Successful Writing at Work

Denise Hensley

295-4244

Writing
A Concise
Handbook

James A. W. Heffernan + John E. Lincoln

As part of Houghton Mifflin's ongoing
commitment to the environment, this text
has been printed on recycled paper.

Successful Writing at Work

Fifth Edition

Philip C. Kolin
University of Southern Mississippi

Houghton Mifflin Company Boston New York

To Kristin, Eric, and Theresa
Julie and Loretta
Tammie
and
MARY

Sponsoring Editor: Jayne M. Fargnoli
Senior Associate Editor: Janet Edmonds
Editorial Assistant: Terri Teleen
Senior Project Editor: Susan Westendorf
Senior Production/Design Coordinator: Jill Haber
Director of Manufacturing: Michael O'Dea
Marketing Manager: Nancy Lyman

Cover photographs: Donovan Reese and Terry Vine for © Tony Stone Images

Printed in the U.S.A.

Library of Congress Catalog Card Number: 97-72505

ISBN: 0-395-87437-8

3 4 5 6 7 8 9 -CW-01 00 99 98

Contents

Preface

Successful Writing at Work is a comprehensive introductory text for use in technical, business, professional, and occupational writing courses. As in the first four editions, the approach in this fifth edition remains practical, emphasizing that communication skills are essential for career advancement and that writing (often as part of a team) is a vital part of almost every job. The fifth edition, however, even more strongly emphasizes the importance and application of the most recent communication technology—such as the Internet, computer graphics, e-mail, and teleconferencing—to writing successfully for and in the world of work. Even the appearance of this redesigned fifth edition underscores the significance of these technologies. TECH NOTES, boxed inserts in each chapter, give students additional information about computer applications in on-the-job writing.

The fifth edition also continues to stress that writing is a problem-solving activity that helps workers meet the needs of their employers, co-workers, customers, and clients. But with this edition's special emphasis on technology, students are shown how to become better problem solvers, and hence better writers, by understanding and using the Internet. In light of these expanding communication resources, *Successful Writing at Work* presents multiple situations and problems that students as business and technical writers will have to address and asks them to consider the rhetorical and technical options available for solving these problems. As earlier editions did, the fifth edition provides students with detailed guidelines for writing and designing clear, well-organized, and readable documents. This edition contains a wide range of examples (many of them annotated and visually varied) drawn from such diverse sources as Internet home pages, e-mail, student papers and reports, letters and memos, proposals, graphics packages, instructions, brochures, news releases, newsletters, and magazine and journal articles. These examples—all dealing with real issues in the world of work—reveal writers as effective, successful problem solvers.

Consistent with this overall view of workplace writing as workplace problem solving, the fifth edition steadfastly continues to emphasize writing as a process. Students will find helpful and concise explanations of the *hows* as well as the *whys* of writing for the world of work. *Successful Writing at Work* helps students develop the crucial skills of brainstorming, researching (through print and on-line sources), drafting, revising, editing, proofreading, and formatting various business and technical documents—correspondence, instructions, summaries, reports, brochures. It also helps them understand why the mastery of such skills is essential to career advancement.

The fifth edition continues to emphasize audience analysis, but this edition stresses even more than its predecessors that writers must often make ethical decisions to meet their readers' needs and fulfill commitments to them. Moreover, given the revolution in information technology, the concept of audience now extends to readers worldwide, whether as co-workers, employers, clients, or representatives of various agencies and organizations. E-mail, letters, brochures and newsletters, instructions, proposals, short and long reports—all are considered from the points of view of the intended audience(s)—that is, from the vantage point of employers, personnel directors, co-workers, and customers, with renewed emphasis on the needs of ESL readers, both in this country and abroad. To engage students fully in job-related writing, the text treats them as professionals seeking advancement at different phases of a business career. For example, they are addressed as employees who must learn to collaborate as part of a writing team of co-workers in Chapter 3; as customer relations representatives addressing the needs of employers and customers with specific requests, problems, and complaints in Chapter 6; as job candidates preparing a variety of documents related to their job search in Chapter 7; as assistants who are asked to summarize a document or a report for a superior in Chapter 11; as competitive businesspeople writing a persuasive proposal to win a contract in Chapter 15; and as presenters at a meeting or conference in Chapter 18.

The organization of the fifth edition reflects the student's own progress in researching and writing for the world of work. The text moves logically and smoothly from consideration of basic concepts in writing (audience, tone, message, purpose, ethical considerations) in Chapter 1 to the overall process of writing (brainstorming, drafting, revising, editing, proofreading, formatting) in Chapter 2. Chapter 3 is devoted to the dynamics of collaborative writing and the procedures groups must go through in order to resolve conflicts and produce successful documents. These three chapters form the foundation on which students can further develop and apply their writing skills. From these introductory chapters, the fifth edition moves sequentially from relatively short and simple assignments (e-mail correspondence, faxes, memos, letters) in Chapters 4 through 7 to longer and more complex business writing (brochures, newsletters, instructions, proposals, reports) in Chapters 8 through 18.

Like the four previous editions, this edition is rich in practical applications, equally useful to readers who have no job experience and to those with years of experience in one or several fields. Another strong feature of the fifth edition is a series of case studies showing how writers use varied resources and rhetorical strategies to solve problems in the business world. An abundance of exercises at the end of each chapter gives students opportunities to practice a variety of writing skills: analyzing the weaknesses and strengths of diverse documents, generating ideas, researching topics, organizing information, drafting, revising and editing, incorporating visuals, designing documents, and working as part of a collaborative writing team.

New Material in the Fifth Edition

The fifth edition has been improved and greatly expanded to make it a more effective tool for the instructor and a more comprehensive and contemporary resource for the student. Throughout the text, new guidelines, examples, figures, case studies, and exercises make the discussion of occupational writing more useful and current. A great deal of information has been added about Internet and other on-line resources in business and technical writing. The following features are new to the fifth edition:

- TECH NOTES help students better understand the ways in which computers, the Internet, and other communication technologies can help them with their work; some TECH NOTES offer advice on applying recent technology (for example, how to send an attachment with e-mail or how to use graphics software to position a table in text) while others elaborate on a topic mentioned in the chapter.
- New material in Chapter 1 on ethical issues in business and technical writing emphasizes the importance of ethics in the writing process, explains how to revise unethical (misleading, exaggerated) writing, and offers possible solutions for ethical dilemmas. The discussion of ethics is not limited to Chapter 1 but continues in other chapters—on e-mail (netiquette), letters, visuals, summaries, reports—with specific guidelines about ethical communication as it pertains to each type of document.
- Chapter 3, on collaborative writing, is entirely new, offering practical advice on the dynamics of team writing. Several collaboration models and a case study on collaborative writing prepare students to be better team members and, therefore, better readers/editors/writers.
- Chapter 4 is also new to the fifth edition. Covering memos, faxes, and e-mail, the chapter introduces students to some of the most frequently written correspondence in the world of work; explains and illustrates both the technology and the protocols students need to understand to produce these documents professionally and concisely; and presents a case study that helps students learn about the format, style, ethics, and rhetorical strategies of this communication medium.
- Chapters 5 and 6 include greatly expanded and updated sections on writing for ESL (English as a Second Language) readers. Examples, guidelines, and a case study on researching, observing, and incorporating a reader's point of view encourage students to respect the cultural traditions and communication needs of diverse audiences.
- This edition supplies expanded coverage of sexist language and ways to avoid it in correspondence and reports.
- New examples throughout Chapter 6 introduce students to the types of problems they are likely to face as writers in the workplace and guide them toward the most effective rhetorical strategies to employ; a new sales letter on promoting ergonomic software, for example, demonstrates the degree of audience analysis required for a mass market promotion.
- A thoroughly revised and updated Chapter 7 on job application letters and résumés contains advice about the variety of resources available to job seekers. Sensitive to the needs of individuals reentering the workplace or changing careers, this chapter devotes additional, valuable space to their situation. The new formats and rhetorical strategies for on-line résumés and electronic job searches are valuable additions to this much-used chapter.
- Responding to the needs of students who are frequently called upon to write or assist in the production of promotional materials for their company or

organization, a new Chapter 8 on news releases, newsletters, brochures, and home pages offers advice, examples, and practical guidelines on preparing these documents. Numerous types of releases, including some from the World Wide Web, are included, along with carefully designed and written examples of a brochure, a newsletter, and a Web home page. This chapter reinforces a prominent theme of the fifth edition—that document design is crucial to the writer's success. No other occupational writing textbook gives students so much detailed and practical information about the construction and rhetoric of home pages.

- A completely revised and substantially updated Chapter 9, on finding and using information resources, including the Internet, introduces students to the most important research tools of the Information Age. A new section on step-by-step research strategies opens the chapter, followed by extended coverage of on-line catalogs, databases, and reference works; the second half of the chapter is devoted exclusively to the Internet, with detailed attention to (and illustrations of) the types of Internet searches, search engines, and Web browsers. The chapter contains a variety of illustrations of home pages, menus, and hyperlinked texts, giving students a thorough but not overwhelming technical introduction to the Internet.
- New material in Chapter 10 on how to document electronic and on-line sources in researched reports offers students clear citation guidelines and plentiful examples. A new section on precautions researchers must follow to verify constantly changing Web sites is also especially useful. A student-written research paper on telecommuting illustrates how to incorporate information from electronic sources and Internet sites and then how to document them.
- A new model article—"Virtual Reality and Law Enforcement"—in Chapter 11, on writing summaries, is used to illustrate, through detailed annotations and extended running commentary, the process of summarizing documents clearly and concisely. New examples of abstracts and summaries also draw upon and highlight new communication technologies to teach careful writing and to introduce interesting and current information into this edition.
- A new Chapter 12, on document design, logically precedes Chapter 13 on visuals and stresses the note that the visual appearance of a document plays in how (and how easily) an audience understands and responds to a piece of writing. Students are introduced to the ABCs of document design, including page layout, levels of headings, type and font size, and design, and are given specific guidelines on mistakes to avoid as they plan and print their work. A before-and-after example of a document illustrates graphically the importance of design in the communication process. This chapter rests upon such central pedagogical bases as the psychology of space and the ways in which design affects readability and retention.
- A thoroughly revised Chapter 13, on designing visuals, pays particular attention to computer-generated visuals and includes additional, updated examples of graphs, tables, and clip art; the emphasis throughout is on the software options available to students configuring appropriate visuals for their written work.

- A new model proposal on notebooks (laptop computers) in Chapter 15 further illustrates a new technology and ways to market it, in writing, to a potential customer.
- Chapter 17, on long reports, contains a new model report on e-cash, or electronic money, using Internet and other contemporary sources.
- New material in Chapter 18 focuses on conferencing and oral reports; a new speech outline on teleconferencing stresses collaboration in the workplace.
- A new appendix—"A Writer's Brief Guide to Paragraphs, Sentences, and Words"—concisely explains and illustrates the most significant and recurrent problems of punctuation, usage, mechanics, and style relevant to writing for the world of work. This appendix is a condensed but not watered-down handbook, useful to the instructor who wants a standard but not intrusive guide for students and for students who need a quick but effective review.

A brief overview of the fifth edition will show how these new materials have been integrated.

An Overview of Part I

Part I deals with the overall writing process. Chapter 1, setting the stage for all occupational writing, identifies the basic concepts of audience analysis, purpose, message, style, tone, and ethics, and relates these concepts to on-the-job writing.

Chapters 2 and 3 continue this important unit on the basic elements of effective writing. Chapter 2, on the writing process at work, introduces students to prewriting strategies, drafting, revising, and editing their written work. Chapter 3 emphasizes the importance of collaborative writing in the world of work and gives students valuable guidelines for being productive, cooperative members of a writing team. This chapter also explores some of the major problems writers face when working together and suggests positive, effective strategies for dealing with problems and resolving them.

An Overview of Part II

Part II deals with business correspondence. Chapter 4 concentrates on memos, faxes, and e-mail, perhaps the most frequent types of writing that students will face in the world of work. Chapter 5 introduces the nuts and bolts of letter writing and focuses on selecting the appropriate format, language, and tone. Chapter 6 examines the rhetorical strategies for producing a variety of business correspondence—complaint, adjustment, order, and sales letters—with additional material on organizational strategies for good-news or bad-news letters, and an expanded discussion of writing for ESL readers. Chapter 7, covering the job search, takes students through the process of preparing a placement file, writing a résumé and organizing it by skill area and/or chronology, and sending it via the Internet as well as through more conventional means, writing a letter of application, anticipating interviewers' questions, and accepting or declining a job offer. For greater teaching flexibility, instructors will find six application letters and six résumés from applicants with varying degrees of experience—helpful models for new and veteran job seekers alike. A new Chapter 8 on writing and designing promotional literature—news releases, brochures, newsletters, and home pages—closes this section and stresses the application of various rhetorical strategies and technical designs to these important customer-centered documents.

An Overview of Part III

Part III, on gathering and summarizing information, occupies a key position in the fifth edition. It helps students acquire the techniques they need to be skilled researchers and accurate summarizers. Chapter 9 takes students on a guided tour of a computerized library, shows them how to locate printed and audiovisual materials, and explains how to profit from searching various computer databases and electronic reference works; the chapter also provides a full introduction to the Internet—its usefulness, organization, search engines, and prominent Web sites in various disciplines. Chapter 10 is devoted to documentation, in particular, the MLA and APA parenthetical methods. Detailed guidelines show students how to document a variety of print and electronic sources, including Web sites, e-mail, listservs, and discussion groups. A reprinted student research paper, "The Advantages of Telecommuting in the Information Age," illustrates several print, electronic, and Internet citations. In Chapter 11 students learn how to write clear and concise summaries and abstracts by seeing how a police officer summarizes an article on virtual reality and law enforcement for a superior.

An Overview of Part IV

In Part IV students get a chance to apply the skills they learned in Part III to more complex writing assignments. The section focuses on key business and technical writing documents—instructions, proposals, and reports. Chapters 12 and 13 form a unit on the related topics of document design and visuals. A new Chapter 12 stresses the significance of document design and gives students practical advice and pertinent examples for making their work more reader-friendly and visually appealing. Chapter 13 supplies practical advice on designing visuals with written commentary; describes how to use visuals in instructions and reports; discusses and illustrates a variety of visuals students can use in their work; and concludes with a greatly expanded and updated section on computer graphics.

Chapter 14 covers writing accurate instructions and selecting the most appropriate language and visuals. Chapter 15 explores three common types of proposals: an internal proposal for an employer, a sales proposal (solicited and unsolicited) for customers, and a research proposal for a teacher. A sample proposal on supplying notebook computers to a small firm is new to the fifth edition. Chapter 16 outlines the principles common to all short reports and then discusses specific types, with detailed coverage of test and laboratory reports and a thorough discussion of

128 E. Leyte

fire station

towards skating rink

Turn Rt off Leyte

asses

the Pinnacle

reports. Finally, students are cautioned about the legal implications of what they write and are shown how to avoid some legal pitfalls.

To make it easily accessible to students, Chapter 17, on long reports, has been revised for this edition to emphasize the *process* of writing such a report. Students are encouraged to see a long report as the culmination of all their work in the course or on a major project at work. The individual parts of such a report are discussed and illustrated in detail, with a fully annotated student-written model report, "E-cash." This paper, together with the Chapter 10 report on telecommuting and the long report on AIDS and health care workers in the Instructor's Guide, gives instructors three complete, documented student research papers from which to teach the long report.

Chapter 18, which stresses the importance of audience analysis in oral communication, offers commonsense advice on preparing briefings and conferences and on generating, organizing, and delivering formal speeches; this chapter includes a new speech outline on the benefits of teleconferencing.

A Writer's Brief Guide

New to the fifth edition is an appendix, "A Writer's Brief Guide," which students can use as a handy, concise reference manual on matters of usage, mechanics, punctuation, style, word choice, sentence structure, and paragraph development.

Acknowledgments

In a very real sense, the fifth edition has profited from a collaboration of various reviewers with the author. I am, therefore, honored to thank the following reviewers who have helped me improve the fifth edition:

John DeSando, Franklin University
Sandra L. Giles, Valdosta State University
Terri Langan, Fox Valley Technical College
Shirley F. Nelson, Chatanooga State Technical
Christopher Patterson, Iona College
Judith N. Serkosky, Madison Area Technical College
Michele Sullivan, Fairfield University
Mary Beth VanNess, University of Toledo
James A. Von Schilling, Northampton Community College

I am also deeply grateful to the following individuals at the University of Southern Mississippi for their help as I prepared this revision. From the Department of English, I thank Steven Barthelme, Amy Dolejs, Marilyn Ford, Laura Hammons, Nancy Hill, Diane Keene, Sandi McBride, Harry McCraw, Gordon Reynolds, and Michael Salada; and Barbara Shoemake (Department of Journalism and Public Relations); Mary F. Lux (Department of Medical Technology); David Goff (Department of Radio, Television, and Film); William Goffe (Department of

Economics and International Business); and university librarians David King and Paul McCarver.

I owe a special debt of gratitude to Cliff Burgess (Department of Computer Science) for his invaluable help with the TECH NOTES and discussions of the Internet throughout this edition.

I also want to thank LaNelle Daniel (Floyd Community College), Colby Kullman (University of Mississippi), George Haber (New York Institute of Technology), Mary Scotto (Kean College), and Dan Jones (University of Central Florida).

Several individuals from business and government also gave me valuable assistance, for which I am very thankful. They include Joycelyn Woolfolk at the Federal Reserve Bank in Atlanta; Michael Sappington of the Bank of Mississippi; Tara Barber at CText; and Lucy Cowan of Tall Pine Farms. I am especially grateful to Sgt. Mannie E. Hall of the U.S. Army, who helped me in innumerable ways as I prepared Chapters 12 and 13.

My thanks to my editors at Houghton Mifflin for their assistance and friendship—Susan Westendorf, Jayne Fargnoli, Linda Bieze, Terry Teleen, Bob Weber, and George Kane. I also thank Henry Rachlin for his professional advice on document design. I am also especially grateful to Janet Edmonds for all her help and encouragement.

I reserve a special thank you for Marge Parish and Al Parish and Dr. Michael O'Neal and Vicki Taylor for their extended support and encouragement.

To all my family I say thanks for your encouragement, patience, and love—Kristin, Eric, and Theresa.

Finally, I am grateful to Tammie Brown for all the joy she has brought into my life.

P. C. K.

To the Student

This book is based on the belief that writing substantially influences your career. Effective writing can help you obtain a job, perform your duties more successfully, and earn promotions for your efforts. Guidelines and examples found in this book emphasize the progress you can make in your career as you acquire effective writing skills. Specifically, *Successful Writing at Work*

- explains the writing process and shows you how planning, drafting, revising, and editing can help you to produce a variety of essential job-related communications
- describes the function and format of these job-related communications
- teaches you how to supply an audience with the information it needs to make decisions and solve problems
- introduces you to various research tools, including the Internet
- prepares you to write a variety of business documents, from simple e-mail messages to longer, more complex proposals and reports

The fifth edition is organized to coincide with your own progress in writing. Part I gives you solid, useful background information to be a successful writer in the world of work. Chapter 1 introduces you to key ideas, strategies, and requirements for writing on the job. Chapter 2 explains and illustrates the process of writing. Chapter 3 explains the advantages and techniques of collaborative writing and the pitfalls to avoid.

Part II, Correspondence, discusses the basics you should know in order to write these major on-the-job documents and so advance in your career. Chapter 4 focuses exclusively on memos and electronic correspondence—faxes and e-mail. Chapter 5 surveys some essential information on style, format, and organization of business letters. Chapter 6 turns to the strategies to be used with specific types of correspondence you will write for your employer and customers. Chapter 7 focuses on how to write a letter of application and to prepare several kinds of résumés, including bullet and electronic (or Internet) résumés. Part II concludes with a chapter that shows you how to write promotional material—news releases, newsletters, brochures, and home pages.

Part III is devoted to helping you gather and summarize information. On your job you will be expected to know how to use a variety of research tools (including the Internet) and strategies. Chapter 9 describes how to find, retrieve, and use research materials, with an emphasis on the Internet and database resources. Chapter 10 is devoted to documenting the information you gather, particularly from on-line sources, and Chapter 11 covers summarizing what you find and know.

Part IV concentrates on effective ways to record your findings in instructions, proposals, and reports. This part helpfully begins with two chapters on designing documents and using visuals. Visuals and graphics are components in all the types of documents discussed in the other Part IV chapters—instructions (Chapter 14), proposals (Chapter 15), short reports (Chapter 16), long reports (Chapter 17), and oral reports, or business presentations (Chapter 18).

A Brief Writer's Guide will assist you in the mechanics of writing paragraphs, composing sentences, and improving your word choice and spelling.

PART I

Backgrounds

CHAPTER 1

Getting Started: Writing and Your Career

What skills have you learned in school or on the job this year? Perhaps you have learned techniques of health care in order to become a nurse, respiratory therapist, or dental hygienist. Maybe you have received training in law enforcement to prepare yourself for work with a crime-detection unit or a traffic-control department. Possibly you have studied or worked in industrial technology, agriculture, computer science, hotel and restaurant management, or forestry. Or maybe you have improved certain skills that will make you a better marketing specialist, salesperson, office manager, computer programmer, or accountant. Whatever your area of accomplishment, the practical know-how you have acquired is essential for your career.

You need an additional skill, however, to ensure a successful career: you must be able to write clearly about the facts, procedures, and problems of your job. Writing is a part of every job. In fact, your first contact with a potential employer is through your letter of application, which determines a company's first impression of you. And the higher you advance in an organization, the more writing you will do. Promotions are often based on a person's writing skills.

The Associated Press reported in a recent survey that "most American businesses say workers need to improve their writing . . . skills." The same report cited a survey of 402 companies that identified writing as "the most valued skill of employees." Still, the employers polled in this survey indicated that 80 percent of their employees need to improve their writing skills. Clearly, writing is an essential skill important to everyone in business—employers and employees alike.

Among the most cost-effective skills you can offer a prospective employer is your writing ability. Businesses pay a premium price for good writing. According to Don Bagin (*Communication Briefings* [May 1995]: 3), most people need an hour or more to write a typical business letter. If an employer is paying someone $30,000 a year, one letter costs $14 of that employee's time; for someone who earns $50,000 a year, the cost of the average letter jumps to $24.

Offices and other workplaces contain numerous reminders of the importance of writing—printers, word processors, floppies, monitors, scanners, keyboards, mouses, modems, fax machines. Why? Writing keeps business moving. It allows individuals working for a company to communicate with one another and with the customers and clients they must serve if the company is to stay in business. Clearly today's most ambitious communication tool is the Internet, the all-purpose information superhighway. Almost every type of written communication can be found on the Internet—e-mail, letters, memos, summaries, instructions, questionnaires, proposals, reports, and much, much more. This book will show you, step by step, how to write these and other job-related communications easily and well.

Chapter 1 presents some basic information about writing and offers some questions you can ask yourself to make the writing process easier and the results more effective. This chapter also describes the basic functions of on-the-job writing and introduces you to one of the most important requirements in the business world—writing ethically.

Four Keys to Effective Writing

Effective writing on the job is carefully planned, thoroughly researched, and clearly presented. Whether you send a routine memo to a co-worker or a special report to the president of the company, your writing will be more effective if you ask yourself four questions.

1. *Who* will read what I write? (Identify your *audience.*)
2. *Why* should they read what I write? (Establish your *purpose.*)
3. *What* do I have to say to them? (Formulate your *message.*)
4. *How* can I best communicate? (Select your *style* and *tone.*)

The questions *who? why? what?* and *how?* do not function independently; they are all related. You write (1) for a specific audience (2) with a clearly defined purpose in mind (3) about a topic your readers need to understand (4) in language appropriate for the occasion. Once you answer the first question, you are off to a good start toward answering the other three. Now let us examine each of the four questions in detail.

Identifying Your Audience

Knowing *who* makes up your audience is one of your most important responsibilities as a writer. In fact, it is important to analyze your audience throughout the composing process.

Look for a minute at the American Heart Association posters reproduced in Figures 1.1, 1.2, and 1.3. The main purpose of all three posters is the same: to discourage individuals from smoking. The essential message in each poster—smoking is dangerous to your health—is also the same. But note how the different details—words, photographs, situations—have been selected to appeal to three different audiences.

The poster in Figure 1.1 emphasizes smoking problems that are especially troublesome to a teenager: red eyes, bad breath, discolored teeth, and unattractive hair. The smiling teen pictured without a cigarette appears to have avoided these problems. The message at the top of the poster plays on two meanings of the word *heart:* (1) smoking can cause heart disease and (2) smoking can be a deterrent to romance. Teenagers are particularly sensitive to the second meaning.

The poster in Figure 1.2 (p. 6) is aimed at an audience of pregnant women and appropriately shows a woman with a lit cigarette. The words at the top and bottom of the poster appeal to a mother's sense of responsibility as the reason to stop smoking, a reason to which pregnant women would be most likely to respond.

Figure 1.3 (p. 7) is directed toward fathers and appropriately shows a small child seated on his father's lap. The situation depicted appeals to a father's wish for his child's happiness. The words in the poster warn that a father who smokes may die prematurely and make his child's life unhappy.

The copywriters for the American Heart Association have chosen appropriate details—words, pictures, and so on—to convince each audience not to smoke. With their careful choices, they successfully answered the question "How can we best communicate with each audience?" As an indication of their skill, note that

FIGURE 1.1 No-smoking poster aimed at teenagers.

© Reprinted with permission of the American Heart Association.

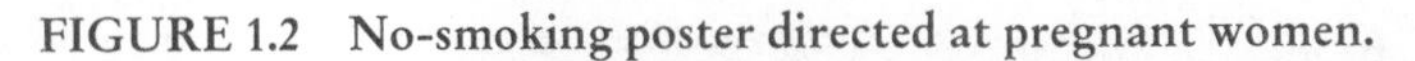

FIGURE 1.2 No-smoking poster directed at pregnant women.

details relevant for one audience (teenagers, for example) could not be used as effectively for another audience (such as fathers).

These three posters illustrate some fundamental points you need to keep in mind when identifying your audience.

- Members of each audience differ in backgrounds, experiences, needs, and opinions.
- How you picture your audience will determine what you say to them.
- Viewing something from the audience's perspective will help you to select the most relevant details for that audience.

Some Questions to Ask About Your Audience

You can form a fairly accurate picture of your audience by asking yourself some questions *before* you write. For each audience for whom you write, consider the following questions.

1. **Who is my audience?** What individual(s) will most likely be reading my work?

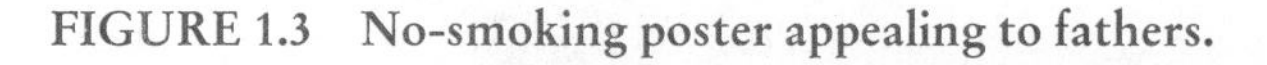

FIGURE 1.3 No-smoking poster appealing to fathers.

If you are writing for individuals at work:

- What is my reader's job title?
- What is my reader's relationship to me? Co-worker? Immediate supervisor? Vice president?
- What kind of job experience, education, and interests does my reader have?

If you are writing for clients or consumers (a very large, sometimes fragmented audience):

- How can I find out about their interest in my product or service?
- How much will this audience know about my company? About me?

2. How many people will make up my audience?

- Will just one individual read what I write (the nurse on the next shift, the desk sergeant) or will many people read it (all the consumers of a product manufactured by my company)?

- Will my boss want to see my work (say, a letter to a consumer in response to a complaint) to approve it?
- Will my letter bear someone else's name, not mine?
- Will I be sending my message to a large group of individuals sharing a similar interest in my topic—such as a usenet or chat-room group on the Internet?
- Will I potentially be communicating with individuals all around the globe via the Internet?

3. How well does my audience understand English?

- Are all my readers native speakers of English?
- Will some of my readers have lesser command of English and require extra sensitivity on my part to their needs as an ESL (English as a Second Language) audience? (See pp. 166–175 for guidelines about communication with this audience.)

4. How much does my audience already know about my writing topic?

- Will my audience know as much as I do about the particular problem or issue, or will they need to be briefed and/or updated?
- Does my audience know little or nothing about the message I am sending them?
- Are my readers familiar with, and do they expect me to use, technical terms and descriptions, or will I have to provide easy-to-understand comparisons and nontechnical summaries?
- Will I need to include detailed visuals or sketches, or will a photograph or simple drawing be enough?

5. What is my audience's reason for reading my work?

- Is reading my communication part of their routine duties, or are they looking for information to solve a problem or make a decision?
- Will they want my writing to describe benefits that another writer or company cannot offer—that is, am I trying to persuade them to buy a product or a service?
- Will they expect complete details, or will a short summary of the main points be enough for their purpose?
- Are they reading my work to take some action affecting a co-worker, a client, or a community official?
- Are they reading something I write because they must (a legal notification, for instance), or are they looking at it just as a courtesy?
- What will my boss want from me—information alone, some analysis and conclusions drawn from the information, or a specific set of recommendations?

6. What are my audience's expectations about my written work?

- Do they want just a piece of e-mail or do they expect a formal letter?
- Will they expect me to follow a certain format and organization?

- Are they looking for a one-page memo or for a comprehensive ten-to-fifteen-page report?
- Should I use a formal tone or a more relaxed and conversational style?

7. **What is my audience's attitude toward me and my work?**

- Will I be writing to a group of disgruntled and angry customers about a very sensitive issue—a product recall or a refusal of credit?
- Will I have to be very sympathetic while at the same time give firm reasons for my company's (or my) decision?
- Will my readers be skeptical—customers whose business I want to attract?
- Will I need to overcome my readers' indifference by arousing their interest and encouraging a response?
- Will my audience be eager and friendly, happy to read what I write?
- Will my readers feel guilty that they have not answered an earlier message of mine or not paid a bill now overdue or not kept a promise or commitment?

8. **What do I want my audience to do after reading my work?**

- Do I want my reader to purchase something from me or my company?
- Do I expect my reader to approve my plan or to send me additional materials or information?
- Do I simply want my reader to get my message and not respond at all?
- Do I expect my reader to get my message, acknowledge it, save it for future reference, or review it and e-mail it to another individual or office?
- Does my reader have to take immediate action or does he or she have several days or weeks to respond?
- Do I want my reader to share my message by e-mail or the Internet?

As your answers to these questions will show, you may have to communicate with many different audiences on your job. If you work for a large organization that has numerous departments, you may have to write to such diverse readers as accountants, office managers, personnel directors, engineers, public relations specialists, marketing specialists, computer programmers, and individuals who install, operate, and maintain equipment. In addition, you will need to communicate effectively with customers about your company's products and services. Each group of readers will have different expectations and requirements; you need to understand these audience differences if you want to supply relevant information.

The advertisement in Figure 1.4 concisely illustrates how the writer for a manufacturer of heavy-duty equipment identified the priorities of five different audiences and selected appropriate information to communicate with each one.

Audience	Information to Communicate
owner or principal executive	The writer appropriately stresses financial benefits: the machine is a "money-maker" and is compatible with other equipment so additional equipment purchases are unnecessary.
production engineer	The writer emphasizes "state of the art" transmissions, productivity, upkeep.

FIGURE 1.4 An advertisement aimed at the needs of five different audiences.

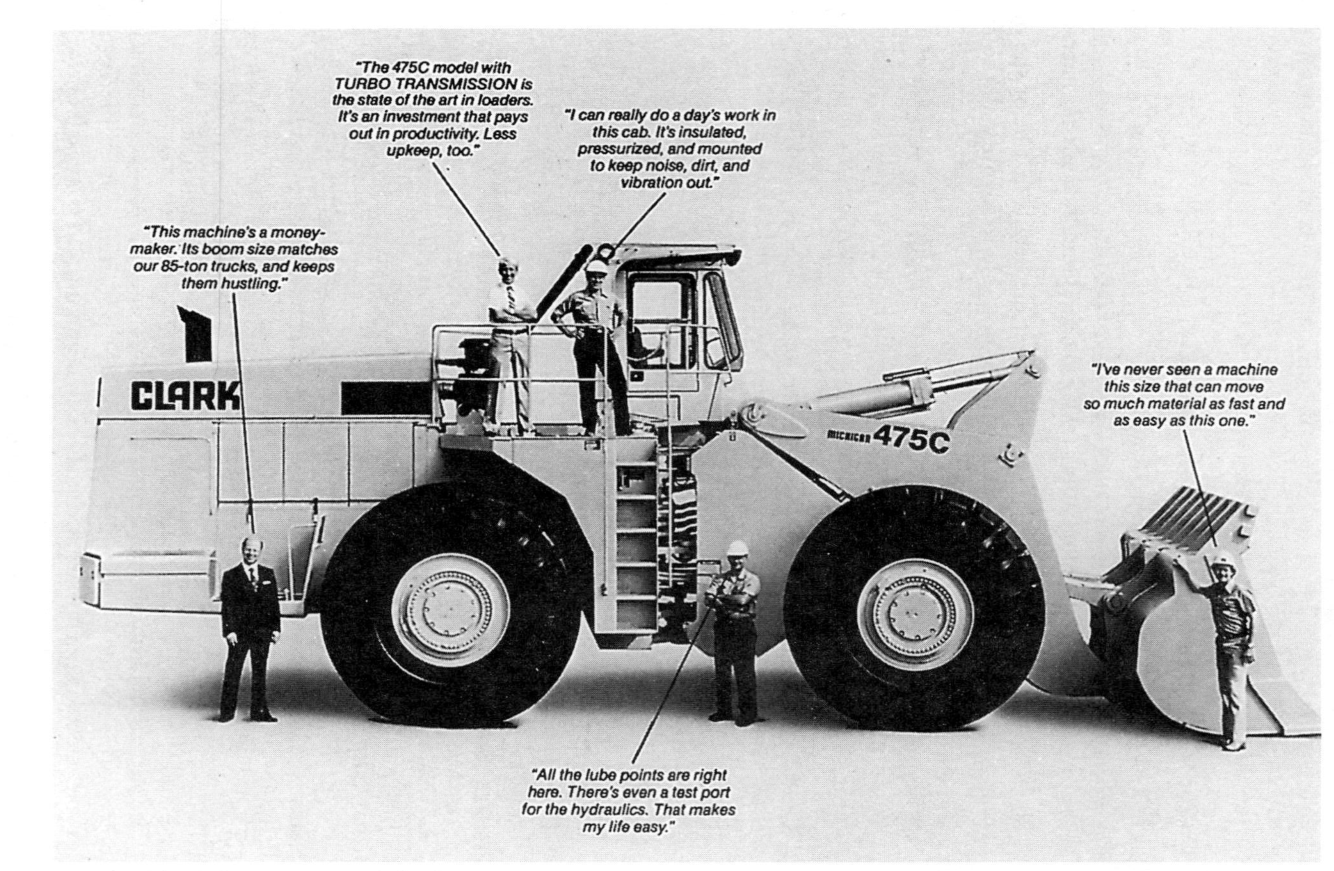

Courtesy of the Clark Company. Reprinted with permission.

operator	The writer focuses on how easy it is to run the machine—the pressurized cab keeps out environmental problems that interfere with a job.
maintenance worker	Since this reader is concerned about such things as "lube points" and "test ports," not costs or operations, the writer correctly selects appropriate information about making the worker's job easier and safer.
production supervisor	The writer emphasizes the speed and efficiency the machine offers, thus zeroing in on this reader's needs and interests.

The lesson of this ad is clear: give each reader the details that he or she needs to accomplish a given job. Sometimes you are not able to identify all the members of your potential audience. In such cases just assume that you have a general audience and keep your message as simple as possible—nontechnical and straightforward.

Establishing Your Purpose

By knowing *why* you are writing, you will communicate better and find writing itself to be an easier process. The reader's needs and your goal in communicating will help you to formulate your purpose. It will help you to determine exactly what you can and must say. With your purpose clearly identified, you are on the right track.

Make sure you follow this most important rule in occupational writing: **Get to the point right away.** At the start of your message, state your goal clearly. Don't feel as if you have to entertain or impress your reader. Don't worry about the way your words sound. It is more helpful to work on your ideas.

> I want new employees to know how to log on to the computer.

Think over what you have written. Rewrite your purpose statement until it states precisely why you are writing and what you want your readers to do or to know.

> I want to teach new employees the security code for logging on to the company computer.

Since your purpose controls the amount and order of information you include, state it clearly at the beginning of every letter, memo, or report. Such an overview will help the reader to follow and act on your communication.

> This e-mail will acquaint new employees with the security measures they must take when logging on to the company computer.

> The following report will give you a detailed account of my progress to date on completing my research this semester.

In the opening purpose statement below, note how the author clearly informs the reader as to what the report will and will not cover.

> As you requested at last week's organizational meeting, I have conducted a study of our use of nontraditional media to advertise our products. This report describes, but does not evaluate, our current practices.

The following preface to a pamphlet on architectural casework details contains a model statement of purpose suited to a particular audience.

> This publication has been prepared by the Architectural Woodwork Institute to provide a source book of conventional details and uniform detail terminology. For this purpose a series of casework detail drawings, . . . representative of the best industry-wide practice, has been prepared and is presented here. By supplying both architect and woodwork manufacturer with a common authoritative reference, this work will enable architects and woodworkers to communicate in a common technical language. . . . Besides serving as a basic reference for architects and architectural drafters, this guide will be an effective educational tool for the beginning drafter-architect-in-training. It should also be a valuable aid to the project manager in coordinating the work of many drafters on large projects.[1]

After reading this preface, readers have a clear sense of why they should use the source book and what to do with the material they find in it.

Formulating Your Message

Your message is the sum of *what* facts, responses, and recommendations you put into writing. A message includes the *scope* and *details* of your communication. The details are those key points you think readers need to know to perform their jobs. Scope refers to how much information you give readers about those key details. Some messages will consist of one or two sentences: "Do not touch; wet paint." "Order #756 was sent this afternoon by Federal Express. It should arrive at your office on March 21." At the other extreme, messages may extend over twenty or thirty pages. Messages may carry good news or bad news. They may deal with routine matters, or they may handle changes in policy, special situations, or problems.

Keep in mind that you will adapt your message to fit your audience. For technical audiences such as engineers or technicians, you may have to supply a complete report with every detail noted or contained in an appendix. For other readers—busy executives, for example—you would be wrong to include such technical details. A short discussion or summary of the financial or managerial significance of these details is what this group of readers prefers.

Consider the message of the following excerpt from a section entitled "Technology in the Grocery Store" included in a consumer handbook. The message provides factual information and a brief explanation of how a grocery store clerk scans an item, informing general consumers about how and why they may have to wait longer in line. It also tells readers that the process, which looks so simple, is far more intricate than they imagine.

Bar Code Readers

Every time you check out at the grocery store, many of your purchases are scanned to record the price. The scanner uses a laser beam to read the bar codes,

---

[1]Reprinted by permission of Architectural Woodwork Institute.

those zebra-striped lines imprinted on packages or canned goods. These codes are fed into the store's computer, which provides the price to go with the product code. The product and its price are then recorded on your receipt.

Scanning an item requires more skill than you might think. To make sure that the scanner accurately reads the bar codes, the clerk has to take into account the following four conditions:

Speed: The clerk has to know how fast to pass the item across the scanner, which is geared to a particular speed. If he or she moves the item too slowly, the bars will look too long or too wide and the computer will reject the item. If the item is moved too fast, the scanner cannot identify the code.

Angle: The clerk needs to know precisely at what angle to pull the item across the scanner. If the angle is wrong there will be insufficient reflection of the laser beam back to the scanner and so it will not be able to read the code. Since it is best for the item to reflect as much of the laser as possible, the clerk should try to hold the code at right angles to the laser.

Distance: Moving the item too close to the scanner is as unproductive as holding it too far away. Either way the code can be out of focus for the scanner reader. Holding the item about 3–4 inches away is best; holding it out more than 8–9 inches ensures that the scanner will not read the code.

Rotation: The code needs to be facing the scanner so that the lines can be read properly.

If your clerk makes a mistake in any one of these calibrations, your wait in line is sure to be longer.

The bar code message is appropriate for a general audience of consumers, who are given neither more nor less information than they need or desire. However, other audiences would need different messages and different details. Individuals at the store who are responsible for entering data into the computer or doing inventory control would need many more detailed instructions on how to program the supermarket's computer to make sure it automatically tells the point-of-sale (POS) terminal what price and product match each bar code.

The product technicians responsible for affixing the bar codes at the manufacturer's plant would require still more detailed and technical information. For example, these individuals would need to be very familiar with the Uniform Product Code (UPC) that specifies bar codes worldwide. They would also have to know about the UPC binary code formulas and how they work—that is, the number of lines, width of spacing, the framework to indicate to the scanner when to start reading the code and when to stop. Such formulas, technical details, and functions of photoelectric scanners clearly would be inappropriate for either the consumer or the grocery store cashier.

Selecting Your Style and Tone

Style

Style is *how* something is written rather than what is written. Style helps to determine how well you communicate with an audience, how well your readers understand and receive your message. It involves the choices you make about

- the construction of your paragraphs
- the length and patterns of your sentences
- your choice of words

You will have to adapt your style to take into account different messages, different purposes, and different audiences. Your words, for example, will certainly vary with your audience. If all your readers are specialists in your field, you may safely use the technical language and symbols of your profession. Your audience will be familiar with such terminology and will expect you to use it. Nonspecialists, however, will be confused and annoyed if you write to them in the same way. The average consumer, for example, will not know what a *potentiometer* is; by writing "volume control on a radio" you will be using words that the general public can understand.

Tone

Tone in writing, like tone of voice, expresses your attitude toward a topic and toward your audience. In general, your tone can range from formal and impersonal (a scientific report) to informal and personal (e-mail to a friend or a how-to article for a consumer).

Tone, like style, is indicated in part by the words you choose. For example, saying that someone is "interested in details" conveys a more positive tone than saying the individual is a "nitpicker." The word *economical* is more positive than *stingy* or *cheap*.

The tone of your writing is especially important in occupational writing, for it reflects the image you project to your readers and thus determines how they will respond to you, your work, and your company. Depending on your tone, you can appear sincere and intelligent or angry and uninformed. Of course, in all of your written work, you need to sound professional and knowledgeable about the topic and genuinely interested in your readers' opinions and problems. The wrong tone in a letter or a proposal might cost you a customer.

A Description of Heparin for Two Different Audiences

To better understand the effects of style and tone on writing, read the following two excerpts. In both, the message is basically the same, but because the audiences differ, so do the style and tone. The two pieces are descriptions of *heparin,* a drug used to prevent blood clots.

Technical/Scientific Style and Tone

The description below appears in a reference work for physicians and other health care professionals and is written in a highly technical style with an impersonal tone.

HEPARIN SODIUM INJECTION, USP
STERILE SOLUTION
Description: Heparin Sodium Injection, USP is a sterile solution of heparin sodium derived from bovine lung tissue, standardized for anticoagulant activity.

Each ml of the 1,000 and 5,000 USP units per ml preparations contains: Heparin sodium 1,000 or 5,000 USP units; 9 mg sodium chloride; 9.45 mg benzyl alcohol added as preservative. Each ml of the 10,000 USP units per ml preparations contains: heparin sodium 10,000 units; 9.45 mg benzyl alcohol added as preservative.

When necessary, the pH of Heparin Sodium Injection, USP was adjusted with hydrochloric acid and/or sodium hydroxide. The pH range is 5.0–7.5.
Clinical pharmacology: Heparin inhibits reactions that lead to the clotting of blood and the formation of fibrin clots both *in vitro* and *in vivo.* Heparin acts at multiple sites in the normal coagulation system. Small amounts of heparin in combination with antithrombin III (heparin cofactor) can inhibit thrombosis by inactivating activated Factor X and inhibiting the conversion of prothrombin to thrombin.
Dosage and administration: Heparin sodium is not effective by oral administration and should be given by intermittent intravenous injection, intravenous infusion, or deep subcutaneous (intrafrat, i.e., above the iliac crest or abdominal fat layer) injection. **The intramuscular route of administration should be avoided because of the frequent occurrence of hematoma at the injection site.**[2]

The writer has made the appropriate stylistic choices for the audience, the purpose, and the message. Physicians and other health care providers reading the description will understand and need the technical vocabulary the writer uses; this audience will also require the sophisticated and lengthy explanations in order to prescribe and/or administer heparin correctly. The author's authoritative, impersonal tone is coldly clinical, which, of course, is also appropriate because the purpose is to convey the accurate, complete scientific facts about this drug, not the writer's or reader's opinions or beliefs. The author sounds both knowledgeable and appropriately objective.

Nontechnical Style and Tone

The following description of heparin, on the other hand, is written in a nontechnical style and with an informal, caring tone. This description is similar to those found on information cards given to patients about the drugs they are receiving in a hospital.

> Your doctor has prescribed for you a drug called *heparin.* This drug will prevent any new blood clots from forming in your body. Since heparin cannot be absorbed from your stomach or intestines, you will not receive it in a capsule or tablet. Instead, it will be given into a vein or the fatty tissue of your abdomen. After several days, when the danger of clotting is past, your dosage of heparin will be gradually reduced. Then another medication you can take by mouth will be started.

The writer of this description also made the appropriate choices for the readers and their needs. Familiar words rather than technical ones are suitable for nonspecialists such as patients. Note also that this audience does not need elaborate descriptions of the origin and composition of the drug. The tone is both personal

[2]Copyright © *Physicians' Desk Reference®* 45th edition, 1991 published by Medical Economics, Montvale, New Jersey 07645. Reprinted by permission. All rights reserved.

and straightforward because the purpose is to win the patient's confidence and to explain the essential functions of the drug, a simpler message than the one for professionals.

The trend today in occupational writing is to make letters, reports, and proposals more natural and personal and less impersonal, formal, or stuffy. But adopting a personal tone does not mean that you should address the reader in a chummy or disrespectful way. Quite the contrary, a business letter or report needs to be personal and professional at the same time.

Characteristics of Job-Related Writing

Job-related writing characteristically serves six basic functions: (1) to provide practical information, (2) to give facts rather than impressions, (3) to provide visuals to clarify and condense information, (4) to give accurate measurements, (5) to state responsibilities precisely, and (6) to persuade and offer recommendations. These six functions tell you what kind of writing you will produce after you successfully answer the *who? why? what?* and *how?* just discussed.

1. Providing Practical Information

On-the-job writing requires a practical here's-what-you-need-to-do-or-to-know approach. One such practical approach is *action oriented.* In this kind of writing, you instruct the reader to do something—assemble a ceiling fan, test for bacteria, perform an audit, or create an Internet home page. Another practical approach of job-related writing is to have someone understand something—why a procedure was changed, what caused a problem or solved it, how much progress occurred on a job site, or why a new piece of equipment should be purchased. Examples of such knowledge-oriented practical writing are a letter sent from a manufacturer to customers to explain a product recall or an e-mail sent to employees telling them about changes in their group health insurance policy.

The following description of Energy Efficiency Ratio combines both the action-oriented and knowledge-oriented approaches of practical writing.

> Whether you are buying window air-conditioning units or a central air-conditioning system, consider the performance factors and efficiency of the various units on the market. Before you buy, determine the Energy Efficiency Ratio (EER) of the units under consideration. The EER is found by dividing the BTUs (units of heat) that the unit removes from the area to be cooled by the watts (amount of electricity) the unit consumes. The result is usually a number between 5 and 12. The higher the number, the more efficiently the unit will use electricity.
>
> You'll note that EER will vary considerably from unit to unit of a given manufacturer, and from brand to brand. As efficiency is increased, you may find the purchase price is higher; however, operating costs will be lower. Remember, a good rule to follow is to choose the equipment with the highest EER. That way you'll get efficient equipment and enjoy operating economy.[3]

---

[3]Reprinted by permission of New Orleans Public Service, Inc.

2. Giving Facts, Not Impressions

Occupational writing is concerned largely with those things that can be seen, heard, felt, tasted, or smelled. The writer uses *concrete language* and specific details. The emphasis is on facts rather than on the writer's feelings or guesses.

The following discussion by a group of scientists about the sources of oil spills and their impact on the environment is an example of writing with objectivity. It describes events and causes without anger or tears. Imagine how much emotion could have been packed into this paragraph by the residents of the coastal states who have watched such spills come ashore.

> The most critical impact results from the escapement of oil into the ecosystem, both crude oil and refined fuel oils, the latter coming from sources such as marine traffic. Major oil spills occur as a result of accidents such as blowout, pipeline breakage, etc. Technological advances coupled with stringent regulations have helped to reduce the chances of such major spills; however, there is a chronic low-level discharge of oil associated with normal drilling and production operations. Waste oils discharged through the river systems and practices associated with tanker transports dump more significant quantities of oils into the ocean, compared to what is introduced by the offshore oil industry. All of this contributes to the chronic low-level discharge of oil into world oceans. The long-range cumulative effect of these discharges is possibily the most significant threat to the ecosystem.[4]

3. Providing Visuals to Clarify and Condense Information

Visuals are indispensable partners of words in conveying information to your readers. On-the-job writing makes frequent use of visuals such as tables, charts, photographs, flow charts, diagrams, and drawings to clarify and condense information. Thanks to various computer software packages, you can easily create and insert visuals into your writing. The use of visuals—including computer visuals—is discussed in detail in Chapter 13.

Visuals play an important role in the workplace. Note how the drawing in Figure 1.5 from the National Safety Council's booklet "Working Safely with Your Computer" can help computer users better understand and follow the accompanying written guidelines. A visual like this, reproduced in an employee handbook or displayed as a poster, can significantly reduce stress and increase productivity.

Visuals are extremely useful in making detailed relationships clear to readers. A great deal of information about the growth and diversity of commercial TV stations is condensed into Table 1.1. Consider how many words a writer would need in order to supply the data contained in the table. Note, too, how easily the numbers can be read when they are arranged in columns. The figures would be far more difficult to decipher them if they were printed like this: Commercial TV stations in operation: 1970, Total 690, VHF 508, UHF 182; 1971, Total 696, VHF 511, UHF 185; and so forth.

---

[4]*The Offshore Ecology Investigation*, Galveston: Gulf Universities Research Consortium, 1975: 4.

FIGURE 1.5 Use of a visual to convey information.

Your working area

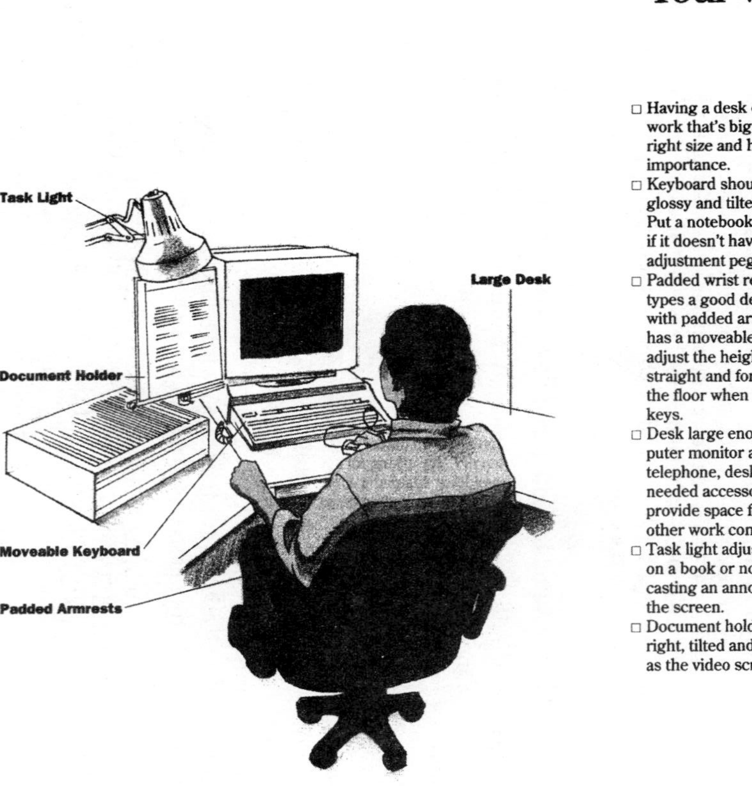

- ☐ Having a desk or other place to work that's big enough, and of the right size and height, is of first importance.
- ☐ Keyboard should be moveable, non-glossy and tilted slightly forward. Put a notebook under the back edge if it doesn't have built-in height adjustment pegs.
- ☐ Padded wrist rests for anyone who types a good deal. Or, use a chair with padded armrests. If the desk has a moveable keyboard tray, adjust the height so that wrists are straight and forearms are parallel to the floor when fingers are on the keys.
- ☐ Desk large enough to hold computer monitor and keyboard, telephone, desk set and all other needed accessories. It also should provide space for writing and doing other work comfortably.
- ☐ Task light adjustable so it can shine on a book or note pad without casting an annoying reflection on the screen.
- ☐ Document holder moveable, upright, tilted and at the same height as the video screen for easy reading.
- ☐ Video monitors at least four feet apart.
- ☐ Muted desk top surface doesn't reflect light into operator's eyes.
- ☐ Screen positioned so it won't reflect light from windows or overhead lights; also so operator won't be distracted by persons walking by.

TABLE 1.1 Commercial TV Stations in Operation, 1970–1996

Year	*Total*	*VHF*	*UHF*
1970	690	508	182
1971	696	511	185
1972	699	510	189
1973	700	511	189
1974	705	513	192
1975	711	513	198
1976	710	513	197
1977	728	517	211
1978	727	516	211
1979	732	516	216
1980	746	517	229
1981	752	519	233
1982	772	524	248
1983	802	526	276
1984	870	536	334
1985	904	539	365
1986	922	541	381
1987	982	547	435
1988	1,017	541	476
1989	1,092	544	548
1990	1,115	563	552
1991	1,128	556	572
1992	1,144	557	587
1993	1,151	556	595
1994	1,157	559	598
1995	1,181	559	622
1996	1,174	554	620

Supplied by the Federal Communications Commission.

In addition to the visuals already mentioned, the following graphic devices (created quickly with the help of a computer) within your letters and reports will make your writing easier for your audience to read and follow.

- Headings, such as **Four Keys to Effective Writing** or **Characteristics of Job-Related Writing**
- Subheadings to divide major sections into parts, such as "Providing Practical Information" or "Giving Facts, Not Impressions"
- Numbers within a paragraph, or even a line, such as (1) this, (2) this, and (3) also this
- Different types of s p a c i n g
- CAPITALIZATION
- *Italics* (easily made by a word processing command or indicated in typed copy by underscoring)
- **Boldface** (darker type for emphasis)
- Scripting (simulating handwriting)

- HYPERTEXT (the use of color, shading, or boldface to mark words or icons linking one text with another on the Internet)
- Asterisks * to * separate * items * or to * note key items
- Lists with "bullets" (raised dots or other symbols like the squares before each entry in this list)

Keep in mind that such graphic devices must be used carefully and with moderation. They should never be used just for decoration or to dress up a letter or report. Used properly, they can help you to organize, arrange, and emphasize your material. They make your work easier to read and recall. For example, headings in a memo, letter, or report make your organizational plan visible to readers and help them preview your ideas. Use bullets or numbers when you have many related points; by setting your points in a list, you make it easier for readers to distinguish, follow, compare, and recall them.

4. Giving Accurate Measurements

Much of your work will depend on measurements—acres, bytes, calories, centimeters, degrees, dollars and cents, grams, percentages, pounds, square feet, units. Readers will look carefully at these measurements. Numbers are clear and convincing. An architect depends on specifications to accomplish a job; a nurse must record precise dosages of medications; a sales representative keeps track of the number of customer visits; and an electronics technologist has to monitor antenna patterns and systems.

The following discussion of mixing colored cement for a basement floor would be useless to readers if it did not supply accurate quantities.

> The inclusion of permanent color in a basement floor is a good selling point. One way of doing this is by incorporating commercially pure mineral pigments in a topping mixture placed to a 1-inch depth over a normal base slab. The topping mix should range in volume between 1 part portland cement, 1 1/4 parts sand, and 1 1/4 parts gravel or crushed stone and 1 part portland cement, 2 parts sand, and 2 parts gravel or crushed stone. Maximum size gravel or crushed stone should be 3/8 inch.
>
> Mix cement and pigment before aggregate and water are added and be very thorough to secure uniform dispersion and the full color value of the pigment. The proportion varies from 5 to 10 percent of pigment by weight of cement, depending on the shade desired. If carbon black is used as a pigment to obtain grays or black, a proportion of from 1/2 to 1 percent will be adequate. Manufacturers' instructions should be followed closely; care in cleanliness, placing, and finishing are also essential. Colored topping mixes are available from some suppliers of ready mixed concrete.[5]

5. Stating Responsibilities Precisely

Job-related writing, since it is directed to a specific audience, must make absolutely clear what it expects of, or can do for, that audience. Misunderstandings waste time and cost money. Directions on order forms, for example, should indicate how and where information is to be listed and how it is to be routed and acted

---

[5]Reprinted by permission from *Concrete Construction Magazine*, World of Concrete Center, 426 South Westgate, Addison, Illinois 60101.

on. The following directions are taken from different business-related communications, showing readers how to perform different tasks and/or explaining why.

- Include agency code numbers in the upper-right corner.
- Items 1 through 16 of this form should be completed by the injured employee or by someone acting on his or her behalf, whenever an injury is sustained in the performance of duty. The term *injury* includes occupational disease caused by the employment. The form should be given to the employee's official superior within 48 hours following the injury. The official superior is that individual having responsible supervision over the employee.
- What is a credit report? A credit report is a record of how you've paid bills with credit grantors such as stores and banks. Credit grantors use credit reports to determine whether or not you will be extended credit. The report identifies you by information such as your name and address, credit accounts, and payment history. Your credit report also includes public record data, such as bankruptcies, court judgments, and tax liens. A list of those who have recently requested a copy of your credit report is also included. A credit report does not contain information on arrest records, specific purchases, or medical records.[6]

Other kinds of job-related writing deal with the writer's responsibilities rather than the reader's. For example, "Tomorrow I will meet with the district sales manager to discuss (1) July's sales, (2) the possibility of expanding our Madison home market, and (3) next fall's production schedule. I will send you a report of our discussion by August 5, 1998." In a letter of application for a job, writers should conclude by asking for an interview and clearly inform a prospective employer when they are available for an interview—for example, weekday mornings, only on Monday and Tuesday afternoons, or any time after February 15.

6. Persuading and Offering Recommendations

Much writing in the business world is directed to employers, customers, and clients to persuade them to (a) buy a product or service or (b) adopt a certain plan of action. You want your audience to act or think in a certain way. To be persuasive, you will need to get your readers' attention, communicate clearly so they understand your message, and make sure they remember it. The very first job-related writing you do will likely be a persuasive letter of application to obtain a job interview with a potential employer, whom you will have to get to recall and select your application from among the hundreds he or she receives. Then, once on the job, you may be required to write persuasive sales letters, reports, proposals, brochures, and even advertisements.

In much job-related writing you will have to convince readers that you (and your company) can save them time and money, increase efficiency, reduce risks, and improve their image. At the same time you may have to show readers why you are better able to deliver on all those promises than your competitors.

---

[6]Reprinted by permission of Associated Credit Bureaus, Inc.

FIGURE 1.6 An advertisement using arguments based on cost, time, efficiency, safety, and convenience to persuade a potential customer to use a service.

SHERIFF
General Medical Services Corp
A subsidiary of
FMI
Federal Medical Industries, Inc. O.T.C.
950 S.W. 12th Avenue, 2nd Floor Suite, Pompano, Florida 33069
(305) 942-1111 FL WATS: 1-800-654-8282
GENERAL MEDICAL WILL STOP THE UNNECESSARY TRANSPORTING OF YOUR INMATES.
We'll bring our X-ray services to your facility, 7 days a week, 24 hours a day.
We can reduce your X-ray costs by a minimum of 28%.
X-ray cost includes radiologists's interpretation and written report.
Same day service with immediate results telephoned to your facility.
Save correctional officers' time, thereby saving your facility money.
Avoid chance of prisoner's escape and possible danger to the public.
Avoid long waits in overcrowded hospitals.
Reduce your insurance liabilities.
Other Services Available: Ultrasound, Two Dimensional Echocardiogram, C.T. Scan, EKG, Blood Lab and Holter Monitor.
General Medical Is Your On-Site Medical Problem Solver

Effective persuasive writing involves using all the skills you will learn about audience analysis, tone, organization, research, and the overall process of drafting, revising, and editing your work. Writing persuasive letters, proposals, and reports requires that you have a clear sense of your audience's needs, priorities, preferences, and even dislikes. You will have to conduct research, provide logical arguments, supply concrete examples or appropriate data, and, especially important, identify the most relevant information for your particular audience.

Notice how the advertisement in Figure 1.6 offers a persuasively worded list of reasons—based on cost, time, safety, efficiency, and convenience—to convince correctional officials that they should use General Medical's services rather than those of a hospital or clinic.

Much job-centered writing also requires you to be highly persuasive when you make a recommendation to your employer. In many of your memos and reports you will have to evaluate various options and products for your employer. Your reader will expect you to offer clear-cut, logical, convincing reasons for your choice. The reader will want to know why and how you arrived at your conclusions.

Below is the summary to one writer's report on why it is better for a company to lease a truck than to purchase one.

> After studying the pros and cons of buying or leasing a company truck, I recommend that we lease it for the following five reasons.
>
> 1. We will not have to expend any of our funds for a down payment, which is being waived.

2. Our monthly payments for leasing the vehicle will be at least $150 less than the payments we would have to make if we purchased the truck on a three-year contract.
3. All major and minor maintenance (up to 36,000 miles) is included as part of our monthly leasing payment.
4. Insurance (theft and damage) is also part of our monthly leasing payment.
5. We have the option of trading in our truck every 16 months for a newer model or trading up every 12 months.

Note the persuasive tone and logical presentation of information this writer has used.

Ethical Writing in the Workplace

On-the-job writing involves much more than conveying facts about products, equipment, costs, and the day-to-day operations of a business. Your writing also has to be ethical. Writing ethically means doing what is right and fair, being honest and just with your employer, co-workers, and customers. Your reputation and character plus your employer's corporate image will depend on your following an ethical course of action.

Many of the most significant bywords in the world of business reflect an ethical commitment to honesty and fairness: *accountability, public trust, good faith effort, truth in lending, fair play, honest advertising, full disclosure, high professional standards, community involvement, social responsibility.*

Unethical business dealings, on the other hand, are stigmatized in *cover ups, shady deals, spin doctors, foul play, misrepresentations, price gouging, bias,* and *unfair advantage.* These are the activities that keep Better Business Bureaus active and make customers angry.

Ethical Requirements on the Job

In the workplace, you will be expected to meet the highest ethical standards by fulfilling the following requirements.

- Complying with all local, state, and federal regulations, especially those ensuring a safe, healthy work environment, products, and/or service
- Adhering to your own profession's code or standard of ethics (for example, certified internal auditors promise to be fair and impartial in all their audits)
- Following your company's policies and procedures
- Honoring guarantees and warranties and meeting customer needs impartially
- Cooperating fully and honestly with any collaborative team of which you may be a member
- Respecting all copyright obligations and privileges

Following the six guidelines above may not be only an ethical requirement; it could also be a legal one. Legal obligations and ethical considerations often merge. For example, doing personal (or outside consulting) work on company time, padding expense accounts, using company equipment for personal use, or accepting a bribe is unethical and illegal. It would also be neglectful and unethical to allow an unsafe product to stay on the market just to spare your company the expense and embarrassment of a product recall. It would be wrong, legally and ethically, to e-mail information about your employer's pending patent plans to the Internet world.

TECH NOTE

Computer ethics, especially when using the Internet, is essential in the world of work. It would be grossly unethical to erase a computer program intentionally, violate a software licensing agreement, or misrepresent (by fabrication or exaggeration) the scope of a database.

In the world of multinational corporations, you also have to make sure you respect the ethics of a host foreign country where your firm does business. Some behavior regarded as normal or routine in the United States might be seen as highly unethical elsewhere. And you should be on your ethical guard not to take advantage of a host country—such as allowing or encouraging poor environmental control because regulatory and inspection procedures are not as strict.

Ethical Dilemmas

Sometimes in the world of work you will face situations where there is no clear-cut right or wrong choice, even with regard to one of the six ethical categories above. You may be involved in an ethical dilemma, or conflict, regarding a decision by your employer, a customer, or a co-worker, or even something you said or did earlier.

Here are a few scenarios, similar to ones in which you may find yourself, that are gray areas, ethically speaking, along with some possible solutions.

- You see an opening for a job in your area but the employer wants someone with a minimum of two years of field experience. You have just completed an internship and had one summer's (12 weeks) experience in the field, which together total almost 7 months. Should you apply for the job describing yourself as "experienced"?

  Yes, but honestly state the types and extent of your field experiences and the conditions under which you had them.

- You work for a company that usually assigns commissions to the salesperson for whom a customer asks. One afternoon a customer visits your store and asks for a salesperson whose name he cannot remember but from whose description you realize is an employee who happens to have the day off. You assist the customer all afternoon, making several long distance calls to locate a particular model and even arrange to have that item shipped overnight to your store so the customer can pick it up in the morning. When it comes time to ring up the sale, should you list your employee number for credit (and the commission) or the off-duty employee's?

You probably should defer crediting the sale to either number until you speak to the absent employee and suggest a compromise—splitting the commission.

- A piece of computer equipment, scheduled for delivery to your customer the next day, arrives with a damaged part. You decide to replace it at your store before the customer receives it. Should you inform the customer?

  Yes, but assure the customer that the equipment is still under the same warranty and that the replacement part is new and also under the same warranty. If the customer protests, agree to let him or her use the computer until a new unit arrives.

As these brief scenarios suggest, concessions and compromises sometimes are absolutely necessary in order to be ethical in the world of work.

Writing Ethically

Your writing as well as your behavior must be ethical. Words, like actions, have implications and consequences. If you slant your words to conceal the truth or gain an unfair advantage, you are not being ethical. In your written work, strive to be careful and straightforward, reporting events and figures honestly without bias or omissions.

As we saw earlier in this chapter, you have to respect your audience, taking their needs and reactions into account. If you lie or exaggerate or minimize, an audience's distrust can come back to haunt you in lost business.

Unethical writing is usually guilty of one or more of the following faults, which can conveniently be listed as the three *M*s— misquotation, misrepresentation, and manipulation. Here are some examples:

1. **Plagiarism** is stealing someone else's words (work) and claiming it as your own. At work, plagiarism is unethically claiming a co-worker's ideas, input, or report as your contribution. In a research report or paper, you are guilty of plagiarism if you use another person's words (or even a rough paraphrase) without documenting the source. Do not think that by changing a few words here and there, you are not plagiarizing. Give proper credit to your source—whether in print, in person (through an interview), or on-line.

2. **Selective misquoting** deliberately omits damaging or unflattering comments to paint a better (but untruthful) picture of you or your company. By picking and choosing words from a quotation, you unethically misrepresent what the speaker or writer originally intended.

Full Quotation: I've enjoyed at times our firm's association with Technology, Inc., although I was troubled by the uneven quality of their service. At times, it was excellent while at others it was far less so.

Selective Misquotation: I've enjoyed . . . our firm's association with Technology, Inc. The quality of their service was . . . excellent.

The dots, called ellipses, unethically suggest that only extraneous or unimportant details were omitted.

3. **Arbitrary embellishment of numbers** unethically misrepresents, by increasing or decreasing percentages or other numbers, statistical or other information. Writers who do this try to stretch the differences between competing plans or proposals to gain an unfair advantage or, on the other hand, express accurate figures in an inaccurate way.

Embellishment: An overwhelming majority of residents voted for the new plan.
Ethical: The new plan was passed by a vote of 53 to 49.

Embellishment: Our competitor's sales volume increased by only 10 percent in the preceding year while ours doubled.
Ethical: Our competitor controls 90 percent of the market, yet we increased our share of that market from 5 percent to 10 percent last year.

4. **Manipulation of data or context,** closely related to #3 above, is the misrepresentation of events, usually to "put a good face" on a bad situation. The writer here unethically uses slanted language or intentionally misleading euphemisms to misinterpret events for readers.

Manipulation: Looking ahead to 2001, the United Funds Group is exceptionally optimistic about its long-term prospects in an expanding global market. We are happy to report steady to moderate activity in an expanding sales environment last year. The United Funds Group seeks to build on sustaining investment opportunities beneficial to all subscribers.
Ethical: Looking ahead to 2001, the United Funds Group is optimistic about its long-term prospects in an expanding global market. Though the market suffered from inflation this year, the United Funds Group hopes to recoup its losses in the year ahead.

In the manipulation example above, the writer minimizes the negative effects of inflation by calling it an "expanding sales environment."

5. **Using fictitious benefits to promote a product or service seemingly promises customers advantages but delivers none.**

False Benefit: Our bottled water is naturally hydrogenated from clear underground springs.
Truth: All water is hydrogenated since it contains hydrogen.

6. **Misrepresenting through distorting or slanted visuals is one of the most common types of unethical communication.** Making a product look bigger, better, or more professional is all too easy with graphics software packages. Other unethical uses of visuals are, for example, making warning or caution statements

(p. 479) the same size and type font as ingredients or directions, or enlarging advertising hype (Double Your Money Back) while reducing major points to small print.

Ethical writing is clear, accurate, fair, and honest. These are among the most important goals of any work communication.

✓ Revision Checklist

At the end of each chapter is a checklist you should review before you submit the final copy of your work, either to your instructor or to your boss. The checklist includes the types of research, planning, drafting, editing, and revising you should do to ensure the success of your work. Regard these checklists as a summary of the main ideas in each chapter as well as a handy guide to quality control. You may find it helpful to check each box as you verify that you have performed the necessary revision/review. Effective writers are also careful editors.

- ❑ Identified my audience—their background, knowledge of English, reason for reading my work, and likely response to my work and me.
- ❑ Made it clear what I want my audience to do after reading my work.
- ❑ Tailored my message to my audience's needs and background—giving them neither too little nor too much information.
- ❑ Pushed to the main point right away; did not waste my reader's time.
- ❑ Selected the most appropriate language, technical level, tone, and level of formality.
- ❑ Did not waste my audience's time with unsupported generalizations or opinions; instead gave them accurate measurements, facts, carefully researched material.
- ❑ Used appropriate visuals to make my work easier for my audience to follow.
- ❑ Used persuasive reasons and data to convince my reader to accept my plan or work.
- ❑ Ensured that my writing is ethical—accurate, fair, honest, a true reflection of the situation or condition I am explaining or describing.
- ❑ Gave full and complete credit to any sources I used, including resource people.
- ❑ Avoided plagiarism and unfair or dishonest use of copyrighted materials—both written and visual, including all electronic media.

Exercises

1. What is your chosen career? Make a list of the types of writing you think you will do, or have already encountered, on the job.

2. Make a list of the kinds of writing you have done in a history or English class or for a laboratory or shop course.

3. Compare your lists for Exercises 1 and 2. How do the two types of writing differ?

4. Bring to class a set of printed instructions, a memo, a sales letter, or a brochure. Comment on how well the printed material answers the following questions.
   a. Who is the audience?
   b. Why was the material written?
   c. What is the message?
   d. Are the style and tone appropriate for the audience, the purpose, and the message? Why?

5. Cut out a newspaper ad that contains a drawing or photograph. Bring it to class together with a paragraph of your own (75–100 words) describing how the message of the ad is directed to a particular audience and commenting on why the illustration was selected for that audience.

6. Pick one of the following topics and write two descriptions of it. In the first description use technical vocabulary. In the second, use language suitable for the general public.
   a. spark plug
   b. blood pressure cuff
   c. carburetor
   d. computer chip
   e. camera
   f. legal contract
   g. electric sander
   h. cyberspace
   i. muscle
   j. protein
   k. compact disc player
   l. e-mail
   m. bread
   n. money
   o. color scanner
   p. soap
   q. calculator
   r. Nintendo game
   s. AIDS
   t. thermostat
   u. trees
   v. food processor
   w. earthquake
   x. recycling

7. Accomplish Exercise 6 above as a collaborative writing project.

8. Select one article from a daily newspaper and one article from either a professional journal or your major field or one of the following journals: *Advertising Age, American Journal of Nursing, Business Marketing, Business Week, Computer, Computer Design, Construction Equipment, Criminal Justice Review, Food Service Marketing, Journal of Forestry, Journal of Soil and Water Conservation, National Safety News, Nutrition Action, Office Machines, Park*

Maintenance, Scientific American. State how the two articles you selected differ in terms of audience, purpose, message, style, and tone.

9. Assume that you work for Appliance Rentals, Inc., a company that rents TVs, microwave ovens, stereo components, and the like. Write a persuasive letter to the members of a campus organization or civic club urging them to rent an appropriate appliance or appliances. Include details in your letter that might have special relevance to members of this specific organization.

10. Evaluate how well the advertisement on p. 30 illustrates the technique of occupational writing described in this chapter. Specifically comment on what analysis of the ad reveals about the copyrighter's analysis of the intended audience. Pay attention to advertising copy (words), the images of people and equipment (visuals), and the situation depicted. Also explain how the ad illustrates the six functions of on-the-job writing.

11. Read the article on pp. 31–32 and identify its audience (technical or general), purpose, message, style, and tone.

12. The following statements contain embellishments, selected misquotations, false benefits, and other types of unethical tactics. Revise each statement to eliminate the unethical aspects.
    - **a.** Storm damage done to water filtration plant #3 was minimal. While we had to shut down temporarily, service resumed to meet residents' needs.
    - **b.** All customers qualify for the maximum discount available.
    - **c.** The service contract . . . on the whole . . . applied to upgrades.
    - **d.** We followed the protocols precisely with test results yielding further opportunities for experimentation.
    - **e.** All our costs were within fair-use guidelines.
    - **f.** Customers' complaints have been held to a minimum.
    - **g.** All the lots we are selling offer easy access to the lake.

13. A co-worker tells you that he has no plans to return to his job after he takes his annual two-week vacation. You know that your department cannot meet its deadline short-handed and that your company will need at least two or three weeks to recruit and hire a qualified replacement. You also know that it is your company's policy not to give paid vacations to employees who do not agree to work for at least three months following a vacation. What should you do? What points would you make in an ethical memo to your boss? What points would you make to your co-worker?

14. Your company is regulated and inspected by the Environmental Protection Agency. In ninety days the EPA will relax a particular regulation about dumping industrial waste. Your company's management is considering cutting costs by relaxing the standard now, before the new, easier regulation is in place. You know that the EPA inspector probably will not return before the ninety-day period elapses. What do you recommend to management?

Courtesy of Trans-Lux Corporation, Norwalk, Conn. Reprinted with permission.

Microwaves

Much of the world around us is in motion. A wave-like motion. Some waves are big like tidal waves and some are small like the almost unseen footprints of a waterspider on a quiet pond. Other waves can't be seen at all, such as an idling truck sending out vibrations our bodies can feel. Among these are electromagnetic waves. They range from very low frequency sound waves to very high frequency X-rays, gamma rays, and even cosmic rays.

Energy behaves differently as its frequency changes. The start of audible sound—somewhere around 20 cycles per second—covers a segment at the low end of the electromagnetic spectrum. Household electricity operates at 60 hertz (cycles per second). At a somewhat higher frequency we have radio, ranging from shortwave and marine beacons, through the familiar AM broadcast band that lies between 500 and 1600 kilohertz, then to citizen's band, FM, television, and up to the higher frequency police and aviation bands.

Even higher up the scale lies visible light with its array of colors best seen when light is scattered by raindrops to create a rainbow.

Lying between radio waves and visible light is the microwave region—from roughly one gigahertz (a billion cycles per second) up to 3000 gigahertz. In this region the electromagnetic energy behaves in special ways.

Microwaves travel in straight lines, so they can be aimed in a given direction. They can be *reflected* by dense objects so that they send back echoes—this is the basis for radar. They can be *absorbed,* with their energy being converted into heat—the principle behind microwave ovens. Or they can pass *through* some substances that are transparent to the energy—this enables food to be cooked on a paper plate in a microwave oven.

Microwaves for Radar

World War II provided the impetus to harness microwave energy as a means of detecting enemy planes. Early radars were mounted on the Cliffs of Dover to bounce their microwave signals off Nazi bombers that threatened England. The word *radar* itself is an acronym for *RA*dio *D*etection *A*nd *R*anging.

Radars grew more sophisticated. Special-purpose systems were developed to detect airplanes, to scan the horizon for enemy ships, to paint finely detailed electronic pictures of harbors to guide ships, and to measure the speeds of targets. These were installed on land and aboard warships. Radar—especially shipboard radar—was surely one of the most significant technological achievements to tip the scales toward an Allied victory in World War II.

Today, few mariners can recall what it was like before radar. It is such an important aid that it was embraced universally as soon as hostilities ended. Now, virtually every commercial vessel in the world has one, and most larger vessels have two radars: one for use on the open sea and one, operating at a higher frequency, to "paint" a more finely detailed picture, for use near shore.

Microwaves are also beamed across the skies to fix the positions of aircraft in flight, obviously an essential aid to controlling the movement of aircraft from city to city across the nation. These radars have also been linked to computers to tell air traffic controllers the altitude of planes in the area and to label them on their screens.

A new kind of radar, phased array, is now being used to search the skies thousands of miles out over the Atlantic and Pacific oceans. Although these advanced radars use microwave energy just as ordinary radars do, they do not depend upon a rotating antenna. Instead, a fixed antenna array, comprising thousands of elements like those of a fly's eye, looks everywhere. It has been said that these radars roll their eyes instead of turning their heads.

High-Speed Cooking

During World War II Raytheon had been selected to work with M.I.T. and British scientists to accelerate the production of magnetrons, the electron tubes that generate microwave energy, in order to speed up the production of radars. While testing some new, higher-powered tubes in a laboratory at Raytheon's Waltham, Massachusetts, plant, Percy L. Spencer and several of his staff engineers observed an interesting phenomenon. If you placed your hand in a beam of microwave energy, your hand would grow pleasantly warm. It was not like putting your hand in a heated oven that might sear the skin. The warmth was deep-heating and uniform.

Spencer and his engineers sent out for some popcorn and some food, then piped the energy into a metal wastebasket. The microwave oven was born.

From these discoveries, some 35 years ago, a new industry was born. In millions of homes around the world, meals are prepared in minutes using microwave ovens. In many processing industries, microwaves are being used to perform difficult heating or drying jobs. Even printing presses use microwaves to speed the drying of ink on paper.

In hospitals, doctors' offices, and athletic training rooms, that deep heat that Percy Spencer noticed is now used in diathermy equipment to ease the discomfort of muscle aches and pains.

Telephones without Cable

The third characteristic of microwaves—that they pass undistorted through the air—makes them good messengers to carry telephone conversations as well as live television signals—without telephone poles or cables—across town or across the country. The microwave signals are beamed via satellite or by dish reflectors mounted atop buildings and mountaintop towers.

Microwaves take their name from the Greek *mikro* meaning very small. While the waves themselves may be very small, they play an important role in our world today: in defense; in communications; in air, sea, and highway safety; in industrial processing; and in cooking. At Raytheon the applications expand every day.

Raytheon Magazine Winter 1981: 22–23. Reprinted by permission.

CHAPTER 2

The Writing Process at Work

In Chapter 1 you learned about the different functions of writing for the world of work and also explored some basic concepts all writers must master. To be a successful writer, you need to

- identify your audience's needs
- determine your purpose in writing to that audience
- make sure your message meets your audience's needs and your established purpose
- use the most appropriate style and tone for your message
- format your work to clearly reflect your message for your audience

Just as significant to your success is knowing how effective writers actually create their work for their audiences. This chapter will give you some practical information about the strategies and techniques careful writers use when they work. These procedures are a vital part of what is known as the *writing process.* This process involves such matters as how writers gather information, how they transform their ideas into written form, and how they organize and revise what they have written to make it suitable for their audiences.

What Writing Is and Is Not

As you begin your study of writing for the world of work, it might be helpful to identify some notions about what writing is and what it is not.

What Writing Is Not

- **Writing is not something mysterious done according to a magical formula known only to a few.** Even if you have not done much writing before, you can learn to write effectively.

- **Writing is not simply a hit or miss affair, left up to chance.** Successful writing requires hard work and thoughtful effort. It is not done well by simply going through an ordered set of steps as if you were painting by number. You cannot sit down for fifteen minutes and expect to write the perfect memo, letter, or short report straight through. Writing does not proceed in some predictable way, in which introductions are always written first and conclusions last.
- **Just because you put something on paper or on a computer screen does not mean it is permanent and unchangeable.** Writing means rewriting, revising, rethinking. The better a piece of writing is, the more the writer has reworked it.

What Writing Is

- Writing is a fluid process; it is dynamic, not static. It enables you to discover and evaluate your thoughts.
- A piece of writing changes as your thoughts and information change, and as your view of the material changes.
- Writing means making a number of judgment calls.
- Writing grows sometimes in bits and pieces and sometimes in great spurts. It needs many revisions; an early draft is never a final copy.

Researching

Before you start to compose any e-mail, memo, letter, or report, you'll need to do some research. Depending on the size of your assignment, your research can be simple or elaborate. Yet whatever the size of the job, research is crucial to obtain the right information for your audience. They expect the information you give them to be factually correct and intellectually significant. The world of work is based on conveying information—the logical presentation and sensible interpretation of facts. You will have to do research to obtain that kind of information.

Don't ever think you are wasting time by not starting to write your report or letter immediately. Actually, you will waste time and risk doing a poor job if you do not find out as much as possible about your topic (and your audience's interest in it).

First, find out as much as you can about the nature of your assignment and your readers:

- audience (expert? technicians? general? routed to other departments?)
- audience's purpose
- kinds of information audience needs and why
- format (e-mail? memo? letter? report?)
- scope (limited—one page? extensive—twenty pages?)

Next, determine the exact kind of research you must do in order to gather and interpret the information your audience needs. Your research can include:

- Interviewing people inside and outside your company
- Doing fieldwork or performing lab studies
- Preparing for conferences to ask the right questions
- Collaborating with colleagues in person or over e-mail
- Distributing a questionnaire and conducting a survey
- Surfing the Net
- Searching abstracts, indexes, reference books, Web sites (see Chapter 9)
- Reading current periodicals, reports, other documents
- Evaluating reports, products, services
- Getting briefings from sales or technical staff
- Contacting customers

TECH NOTE

The computer is an invaluable resource during your research phase; in fact, it has revolutionized the way information is disseminated. It can help you search efficiently and thoroughly through thousands of databases for key articles and reports in just a matter of seconds. Many reference works—encyclopedias, handbooks, abstracts—are also available on-line (see Chapter 9). The resources of the Internet are limitless. Depending on your library's or company's software and Web browsers, you can easily access these databases, reference works, and the Net itself. Computerized information retrieval will give you the most current data, about the stock market, technological advances, or world events.

You can also use your computer to research appropriate company records—e-mail and memos from your boss and co-workers, previous communications with customers, and company ads, brochures, newsletters, reports, catalogs, and budgets. By becoming familiar with the history of communications about a specific problem or with a particular customer, you will be better able to document and assess your topic. Such in-house research via your computer will give you the necessary edge to succeed. You can also call up previously prepared documents (boilerplates) to insert into your document later.

As you do your research, take notes at your terminal and copy them into new files for use when you outline and draft. You can also transfer quotations, graphs and other visuals, statistical data, and ideas and opinions from your notes into a file for later use in writing your document. During your research, too, you can create a Works Cited file, listing all the sources you consulted, and then include it in the final version of your document.

Research is not confined to the beginning of the writing process; it goes on throughout. As you start to tackle your subject, take a little extra time during this formative period to think about the information you have gathered so you can most effectively adapt it for your audience.

Case Study: The Writing Process

Office manager Melissa Hill asked Gorden Reynolds to recommend ways to improve office efficiency and customer relations. In preparing his report, Reynolds knew he had to do some research. To make a recommendation, he needed to find out information about different kinds of communication technologies. As he read articles in various business journals and magazines and home pages on the Web, he learned about the various laser printers on the market.

As he studied the literature about these printers and their vendors, he realized that several offered very strong benefits for his company in the areas of concern to his boss. He visited a few dealers in his city and, after seeing some demonstrations, was convinced that purchasing a LaserJet printer was feasible and economically wise. The results of all his research would be reflected in his final report.

Planning

At this stage in the writing process, your goal is to get something—anything—down on paper or on your computer screen. For most writers, many of whom are fearful of writer's block, getting started is the hardest part of the job. But you will feel more comfortable and confident once you begin to see your ideas before your eyes. It is always easier to clarify and criticize something you can see.

TECH NOTE

Use your word processing program to plan your document either by creating an outline or a brainstormed list. Don't jump right into the drafting stage, however tempting a blank computer screen might be. A computer can help you to generate items for your outline or list quickly. You might even use one of the software packages designed to generate and organize information so you can identify ideas quickly. With a word processing program you can add, delete, rearrange, combine, expand, and organize points into categories and subcategories. Be sure to save your outline or brainstormed list and put it into a new file that you can call up when you begin drafting your document.

Getting started is also easier if you have researched your topic because you have something to say and to build on. Each part of the process relates to and supports the next. Careful research prepares you to begin writing.

Still, getting started is not easy. You can take advantage of a number of widely used strategies that can help you to develop, organize, and tailor the right information for your audience. Use any one of the following techniques, alone or in combination.

1. Clustering. In the middle of a sheet of paper, write the word or phrase that best describes your topic, and start writing any words or phrases that come to mind about this word. As you write, circle each word or phrase and then connect it to the word from which it sprang. Note the clustered grouping in Figure 2.1. The writer used this strategy to get started on a report encouraging a manager to switch to flex time. (Flex time is a system in which employees may work on a flexible time schedule within certain limits.) The resulting diagram is still an incomplete picture of the final report, but it does give the writer a rough sense of some of the major divisions of the topic and where they may belong in the report.

2. Brainstorming. At the top of a sheet of paper or your computer screen, describe your topic in a word or phrase and then list any information you know or found out about that topic—in any order and as quickly as you can. Brainstorming is like thinking aloud except that you are recording your thoughts. As you brainstorm, don't stop to delete, rearrange, or rewrite anything, and don't dwell on any one item. Don't worry about spelling, punctuation, grammar, or whether you are using words and phrases instead of complete sentences. Keep the ideas flowing. The result may well be an odd assortment of details, comments, and opinions.

After ten to fifteen minutes, stop and take a short break. When you come back to your list, you will no doubt want to make some changes. Some of your points will be irrelevant, so strike them. You may add some ideas, or combine or rearrange others as you start to develop them in more detail. Your list is not final by any means: you have just begun to mine the raw ore.

Figure 2.2 (p. 39) shows Gordon Reynolds's initial brainstormed list. After he began to revise it, he realized that some items were not relevant for his audience (6, 8, and 13) and that others were pertinent but needed to be adapted for his reader (5). He also recognized that some items were repetitious (1, 2, and 11). Further investigation revealed that his company could purchase a laser printer for far less than his initial high guess (17). As Reynolds continued to work on his list, other important points came to mind that were not part of his original brainstorming.

3. Outlining. For most writers, outlining is the easiest and most comfortable way to begin planning their report or letter. Outlines can go through stages, so don't worry if your first attempt is brief and messy. It does not have to be formal (with roman and arabic numerals), complete, or pretty. It is intended for no one's eyes but yours. Use your preliminary outline as a quick way to sketch in some ideas, a convenient container into which you can put information. You might simply jot down a few major points and identify a few subpoints.

FIGURE 2.1 Clustering on the topic of flex time.

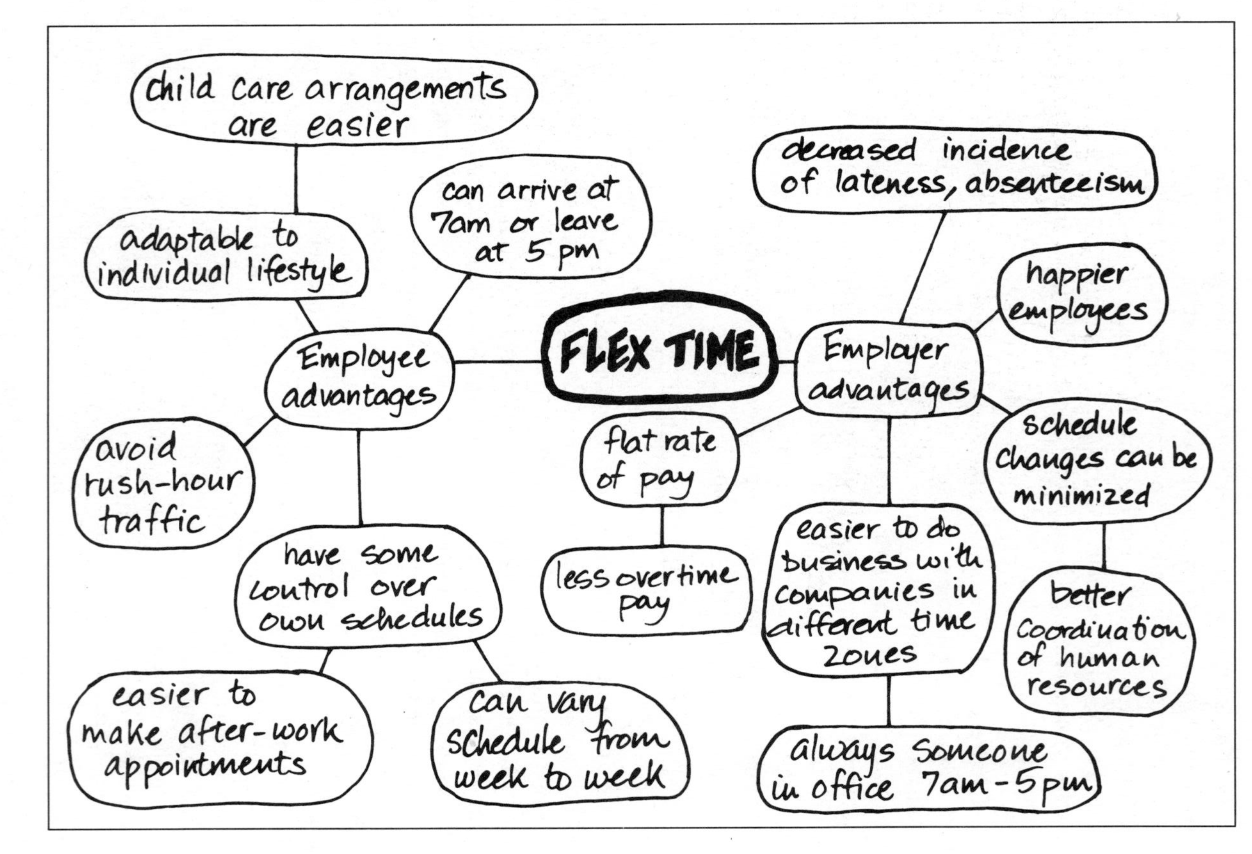

FIGURE 2.2 Gordon Reynolds's initial, unrevised brainstormed list.

1. laser printers work better
2. laser printers faster and quieter
3. would blend into the office better
4. better print quality and graphics
5. more font options
6. same color as dot matrix printers
7. would use ordinary office paper
8. laser optics—real contribution to technology
9. compatible with new software
10. help get the most out of our inhouse documents
11. laser printers reliable and energy efficient
12. can print envelopes
13. stock is doing better on Wall Street
14. would be easy to change over
15. would drastically reduce our cost for copying
16. would help us do our work better
17. top-of-the-line can be bought for a little over $1000

FIGURE 2.3 Gordon Reynolds's early outline after revising his brainstormed list.

I. Convenience

A. One cable hookup like any other printer

B. Easy to install and operate

C. Can be configured easily through our operating system

D. Can handle the output for the entire office

E. Good print quality

F. Uses regular, not perforated, paper

G. Less noise

II. Time/Efficiency

A. Fast – 12 to 20 pages per minute

B. Can print envelopes

C. Greater graphics capability –75 scalable true type fonts

D. 50,000 page monthly duty cycle means less maintenance

III. Money

A. Save on service calls

B. Can handle the workload of the entire office

C. Save cost of copies by making our own

D. Will be compatible with future upgrades of the office PCs

As you continue to work on your outline, you will discover key relationships about major points and supporting ideas and different ways to develop them. You will also be able to delete and add details. If you have done your research, you should find that with each successive outline you are coming closer and closer to drafting, your next stage.

As with a brainstormed list, leave your outline to cool for a while. When you return you can juggle points and add or delete others.

Case Study: Outline

Note that Gordon Reynolds organized his revised brainstormed list into an outline (Figure 2.3). There is no one right way to get started. If you are not sure which way works best for you, experiment with a few techniques, as Reynolds did.

Drafting

If you have done your planning carefully, you will find it easier to start your first draft. When you draft, you convert the words and phrases from your outlines, brainstormed lists, or clustered groups into paragraphs. Think of your earlier jottings as the material out of which the basic building blocks (paragraphs) of your drafts will come. During drafting, as elsewhere in the writing process, you will see some overlap as you look back over your lists or outlines to create your draft(s).

Keep in mind that drafting is still preliminary to the creation of your final copy. Don't expect to wind up with a polished, complete version of your paper after working on only one draft. In most cases, you will have to work through many drafts, but each draft should be less rough, more acceptable, than the preceding one.

TECH NOTE

When you draft your document, the computer will help you to get some writing on the monitor quickly. A word processing program keeps pace with the speed of your thought process and allows you to concentrate on your writing rather than on the more technical aspects of creating a document. You do not have to stop to check spelling or punctuation; you can return to these tasks later. And if you run into a snag on any draft, you do not have to lose momentum by stopping to fix it. With a single stroke of the keyboard or mouse, the word processor will flag items to which you want to return or about which you need more information.

Moreover, you can manipulate line spacing so that if you are not ready to expand a point or qualify an idea, you can leave extra space for later additions or highlight a sentence or word to flag it for more attention.

You can insert notes to yourself to supply information or qualification or even documentation. The computer lets you send helpful signals to yourself, reminding you that your draft is a document in the making.

Thanks to word processing software, therefore, you can create, expand, delete, move, save, and retrieve different drafts, or even portions of one or more drafts. That way you can go back to see which version (of a draft) you like best or what might be salvaged from one version and incorporated into another. For that reason it is best to press the save command after every page, or at regular intervals of every two to three minutes. You won't lose your work that way. When you save drafts (or revisions), your file names should always clearly reflect their contents (Draft 1, Works Cited).

You can print out a draft as often as you like to see what it looks like; you do not have to worry about the time-consuming drudgery of having to rekey successive drafts because of changes. Drafting at your terminal thus gives you maximum flexibility and reduces the risk of mistakes and omissions because of a messy paper copy. During drafting some writers in fact move back and forth between hard copy (the paper printout) and the text on the screen.

Function of Drafting

The purpose of drafting is to get your main points down in the most logical order for your readers. The most common reader complaint is that key information is hard to find. Effective drafting means making big decisions about (1) what you say, and (2) where you say it. Accordingly, during drafting you need to pay special attention to the content and organization of your work.

Key Questions to Ask as You Draft

As you work on your drafts, ask yourself the following questions about your content and organization.

- Is this the best way to start?
- Am I giving my readers too much or too little information?
- Is my information too technical for my audience?
- Does this point belong where I have it, or would it more logically follow or precede something else?
- Is this point necessary and relevant?
- Am I repeating myself?
- Have I contradicted myself?
- Have I ended appropriately for my audience?

To answer these questions successfully, you may have to continue researching your topic and reexamining your audience's needs. But in the process new and even better ideas will come to you, and ideas that you once thought were essential may in time appear unworkable or unnecessary.

Guidelines for Successful Drafting

Following are some suggestions to help your drafting go more smoothly and efficiently.

- Select a comfortable place to write. Use a pen, typewriter, PC, or laptop computer—whatever gives you the most satisfying feeling about creating. Take another look at Figure 1.5 to see proper posture.
- In an early draft, write the easiest part first, regardless of where in the paper it may finally end up. Some writers feel more comfortable drafting the body (or middle) of their work first.
- As you work on a later draft, write straight through. Do not worry about spelling, punctuation, or the way a word or sentence sounds. Save these concerns for later stages.
- Allow enough time between drafts so that you can evaluate your work with fresh eyes and a clear mind.
- Get frequent "outside" opinions. Show, e-mail, or fax a draft to a fellow student, a co-worker, or maybe an employer for comment. A new pair of eyes will see things you missed.
- Start considering what visuals might enhance the quality of your work and where they might best be positioned.

Case Study: Draft

Figure 2.4 shows one of the several drafts Gordon Reynolds prepared. Because he wisely recognized that his outline was not final, he continued to work on it during the drafting stage. Note that he has added an introduction and conclusion that were not part of his outline (Figure 2.3). These are necessary parts of Reynolds's report if he is to accomplish his purpose of convincing Melissa Hill to purchase a laser printer. Even so, Reynolds recognized that his draft was still not ready for his boss to see, so he showed it to a co-worker for suggestions.

From discussions with his co-worker, and after further studying his draft, Reynolds realized that he had placed one of the most important considerations for his audience (savings) last. In his final version, shown in Figure 2.5 (pp. 46–47), he moved this section to the beginning of his memo because he realized that Melissa Hill would be most concerned about costs. Reynolds thus paid attention to his audience's priorities and needs. He also added headings and bulleted lists to help his reader find information. The design of his earlier draft in Figure 2.4 did not assist his readers in finding information quickly and did not reflect a convincing organization of his proposal.

FIGURE 2.4 An intermediate draft of Gordon Reynolds's report.

To: Melissa Hill, Office Manager
From: Gordon Reynolds
Date: September 4, 1998
Subject: Improving Efficiency

As you requested, I have been researching what to do about improving office efficiency and customer relations. The most beneficial and immediate solution I have found so far is to replace our old printers with a new laser printer. More and more businesses today have these printers, and they are fast becoming a part of many companies' information technology systems. With the advances in technology of the past few years it makes good sense to replace our TechWorld printers with one **BaxterLaserJet GTP**. Today, laser printers are faster, quieter, and more efficient than ever.

A laser printer is easy to set up and operates just like our old printers, and can be configured through our operating system. Among the models of laser printers I have seen, most are about the same size as one of our old printers. It should be easy, too, for our staff to learn to operate the new laser printer. Because a laser printer would be faster and quieter, one could easily handle the output of our entire office. Furthermore, the print quality and the fact that it uses regular office paper would make it just that much more convenient. From several demonstrations I have seen, I am truly impressed.

A laser printer is fast and efficient, printing 12-20 pages a minute with little or no noise. A laser printer is also more versatile because it can print on different types of paper and even envelopes. It would be so much easier for large mailings we routinely have just before the holidays.

Continued

FIGURE 2.4 (Continued)

page 2

With 75 scalable true type fonts, we can produce top-quality documents in-house, saving us the trouble of finding an outside contractor to do the work. And finally, with a 50,000 page monthly duty cycle, a laser printer would be low-maintenance.

The greatest benefit to our company is the money a laser printer will save. Generally speaking, a laser printer runs about $1100 to $1300, depending on the model. The **Baxter LaserJet GTP** can be purchased direct from USA Computers for $1023 plus shipping and handling, not much more than a good fax machine. Because a laser printer can go longer without maintenance, we can save on service calls. Also, USA Computers offers a two year warranty. A new laser printer will also be compatible with future upgrades of our PC's, should we decide to do so.

Revising

Revision is an essential stage in the writing process. It requires more than giving your work one more quick glance. Do not be tempted to skip the revision stage just because you have written the required number of words or sections or because you think you have put in too much time already. Revision is done *after* you produce a draft that you think conveys the appropriate message for your audience. The quality of your letter or report depends on the revisions you make now.

Allow Enough Time to Revise

Like planning or drafting, revision is not done well in one big push. It evolves over a period of time. Because revision is so important, allow yourself enough time to do it carefully. Avoid drafting and revising in one sitting. If possible, wait at least a day before you start to revise (or, in the busy work world, waiting a couple of hours may suffice). In the meantime, you might ask a co-worker or friend familiar with your topic to comment on your work, as Gordon Reynolds did. Plan to read your revised work more than once.

FIGURE 2.5 The final copy of Gordon Reynolds's report.

To: Melissa Hill, Office Manager
From: Gordon Reynolds
Date: September 4, 1998
Subject: Purchasing a New Laser Printer

As you requested, I have investigated some ways to improve our office efficiency and customer relations. The best solution I have found is to replace our two TechWorld dot matrix printers with a new laser printer. With advances in printer technology in the past few years (I enclose a copy of a review article "Share and Share A Lot" from *Computer World Magazine* [June 1998]: 121-29), it makes good sense to replace our printers with a **Baxter LaserJet GTP**. Laser printers are economical, more efficient, and will improve our document design.

Cost

The greatest benefit of a laser printer is the money it will save. We can purchase it from USA Computers for $1039 plus shipping and handling, totaling $1074. Purchasing it allows us to easily recoup that cost in just a month or two because we will

- have less maintenance, saving at least $200 per month in service calls
- receive a two year warranty with guaranteed overnight service, which should decrease down time
- save an additional $50 to $100 each month by not having to buy special perforated paper
- be able to trade in dot matrix printers for $75 each

Continued

FIGURE 2.5 (Continued)

Page 2

Efficiency

A laser printer would be easy to set up and is compatible with our current computer system. Other key features include

- ample paper capacity (500 sheets)
- enhanced memory that manages multiple jobs easily, accepting and printing them in the order sent
- ability to print from 12 to 20 pages per minute with little or no noise
- ability to handle the output of our entire office, including our large quarterly mailings

Quality

The **LaserJet** prints high quality documents because it

- has 1200 dpi resolution
- exhibits same density on 1000th as on first
- produces superior graphics
- functions with different types of paper (ledger or letterhead)
- can print envelopes
- offers 75 scalable true type fonts
- can be used for inhouse documents such as our own reports and the monthly newsletter, saving us the trouble and expense of using an outside contractor

I recommend that we order a **Baxter LaserJet GTP** from USA Computers. It will unquestionably save us money, improve office efficiency, and help us to project a much more professional corporate image.

Revision Is Rethinking

When you revise, you **re**see, **re**think, and **re**consider your entire document. You ask questions about the major issues of content, organization, and tone. Revision involves going back and repeating earlier steps in the writing process. You can also get help **re**seeing from colleagues through a collaborative effort (see Chapter 3).

When effective writers revise their work, they go through their drafts several times to check for content, organization, and tone. Note the extent of changes Gordon Reynolds made between his draft in Figure 2.4 and the final, revised copy of his memo in Figure 2.5.

TECH NOTE

When you are revising, a word processor assists you in refining and refocusing your thoughts. Using a word processor can actually encourage you to make revisions because they can be accomplished so easily and quickly. You are relieved of the burden of rekeying what you want to save each time you make a change or spot an omission or error. Since you can add, delete, qualify, and rearrange words, sentences, and paragraphs, or even transform the appearance of your material, you can experiment with a number of different versions of your work without having to key in each one separately. Just remember to save what you have done each time.

At the revision stage you can insert visuals exactly where you want them and format your document to include headings, different fonts, and other design elements (see Chapter 12).

Various software packages allow you to split your monitor screen into two or four parts (windows) to view different versions of a text or several pages of the text simultaneously. Or you can view one page while working on another. Moreover, you can go to the print mode and enter "print preview" to see each page layout. That way you can judge if your document falls under or exceeds the page limits your boss or instructor has set for you. From the broader perspective of your overall document, you will be able to make global revisions, rather than just line changes.

Revision means asking again the questions you have already asked and answered during the planning and drafting stages. In the process you will discover gaps and points to change and errors to correct in your draft. Revision gives you a second (or third or fourth) chance to get things right for your audience and to clarify your purpose in writing to them.

Don't Forget Your Audience

As you revise your draft to make it better, ask the same questions about your audience that we studied in Chapter 1 (pp. 6–9). Once more you will inspect your draft to make sure that it meets your audience's priorities. Your audience's reason for reading your work will determine the three big issues that you should be concerned with when you revise—content, organization, and tone.

Key Questions to Ask as You Revise

Content

1. Is it accurate? Are my facts (figures, names, dates, costs, references, statistics) correct?
2. Is it relevant for my audience and purpose? Have I included information that is unnecessary, too technical, not appropriate? Does every detail belong in this communication with my readers?
3. Have I included sufficient information for my readers' purpose? Have I given enough evidence to explain things adequately and to persuade my readers? (Too little information will make readers skeptical about what you are describing or proposing, so they may question your conclusions.) Have I left anything out? Do I need to clarify or explain more for my readers' purpose?

Organization

1. Is the arrangement of my information clear and straightforward? Is it easy to follow?
2. Have I clearly identified my main points and shown my readers that they are important?
3. Is everything proportionate to my purpose and my readers' needs, or have I spent too much (or too little) effort on one section? Do I repeat myself? What can be cut?
4. Is everything in the right, most effective, order? Do I need to regroup sections or paragraphs of my document? Should anything be switched or moved closer to the beginning or the end of my document?
5. Have I grouped related items in the same part of my report or letter, or have I scattered details that really need to appear together in one paragraph or section?
6. Is my work logical? Do my conclusions follow from the evidence I present? Are my recommendations valid and based on the conclusions I draw?

Tone

1. How do I sound to my readers? Am I professional and sincere, or arrogant or unreliable? What attitude do my words or expressions convey?
2. How will my readers think I perceive them? Will they know I believe they are honest and intelligent, or have I used words and details that seem to question their judgments, professionalism, or intelligence?

Case Study: The Revision Process

Mary Fonseca, an employee at Seacoast Labs, was asked by her supervisor to prepare a short report for the general public on the lab's most recent experiments. One of her later drafts begins with the paragraphs in Figure 2.6.

As Fonseca worked on her ideas and thought about her audience's needs, she realized that this opening section of her draft presented some problems that she would need to resolve. As she started to revise, she asked herself questions like those on the previous pages about the content, order, and tone of her work. Answering these questions in turn led her to make a number of major changes that often occur at the revision stage.

Fonseca knew that her first paragraph lacked focus. It was not carefully organized, jumping back and forth between drag on a ship and on an airplane. Consequently, she decided to delete information about planes, especially since Seacoast Labs did not work in this area. She also decided that the information on the effect of drag on a ship was so important that it deserved a separate paragraph. So she removed the details on that topic from this long first paragraph and started

FIGURE 2.6 Opening paragraphs of Mary Fonseca's draft.

Drag is an important concept in the world of science and technology. It has many implications. Drag occurs when a ship moves through the water and eddies build up. Ships on the high seas have to fight the eddies, which results in drag. In the same way, an airplane has to fight the winds at various altitudes at which it flies; these winds are very forceful, moving at many knots per hour. All of these forces of nature are around us. Sometimes we can feel them, too. We get tired walking against a strong wind. The eddies around a ship are the same thing. These eddies form various barriers around the ship's hull. They come from a combination of different molecules around the ship's hull and exert quite a force. Both types of molecules pull against the ship. This is where the eddies come in.

Scientists at Seacoast Lab are concerned about drag. Dr. Karen Runnels, who joined Seacoast about three years ago, is the chief investigator. She and her team of highly qualified experts have constructed some fascinating multilevel water tunnels. These tunnels should be useful to ship owners. Drag wastes a ship's fuel.

to develop them separately. In the process, she discovered that her explanation of molecules, eddies, and drag was unclear; it needed to be reorganized and made more reader-friendly.

Often additional ideas come to you even as you revise your work. As Fonseca continued reading about the topic, she came across an interesting analogy that she felt was appropriate for her audience, so she included it. Working on this explanation, she had to decide how to organize her information. After some deliberation, she decided to follow a cause and effect pattern—drag is the cause and the effect is how it works to slow a ship.

So much for a new second paragraph. But Fonseca was then left with the job of finding an opening for her short report. She wanted to start with something that would both introduce the topic and encourage her layperson audience to continue reading.

Buried in the original paragraph from her last draft was an idea she found worth developing—that we cannot always *see* the forces of nature but we can *feel* them. So rather than beginning with a wooden statement about the scientific concept of drag (which she realized did not have to be scientifically defined for her audience), Fonseca revised her draft to start with a simple but visual example about an individual walking against the wind. Through continued revisions she created two useful, carefully organized paragraphs out of the one long opening paragraph of her last draft.

Throughout the revision process Fonseca kept her readers' needs in mind. Working on her original second paragraph, she realized that it said very little about the lab's experiments with drag. In fact, what she had initially written might even confuse readers about how and why the tunnels were useful. Once Fonseca put herself in her readers' place, however, she asked the exact question that helped her to develop the paragraph: "What specific things are they doing at Seacoast Labs to reduce drag?"

Generating a precise starting sentence from this question made all the difference. Focusing now on the lab's "ways to reduce drag on ships," Fonseca hit upon a focused idea that guided her in choosing specific examples about experiments at Seacoast. Instead of leaving in interesting but unnecessary information (such as that Dr. Runnels has been at the lab for three years), Fonseca sought concrete illustrations and explained their significance to her audience. She also discovered that the last sentence in her original second paragraph did not belong in the new revised paragraph on experiments. The cost of drag was much more appropriately placed in her new opening paragraph.

Concerned about the design of her revised document, Fonseca recognized that though it was better written than her draft, it looked no different. Her page appeared crowded, uninviting, and offered no signposts to direct her readers. Accordingly, she added two headings and double-spaced the text. The result was a much more professional-looking document.

Fonseca's final revision is shown in Figure 2.7. Through revising, she transformed two poorly organized and incomplete paragraphs in three carefully separated yet logically connected paragraphs.

FIGURE 2.7 A revision of Mary Fonseca's draft in Figure 2.6.

What is drag?

We cannot see or hear many of the forces around us, but we can certainly detect their presence. Walking or running into a strong wind, for example, requires a great deal of effort and often quickly leaves us feeling tired. When a ship sails through the water, it also experiences these opposing sources known as drag. Overcoming drag causes a ship to reduce its energy efficiency, which leads to higher fuel costs.

How drag works

It is not easy for a ship to fight drag. As the ship moves through the water, it drags the water molecules around its hull at the same rate the ship is moving. Because of the cohesive force of these molecules, other water molecules immediately outside the ship's path get pulled into its way. All these molecules become tangled rather than simply sliding past each other. The result produces an eddy, or small circling burst of water around the ship's hull, that intensifies the drag. Dr. Charles Hester, a noted engineer, explains it using an analogy: "When you put a spoon in honey and pull it out, half the honey comes out with the spoon. That's what is happening to ships. The ship is moving and at the same time dragging the ocean with it."

At Seacoast Labs, scientists are working to find ways to reduce drag on ships. Dr. Karen Runnels, the principal investigator, and a team of researchers have constructed water tunnels to simulate the movement of ships at sea. The drag a ship encounters is measured

Continued

FIGURE 2.7 (Continued)

from the tiny air bubbles emitted in the water tunnel. Dr. Runnel's team has also developed the use of polymers, or long carbon chain molecules, to reduce drag. These polymers act like a slimy coating for the ship's hull to help it glide through the water more easily. When asbestos fibers were added to the polymer solutions, the investigators measured a 90 percent reduction in drag. The team has also experimented with an external pump attached to the hull of the ship, which pushes the water away from the ship's path.

Editing

Editing means getting the final copy ready for your audience. This last stage in the writing process might be compared to detailing an automobile—the preparation a dealer goes through to ready a new car for prospective buyers. Editing is done only after you are completely satisfied that you have made all the big decisions about content and organization—that you have said what you wanted to, where and how you intended, for your audience.

When you edit, you will check your work for

- sentences
- word choices
- punctuation
- spelling
- grammar and usage

As with revising, don't skip or rush through the editing process, thinking that once your ideas are down, your work is done. Your style, punctuation, spelling, and grammar matter a great deal to your readers. If your work is hard to read or contains mistakes in spelling or punctuation, your readers will think that your ideas, your research, and your organization are also faulty.

The following sections will help you understand what to look for when you edit your sentences and words. The appendix, Writer's Guide, contains helpful suggestions on correct spelling and punctuation. You can also benefit from various computer software programs such as spell-checkers and style guides.

TECH NOTE

Numerous software programs make editing on-line efficient and easy. These programs flag errors in spelling, punctuation, usage, word choice, and sentence length (readability). Use your spell check to identify and correct misspelled words; with a single search command you can correct all instances of one misspelling found in your document. If there is a group of words that you frequently misspell, add them to your on-line dictionary. It will be worth the time and effort. Also, include in your dictionary any proper names, brand names, technical terms, or concepts that you use often. Other programs will, by using the search command, help you eliminate wordiness by deleting excessive words (for example, *due to the fact that*) or by highlighting overused or misused words and suggesting alternatives.

When you edit on-line, be careful that you do not focus only on the twenty-four lines you can see on the screen and neglect the larger organization of your document. Scroll or print out the text to check for editing problems throughout the document or to observe how a change in one place affects (contradicts, duplicates, weakens) something earlier or later.

When you finish editing, always make a back-up and a hard copy. Your document could be lost because of a power surge, someone mistakenly erasing your disk, or problems reading the disk at a later date.

Editing Guidelines for Writing Lean and Clear Sentences

Here are three of the most frequent complaints readers voice about poorly edited writing in the world of work:

- The sentences are too long. I could not follow the writer's meaning.
- The sentences are too complex. I could not understand what the writer meant the first time I read the work; I had to reread it several times.
- The sentences are unclear. Even after I reread them, I am not sure I understood the writer's message.

Wordy, unclear sentences frustrate readers and waste their time as well. Writing clear, readable sentences is not always easy. It takes effort, but the time you spend editing will pay off in rich dividends for you and your readers.

If you follow the nine guidelines below during the editing phase of your work, you will be better able to write easy-to-read—lean and clear—sentences. As you study these guidelines, keep in mind that readability depends on

- the length of your sentences
- the order in which you list information
- the way in which you signal the relationships among your sentences

1. Avoid Needlessly Complex or Lengthy Sentences

How long should a sentence be? Most readers have very little trouble with sentences ranging from eight to fifteen words. On the other hand, readers find sentences over twenty words much more difficult. As a general rule keep your sentences under eighteen words. The longer your sentence is, the more difficult it will be to understand.

Do not pile one clause on another. Instead, edit one overly long sentence into two or even three more manageable ones.

Too long: The planning committee decided that the awards banquet should be held on May 15 at 6:30, since the other two dates (May 7 and May 22) suggested by the hospitality committee conflict with local sports events, even though one of those events could be changed to fit our needs.

Edited for easier reading: The planning committee has decided to hold the awards banquet on May 15 at 6:30. The other dates suggested by the hospitality committee—May 7 and May 22—conflict with two local sports events. Although the date of one of those sports events could be changed, the planning committee still believes that May 15 is our best choice.

2. Combine Short, Choppy Sentences

Don't shorten long, complex sentences only to turn them into choppy, simplistic ones. A memo or letter written exclusively in short, staccato sentences sounds immature and makes for boring reading.

TECH NOTE

The computer will help you increase your writing productivity, improve your writing skills, and design more professional-looking documents. But it will not do your writing and thinking for you. A software package may help you organize your ideas, but *you* must first do research to discover what ideas are relevant and convincing for your audience. The computer enables you to produce more writing, but, again, *you* are the one who must select the right words with appropriate tone and put them into readable sentences and logically organized paragraphs. A computer cannot tell you what points will please or alienate your audience. Your computer can greatly assist you in revising your work, but *you* must decide what must be changed, modified, retained, or moved, and why and how. A software program can tell you your sentences are too long, but changing every one into a short (ten-word) statement will make your work choppy and less pleasing to read. And never be lulled into thinking that a clear, professionally printed document will hide or make up for incorrect grammar, irrelevant content, or poor organization. Your computer is an efficient writing tool, not a substitute writer for you.

Effective editing blends short sentences with longer ones to achieve variety and to reflect logical relationships. For example, a sentence containing a subordinate clause followed by an independent clause may signal a cause-to-effect relationship to readers. A sentence with a series of parallel independent clauses points to the equality of the ideas spelled out in these clauses. A short sentence at the end of a paragraph can emphatically summarize a main idea.

When you find yourself looking at a series of short, blunt sentences, as in the following example, combine them where possible and use connective words similar to those italicized in the edited version.

Choppy: Medical secretaries have many responsibilities. Their responsibilities are important. They must be familiar with medical terminology. They must take dictation. Sometimes physicians talk very fast. Then the secretary must be quick to transcribe what is heard. Words could be missed. Secretaries must also prepare final reports. This will take a great deal of time and concentration. These reports are copied and stored properly for reference.

Edited: Medical secretaries have many important responsibilities. *These* include transcribing physician's orders using correct medical terminology. *When* physicians talk rapidly, secretaries will have to transcribe accurately *so* that no words are omitted. *Among the most demanding* of their duties are preparing final transcriptions *and then* making copies of them and storing those copies properly for future reference.

3. Edit Sentences to Tell Who Does What to Whom or What

The clearest sentence pattern in English is the subject-verb-object **(s-v-o)** pattern.

s v o

Sue mowed the grass.

s v o

Today's newspaper contains a special supplement on our school.

Readers find this pattern easiest to understand because it provides direct and specific information about the action. Hard-to-read sentences obscure or scramble information about the subject, the verb, or the object. Unedited sentences bury the subject in prepositional phrases (see Appendix, p. 685) in the middle or the end of the sentence, smother the verb in phrases, or allow an object to act like a subject.

Not all clear sentences, however, follow the subject-verb-object pattern. You might use a subordinate clause in addition to the subject, verb, and object in the independent clause.

To edit your sentences to tell readers clearly what's going on, follow these steps. (a) Identify the subject—the person, place, or concept that controls the main action. (Avoid using vague words such as *factors, conditions, processes,* or *elements* for subjects.) (b) Select an action-packed verb that shows what the subject does. (c) Specify the object that is acted upon by the real subject through the verb.

In the following unedited sentences, subjects are hidden in the middle rather than being placed in the most crucial subject position.

Unclear: An assessment of the market helped our company design its new food blender. (The main action is *designing*. Who did it?)
Edited: Our company designed its new food blender by assessing the market.

Unclear: The control of the ceiling limits of glycidyl ethers on the part of the employers for the optimum safety of workers in the workplace is necessary. (Who is responsible for taking action? What action must they take? For whom is such action taken?)
Edited: For the workers' safety, employers must control the ceiling limits of glycidyl ethers.

4. Arrange Information Logically Within Sentences

The order in which you list information in a sentence can help or hinder a reader in understanding your message. You cannot list details in random order. Edit your sentences to make sure readers receive information in the most logical, helpful sequence. The content of the sentence will help you choose the best pattern: chronological, cause to effect, action to reaction, and the like.

Not Logical: They shut off the computer once they finish with their program.
Edited: Once they finish their program, they shut off the computer. (Since finishing the program precedes shutting the computer off, give readers the information in that order.)

5. Use Strong, Active Verbs Rather than Verb Phrases with Verbs Disguised as Nouns

In trying to sound important, many bureaucratic writers avoid using simple, graphic verbs. Instead, these writers add a suffix *(-ation, -ance, -ment, -ence)* to a direct verb *(determine)* to make a noun *(determination)* and then couple the new noun with *make, provide,* or *work* to produce a weak verb phrase (for example, *provide maintenance of* instead of *maintain, work in cooperation with* instead of *cooperate).* Such verb phrases imprison the active verb inside a noun format and slow a reader down.

Note how the edited versions below rewrite weak verb phrases.

Weak: The officer made an assessment of the damages the storm had caused.
Strong: The officer assessed the damage the storm had caused.

Weak: The city provided the employment of two work crews to assist the strengthening of the dam.
Strong: The city employed two work crews to strengthen the dam.

6. Avoid Piling Up Modifiers in Front of a Noun

It can be hard for readers to grasp your message when you string a series of modifiers in front of a noun. Modifiers are subordinate units (see Appendix, pp. 689–690) that clarify and qualify a noun. Putting too many of them in the reader's path to the noun will confuse the reader, who cannot decipher how one modifier relates to another modifier or to the noun. To avoid this problem, edit the sentence to place some of the modifiers in prepositional phrases after or before the nouns they modify.

Crowded: The ordinance contract number vehicle identification plate had to be checked against inventory numbers.
Spaced: The ordinance contract number on the vehicle identification plate had to be checked against the inventory numbers.
Crowded: The vibration noise control heat pump condenser quieter can make your customer happier.
Spaced: The quieter on the condenser for the heat pump will make your customer happier by controlling noise and vibrations.

7. Replace a Wordy Phrase or Clause with a One- or Two-Word Synonym

Wordy: The college has parking zones for different areas for people living on campus as well as for those who do not live on campus and who commute to school.
Edited: The college has different parking zones for resident and commuter students. (Twenty words of the original sentence—everything after "areas for"—have been reduced to four words: "resident and commuter students.")

8. Combine Sentences Beginning with the Same Subject or Ending with an Object That Becomes the Subject of the Next Sentence

Wordy: I asked the inspector if she were going to visit the plant this afternoon. I also asked her if she would come alone.
Edited: I asked the inspector if she were going to visit the plant alone this afternoon.
Wordy: Homeowners want to buy low-maintenance bushes. These low-maintenance bushes include the ever-popular holly and boxwood varieties. These bushes are also inexpensive.
Edited: Homeowners want to buy such low-maintenance and inexpensive bushes as holly and boxwood. (This revision combines three sentences into one, condenses twenty-four words into fourteen, and joins three related thoughts.)

9. Avoid Unnecessary That/Which *Clauses*

That/which clauses using some form of the verb *to be (is, are, were, was)* are infamous for adding words but not meaning to your sentences. They are popular with bureaucratic writers, who like to draw out an idea beyond the number of words needed. Edit them by eliminating *that/which* clauses or using an adjective to represent the clause.

Wordy: The pain medication that was prescribed by the doctor was very helpful to her father.
Edited: The pain medication prescribed by the doctor helped her father. (Note how the editing reduces "was helpful to" to "helped" to further save words and time.)
Wordy: The organizational plan that was approved last week contains a number of points, all of which are considered to be especially important for new employees.
Edited: The organizational plan approved last week contains important points for new employees.

Editing Guidelines for Cutting Out Unnecessary Words

Too many people in business and industry think the more words, the better. Nothing could be more self-defeating. Your readers are busy; unnecessary words slow them down. Make every word work. When a word takes up space but gives no meaning, cut it. Cut out any words you can from your sentences. If the sentence still makes sense and reads correctly, you have eliminated wordiness.

The phrases on the left should be replaced with the precise words on the right.

Wordy	Concisely Edited
at a slow rate	slowly
at an early date	early
at the point where	where
at this point in time	now
be in agreement with	agree
bring to a conclusion	conclude, end
bring together	combine, join
by means of	with
come to terms with	agree, accept
due to the fact that	because
express an opinion that	believe
feel quite certain about	believe
for the length of time that	while
for the period of	for
for the purpose of	to
in an effort to	to
in such a manner that	so
in the area / case / field of	in
in the event that	if
in the neighborhood of	approximately
look something like	resemble
serve the function of	function as
show a tendency to	tend
take into consideration	consider
take under advisement	consider
take place in such a manner	occur
with reference to	regarding, about
with the result that	so

Another kind of wordiness comes from using redundant expressions. Being redundant means that you say the same thing a second time, in different words. "Fellow colleague," "component parts," and "corrosive acid" are phrases that contain this kind of double speech; a fellow *is* a colleague, a component *is* a part, and acid *is* corrosive. Redundant expressions are uneconomical and are often clichés. The suggested changes on the right are preferable to the redundant phrases on the left.

Redundant	Concise
absolutely essential	essential
advance reservations	reservations
basic necessities	necessities, needs

Redundant	Concise
close proximity	proximity, nearness
each and every	each, every, all
end result	result
eradicate completely	eradicate
exposed opening	opening
fair and just	fair
final conclusions / final outcome	conclusions / outcome
first and foremost	first
full and complete	full, complete
grand total	total
null and void	void
passing fad	fad
personal opinion	opinion
prerecorded	recorded
over and done with	over
tried and true	tried, proven

Watch for repetitious words, phrases, or clauses within a sentence. Sometimes one sentence or one part of a sentence needlessly duplicates another.

Redundant: The post office hires part-time help, especially around the holidays, to handle the large amounts of mail at Christmas time.

Edited: The post office often hires part-time help to handle the large amounts of mail at Christmas time. ("Especially around the holidays" means the same thing as "at Christmas time.")

Redundant: The fermenting activity of yeast is due to an enzyme called zymase. This enzyme produces chemical changes in yeast.

Edited: The fermenting activity of yeast is due to an enzyme called zymase. (The second sentence says vaguely what the first sentence says precisely; delete it.)

Redundant: To provide more room for employees' cars, the security department is studying ways to expand the employees' parking lot.

Edited: The security department is studying ways to expand the employees' parking lot. (Since the first phrase says nothing that the reader does not know from the independent clause, cut it.)

Adding a prepositional phrase can sometimes contribute to redundancy. The italicized words below are redundant because of the unnecessary qualification they impose on the word they modify. Be on the lookout for the italicized phrases and delete them.

audible *to the ear*
bitter *in taste*
fly *through the air*
orange *in color*
rectangular *in shape*
second *in sequence*
short *in duration*
hard *to the touch*
honest *in character*
light *in weight*
soft *in texture*
tall *in height*
twenty *in number*
visible *to the eye*

Certain combinations of verbs and adverbs are also redundant. Again, the italicized words below should be deleted.

advance *forward*	lift *up*
burn *up*	merge *together*
cancel *out*	open *up*
circle *around*	plan *ahead*
commute *back and forth*	prove *conclusively*
combine *together*	refer *back*
continue *on*	repeat *again*
drop *down*	reply *back*
funnel *through*	revert *back*
join *together*	write *down*

Figure 2.8 shows a memo that Trudy Wallace wants to send to her boss about installing cellular phones in the company cars. Wallace attempted to get her ideas down on paper without worrying about finding the most concise and precise words. Her unedited work is bloated with unnecessary words, expendable phrases, and repetitious ideas.

After careful editing, Wallace was able to streamline her memo to Lee Chadwick and eliminate the wordiness. Note how, in Figure 2.9, she pruned wordy expressions, combined sentences to cut out duplication, and found a two-word phrase, "phone tag," to replace the unnecessarily complex description of calling clients back. This revised version is only 102 words, as opposed to 266 words in the draft. Not only has Wallace shortened her message but she has made it easier to read.

Editing Guidelines to Eliminate Sexist Language

Editing involves far more than just making sure your sentences are readable and free from wordiness. It also reflects your professional style—how you see and characterize the world of work and the individuals in it, not to mention how you want your readers to see you. Your words should reflect a high degree of ethics and honesty, free from bias and offense. Prejudice has no place in business and technical writing, or in any other type of writing or speaking for that matter.

Sexist language offers a distorted view of our society and discriminates in favor of one sex at the expense of another, usually women. Accordingly, you should avoid sexist language for a variety of reasons:

- You will offend female readers by depriving them of their equal rights.
- You run the risk of readers branding you as sexist and biased.
- You decrease the effectiveness and persuasiveness of any points you are trying to make.
- You will cost your company business.

To avoid sounding prejudiced, eliminate sexist terms and attitudes as you edit your work.

Sexist language unfairly assigns responsibilities, jobs, or titles to individuals on the basis of sex. It is often based on sexist stereotypes that depict men as superior to women. For example, calling politicians *city fathers* or *favorite sons* follows the stereotypical picture of seeing politicians as male. Such phrases discriminate

FIGURE 2.8 Wordy, unedited memo.

TO: Lee Chadwick
FROM: Trudy Wallace T.W.
DATE: May 15, 1998
SUBJECT: Installing cellular phones

Due to the amount of time our sales force spends traveling the roads each day, it strikes me as beneficial to look into the distinct possibility of installing cellular car phones in our company cars. Such an installation would benefit our sales force in a variety of multiple ways. The sales force could increase their efficiency and morale with the installation of these car phones. With the aid of a cellular phone we could bring together our customers and our sales force a lot easier. Rather than wasting an amount of time in the neighborhood of 40 to 50 minutes each day tracking down phones on the road, our sales people could have a shortened period of time to respond using their cellular phones in their cars. The response rate of returning a call could be markedly reduced and dropped down. Moreover, by having cellular phones in their cars sales people would minimize the problems of returning calls to people and then finding out they are out and then having to call them back. It is difficult to catch people this way. Cellular phones would increase both the convenience and the ease by which we operate our business. I think it would be absolutely essential to the ongoing operation of our company's business today to respond fully and completely to the possibility such a proposal affords us. It would therefore appear safe to conclude that with reference to the issue of cellular phones that every means at our disposal should be brought to bear on including such phones in our company cars.

FIGURE 2.9 The memo in Figure 2.8 edited for conciseness.

TO: Lee Chadwick
FROM: Trudy Wallace T.W.
DATE: May 15, 1998
SUBJECT: Installing cellular phones

Because our sales people spend so much time on the road, I think we should install cellular phones to increase employee efficiency and improve morale. Cellular phones would help our sales people communicate with their clients a lot easier and faster. They would not waste 40 to 50 minutes each day looking for phones to call clients. And they would save even more time by not having to play phone tag.

I think installing cellular phones is a wise investment, and so with your approval, I will obtain more information from suppliers to prepare a formal proposal to request bids.

against women who do or could hold public office at all levels of government. Not all bosses are male, either.

Sexist language also prejudiciously labels some professions as masculine and others as feminine. For example, sexist phrases assume engineers, physicians, and pilots are male (*he, his,* and *him* are often linked with these professions in sexist descriptions) while social workers, nurses, and secretaries are female *(she, her),* although members of both sexes work in these professions. Sexist language also wrongly points out gender identities when such roles do not seem to follow biased expectations—*lady lawyer* and *male nurse; female surgeon* or *female astronaut.* These offensive distinctions reflect prejudiced attitudes that you should eliminate from your writing. Using these phrases is just as sexist as saying, "There's a girl in our office who is as good at selling as a man."

Always prune the following sexist phrases: *gal Friday, little woman, lady of the house, masterpieces, the best man for the job, the weaker sex, woman's work, working wives,* and *young man on the way up.* These sexist terms will not only offend but also exclude many of the members of the audience you want to reach.

Ways to Avoid Sexist Language

1. Replace Sexist Words with Neutral Ones

Neutral words do *not* refer to a specific sex; they are genderless. The sexist words on the left can be replaced by the neutral nonsexist substitutes on the right.

Sexist	Neutral
authoress	writer, author
businessman	businessperson
chairman	chair, chairperson
craftsman	skilled worker
fireman	firefighter
foreman	supervisor
janitress	cleaning person
landlord, landlady	owner
mailman/postman	mail carrier
man-hours	work-hours
mankind	humanity
manmade	synthetic, artificial
manpower	strength, power
man to man	candidly
men	human beings, people
modern man	modern society
policeman	police officer
repairman	repair person
salesman	salesperson, clerk
stewardess	flight attendant
woman's intuition	intuition
workman	worker

2. Watch Masculine Pronouns

Avoid using the masculine pronouns *(he, his, him)* when referring to a group that includes both men and women.

Every worker must submit his travel expenses by Monday.

Workers may include women as well as men, and to assume that all workers are men is misleading and unfair to women. You can edit such sexist language in several ways.

a. Make the subject of your sentence plural and thus neutral.

Workers must submit their travel expenses by Monday.

b. Replace the pronoun *he* with *the* or drop it altogether.

Every worker must submit travel expenses by Monday.

Every employee is to submit the work activity report by Monday.

c. Use *his or her* instead of *his*.

Every worker must submit his or her travel expenses by Monday.

d. Reword the sentence using the passive voice.

All travel expenses must be submitted by Monday.

Moreover, in some contexts exclusive use of the masculine pronoun might invite a lawsuit. For example, you would be violating federal employment laws prohibiting discrimination on the basis of sex if you wrote the following in a help-wanted advertisement for your company.

> *Each applicant must submit his transcript with his application. He must also supply three letters of recommendation from individuals familiar with his work.*

The language of this ad implies that only men can apply for the position.

3. Eliminate Sexist Salutations

Never use the following salutations when you are unsure of who your readers are:

- Dear Sir
- Gentlemen
- Dear Madam

Any woman in the audience will surely be offended by the first two greetings above and may also be unhappy with the pompous and obsolete *madam.* It is usually best to write to a specific individual, but if you cannot do that, direct your letter to a particular department or office: *Dear Warranty Department* or *Dear Selection Committee.*

Be careful, too, about using the titles *Miss, Mr.,* and *Mrs.* Sexist distinctions are unjust and insulting. It would be preferable to write *Dear Ms. McCarty* rather than *Dear Miss or Mrs. McCarty.* A woman's marital status should not be an issue. Review pp. 147–149 about the acceptable salutations in your letters.

4. Never Single Out a Woman's Physical Appearance

> *The manager is a tall blonde who received her training at Mason Technical Institute.*

Such sexist physical references negatively draw attention to a woman's gender. Sexist writers would not describe a manager who happened to be male that way.

✓ Revision Checklist

- ❑ Investigated the research, drafting, revising, and editing benefits available through home or work computer.
- ❑ Researched my topic carefully to obtain enough information to answer all my readers' questions. Used such appropriate means as library research, on-line data search, interviews, questionnaires, personal observations, or a combination of methods.
- ❑ Before writing, determined how much and what kind of information is needed to complete writing task.

- ❑ Spent enough time planning—brainstorming, outlining, clustering, or a combination of these techniques. Produced enough substantial material from which to shape a draft.
- ❑ Prepared enough drafts to decide upon major points in message to readers. Was willing to make major changes and deletions if necessary in drafts.
- ❑ Revised drafts carefully to successfully answer reader questions about content, organization, or tone.
- ❑ Made time to edit work so that style is clear and concise and sentences are readable and varied. Checked words to make sure they are spelled correctly and appropriate for audience.
- ❑ Eliminated sexist language.

Exercises

1. Below is a writer's initial brainstormed list on stress in the workplace.

leads to absenteeism
high costs for compensation for stress-related illnesses
proper nutrition
numerous stress-reduction techniques
good idea to conduct interviews to find out levels, causes, and extent of stress in the workplace
low morale caused by stress
higher insurance claims for employees' physical ailments
myth to see stress leading to greater productivity
various tapes used to teach relaxation
environmental factors—too hot? too cold?
teamwork intensifies stress
counseling
work overload
setting priorities
wellness campaign
savings per employee add up to $2,500 per year
skills to relax
learning to get along with co-workers
need for privacy
interpersonal communication
employee's need for clear policies on transfers, promotion
stress management workshops very successful in California
physical activity to relieve stress

affects management
breathing exercises

Revise this brainstormed list, eliminating any repetition and combining any related items.

2. Prepare a suitable outline from your revised list in Exercise 1 for a report to a decision maker on the problems of stress in the workplace and the necessity of creating a stress management program.

3. From the revised brainstormed list in Exercise 1, write a short memo to a decision maker about how the problems of stress negatively affect workplace production.

4. Write a short report (2–3 pages) to the manager of the small company you work for to convince her to establish a system of flex time. Prepare an outline based on the clustered items in Figure 2.1 (p. 38). Add, delete, or rearrange anything in this clustered grouping to complete your outline. Submit your final outline along with your report to your instructor.

5. Compare the draft of Gordon Reynolds's report in Figure 2.4 (pp. 44–45) with the final copy of his report in Figure 2.5 (pp. 46–47). What kinds of changes did he make? Were they appropriate and effective for his audience and purpose? Why or why not?

6. Assume you have been asked to write a short report (similar to Gordon Reynolds's in Figure 2.5) to a decision maker (the manager of a business you work for or have worked for; the director of your campus union, library, or security force; a city official) about one of the following topics.
   a. computer software
   b. Internet resources
   c. security lighting
   d. food service
   e. insurance plans
   f. public transportation
   g. sporting events/activities
   h. training programs
   i. morale
   j. hiring more part-time student workers

   Do some research and planning about one of these topics and the audience for whom it is intended by writing your answers to the following questions.

   - What is my precise purpose in writing to my audience?
   - What do I know about the topic?
   - What information will my audience expect me to know?
   - Where can I obtain relevant information about my topic to meet my audience's needs?

7. Using one or more of the planning strategies discussed in this chapter (clustering, brainstorming, outlining) generate a group of ideas for the topic you

chose in Exercise 6. Work on your planning activities for about 15–20 minutes, or until you have about 10–15 items. At this stage do not worry about how appropriate these ideas are or even if some of them overlap. Just get some thoughts down on paper.

8. Go through the list you prepared in Exercise 7 and eliminate any entries that are inappropriate for your topic or audience or that overlap. Try to see how many of them you might expand or rearrange into categories or subcategories. Then create an outline similar to the one in Figure 2.3 (p. 40).

9. Using your outline in Exercise 8, prepare some drafts of your memo report. Submit at least two drafts to your instructor.

10. Revise your drafts as much as necessary to create the final copy of your report.

11. In a few paragraphs, explain to your instructor the changes you made between your early drafts and your revised drafts. Explain why you made them. Concentrate on major changes—adding and moving paragraphs—as well as matters of style, tone, and even format.

12. In another memo—addressed to your instructor—describe any problems that bothered you at various phases of working on your report. Also point out what planning, drafting, and revising strategies worked especially well for you.

13. The following paragraphs are wordy and full of awkward, hard-to-read sentences. Edit these paragraphs to make them more readable and concise by using clear and concise words and sentences.
    a. It has been verified conclusively by this writer that our institution must of necessity install more bicycle holding racks for the convenience of students, faculty, and staff. These parking modules should be fastened securely to walls outside strategic locations on the campus. They could be positioned there by work crews or even by the security forces who vigilantly patrol the campus grounds. There are many students in particular who would value the installation of these racks. Their bicycles could be stationed there by them, and they would know that safety measures have been taken to ensure that none of their bicycles would be apprehended or confiscated illegally. Besides the precaution factor, these racks would afford users maximized convenience in utilizing their means of transportation when they have academic business to conduct, whether at the learning resource center or in the instructional facilities.
    b. On the basis of preliminary investigations, it would seem reasonable to hypothesize that among the situational factors predisposing the Smith family toward showing pronounced psychological identification with the San Francisco Giants is the fact that the Smiths make their domicile in the San Francisco area. In the absence of contrariwise considerations, the Smiths' attitudinal preferences would in this respect interface with earlier behavioral studies. These studies, within acceptable parameters, correlate the fan's domicile with athletic allegiance. Yet it would be counterproductive to establish domicility as the sole determining factor for the Smiths'

preference. Certain sociometric studies of the Smiths disclose a factor of atypicality which enters into an analysis of their determinations. One of these factors is that a younger Smith sibling is a participant in the athletic organization in question.

14. Below are very early drafts of memos that businesspeople have sent to their bosses or fellow workers. Revise and edit each draft, referring to the revising and editing checklists. Turn in your revision and the final, reader-ready copy. As you revise, keep in mind that you may have to delete and add information, rearrange the order of information, and make the tone suitable for the reader. As you edit, make sure your sentences are clear and concise and your words well chosen.

a. TO: All workers
FROM: B.J. Blackwell
SUBJECT: Parking
DATE: September 29, 1999

The parking violations around here have gotten very very bad. And the administration is provoked and wants some action taken. I don't blame them. I have been late for meetings several times in the last month because inconsiderate folks from other divisions have parked their cars in our zone. That just is not fair, and so I must not be the only one who is upset. No wonder the management finds things so bad they have asked me to prepare this memo.

A big part of the problem it seems to me is that employees just cannot read signs. They park in the wrong zones. They also park in visitors' spots. The penalties are going to be stiff. The administration, or so I was led to believe, is thinking of fining any employee who does not obey the parking policies. I know for a fact that I saw someone from the research department pull right into a visitor parking area last week just because it was 8:55 and he did not want to be late for work. That gives our business a bad name. People will not want to do business with us if they cannot even find a parking spot in the area that the company has reserved for them.

Ms. Watson has laid the law down to me about all this and told me to let each and every one of you know that things have to improve. One of the other big problems around here is that some employees have even parked their cars in loading zones, and security had to track them down to move.

As part of the administration's new policy, each employee is going to be issued a company parking policy and will have to come in and sign for it verifying that he received it. I think things really have gotten out of hand and that some drastic action has to be taken. We will all have to shape up around here.

b. TO: Betty Jones, Director
FROM: Tom Cranford
RE: Vacation request for vacation from June 12-24
DATE: April 21, 2000

I have been a highly productive employee and so I do not think that it is out of line for me to make this request. I have put in overtime and even done others' work in the department while they were away. So, I think that it is fair and just, and I can see no reason why I should not be allowed to take my vacation during the last two weeks of June.

Let me explain some of the reasons. I could have others watch my desk and do the work. I have helped them out, too, and they know it. I have been remarkably dependable. I have been readily accessible whenever their vacations have come around, and so I know it can be done.

I fully realize that this is the busiest time of the year for our company and that vacations are not usually granted during this season. But I do have personal reasons which I think should be honored/respected. Peak business times are major. I understand this, and I do hope an exception will be made in my case. After all, I do have the on-the-job training that other companies would reward. Thanks very much.

c. TO: All Employees
FROM: George Holmes
DATE: February 21, 1997
RE: Travel

Every company has its policies regarding travel and vouchers. Ours strike me as important and fairly straightforward. Yet for the life of me I cannot fathom why they are being ignored. It is in everyone's best interest. When you travel, you are on company time, company business. Respect that, won't you. Explain your purpose, keep your receipts, document your visits, keep track of meals.

If you see more than one client per day, it should not be too hard or too much to ask you to keep a log of each, separate, individual visit. After all, our business does depend on these people, and we will never know your true contributions on company trips unless you inform us (please!) of whom you see, where, why, and how much it costs you. That way we can keep our books straight and know that everything is going according to company policy.

Please review the appropriate pages (I think they are pages 23-25) about travel procedures. Thanks. If you have questions, give me a call, but check your procedures book or with your office/section manager, first. That will save everyone more time. Good luck.

15. Find a piece of writing—a memo, letter, brochure, or short report—that you believe was not carefully drafted or revised. In a short memo or e-mail, point out to your instructor what is wrong with this piece of writing—for example, not logically organized, inappropriate tone, incomplete or too technical information. Attach a copy of the poor example to your memo.

16. Revise the piece of poor writing you analyzed in Exercise 15. Submit your improved version to your instructor.

CHAPTER 3

Collaborative Writing at Work

In the workplace you will not always have to write alone, isolated from co-workers or managers. Much of your business writing time may be spent working as part of a team. You will prepare a document with other employees or managers who will collaborate or, at the very least, review your work and help improve it.

Collaborative writing occurs when a group of individuals—as few as two to as many as seven or more—

- Combine their efforts to prepare a single document
- Share authorship
- Work together for the common good, maybe even the survival, of your department, company, or agency

The trend of the late 1990s is toward increased collaboration. One survey estimates that in the world of work 90 percent of all businesspeople spend some time writing as part of a collaborative team. Simply put, then, collaborative writing is one very real way you can expect to communicate in the business world.

Teamwork Is Crucial to Business Success

Communications and quality experts emphasize that the success of a business depends on how well people interact as a team. Working as part of a writing team is an essential responsibility in a technical world where each employee is connected to co-workers and managers via computer modems and where customers are connected to almost every company through the worldwide marketplace of the Internet. Collaboration is networking, and collaborative writing is, therefore, a vital part of the global network where individuals depend upon each other's expertise, experience, and viewpoints to benefit their company and customers.

The heart of collaboration is being a team player, a highly valued skill in the workplace. Being a collaborative, or team, player means you interact successfully on an interpersonal level; it means you know how to talk to people to get information, provide feedback, give and take constructive criticism, raise important and relevant questions, get assistance from resource experts in other departments and fields, and, most significant, put the good of your company above your ego. Collaboration helps get writing done more easily and more efficiently.

Chapter 3 introduces you to successful ways to collaborate with co-workers in your office as well as people with whom you'll do business around the globe. In this chapter you'll receive practical advice on how to write in and for a group and how to solve communication problems within a group setting. It takes much work and skillful negotiation to be a member of an effective collaborative writing group, but it is worth the effort.

Collaboration takes place in preparing many types of writing—from brochures to technical manuals, from proposals to long reports. Collaboration can be done in a variety of ways—face to face, one on one, over the telephone, through faxes, and via e-mail and the Internet. In the last few years, several companies have marketed sophisticated software packages called *groupware* that facilitate collaborative writing efforts through electronic conferencing (see Figure 3.8).

Advantages of Collaborative Writing

Collaborative writing teams benefit both employers and employees. Specific advantages of collaboration include the following:

1. Builds on collective talents. Because no one individual has all the answers, a writing team profits from the diverse expertise and talents of individual members. A group of writers and researchers together can do a much more thorough job than can one writer working without the advantage of outside help or commentary. One team member may have strengths in statistical research, another in writing, and a third in document design and graphics. Their combined skills will produce a much more informed and professional document.

2. Provides productive feedback. A company profits by pooling the professional resources of its workforce, getting the benefit of diverse viewpoints, insightful criticism, and immediate feedback. Members can offer helpful critiques of each other's suggestions, drafts, and revisions. Constructive discussion clarifies issues, sorts out appropriate from inappropriate solutions, and leads to better, more persuasive recommendations for a customer or an employer. Thanks to collective planning and decision making, the group can identify and solve major problems and survey more opinions than an individual writer could. Employees working as a team are also much more likely to incorporate all the relevant points an audience needs. If coordinated properly, group interaction will net much more information, and in a shorter time, than could an individual writer working alone.

3. Increases productivity and can save time. When a group has planned its strategies carefully, collaboration actually cuts down on the number of meetings and conferences, saving the company and employees valuable time. Also, with group checks and balances, the likelihood of wasting resources on false leads and irrelevant issues is reduced.

4. Assures overall writing effectiveness. The more people involved in developing a document, the greater its chances for thoroughness and cohesion. Guided by shared principles of style, a collaborative team can better guarantee the uniformity and consistency of a piece of writing than can two or three individuals each writing a different section of a document. Document Manager Tara Barber's remarks in Figure 3.1 about shared editorial responsibilities reinforce the point that effective collaboration can uncover inconsistencies in organization, format, and style. Several pairs of eyes see more clearly and objectively than just one.

5. Offers psychological benefits. Collaboration contributes significantly to employee confidence and morale. Workers are more productive when they know that their opinions are solicited and acted upon. By sharing information and offering suggestions, members feel that they are helping others while simultaneously fulfilling themselves. Furthermore, working as part of a team relieves individuals of some job-related stress. Knowing that he or she is not carrying sole responsibility for planning, drafting, and writing a document lightens the pressure and may be an incentive to work more effectively. A division of labor makes deadlines less fearsome, too.

6. Contributes to customer service and satisfaction. Collaborative writing teams can assess and describe a product effectively by pooling their knowledge. Their collective judgment of an audience's needs helps their company better meet those needs. Moreover, a collaborative team is much more likely—through discussions, interactions, even disagreements—to anticipate a customer's requests or complaints and thereby respond to or resolve them. A collaborative environment fosters greater opportunities for originality. In the give and take of discussion, groups are challenged to move beyond ordinary methods to construct more creative, compelling presentations that ultimately lead to greater sales.

Collaborative Writing and the Writing Process

The writing process described in Chapter 2 also applies to collaborative writing. Groups use the same strategies and confront the same problems individual writers do. Like the individual writer, writing teams brainstorm, plan, research, draft, revise, and edit. Like the individual writer, too, a team moves through the writing process only by identifying its audience, following a common purpose, defining the problem, and deciding on the best ways to solve that problem. Effective collaborative writing merges with individual writing at times. That is, although collaboration involves group interaction, it also allows for independent time so individual members can perform tasks on their own that will contribute to the team's success.

Below is a brief rundown of how groups move through the writing process.

1. Groups must plan before they can write. This planning phase includes brainstorming on the group scale with each member contributing to the overall discussion. Members should be encouraged to express ideas freely and without fear of criticism at an early planning session. Groups will perform outlining and looping at this stage as well. (Review pp. 37–41.) During planning, the group, like the individual writer, should identify its audience and determine the audience's purpose in reading the document. In this early phase, the group must also establish the ground rules by which it will operate, including making individual assignments, establishing schedules, and selecting a facilitator.

2. Groups do research. Researching entails more than chitchat or a casual pooling of undocumented opinions. It requires searching, interviewing, and reading. Some tasks may be undertaken by individuals working alone to prepare for the next group session. Each member may be assigned to research one part of a subject, and all will be expected to contribute their research.

TECH NOTE

As part of your research, you may have to e-mail someone in another department or division of your company for information to incorporate into your report. Many documents include boilerplates, or specific contractual statements, that you can pick up directly from another company document without having to create, revise, or edit new text. With the help of scanners, you can also obtain research graphics from other company divisions to include in your work.

Never hoard information; always share for the benefit of the group. As a team member, you may even be called on to distribute copies of documents you retrieved from the Internet or other sources or to report on an interview you conducted.

3. Groups prepare drafts. While it is possible for a group to draft a document together (see "Cowriting Model," p. 91), requiring everyone to sit down together and write word for word is not always feasible or justifiable. Individuals are more likely to draft sections of a document on their own and then present their work for group discussion and revision.

4. Groups revise and edit. These phases of the writing process can benefit from group interaction. A group can collectively spot and resolve problems—omissions, difficulties in organization (a section out of order), length, inconsistencies in content, and so on. Discussion also helps groups agree on final style and document design, provided one individual's personal preferences do not override everyone else's or accepted rules of usage. It often makes sense to appoint someone who has more writing ability to give the group document a final editing for style.

Each stage of the writing process just outlined may be modified to account for group activities. Flexibility is as important to a group of writers as it is to an individual writer.

A Case Study in Collaborative Editing

Figure 3.1 contains a memo written by Tara Edwards Barber, the Documents Manager of CText, Inc., a large firm in Ann Arbor, Michigan, that develops software for the publishing industry, describing the writing process followed by her team of collaborative writers. Communicated over the Internet, Barber's observations—based on many years of experience in corporate writing—shed light on the practical, day-to-day process of group writing and editing. Barber's team routinely interacts with her and one another in the writing process, especially at the editing stage. Group harmony means documents are completed successfully and on time. Study Barber's approach: she effectively works *with* and *not against* her staff. She manages the collaborative effort without being heavy-handed or squelching individual creativity. Her strategy is a good one to follow.

Guidelines for Successful Group Writing

The success of a team-written document depends on (1) the cooperation of team members, (2) the information the team gathers, and (3) the ability of the team to adapt that information to meet the audience's needs and the employer's goals. Making sure a team functions smoothly and productively benefits a company. In fact, many firms hire consultants to assist them in forming the most compatible and creative groups possible.

Following the ten guidelines below will increase your chances of success when you write in a group.

1. Know the individuals in your group. Establish rapport with your team. If you do not know them, introduce yourself before you start working on a document. Even if you know them, touch base to assure them that you are looking forward to working with them. Learn as much as you can about their schedules, backgrounds, special competencies, experiences within your organization, and even their pet peeves. By being concerned and friendly, you create an environment that enhances productivity.

2. Do not regard one person on the team as more important than another. Favoritism leads to hard feelings and decreases the group's productivity. Instead, adopt the attitude that the group succeeds or fails as a whole. Think collectively. Everyone's input is necessary to group effort and success. When individuals regard themselves as the most important or indispensable member of the team, group harmony is jeopardized. A unified team accomplishes more than a collection of disgruntled individuals.

3. Set up a preliminary meeting to establish guidelines. Unless members share the same vision and objectives, they may work at cross purposes and head in different directions. The group should discuss the objectives, audience, scope, format, and importance of the document at an introductory meeting. It is helpful at this first meeting to identify and discuss any directions and/or directives from management and to set up priorities. Such a preliminary meeting is crucial to the

FIGURE 3.1 **Collaborative editing: Advice from a pro.**

Date: 7 Mar 1996 13:35:15–0600
From: tara@barber.ctext.com
Newsgroups: bit.listserv.techwr-1

I would like to describe some ways to edit constructively to help new writers improve.

I agree that the more you make collaborative, the better results you will get. Here's how we do it in my department. Although all my writers are experienced, they come from such different backgrounds that some of the ways I have coordinated editing activities are similar to how I would interact with new writers. It's my job to make sure all the pieces fit.

When I first took this job, there were several documentation styles being used for the company's manuals. But the department at that time was very small and so the senior writer and I sat down and prepared a cohesive and consistent style manual. We experienced some tension, but by compromising, we were able to iron out our differences. We developed a style guide and a set of manual conventions that gave new writers precise and helpful guidelines and procedures to follow from the start. The new manual helped us to eliminate glaring editing problems. That way we do not worry about inconsistencies in spelling, capitalization, formats, headings, or reference documentation.

But the department has grown since that time, and on at least two occasions all of us have sat down as a department to reassess the style guide and the manual. By doing this, we are able to get input from everyone and to allow for new ideas. When a problem arises now, we are comfortable addressing it as a group.

In addition to the usual subject-matter expert reviews, all of us review each other's materials. This review helps us remain familiar with each other's projects and homogenizes our writing styles, which keeps our documentation consistent. It's also a great way to get a good "clean-eyes edit." And, happily, a writer will offer an original idea about how

Continued

FIGURE 3.1 (Continued)

Page 2

to approach a problem that eliminates a potential editing dilemma. We all win.

In my role as documentation manager, I try to edit early and late, but not in the middle. Our materials usually go through several edits before they're printable (usually due to the changes in software). I try to look over early material in the form of outlines and first drafts, to make sure everything looks on track and follows our house style and procedures. I always edit final drafts because I am responsible for the output of my department. But I try not to do too much in the middle unless there is a problem. Peer review works well during this intermediate phase, and I don't want to step on individual creativity.

If I have to make comments, I make them at various levels. And here I'm not talking about catching grammar mistakes, spelling errors, typos, misnumberings, etc. Instead,

a) I point out problems, or sections that seem confusing, but let the writer suggest fixes.
b) I make suggestions and provide examples–more than one if I can.
c) I actively work with the writer to develop ways around a problem. Sometimes this means coming up with a whole new way of approaching the documentation, which is then addressed at our next style meeting. Sometimes it also works to get the rest of the department involved in a brainstorming session.
d) If nothing else works, I play the heavy manager and say "do it this way because I say so." I try to avoid this, however, if I possibly can.

Working as a constructive team, we rarely have editing problems that we can't solve, and everyone is happy with the comments and the final product. And our customers find the documents usable and valuable. You can't really ask for more than that.

Tara Barber
Documentation Manager
CText, Inc.

Reprinted with the permission of Tara Barber.

success of all subsequent meetings. Each member should take this opportunity to ask questions or share comments and concerns before research and drafting begin.

4. Agree on the group's organization. Here are some questions to consider.

- Is the group to appoint a leader? If so, how is this person to be chosen?
- Will the leader be the most experienced or skilled writer? The member who has been with the company longest?
- What are the leader's responsibilities?
- Will the leader have the authority to settle disputes, break deadlocks, or make final decisions?

An effective group leader must be skillful at initiating discussions, encouraging team members, compromising for the sake of consensus, and generally presiding over meetings. The leader is the group's gatekeeper, or moderator.

The group must also consider how to communicate with management and decide whether its work should be shared with individuals outside the group (such as co-workers from another department). Members will also have to decide how and how often to communicate between meetings. For example, since one member's work might affect another's, it will be important to be able to notify other team members expeditiously. Finally, the group must decide whether it will appoint a recorder to distribute minutes or documents from meetings.

5. Identify each member's responsibilities, but allow for individual talents and skills. The division of labor should be fair in terms of workload and should reflect the group's diverse talents. Members should be aware of one another's strengths—in document design, graphics, software, marketing, editing. Don't underestimate or overestimate the time necessary for each member's (or the entire group's) tasks. Individuals should feel comfortable with their assigned duties, whether gathering data, drafting a section, editing, verifying documentation, or designing graphics. Because some jobs are more demanding and may take longer than others, one member may need extra time or help from an outside expert (statistician, engineer). Make sure members have opportunities to voice difficulties and to request help.

6. Establish the times, places, and length of group meetings. The group must decide on a calendar of scheduled meetings and ensure that each member has a copy. When making its schedule, the group should consider likely conflicts and other obligations members may have (forthcoming business trips, conventions). The leader, or the recorder, should remind members of meetings and notify them of any problems that arise between meetings. Accordingly, the leader needs to be available between meetings if members have questions. A facilitator can keep the group on task during meetings by seeing to it that the time limits for each segment are adhered to—within reason.

7. Follow an agreed-on timetable, but leave room for flexibility. The group should estimate a realistic time necessary to complete the various stages of their work—when drafts are due or when editing must be concluded, for example. A project schedule based on that estimate should then be prepared. Everyone in the group must be aware of the short- and long-range deadlines. The group's timetable, with major milestones (dates when key parts are to be completed) boldfaced or highlighted, should be sent to each member. Either collectively or through its leader or recorder, the group should also build some failsafe time into the schedule. **Don't forget: Projects always take longer than initially planned.** Prepare for a possible delay at any one stage—say, research is not completed because materials are not available, or someone fails to complete a task because of illness or a transfer. The group may have to submit progress reports (see pp. 589–594) to its members as well as to management.

8. Provide clear and precise feedback to members. Feedback is the most essential ingredient in group dynamics; it should be relevant, intelligent, and timely. The group needs to decide on how to give feedback—in a face-to-face meeting, privately one on one, via the computer, or a combination of these. Don't come to a meeting without having done your homework. Also, you are not doing yourself or the group a favor by simply responding "OK" or "Looks good" on a draft. Skimming helps no one. Be precise and helpful. Nothing frustrates a writer more than having a reader offer superficial concluding comments—for example, "needs improvement," "lacks focus," "does not flow." Criticize a draft by supplying specific reasons why you find something wrong, incomplete, or misleading. Then be ready to offer detailed advice on where and how to fix the shortcoming. If something is effective, state why you think so, but be honest. Holding back valid criticism hurts group effort.

9. Be an active listener. Good listeners are active participants in a group discussion, not passive observers. Listen to what members say and resist the temptation to interrupt. Develop what some communication consultants call a "third ear," listening to the person's words as well as the meanings and feelings behind those words. Try to identify the themes of a discussion. By following the ideas of an argument, you'll understand the big picture. But don't tune out the technical details. You need to hear the smaller elements as well as the larger framework of a dialogue to write effectively with and for the group.

10. Use a standard reference guide for matters of style, documentation, and format. Establishing guidelines about word processing programs, style, documentation, format, graphics, and so forth makes the group's task easier. Many large companies have policy or style manuals that their teams must use, as Tara Barber points out in Figure 3.1. If such a document is not available, your group should select a manual or other reference work to which members should adhere, such as the APA (American Psychological Association) guide, *The Chicago Manual of Style,* or another.

Sources of Conflict in Group Dynamics and How to Solve Them

The success of collaborative writing depends on how well the team interacts. Discussion and criticism are essential to discover ideas, results, and solutions. Members must build upon one another's strengths and eliminate or downplay weaknesses. Inevitably, members have different perspectives or viewpoints out of which conflicts will arise.

"Conflict" in the sense of conflicting opinions—a healthy give and take—can be positive if it alerts the group to problems (inconsistencies, redundancies, incompleteness) and provides ways to resolve them. A conflict can even help the group generate and refine ideas, thus leading to a better organized and written document.

Establishing Group Rules

But when conflict translates into ego-tripping and personal attacks, nothing productive emerges. Everyone in the group must agree beforehand on three iron-clad working policies of group dynamics: (1) individuals must seek and adhere to group consensus, (2) compromise may be advisable, even necessary, to meet a deadline, and (3) if the group decides to accept compromise, the group leader's final decision on resolving conflicts must be accepted.

Following are some common problems in group dynamics, with suggestions on how to avoid or solve them.

Common Problems; Practical Solutions

1. Resisting constructive criticism. No one likes to be criticized, yet criticism can be vital to the group effort. Collaboration requires being open to suggestions. But when an individual responds to criticism with anger and threatens to disrupt a meeting or lead the group away from its main goals, serious miscommunication results. Some individuals insist on "their way or no way" and so want to force everything through their very narrow, limiting perspective. They can become hostile to any change or revision, no matter how big or small.

Solutions: When emotions become too heated, the group leader may wisely move the discussion to another section of the document or to another issue and allow some cooling-off time.

2. Failing to give constructive criticism. Just as it is counterproductive to reject criticism, it is also tiresome to saturate a meeting with nothing but negatives. You will block communication if you start criticizing with words such as "Why don't you try," "What you need is," "Don't you realize that," or "If you don't . . ."

Solutions: When you find it necessary to criticize an idea, link your reaction to the team's overall goals. Diplomatically remind the individual of those goals and point to ways in which revision (criticism) furthers them. Identify a specific section in the document, explain the problem, and offer a helpful revision. If you cannot devise a suitable remedy, ask the group for suggestions or, at least, for confirmation of your criticism. Stressing that "we are all in this together" may defuse some anger. Never attack an individual; criticism should never be a personal insult but a plan to improve the document for the group's benefit. Mutual respect is everyone's right and obligation. Take a "let's work on this together" attitude rather than disrupt group harmony. Be objective, constructive, and cooperative.

3. Refusing to participate. This is a lethal problem in group dynamics. Withholding your opinions hurts the group efforts; identify what you believe are major problems and give the group a chance to consider them.

Solutions: When one individual in the group cannot, or will not, take a stand, the leader may be forced to say, "Even though you have not expressed your preference, we need you to make a choice anyway." If you don't feel sure of yourself or your points, talk to another member of the group before a meeting to "test" your ideas or to see if he or she reacts the same way.

4. Interrupting with incessant questions. Some people interrupt a meeting so many times with questions that all group work stops. Sometimes the individual questioner is simply trying to exercise control by asking irrelevant questions.

Solutions: When that happens, a group leader may turn the tables on the person by remarking, "We appreciate your interest, but would you try an experiment, please, and attempt to answer your own questions." If the person claims not to know, the leader might then say, "If you can't come up with an answer now, why don't you think about it for a while and then get back to us." If these tactics fail, the leader may have to confront the disrupter privately after the meeting.

5. Inflating small details out of proportion. Some individuals waste valuable discussion and revision time by dwelling on relatively insignificant points—the choice of a single word, an optional comma—and overlook larger problems in content and organization. Nitpickers can derail the group.

Solutions: If there is consensus about a matter, leave it alone and turn to more important issues. Bring the group back to the big picture.

6. Dominating a meeting. Developing interpersonal skills means sharing and responding, not taking over. Sometimes a group member is so aggressive and focused on individual achievement that he or she seems to occupy the floor every minute.

Solutions: A couple of intervention strategies can work when one person monopolizes a discussion. The leader may say, "We've been hearing from primarily one or two people; now we need to hear from the rest of the group." Or more directly the leader might point out, "You've made some interesting observations. Now let's give others a chance to offer their comments." The leader may have to take this person aside to remind him or her of the others' right to speak. Otherwise, dialogue is impossible. Some groups operate democratically with the "one-minute rule." Each member has a minute to voice objections, suggest revisions, report on progress, and so on, and does not get the floor again until everyone else has had a chance to speak. Holding members to a minute of "air time" can also force them to be clear and concise.

7. Being too deferential to avoid conflict. This problem is just the opposite of that described in #1 above. You will not help your group by being a "yes person" simply to appease a strong-willed member of the group. Saying that an idea or plan is excellent, when you know that it is flawed and contradictory, will only increase your team's workload. Being too deferential is as counterproductive as being too assertive.

Solutions: Feel free to express your opinions politely, and if tempers begin to flare, call in the group leader or seek the opinions of others on the team.

8. Not finishing on time or submitting an incomplete document. Meeting deadlines is the group's most important obligation to one another and to the company. Deadlines exist for various stages of a document as well as for the most important deadline of all—the date when final work must be submitted to the boss or to the customer. When some members are not involved in the planning stages or when they skip meetings, deadlines are invariably missed. If you miss a meeting, get briefed by an individual who was there. Repeated absences seriously violate group work. A deadline can also be jeopardized when a member does not clearly understand his or her assignment and therefore risks duplicating or delaying what someone else in the group has been asked to do.

Solutions: The group leader may institute networking through e-mail to announce meetings, keep members updated, or provide for ongoing communication and questions.

Models for Collaboration

There are as many types of collaborative writing activities and methods as there are companies. Writers interact with other writers, editors, and outside specialists in a variety of ways. The process can range from relatively simple phone calls or e-mail to a much more extensive network of checks and balances, revisions and refinements. The scope, size, and complexity of your document as well as your company's organization will determine what type of (and how much) collaboration is necessary. Management's participation in and instructions to a group also affect the way it is organized and functions.

A shorter assignment (say, a memo) will not require the same type of group structure and participation as would a policy handbook, a technical proposal, or a long report. At some large companies, for example, a staff of professional editors revises the final draft prepared by a departmental team. The more important and more complex a document is, the more extensive collaboration will be.

The following sections describe some possible models for collaboration that are used in the world of work.

Cooperative Model

This is one of the simplest and most expedient ways to write in the business world. An individual writer is given an assignment and then goes through the writing process (see Chapter 2) to complete it. Along the way, he or she may show a draft to a peer to get feedback or to a supervisor for a critique. This is what Randy Taylor did in Figure 3.2. He showed a draft of a letter to a potential client to his boss Felicia Krumpholtz who made changes in content, wording, and format and then sent it back to Taylor. Following his boss's suggestions, Taylor then created the revised letter in Figure 3.3. If Taylor had sent his first draft, the customer would hardly have been impressed with the company's professionalism and might have reconsidered placing an order. Following Krumpholtz's revisions, though, Taylor made his letter much more effective. Careful writers always benefit from constructive criticism. Even though Felicia Krumpholtz helped Randy Taylor improve his work, strictly speaking it was not a case of group writing. Although the two interacted, they did not share the final responsibility for creating the letter.

In the course of preparing a longer, more detailed piece of work, a writer may interview technical experts and lawyers, prepare an early draft and show it to

FIGURE 3.2 A draft of Randy Taylor's letter edited by his supervisor Felicia Krumpholtz.

November 18, 1997

Terry Tatum
Manager
Consolodated Industries [i]
Houston, TX [add zip code]

Dear ~~Sir,~~ [Terry Tatum:]

~~I am taking the opportunity of answering your request~~ [Thank you for asking] for a price list of Servitron products. Servitron has been in business in the Houston area for more than ~~twenty-two~~ [22] years and we offer unparalleled equipment and service to ~~any customer~~ [Consolidated Industries]. Utilizing [can] a Servitron will give you both efficiency and economy. [boldface]

Whatever model Servitron you choose carries with it a full one-year warranty [boldface] on all parts and labor. After the expiration date of your warranty you ~~should~~ [might] purchase our service contract for $75,000 a year.

~~Here are the models Servitron offers~~

Zephyr	81072	$ 459.95
Colt	86085	$629.95
Meteor	88096	$769.95

Depending on your needs, one of these models should be right for you.

If I might be of further assistance to you, please call on me. I am also enclosing a brochure giving you more information, including specifications, on those Servitron products. [add phone number and e-mail] [reverse the order of these two sentences]

~~Truly,~~ [Sincerely yours,]

Servitron [leave 4 spaces]
Randy Taylor [sign your name]
Sales Associate

FIGURE 3.3 The edited, final copy of Figure 3.2.

November 18, 1997

Terry Tatum
Manager
Consolidated Industries
Houston, TX 77005-0096

Dear Terry Tatum:

Thank you for asking for a price list of Servitron products. Servitron has been in business in the Houston area for more than 22 years and we can offer unparalleled equipment and service to Consolidated Industries. Using a Servitron will give you both **efficiency** and **economy.**

Depending on your needs, one of these models should be right for you.

Model Name	Number	Price
Zephyr	81072	$459.95
Colt	86085	$629.95
Meteor	88096	$769.95

Whatever model Servitron you choose carries with it a full **one-year warranty** on all parts and labor. After the expiration date of your warranty you might purchase our service contract for $75.00 a year.

I am also enclosing a brochure giving you more information, including specifications, on those Servitron products. If I might be of further assistance to you, please call me at (904)555-1689 or e-mail me at rtaylor@servitron.com.

Sincerely yours,

SERVITRON

Randy Taylor

Randy Taylor
Sales Associate

co-workers to gather their opinions, and then submit the document to a midlevel supervisor who may further revise it. Cooperative writing benefits employees and employers.

Sequential Model

In this plan, each individual in a group is assigned a specific, nonoverlapping responsibility—from brainstorming to revising—for a section of a proposal, report, or other document. There is a clear-cut, rigid division of labor. If four people are on the team, each will be responsible for his or her part of the document. For example, one employee may write the introduction, another the body of the report, another the conclusions, and the fourth, the group's recommendation.

Team members may discuss their individual progress and even choose a coordinator to oversee the progress of their work. They may even exchange their work for group review and commentary. When each team member finishes his or her section, the coordinator then assembles the individual parts to form the report.

Functional Model

The division of labor in this collaborative model is assigned not according to parts of a document but by skill or job function of the members. For example, a four-person team may be organized as follows:

- **A leader** schedules and conducts meetings, assists team members, issues progress reports to management, solves problems by proposing alternatives, and generally coordinates everyone's efforts to keep the project on schedule.
- **A researcher** collects data, conducts interviews, searches the literature, and administers tests. This researcher gathers and classifies information and then prepares notes on the work.
- **A designated writer/editor,** who receives the researcher's notes, prepares outlines and drafts and circulates them for corrections and revisions.
- **A graphics expert** obtains and prepares all visuals, specifying why, how, and where visuals should be placed and might even suggest that visuals replace certain sections of text; the graphics expert may also be responsible for the design (layout) of the document.

This organizational scheme fosters much more group interaction than does the sequential model.

Figure 3.4 illustrates the workings of a functional model. It describes the behind-the-scenes joint effort that went into Joycelyn Woolfolk's preparation of a proposal for her boss about starting a professional journal. Woolfolk wanted her employer to authorize this new journal, which would incorporate a newsletter her office currently prepares. A publications coordinator for a large, regional health maintenance organization (HMO), Woolfolk supervises a small staff and reports directly to the public affairs manager, who in turn is responsible to the vice president of the regional office. The vice president is ultimately accountable to the president and board of directors.

As you will see from reading Woolfolk's scenario, she assigned specific duties to her five-person team. Then, based on their initial research and documentation,

FIGURE 3.4 An account of how one proposal originated and was collaboratively prepared following a functional model.

The idea to start a regional magazine was first expressed in passing by our vice president, a member of our senior management, who is interested in getting more and higher-level visibility for our regional office. Several other regional offices in our company have recently put out fairly attractive magazines, and one office in particular has earned a lot of good publicity.

The public affairs manager (my boss) and I quickly picked up on the vice president's hint and began to formulate ways to investigate the need for such a publication and ways to substantiate our recommendation. For several weeks the public affairs manager and I had a number of conversations addressing specific points, such as the kind of documentation our proposal would need, what our resources for researching the question were, what our capabilities would be for producing such a publication, and so on. We were guided by the twofold goal of getting the vice president's approval and, ideally, meeting a genuine market need.

After discussions with my boss, I met with members of my staff to ask them to do the following tasks:

1. review existing HMO publications and report on whether there was already a regional magazine for the Northwest
2. develop, administer, and analyze a readership survey for current subscribers to our newsletter, which would potentially be incorporated into the new magazine
3. formulate general design concepts for the magazine that we can implement without increasing staff, while still producing the quality magazine the vice president wants
4. prepare a budget for projected costs
5. consult with experts on our staff (actuaries, physicians, nurses) about topics of interest

I requested memos and other written documentation from my staff members about most of these tasks, and then I used that information to draft the proposal that eventually would go to the vice president,

Continued

FIGURE 3.4 (Continued)

Page 2

and perhaps even to the presidents office. And I communicated with my staff often, through e-mail and personal meetings.

I revised my draft several times and, as a courtesy, had my staff look over each draft for feedback and proofreading. Based on their comments, I made further revisions and did careful editing. My boss reviewed my proposal and also revised and edited it in minor ways innumerable times. Ultimately, the proposal will go out as a memo from me to my boss, who will then send it under her name to the vice president.

The copy of my proposal may be further revised in response to the vice president's comments when she gets it. She may use some portion or all of it in another memo from her to the president, or she may write her own proposal to the president supporting her argument with the specifics from my proposal. The vice president will word the proposal to fit the expressed values and mission of our regional office as well as provide information that addresses budgetary and policy concerns for which her readers (our company president and board members) have final responsibility.

This is how a proposal started and where it will eventually end.

she drafted a document for her staff and her boss to read. As Woolfolk's functional approach shows, collaboration can move up and down the chain of command, with participation at all levels.

Woolfolk's plan is a common one in business. Often, individual employees will pull together information from their separate functional areas (such as finance, marketing, sales, and transportation) and someone else will put this information into a draft that others read and revise until the document is ready to send to the boss. By the time the boss reviews the document, it has been edited and revised many times and by many different individuals.

Integrated Model

In this model, all members of the team are engaged in planning, researching, and revising. Each shares the responsibility of producing the document. Members participate in every stage of the document's creation and design, and the group goes back to each stage as often as needed. This model offers intense group interaction. However, even though individual writers on the team may be asked to draft different sections of the document, all share in revising and editing that document. Depending on the scope of the document and company policy, the group may also go outside the team to solicit reviews and evaluations from experts.

Figures 3.5, 3.6, and 3.7 show the evolution of a memo that follows an integrated model of collaboration. This memo informed employees that their company was beginning a recycling program. Alice Schuster, the vice president of Fenton Industries, an appliance manufacturing company, asked two employees in the personnel department—Abigail Chappel and Manuel Garcia—to prepare a memo ("a few paragraphs" is how Schuster put it) to be sent to all Fenton employees. Schuster had an initial conference with Chappel and Garcia, at which she stressed that their memo had to convey Fenton's commitment to conservation and that, as part of this commitment, the employees had to practice recycling. Chappel and Garcia thus shared the responsibility of convincing co-workers of the importance of recycling and educating them about practicing it. Chappel and Garcia also had the difficult job of writing for several audiences simultaneously—the boss, whose name would not appear on the memo, other managers at Fenton, and the Fenton workforce itself.

Figure 3.5 is the first draft that Chappel and Garcia collaborated on and then presented to the manager of the personnel department—Wells McCraw—for his comments and revisions. As you can see from McCraw's remarks, written in ink, he was not especially pleased with their first attempt and asked them to make a number of revisions. As a careful reader (conscious of the memo's audience), McCraw found Chappel and Garcia's paragraphs to be rambling and repetitious—the writers were unable to stick to the point. Specifically, he pointed out that they included too much information in one paragraph and not enough in others. He also directed their attention to factual mistakes, irrelevant and even contradictory comments, and essential information they had omitted. McCraw also offered some advice on using visual devices (see Chapter 1, pp. 17–20) to make their information more accessible to readers.

Figure 3.6 shows the next stage in the collaboration. In this version of the memo, prepared through several revisions over a two-day period, the authors incorporated McCraw's suggestions as well as several changes of their own. It was this revision that they submitted to Vice President Schuster, who also made some comments on the memo. Schuster's suggestions—all valid—show how different readers can help writing teams meet their objectives. With Schuster's input and continuing to revise the memo on their own, Chappel and Garcia submitted the final, revised memo found in Figure 3.7 to the vice president a few days later. This final copy received Schuster's approval and was then routed to the Fenton staff.

Thanks to an integrated model of collaboration and careful critiques by McCraw and Schuster, Chappel and Garcia successfully revised their work. Note that in the process of revising their memo they had to make major changes from an early draft through several versions to the final copy. These changes—shortening and expanding paragraphs, adding and deleting information, and refocusing their approach to meet the needs of their audience—are the revisions a collaborative writing team, like an individual writer, should expect to make. Effective team effort and shared responsibility were at the heart of Chappel and Garcia's assignment.

FIGURE 3.5 Early draft of the Chappel and Garcia collaborative memo, with revisions suggested by Wells McCraw.

FENTON INDUSTRIES

TO: All Employees
FROM: Abigail Chappel; Manuel Garcia
RE: Starting a Recycling Program
DATE: February 10, 1998

This ¶ is too long Too many topics – costs, protecting the environment keep it short – say what we are doing and why

An in-house study has shown that Fenton sends approximately 26,000 pounds of paper to the landfill. The landfill charge for this runs about $2,240, which we could save by recycling. Fenton Industries is conscious of our responsibility to save and protect the environment. Accordingly, starting March 1 we will begin a paper recycling program. Our program, like many others nationwide, seeks first of all to employ the latest degradable technology to safeguard the air, trees, and water in our community. It has been estimated that of the 250 million tons of solid waste, three quarters of goes to landfills. These landfills across the country are becoming dangerously overcrowded. Such a practice wastes our natural resources and endangers our air and drinking water. For example, it takes ~~10~~ trees to make one ton of paper, or roughly the amount of paper Fenton uses in four weeks. If we could recycle that amount of paper, we would save those trees. Recycling old paper into new paper involves less energy than making paper from new trees. Moreover, waste sent to landfills can, once broken down, leach, seep into our water supply, and contaminate it. The dangers are great.

start off with this key idea

word "it" left out

check your facts; I think it is closer to 16–17

Delete– not relevant to our purpose

No cap

By recycling, we will not be sending so much to the Springfield Landfill and so help alleviate a dangerous condition there. We will keep it from overflowing. Fenton will also be contributing to transforming waste products into valuable reusable materials. Recycling paper in our own office shows that we are concerned about the environmental clutter. By having a paper recycling program we will establish our company's reputation as an environmentally conscious industry and enhance our company's image.

Add the fact about our saving trees in this ¶

Delete– makes us look bad

Start ¶ with this point

Fenton is primarily concerned with recycling paper. The 200 old phone books that otherwise would be tossed away can get our recycling program off to a good start.

When? How? Implications for saving/costs?

Give some examples

We encourage you to start thinking about the kinds of paper around your office/workspace that needs to be earmarked for recycling. When you start to think about it, you will see how much paper we as a company use.

Continued

FIGURE 3.5 (Continued)

not into your wastebaskets

Starting the last week of February, paper bins will be placed by each office door inside the outer wall. These bins will be green–not unsightly and blending with our decor. Separate your waste paper (white, colored, and computer) and put it into these bins. You do not need to remove paper clips and staples, but you must remove rubber bands, tape, and sticky labels. They will be emptied each day by the clean-up crew. There will also be large bins at the north end of the hallway for you to deposit larger paper products. [The crucial point is that you use these specially marked bins rather than your wastebasket to deposit paper.]

Delete – repetitious

¶ lacks effective "call to action"

Fenton Industries will deeply appreciate your cooperation and efforts. Thanks for your cooperation.

This memo needs more work

–add at least one ¶ on how recycling will save us money
–make directions clearer and easier to follow; try using numbered steps
–end on a more upbeat note; tell employees about the benefits coming to them for recycling – i.e., the office fund/our contribution to their favorite charity.

Cowriting Model

According to this model, everyone on the team actually drafts the document together, word for word. Each may work at a different computer terminal, but all their effort is focused on the same section of the document at the same time. This model offers the highest degree of collaboration and might be compared to a committee in which everyone has a direct say (or hand) in every phase of the document. While cowriting may work for a relatively short document such as a memo, it is rarely used in business because it is neither cost-effective nor practical. Cowriting collaboration is extremely labor intensive, as it ties up all members of the team to draft the document.

Collaborating On-Line

Collaborating on-line takes writers into a different communication environment. Individuals in the world of work are meeting much more frequently in cyberspace to get a job done. One of the most frequently used services on the information superhighway—the Internet—is electronic mail, or e-mail. You will learn more

FIGURE 3.6 Revision of the Chappel and Garcia memo with changes suggested by Vice President Schuster.

FENTON INDUSTRIES

TO: All Employees
FROM: Abigail Chappel; Manuel Garcia
RE: Starting a Recycling Program
DATE: February 10, 1998

To save money and protect our environment, Fenton Industries will begin a waste paper recycling program on March 1. Recycling has become an environmental necessity. It has been estimated that three quarters of the 250 million tons of solid waste dumped annually in America could be recycled. Our program will employ the latest recycling technology to safeguard trees, air, and water.

say strengthen or continue

¶ needs more infor- mation

This new program will ~~establish~~ Fenton's reputation as an environmentally conscious company. Fenton now uses one ton of paper every four weeks. This represents 17 trees that can be saved just by recycling our paper waste. Recycling also means we will send less waste to the landfill, alleviating problems of overfill and reducing the potential for contamination of the water supply. Waste sent to large landfills can leach and seep into water systems. By recycling paper, Fenton will also reduce the risk of long-term environmental pollution.

Add that we no longer use Styrofoam and that our suppliers use only biodegradable products

Recycling will also save us money. Right now Fenton pays $180 per month to dump 2,000 pounds (one ton) of paper waste at the landfill. However, scrap paper is worth $100 per ton. Recycling will generate $100 each month in new revenue while eliminating the $180 dumping expense.

Ultimately, the success of our project depends on being aware of the variety of office paper suitable for recycling. Paper products that can be recycled include newspapers, scrap paper, computer printouts, letters, envelopes (without windows), shipping cartons, old phone books, and uncoated paper cups.

This sentence more logically goes in ¶ 3

Put in itemized or bulleted list to standout better

Continued

FIGURE 3.6 (Continued)

Do not start new ¶ here; keep as part of previous ¶

Recycling 200 or so old phonebooks each year alone will save 22 cubic yards of landfill space (or close to $100) and will generate $25 in scrap paper income for our company. The old 1997 phone books, which will be replaced by new 1998 ones on March 1, will give us an excellent opportunity to start recycling.

Here are some easy to follow directions to make our recycling efforts effective:

Boldface these words

1. Starting the last week in February, paper bins will be placed inside each office door. Put all waste paper into these bins, which will be emptied by maintenance.
2. Place larger paper products–such as cartons or phone books–in the green bigger paper bins at the end of each main corridor.
3. Put white, colored, computer printout, and newspapers into separate marked bins. Remove all rubber bands, tape, and sticky notes.

Thanks for your cooperation. To show our appreciation for your help, 50 percent of all proceeds from the recycled paper will go into the general office fund and the other 50 percent will be given to the office's favorite charity. The benefits of our new recycling program will more than outweigh the inconvenience it may cause.

Add another sentence to this ¶ on how a safer environment will benefit our company and the employees, too

about this type of correspondence in Chapter 4. Here we will examine how e-mail can improve collaboration.

TECH NOTE

E-mail involves the exchange of information between and among computers (with the benefit of modems) around the globe. E-mail enhances any collaboration from on-line conversations in **real time** (both parties are speaking to each other at the same time) to electronic conferences in which both or all parties may be in nonreal (different) time frames; for example, you send a message at 10:00 A.M. EST and your respondent answers at 10:00 P.M. CST.

FIGURE 3.7 Final copy of the memo prepared by Chappel and Garcia using an integrated model of collaboration.

FENTON INDUSTRIES

TO: All Employees
FROM: Abigail Chappel; Manuel Garcia
RE: Starting a Recycling Program
DATE: February 10, 1998

To save money and protect our environment, Fenton Industries will begin a waste paper recycling program on March 1. For the program to succeed, all of us need to recycle paper. Recycling has become an environmental necessity. It has been estimated that three quarters of the 250 million tons of solid waste dumped annually in America could be recycled. Our program will employ the latest recycling technology to safeguard trees, air, and water.

Recycling will save us money. Right now Fenton pays $180 per month to dump 2,000 pounds (one ton) of paper waste at the landfill. However, scrap paper is worth $100 per ton. Recycling will generate $100 each month in new revenue while eliminating the $180 dumping expense.

This new program will also strengthen Fenton's reputation as an environmentally conscious company. Three years ago, we stopped using Styrofoam products and asked suppliers to use biodegradable materials for all our shipping containers. Fenton now uses one ton of paper every four weeks. This paper represents 17 trees that can be saved just by recycling our paper waste. Recycling also means we will send less waste to the landfill, alleviating problems of overfill and reducing the potential for contamination of the water supply. Waste sent to large landfills can leach and seep into water systems.

Ultimately, the success of our project depends on all of us being aware of the variety of office paper suitable for recycling. Paper products that can be recycled include:

- newspapers
- letters, envelopes (without cellophane windows)
- old phone books
- uncoated paper cups
- scrap paper
- computer printouts
- junk mail (only black print on white paper)
- shipping cartons

Continued

FIGURE 3.7 (Continued)

Page 2

The old 1997 phone books, which will be replaced by new 1998 ones on March 1, will give us an excellent opportunity to start recycling.

Here are some easy to follow directions to make our recycling efforts effective:
1. Starting the last week in February, put all waste paper into **paper bins** which will be placed **inside each office door**. These bins will be emptied by maintenance.
2. Place **larger paper products**–such as bulky cartons–in the **green** bigger paper bins at the end of each main corridor.
3. Remove all rubber bands, tape, and sticky notes and put white, colored, computer printout, and newspapers into separate marked bins.

Thanks for your cooperation. To show our appreciation for your help, 50 percent of all proceeds from the recycled paper will go into the general office fund and the other 50 percent will be given to the office's favorite charity. The benefits of our new recycling program will more than outweigh the inconvenience it may cause. A safer environment–and a more cost-effective way to run our company–benefits us all.

Thanks to e-mail, employees communicate more and better with each other and with their customers and management. E-mail reaches a much broader audience than a telephone call or a fax; a company hooks into a worldwide network through e-mail. E-mail also allows individuals to send more messages more quickly and efficiently than any other form of communication. Using e-mail, employees can speed the flow of collaborative information, especially using software designed to enhance teamwork in business. Going far beyond the potential of basic e-mail, Lotus Notes (Figure 3.8) is an example of software that can be linked with the programs and databases already in use by the organization, allowing it to customize employees' collaborative tasks, scheduling, and networks easily and seamlessly. E-mail will not completely eliminate the need for a group to meet in person to discuss thorny issues, to clarify subtleties, or to build group harmony. But face-to-face communications, though essential, are frequently supplemented with on-line collaboration.

FIGURE 3.8 An advertisement for Lotus Notes emphasizing groupware.

It Is Easy To Get Started With Notes

Notes gives you your choice of functionality.

Lotus Notes provides you with three licensing options for building a unified groupware infrastructure: Lotus Notes Mail, Lotus Notes Desktop, and Lotus Notes.

Lotus Notes Mail.

Notes Mail is a state-of-the-art, client/server messaging system. Notes Mail includes the cc:Mail user interface, OLE 2.0 support, platform independent viewers, collapsible sections, the InterNotes Web Navigator, task management, document libraries, personal journal, phone messages, and unparalleled mobile user support.

Lotus Notes Desktop.

Lotus Notes Desktop is a run-time Notes client with the ability to run any Notes application. Notes Desktop enables you to extend your most strategic applications to everyone in your workgroup. Includes Notes Mail functionality and sample application templates. Notes Desktop is an affordable solution ideal for Notes users who need to access customized applications.

Lotus Notes.

Lotus Notes gives you the full-function power to create custom applications for improving the quality of everyday business processes. Includes an application development environment, system administration capabilities, and Notes Mail functionality.

Build your Notes system as you build your business.

Start with the core capabilities of Lotus Notes and build strategic business process applications for your workgroup. Tailor your application development, database and messaging capabilities as your needs change and your workgroup expands.There's a rapidly-growing industry of value-added software products and services based on Lotus Notes. Choose from hundreds of off-the-shelf applications or work with one of the 11,000 Lotus Business Partners to customize an application appropriate for your company.

GROUP SCHEDULING IS EASY WITH LOTUS ORGANIZER AND NOTES

Lotus Organizer® uses existing Notes and cc:Mail directories to set up group meetings and address meeting notices. There is no faster or easier way to manage your workload, stay in touch with business contracts, and schedule meetings, than with Lotus Organizer and Notes.

Excerpted from "It's Easy to Get Started with Notes," from "The Power of People Working Together: Now Is the Time for Lotus Notes" © 1996 Lotus Development Corporation. Used with permission of Lotus Development Corporation. Lotus and Lotus Notes are registered trademarks and Notes is a trademark of Lotus Development Corporation.

Advantages of Collaborating On-Line

Collaborating via e-mail offers significant advantages to employers and workers. As you develop your on-line communication skills, here are some of the benefits you can count on:

1. Eliminating time barriers. Communicating by e-mail, team members do not have to worry about tight schedules, conflicts, delays, last-minute appointment changes, or out-of-town business trips interfering with their own or another's participation. Members can send and receive messages and commentary via e-mail at any time from anywhere. Employers and employees can take advantage of a handless clock.

2. Removing geographic limitations. With e-mail, a team member does not have to be physically present in order to contribute to group work. Team members can be in different cities, even different hemispheres. A collaborative team can be composed of members from all over the globe, which benefits international firms that want to involve employees from foreign offices in decision-making processes. Team members who live in the same city can get their work in on time even if a crippling winter storm prevents them from leaving their homes. With a personal computer members can send and receive e-mail within the group at any time.

3. Increasing the amount and quality of feedback. E-mail allows participants to exchange ideas with individuals from around the globe and to read about the latest research. Moreover, e-mail encourages users to ask questions about any topic and to receive helpful, informative replies. Rather than merely distributing information, collaborating on-line inspires individuals to analyze and solve problems more quickly. E-mail can thereby increase the level and even the objectivity of feedback in collaboration.

4. Generating more information through on-line exchanges. Many writers are willing to take more chances via e-mail than they would be in person. E-mail collaboration encourages some writers to risk innovative suggestions and proposals they might fear to voice in face-to-face meetings. By reading and responding to drafts and revisions at times most convenient for them, team members are given more time to pursue research and study their replies and thus to offer much more carefully considered responses.

5. Allowing team members to work on a project with less pressure. Collaborating via e-mail can make criticism easier to take and to give without the intimidating presence of an especially difficult colleague. Moreover, on-line collaboration lessens the likelihood of interruptions, miscommunications, or unpleasant confrontations.

6. Broadening the range of participation in the collaborative process. E-mail encourages the most flexible and extensive kinds of group organization possible. Supervisors within a writer's department and from related areas as well as customers and vendors or suppliers can join in the discussion and revision to improve the quality and service a company offers.

7. Lowering the cost of communicating in the world of work. The expense a company incurs to put several workers on-line to draft and revise documents is far less than if these employees, from different plants, must meet in person to perform the same writing tasks. An employer is also able to take advantage of e-mail rates, which are much more economical than phone or fax charges.

The Collaborative Dynamics of E-Mail Technology

E-mail is widely used in the business world, and its impact is sure to grow. But it is not just for short exchanges. E-mail plays a vital role in allowing individuals to write and edit all types of documents—letters, memos, reports, proposals. Like a conference call or a videoconference (see Chapter 18), e-mail helps a group to share information and responses and to communicate with one another about them.

Using E-Mail to Write Collaboratively

E-mail technology is central to groupware, allowing for a variety of types of collaboration. Here are just two of the most likely ways a group can create and edit a document using e-mail. The first method is as follows:

1. Each member of the writing team drafts his or her assigned section of a document on a computer.
2. Then each writer e-mails that section to everyone in the group for review and comments.

TECH NOTE

Attaching a word processing file to an e-mail message is easy. Save the document from your work processing file as a text or ASCII file. Then get into your e-mail program and choose the "send" command. Before you send your e-mail, though, you will have to choose from your option menu "attach." Then indicate what file you want to attach. Your program will automatically pull up that file and attach it to your e-mail. Then press "send," and your document, plus attachment, is on its way to the reader or readers you have specified.

3. All the individuals in the group then receive and react to each writer's draft and e-mail their comments—corrections, suggestions for revisions, additions, whatever—back to the individual contributors.

TECH NOTE

Once you receive an e-mail with an attached document, you can download it into a file or you can make a hard copy. Some individuals like to read and edit on-screen, while others prefer a hard copy manuscript.

4. Each writer revises his or her draft based on the group's comments and sends the revised draft to a team leader or coordinator.
5. The team coordinator receives all the revised drafts and puts them together on one file, making sure that all changes are accurate and appropriate and that the style and format of the assembled final document are consistent.

Essentially, this is the sequential model of collaboration (see p. 86) using e-mail technology.

A second method of e-mail collaboration is as follows:

1. Each member of the writing team actually revises the document he or she receives and then sends that revision document on to a leader or coordinator rather than back to the original writer.
2. It is the coordinator's responsibility to select from among the revisions which ones to include in the final draft and which ones to leave out.

It is essential for the team leader to link revisions with specific individuals. Current technology allows for the identification of multiple versions by multiple authors of the same document so that the leader will know which group member made a particular change.

Case Study: Collaborating via E-Mail

The following shows a series of e-mail exchanges among a writing team trying collaboratively to write a single document, a report on expanding a hospital parking facility. In this string of exchanges you will see the dynamics of collaboration made possible through e-mail technology. As you read the participants' comments, identify each person's concerns and how other team members react to them. Who is leading this collaborative effort? Whose ideas most influence the final report? Who raises an ethical issue and what is it? In broad terms, how will the revised document differ from the first draft? Finally, if these co-workers did not have e-mail technology, how would they go about preparing their collaborative report?

Four on-line collaborative writers revise a report via e-mail.

Date: Wed. 10 April 1998 16:55:00EST
Reply to: Ramon Calderez, Alex Latriere, Loretta Bartel
Sender: Nicole Goings
Subject: Report on Expanded Hospital Parking

Now that we have a pretty clear outline and a first draft of the report before us, I think we all need to try to flesh it out. Thanks for reviewing the attached document to see how it hangs together. My initial reaction is that the draft needs reorganization and more attention to detail.

Date: Wed. 10 April 1998 18:01:00EST
Reply to: Nicole Goings, Alex Latriere, Loretta Bartel
Sender: Ramon Calderez
Subject: Report on Expanded Hospital Parking

Thanks for the draft. One thing I noticed right away was that our opening is not very strong or convincing. There is too little sense of the overall reason for the hospital investing $3.2 million in expanding parking facilities. I have rewritten the opening, as you will see, and tried to link the currently inadequate parking facilities (a detriment) to the overall growth of patient care (our strong suit). Then I tried to emphasize how responsive Bloomington Memorial has been to the needs of visitors and the patients they have come to see.

Let me know what you all think.

Date: Thurs 11April 1998 8:23:00EST
Reply to: Nicole Goings, Alex Latriere, Ramon Calderez
Sender: Loretta Bartel
Subject: Report on Expanded Hospital Parking

I agree with Ramon and think the revised introduction will work much better, but aren't we being too dramatic and not very pro-Bloomington Memorial by using the last sentence of his in the second paragraph—"new parking facilities will prevent visitors from walking long, bone-soaking distances in the rain"–and so I cut it and used a different closing sentence.

Date: Thurs. 11 April 1998 8:54:00EST
Reply to: Nicole Goings, Ramon Calderez, Loretta Bartel
Sender: Alex Latriere
Subject: Report on Expanded Hospital Parking

Loretta's change is o.k. but I want to point out two much more important revisions we need to make. One, it is unfair to say we are adding 500 new parking spaces. The exact number is 417. I am more comfortable with saying "more than 400" or just giving the exact number.
Another much larger issue is rerouting traffic on Wentworth Avenue once the new parking lot is complete. Since Wentworth has been a two-way street for a long time–at least for the last 11 years that I have lived in Bloomington–I think we need to prepare readers better for this change. In short, we should add another paragraph under the section now labeled "Increased Traffic Flow" and devote it entirely to the Wentworth Avenue change.

I inserted one or two points at the end of that section that I think would make a coherent paragraph.

Date: Thurs. 11 April 1998 11:08:00EST
Reply to: Alex Latriere, Ramon Calderez, Loretta Bartel
Sender: Nicole Goings
Subject: Expanded Parking Report

All right, bravo–I hear you have taken your ideas to draft a new paragraph on Wentworth and have also done some slight editing to make the transition to this topic a little smoother. What do the rest of you think?

Continued

(Continued)

Date: Thurs 11April 1998 13:45:00EST
Reply to: Nicole Goings, Ramon Calderez, Alex Latriere
Sender: Loretta Bartel
Subject: Expanded Parking Report

Alex's suggestion and Nicole's additional paragraph work very well together.

Date: Wed. 10 April 1998 14:23:00EST
Reply to: Nicole Goings, Alex Latriere, Loretta Bartel
Sender: Ramon Calderez
Subject: Expanded Parking Report

Yes, a good job. The more I looked at the section on "Entry Points" the more I was troubled by including the topic of handicap access under this heading. Given the fact that the new parking facility will also involve widening the entrance to the ER, thus making 11 additional handicap spots available is, in my mind, worthy of a separate section in the report. Accordingly, I think we should take handicap access out of the "Entry Points " section and create a new, even if small, section on "Handicap Access." I have transferred that information, therefore, to the new section I want to see follow the section on "Entry Points."

Date: Thurs. 11 April 1998 17:17:00EST
Reply to: Ramon Calderez, Nicole Goings, Alex Latriere
Sender: Loretta Bartel
Subject: Report on Expanded Parking

I realize we want to emphasize the new technology that the hospital is putting into this parking facility, but since the report is going to the board of directors and to other general readers, I think we need to cut back on the descriptions of this technology.

What I want to eliminate–or at least tone down–is all the details on stress points, pre-cast concrete, and the low slope vehicular access ramps. I think we can still include some information about the safety and engineering benefits but since we are also making the architect's site plans available, couldn't we tighten and shorten this section. I have tried to takes some things out. Have I taken out too much? Not enough? Let me hear from you!

Date: Fri. 12 April 1998 9:49:00 EST
Reply to: Loretta Bartel, Ramon Calderez, Alex Latriere
Sender: Nicole Goings
Subject: Report on Expanded Parking

Thanks and more thanks Loretta. I spoke with Lee Bukowski, the hospital architect, last evening about your changes. You are on target and so I have left your revisions stand. BUT....I do think we need to retain some information about the access ramps and covered areas surrounding them.

Date: Fri. 12 April 1998 13:22:00EST
Reply to: Nicole Goings, Loretta Bartel, Alex Latriere
Sender: Ramon Calderez
Subject: Report on Expanded Parking

Let's not forget the sacrifices our staff are making. They're being forced to give up their parking spaces during construction so visitors will have closer access. I think we should develop the idea even more that the staff share and exemplify the hospital's dedication to service.

Continued

(Continued)

Date: Fri. 12 April 1998 14:56:00 EST
Reply to: Ramon Calderez, Loretta Bartel, Alex Latriere
Sender: Nicole Goings
Subject: Report on Expanded Parking

You are absolutely right, Ramon, but certainly we can say it in fewer words. See if you like my revisions of your staff contributions.

Date: Mon. 15 April 1998 10:23:00EST
Reply to: Loretta Bartel, Nicole Goings, Alex Latriere
Sender: Ramon Calderez
Subject: Expanded Parking Report

Sure, Nicole, your changes work very well. Any way we can polish this document we should.

It occurred to me, though, that we need some visual conformation of all the hospital's efforts. I e-mailed the archives earlier this morning and they found the photograph of an aerial view of the hospital's original parking lot in 1971. I think we should incorporate that somewhere in the introduction to show the growth we want to stress and then use the artists drawing of what the new parking lot will look like to begin the section "Expanded Parking Facilities Planned." I have scanned (and now attach) both documents to give you an idea of what I have in mind.

Date: Mon. 15 April 1998 14:04:00EST
Reply to: Ramon Calderez, Alex Latriere, Loretta Bartel
Sender: Nicole Goings
Subject: Report on Expanded Parking

Ramon, you deserve a pat on the back. Yes, the visuals definitely work. I wonder, too, if we might not retitle the section "Safety and Security" that we all worked on last week to "A User Friendly Facility" and then use "Safety" and "Security" as subheadings in that section. I rather like the idea of "User Friendly" to suggest both the technology and our goal to stress Bloomington Memorial as a friendly facility.

I think we have done a good job in revising and editing the report. I want to give it to Christine Murphrey tomorrow for her approval before it goes to the Board.

Let me know before 4:30 today if you have any further suggestions or revisions.

Thanks for all your help. I'm going to acknowledge and thank each of you for your excellent work in my cover letter to Director Murphrey.

✓ Revision Checklist

- ❑ Tried to be a team player by putting the success of my group over the needs of my own ego.
- ❑ Followed necessary steps of the writing process to take advantage of team effort and feedback.
- ❑ Attended all group meetings and understood responsibilities of the group and my own obligations.
- ❑ Finished research, planning, and drafting expected of me.
- ❑ Conducted necessary interviews and conferences to gather and verify information.
- ❑ Shared my research, ideas, and suggestions for revision through constructive criticism.
- ❑ Participated honestly and politely in discussions with colleagues.
- ❑ Treated members of my team with respect and courtesy.
- ❑ Was open to criticism and suggestions for change.
- ❑ Read colleagues' work and gave specific and helpful criticism and suggestions.
- ❑ Kept matters in proper perspective by not being a nitpicker and by not interrupting with extraneous points or unnecessary questions.
- ❑ Sought help when appropriate from relevant subject matter experts and from co-workers.
- ❑ Secured responses and approval from management.
- ❑ Took advantage of e-mail to disseminate information, to communicate with my collaborative team, and to keep my own research current.
- ❑ E-mailed messages in clear, diplomatic, and correct sentences.
- ❑ Attached pertinent documents e-mailed to collaborative team.
- ❑ Answered e-mail questions and responded to requests promptly and completely.

Exercises

1. Assume you belong to a three- or four-person editing team that functions the way Tara Barber's does in Figure 3.1. Each member of your team should bring in four copies of a paper done for this course or another one. Exchange copies with the other members of your team so that each team member has everyone else's papers to review and revise. For each paper you receive, comment on the style, organization, tone, and discussion of ideas as Wells McCraw did in Figure 3.5.

2. With members of your collaborative team (selected by your teacher or self-appointed) select four different brands of the same leading product (such as software package, a web browser, a CD player, a microwave, a VCR, a power tool, or other item). Each member of your team should select one of the brands and prepare a two-page memo report (see pp. 110–118), evaluating it for your instructor according to the following criteria:

   - convenience
   - performance
   - technical capabilities/capacities
   - appearance
   - adaptability
   - price
   - weaknesses/strengths compared with competitors' models

   Each team member should then submit a draft to the other members of the team to review. At a subsequent group meeting, the group should evaluate the four brands based on the team's drafts and then together prepare one final recommendation report for your instructor.

3. Your company is planning to construct a new office and you, together with other employees from your company, have been asked to serve on a committee to make sure that plans for the new building adhere precisely to the Americans with Disabilities Act, passed in 1993. According to that act, it is against the law to discriminate against anyone with impairments that limit "major life activities"—walking, seeing, speaking, working.

   The law is expressly designed to remove architectural and physical barriers and to make sure that plans are modified to accommodate those protected by the law (for example, wider hallways to accommodate individuals in wheelchairs). Other considerations include choosing appropriate nonstick floor surfaces (reducing the danger of slipping), placing water fountains low enough for use by individuals in wheelchairs, and installing doors that require just the right amount of pressure to open and close.

   After studying the plans for the new building, you and your team members find several problem areas. Prepare a group-written report advising management of the problems and what must be done to correct them to comply with the law. Divide your written work according to areas that need alteration—doors, floors, water fountains, restroom facilities. Each team member should bring in his or her section and then the group should edit and revise these sections and prepare the final report for management.

4. A new president will be coming to your college in the next month, and you and five other students have been asked to serve on a committee that will submit a report about campus safety problems and what should be done to solve them. You and your team must establish priorities and propose guidelines that you want the new administration to put into practice. After two very heated meetings, you realize that what you and two other students have considered

solutions, the other half of your committee regards as the problems. Here is a rundown of the leading conflicts dividing your committee:

- **speed bumps**—half of the committee likes the way they slow traffic down on campus but the other half says they are a menace because they jar car CD players
- **sound pollution**—half of your team wants Campus Security to enforce a noise policy preventing students from playing loud music while driving on campus but the other half insists this would violate students' rights
- **van and sports utility vehicle parking**—half of the committee demands that vans and sports utility vehicles park in specially designated places because they block the view of traffic for any vehicle parked next to them; the other members protest that people who drive these vehicles will be singled out for less desirable parking places on campus

Clearly your committee has reached a deadlock and will be unproductive as long as these conflicts go unresolved. Based on the above scenario, do the following:

a. Have each student on the committee write (or e-mail, if available) the other five students suggesting a specific plan on how to proceed—how can the group resolve their conflicts. Prepare your e-mail message and send it to the other five committee members and to your instructor. What's your plan to get the committee moving toward writing the report to the incoming president?

b. Assume that you have been asked to convince the other half of the committee to accept your half's views on these three areas—speed bumps, noise control, parking. Send the three opposition students a memo or e-mail persuading them to your way of thinking. Keep in mind that your message must assure them that you respect their point of view.

c. Assume that the committee members reach a compromise after seeing your plan put forth in (a) above. Collaboratively draft a three-page report to the new president.

d. Collaboratively draft a letter to the editor of your student newspaper defending your recommendations to the student body and explaining how the group resolved its difficulty. This is a public statement that the group felt it was important to write; you will have to choose your words very carefully to win campus-wide support.

5. You work for a hospital laboratory and your lab manager, under pressure from management to save money, insists that you and the three other medtechs switch to a different brand of vacuum blood drawing tubes. You and your colleagues much prefer the brand of tubes you all have been using for years. Moreover, the price difference between the two brands is very small. As a group project, prepare a memo to the business manager of the hospital explaining why the switch is unnecessary, unwise, and unpopular. Then prepare another collaboratively written memo to your lab manager. But be sensitive to each reader's needs as you diplomatically explain the group's position.

PART II

Correspondence

CHAPTER 4

Writing Memos, Faxes, and E-Mail

Memos, faxes, and e-mail are the types of writing you can expect to prepare most frequently on the job. These three forms of business correspondence are quick, easy, and effective ways for a company to communicate internally as well as externally. You will find yourself preparing one or more of these types of writing each day to co-workers in your department, to colleagues in other departments and divisions of your company, and to decision makers at all levels. You can expect to send memos and e-mail to co-workers anywhere in the world.

What Memos, Faxes, and E-Mail Have in Common

Each of these three forms of writing is streamlined for the busy world of work. Memos are far less formal in tone than letters; e-mail can be even more informal than a memo. Faxes, memos, and e-mail also require you to follow different formats than you do for letters. As you will see, these formats are designed to expedite communication between you and your readers. E-mail, faxes, and memos can be composed and sent to various sites electronically from your PC.

The overall function of memos, faxes, and e-mail is to give busy readers information fast. Each type of document should be clear, direct, and narrowly focused. Memos, faxes, and e-mail are **not** designed for long messages but, instead, for shorter ones (usually a page or two). While these messages can be about any topic in the world of work, most often they focus on the day-to-day activities and operations at your company—sales and product information, policy and schedule changes, progress reports, orders, personnel decisions, and so forth.

Don't think, though, that because memos, faxes, and e-mail are routine that writing them will not demand a great deal of your thought and time. Each of these communications is a vital piece of documentation that requires your best written work. While some individuals believe we are moving toward a paperless office, a company will still want to see a paper or electronic trail with memos, faxes, and e-mail documenting what has been done, when, and by whom. Your success as an employee can depend as much on your preparing a readable and effective memo, fax, or e-mail message as it will on your technical expertise.

Memos

Memorandum, from which the noun *memo* comes, is a Latin word signifying "something to be remembered." The Latin meaning points to the memo's chief function: to record information of immediate importance and interest in the busy world of work. Memos are in-house correspondence sent up and down the corporate ladder—from managers to employees and from employees to managers. They are also sent to and from co-workers. Memos allow a business or agency to communicate with itself in its day-to-day operations. They can be handwritten, faxed, or sent through e-mail.

Functions of Memos

Memos have a variety of functions, including

- announcing a company policy or plan
- making a request
- explaining a procedure or giving instructions
- clarifying or summarizing an issue
- alerting readers to a problem
- confirming the outcome of a conversation
- reminding readers about a meeting, policy, or procedure
- providing documentation necessary for business
- offering suggestions or recommendations

Memos are valuable written records used for a variety of purposes. They are used for short reports; see Chapter 16 for examples of field, trip, or progress reports in memo format. Many internal proposals (pp. 556–560) also are written as memos.

Memo Format

Memos vary in format. Some companies use standard, printed forms (Figure 4.1), while others have their names (letterhead) printed on their memos, as in Figure 4.2. Memos can be printed on 8 1/2" × 11" sheets of paper or on half-sheets. The smaller option is useful for shorter communications to encourage the writer to be concise. You can also make your own memos by including the necessary parts discussed below. You can also send a memo in an e-mail.

As you can see from looking at Figures 4.1 through 4.4, memos look different from letters. They are more streamlined and less formal. Because they are sent to individuals within your company, memos do not need the formalities necessary in business letters, such as an inside address, salutation, complimentary close, or signature line, as discussed in the next chapter. (See pp. 147–151.)

FIGURE 4.1 Standard memo without letterhead.

MEMO

TO: All RNs
FROM: Margaret Wojak, Director of Nurses M.W.
Date: August 16, 1999
Subject: RN Identity Patches

Effective September 1, 1999, all RNs will be asked to wear an identity patch in addition to any other means of identification (name badge, cap). You should sew your patch on the upper right arm of lab coats or uniforms so that staff and patients can easily identify you as an RN.

You may obtain an authorized Memorial Hospital identity patch for two dollars at the Health Uniform Shop directly across the street from the hospital on Ames Street.

Please write me at my e-mail address mwojak@memorial.com or call me at Extension 3106 if you have any questions.

Basically, the memo consists of two parts: the identifying information at the top and the message itself. The identifying information includes these easily recognized parts: the **To, From, Date,** and **Subject** lines.

TO: Aileen Kelly, Data Processing Manager
FROM: Stacy Kaufman, Operator, Level II
DATE: January 31, 2000
SUBJECT: Progress report on the fall schedule

On the **To** line write the name and job title of the individual(s) who will receive your memo or a copy of it. If your memo is going to more than one reader, make sure you list your readers in the order of their status in your company or agency, as Mike Gonzalez does in Figure 4.3: the vice president's name appears before the public relations director's. If you are on a first-name basis with the reader, just use his or her first name. Otherwise, include the reader's first and last names.

In some companies memos are sent to everyone whose name is on a distribution list. For example, your name might be on the list for receiving all company

FIGURE 4.2 Printed memo on letterhead stationery.

GREENWOOD CORP.
56 North Jones
Canton, Ohio 45307-0299
phone: (216)555-4232 fax: (216)555-1172
e-mail: rblack@gwc.com

TO:	Marge Adcox	DATE:	November 23, 1999
FROM:	Roger Blackmore RB	SUBJECT:	Review of Management Training Seminar

I attended the Management Training Seminar (November 10-20) and learned about several training techniques that could easily be incorporated into our monthly orientation programs.

Here is a review of these major points made by the director, Jackie Lowery.

1. The individual conducting the training sessions should always talk in a loud voice so that trainees can hear him or her.
2. The main purpose of the training session should always be announced so that trainees can focus on a specific set of topics.
3. Instructors should allow at least a ten-minute break for each ninety-minute session. A brief break will help increase trainees' attention spans and promote better learning.
4. The easiest tasks should be assigned first; more complicated ones should follow.
5. Trainees should be divided into small groups to encourage collaborative exchanges.
6. The instructor should provide feedback to the trainees at the end of each major section.

Could we meet in the next day or two to discuss implementing these points? I would like to get your reaction to them as well.

information on a given project or weekly reports from certain departments. Your name may appear on a number of lists and some of your memos may be distributed to several people. Don't send copies of your memos to individuals who don't need them, however. You will only increase the paper inflation or electronic traffic in your office.

On the **From** line write your name (first name only if your reader refers to you by it) and your job title (unless it is unnecessary for your reader). Some writers put their handwritten initials after their typed name to verify that the message comes from them.

On the **Subject** line write the purpose of your memo. The subject line serves as the title of your memo. Be precise so that readers can file your memo correctly. Vague subject lines such as "New Policy," "Operating Difficulties," or "Shareware" do not identify your message precisely and may suggest that your message is not carefully restricted or developed. "Shareware," for example, does not tell readers if your memo will discuss new equipment, corporate arrangements, or vendors; offer additional or fewer benefits; or warn employees about abusing the system.

On the **Date** line do not simply name the day of the week—Monday. Give the full calendar date—June 8, 1998.

TECH NOTE

A template is a predesigned word processing form created either by a company (Microsoft, Corel, WordPerfect, for example) or an individual (such as you or me) for use as a basis for all subsequent similar documents in order to save time. A memo template is a predesigned form that you can use every time you send a memorandum. You simply open the template and fill in the blanks instead of sitting down and creating a document from scratch. Below is an example of a memo template. The document heading always remains the same. The date is inserted automatically, and the only entries you must add are the reader, the subject, and—of course—the body of the memo.

Memorandum

TO:
FROM: Lucy Cowan
DATE: October 2, 1999
RE: (Enter subject here.)

****

(Body of text goes here.)

Memo Protocol and Company Politics

Memos are important tools for any company or agency. They reflect a company's politics, policies, and organization. Memos are sent down the administrative ladder from presidents, vice presidents, managers, and so on to employees, and memos are sent up the ladder, too, from employees to their supervisors. Workers also send memos to one another. Figures 4.1 and 4.4 illustrate memos sent from the top down; Figure 4.2 contains a memo from one worker to another; and Figure 4.3 shows a memo sent from an employee to management.

FIGURE 4.3 A memo that uses headings to highlight organization.

Ramco Industries
"Where Technology Shapes Tomorrow"
ramco@gem.com http://www.Ramcogem.com

TO: Rachel Mohler, Vice President
Harrison Snowden, Public Relations
FROM: Mike Gonzalez MG

Subject: Ways to Increase Ramco's Community Involvement
Date: March 3, 1999

At our planning session in early February, our division managers stressed the need to generate favorable publicity for our new Ramco plant in Mayfield. Knowing that such publicity will highlight Ramco's visibility in the Mayfield community, the company's image might be enhanced in the following ways:

CREATE A SCHOLARSHIP FUND

Ramco would receive favorable publicity by creating a scholarship at Mayfield Community College for any student interested in a career in technology. A one-year scholarship at Mayfield Community College would cost $4,800. The scholarship could be awarded by a committee composed of Ramco's executives and administrators. Such a scholarship would emphasize Ramco's support for industrial education at a local college.

OFFER FACTORY TOURS

Guided tours of the Mayfield facility would introduce the community to Ramco's innovative technology. The tours might be organized for community and civic groups. Individuals would see the care we take in production, equipment choice, and the speed with which we ship our products. Of special interest to visitors would be Ramco's use of industrial robots alongside Ramco employees. Since these tours would be scheduled well in advance, they would not conflict with our production schedules.

PROVIDE GUEST SPEAKERS

Many of our employees would be excellent guest speakers at social and educational meetings in Mayfield. Possible topics include the technological advances Ramco has made in designing and engineering and how these advances help consumers.

REQUEST

Thanks for giving me your comments as soon as possible. If we are going to put one or more of the suggestions into practice before the plant opens, we'll need to act before the end of the month.

FIGURE 4.4 Memo with a clear introduction, discussion, and conclusion.

MEMORANDUM

Dearborne Equipment Company

TO: Machine Shop Employees
From: Janet Hempstead JH
Shop Supervisor

Date: September 27, 2000
Subject: Brake Machines

During the past two weeks I have received a number of reports that the brake machines are not being cleaned properly after each use. Through this memo I want to emphasize and explain the importance of keeping these machines clean for the safety of all employees.

When the brake machines are used, the cutter chops off small particles of metal from the brake drums. These particles settle on the machines and create a potentially hazardous situation for anyone working on or near the machines. If the machines are not cleaned routinely before being used again, these metal particles could fly into an individual's face when the brake drum is spinning.

To prevent accidents like this from happening, please make sure you vacuum the brake machines after each use.

You will find two vacuum cleaners for this purpose in the shop–one of them is located in work area 1-A and the other, a reserve model, is in the storage area. Vacuuming brake machines is quick and easy: it should take no more than a few seconds. I am sure you will agree that this is a small amount of time to make the shop safer for all of us.

Thanks for your cooperation. If you have any questions, please call me at Extension 324 or come by my office.

204 South Mill St. South Orange, NJ 02341-3420 (609) 555-9848 JHEMP@dearco.com

Most companies have their own memo protocol—accepted ways in which in-house communications are formatted, organized, written, and routed. In fact, some companies offer protocol seminars on how employees are to prepare memos. In the corporate world, protocol determines where your memo will go. For example, it would be presumptuous to send copies of all your memos to the vice president. You would offend your immediate supervisor, who would think that you are trying to avoid going through proper channels.

Conversely, a vice president may want a certain memo to be distributed to the staff by another, lower administrator to ensure that the staff is being informed but that responses are not sent back up to the vice president but instead go through a lower administrative channel. Familiarize yourself early in your employment with how your company wants memos (or even e-mail) to be directed.

You may be asked to write a memo for the vice president or supervisor to sign (see Jocelyn Woolfolk's account in Figure 3.4). Your writing skills may, therefore, need to include the ability to write in another person's voice to some degree. Moreover, as you learned in Chapter 3, it is common in offices for two or more people to collaborate in preparing memoranda just as they collaborate to prepare reports and proposals. (See Figures 3.5–3.7.)

Memo Style and Tone

The style and tone of your memos will be controlled by your audience within your company or agency. Writing to a co-worker whom you know well, you can adopt a casual, conversational tone. You want to be seen as friendly and cooperative. In fact, to do otherwise would make you look self-important, stuffy, or hard to work with. Consider the friendly tone appropriate for one colleague writing to another in Roger Blackmore's memo to Marge Adcox in Figure 4.2.

When writing a memo to an manager, though, you want to use a more formal tone than when communicating with a co-worker or peer. Your boss will expect you to show a more respectful, even official, posture. Here are two ways of expressing the same message, the first more suitable when writing to a co-worker and the second more appropriate for a memo to the boss.

> Co-worker: I think we should go ahead with Alison's plan for reorganization. It seems like a safe option to me, and I don't think we can lose.
>
> Boss: I think that we should adopt the organizational plan developed by Alison Pierson. Her recommendations are carefully researched and persuasively answer all the questions our office has about the plan.

Finally, keep in mind that your employer and co-workers deserve the same clear and concise writing and attention to the "you attitude" (see pp. 154–160) that your customers do. Memos require the same care and follow the same rules of effective writing as letters do.

Strategies for Organizing a Memo

Your memos need to be drafted and organized so that readers can find information quickly and act on it promptly. For longer, more complex communications, such as the memos in Figures 4.2 through 4.4, the message of your memo might be divided into three parts: (1) introduction, (2) discussion, and (3) conclusion.

Introduction

In the introduction of your memo, do the following:

- Tell readers clearly about the problem, procedure, question, or policy that prompted you to write.
- Link the first sentence of your memo to the subject line.
- Explain briefly any background information the reader needs to know.
- Be specific about what you are going to accomplish in your memo.

Note, for example, the ways in which the writers of Figures 4.2 and 4.4 tell readers why a list of items is provided and why recommendations are included. Do not hesitate to come right out and say: "This memo explains new e-mail security procedures" or "This memo summarizes the action taken at the industrial site near Evansville to reduce air pollution."

Discussion

In the discussion section (the body) of your memo, help readers in these ways:

- Inform them why a problem or procedure is important; who will be affected by it; and what caused it and why.
- Indicate why changes are necessary.
- Give them precise dates, times, locations, and costs.

See how Janet Hempstead's memo in Figure 4.4 carefully describes an existing problem and explains the proper procedure for cleaning the brake machine.

Conclusion

In your conclusion, state specifically how you want the reader to respond to your memo. To get readers to act appropriately, you can do one or more of the following in your conclusion:

- Ask readers to call you if they have any questions.
- Request a reply—in writing, over the telephone, via e-mail, or in person—by a specific date.
- Provide a list of recommendations that the readers are to accept, revise, or reject.

Throughout your memo use organizational markers, such as headings, to make information easy for readers to follow. When you have a large number of related points, as the writer of the memo in Figure 4.2 does, number them so that the reader can comprehend them more readily. Bulleted items and numerals help you organize information and also assist your readers in following your work. Lists in a memo point out comparisons and contrasts easily. Underlining or boldfacing key sentences also helps readers by emphasizing important points. But do not abuse this technique by underlining or boldfacing too much. Draw attention only to points that contain summaries or draw conclusions.

Memo reports, such as Figure 4.3, use headings to separate information for readers so that they can find it more quickly. Headings (often discovered from brainstorming and polished and refined through revision) will also help you organize information as you write.

These organizational markers can also be used in the faxes and e-mail you send.

Faxes

Fax (facsimile) machines are widespread in the world of work; almost every business and home office has a fax machine. A fax machine can send copies of letters, memos, reports, graphs, blueprints, and artwork over ordinary phone lines. Fax technology is so advanced that you can send graphics that resemble actual photos. You will reduce costs if you send faxes at times when phone rates are lower—evenings and weekends. Some fax machines are programmed to offer speed dialing, allowing a business to send the same fax to numerous customers in a relatively short time. Newsletters and time-sensitive reports are often faxed to subscribers more quickly and cheaply than if these documents were sent through the U.S. Postal Service. Subscriptions and responses to ads and surveys are increased when readers can fax a reply.

When you send a fax, be sure to include a facsimile cover sheet (Figure 4.5). A cover sheet indicates the person(s) sending and receiving the fax, their addresses, phone, and fax numbers, and the total number of pages being faxed. This last information is essential so that the recipient will know when the transmission is complete and can alert the sender of any interruption in transmission.

Be aware that, unless the recipient has her or his own secured fax machine, your confidentiality is not easily protected when communicating by fax. If your fax is sent to a machine available to the entire office staff, anyone can read it.

Fax Guidelines

When you send a fax, observe the following guidelines.

1. Be careful about sending anything longer than three to four pages. You can tie up the recipient's phone line. Call before you fax to see whether the recipient will allow you to fax a longer document or would prefer that you send it another way.
2. Make sure your fax is clear, but always include your phone and fax numbers in case the recipient needs to verify your message or has questions about it.
3. Avoid writing any comments in the margins or at the very top or bottom of a fax. Your notes might be cut off or blurred in transmission.
4. While it may be permissible to send a fax to a company after business hours, do not risk sending one after hours to a customer's home unless he or she has agreed or requested it.

FIGURE 4.5 A fax cover sheet.

Westwood Communications Inc.
4277 Old Trail Road
El Paso, TX 79968

FAX COVER SHEET

Date: May 16, 2000

PLEASE DELIVER THE FOLLOWING PAGE(S) TO:

NAME: Deborah Shapiro

COMPANY/DEPARTMENT: Marketing Department

FAX NUMBER: (502)555-8449

THIS FAX IS BEING SENT BY:

NAME/DEPARTMENT: Malcolm T. Belleau, Design Dept.

NUMBER OF PAGES: 3 INCLUDING COVER SHEET

TIME SENT: 10:30 AM X PM ____

IF YOU DO NOT RECEIVE ALL PAGES CLEARLY,
PLEASE CALL (915)555-3200

E-Mail

What Is E-Mail?

E-mail, or electronic mail, is one of the most popular basic features of the Internet. E-mail has become an essential communication tool found in almost every business and home office. Millions, perhaps billions by the year 2005, of consumers also use e-mail to communicate with businesses each day. Everyone who has computer access to an Internet service provider can send and receive e-mail. As with a fax, the recipient of e-mail does not need to be present to receive the message. But there are major differences between the two media.

TECH NOTE

As long as a fax phone line is open, a fax will be transmitted in a matter of minutes. A fax may not be as fast as e-mail. Of course, e-mail is delivered more slowly if the traffic on the Internet is heavy or if a server is down.

With e-mail you send and receive messages through your personal computer, which is linked via a phone line and modem to a worldwide communication network. Functioning like an electronic mailbox, your computer can receive and store messages while you are out and then inform you that they are waiting. You can then read these messages on your screen, make hard (printed) copies if necessary (see Figures 4.6 through 4.9), forward them, file messages on disk for future reference, or delete them. Through your computer you can communicate quickly and efficiently with hundreds of people in your office, co-workers at branch offices, or employees and customers across the country or around the world. For example, you can e-mail the president of the United States at *president@whitehouse.gov* or the CEO of IBM at *ceo@ibm.com.*

Advantages of E-Mail

E-mail is fundamentally different from paper-based communications. It has many advantages over conventional snail mail, as the U.S. Postal Service is sometimes called. Among them, e-mail is

1. **Quick.** Traveling at the speed of sound, e-mail can usually reach its destination within minutes, if there is not a Net slowdown. You don't have to wait for postal or overnight deliveries. E-mail immensely expedites any kind of domestic or international communication.

2. Convenient. You can send e-mail anytime and know that it will be delivered even if the receiver is away from his or her computer. When the individual returns, your message will be waiting conveniently for him or her. Consequently, you don't have to worry about repeatedly exchanging phone messages or hearing busy signals when sending a fax. Moreover, you can send your e-mail via a notebook computer even when you are away from your office—sitting in a hotel lobby, riding in a car, or even flying in an airplane—as long as you have a phone line to which you can connect the modem.

3. Cost-effective. E-mail does not require stamps, envelopes, or even paper (assuming you don't need a hard copy for your records). It is also cheaper than long-distance telephone calls or faxes. Since no long distance charges are billed on the Net, it costs no more to send the same message across town than it does around the world. You will be charged only for the time required to transmit.

4. Efficient. Sending and receiving e-mail makes it easier to conduct business than by communicating through conventional channels. Recipients can read and reply at their convenience (when there is a window in a hectic business day). They can ask questions or provide clarifications immediately. Even when they are busy, they can e-mail you that your message was received and they will answer it later.

5. Conducive to collaborative writing. E-mail enables many individuals to "talk" with each other simultaneously. With a click of your "reply to all" button, you can e-mail everyone on your team who received a particular message. You might send the same message by e-mail to a variety of individuals, or a variety of messages to just one or two people. They can read, comment on, incorporate, and revise your document and then send the revised version back to you or to everyone else in the group. (Review the collaborative e-mail exchange on pp. 100–102). Correspondingly, you can reply to messages from a team member and, with his or her permission, share your comments with the group. You can also incorporate changes from a team member's e-mail message directly into your draft. See pages 98–99 for additional uses of e-mail in collaborative writing.

Specific Business Applications of E-Mail

E-mail is the most informal type of business correspondence, far more so than a printed memo or letter. Basically, e-mail is chatty and casual. Yet this informality is an asset in interoffice communications. E-mail is used extensively within organizations and their various branches—from one individual to another and from one department to another—to communicate efficiently and concisely. The main office of a company, for example, may e-mail its branches every day—several times a day, in fact—about orders, customer requests, complaints, technical problems, travel schedules, sales figures, price changes, or security matters. A company can also e-mail vendors, customers, and anyone else in the world of work.

Think of e-mail as a polite, informative telephone conversation—friendly, to the point, but always accessible. E-mail allows individuals to conduct business professionally yet with a minimum of wasted time. E-mail is easy and immediate.

But e-mail's accessibility does not mean you can forget about the principles of clear, correct, and effective writing. The same standards that apply to letters and memos—punctuation, spelling, accuracy—also are necessary for your e-mail. Nor does e-mail relieve you of your obligation to be courteous and audience-centered. Make your e-mail user-friendly.

Do not use e-mail to send long, complicated documents or messages, with lots of attachments. Most e-mail traffic is taken up with short documents—a few paragraphs, rarely more than a page. Keep in mind that if your e-mail exceeds one computer screen, your readers will lose time scrolling your message.

The number of messages that can be sent over e-mail is limitless. You can send pictures, spreadsheets, even soundbites and video clips via on-line transmissions, but realize that downloading these documents takes time.

TECH NOTE

Your e-mail program will tell you whether or not an attachment has been sent, but be aware that attachments may arrive in an unusable or partly usable form. Often the sender and the receiver need to use the same or compatible software to open and manipulate each other's documents, especially if the files contain graphics, databases, or spreadsheets. Even simple word-processing documents may arrive "mangled" if the sender and the receiver use different operating systems or incompatible programs. It's a good idea first to attach a test document and to ask your reader to send you one in return. If this experiment fails with a word-processing document, launch your word processor, and save the document again—this time as a "text file." Although some formatting may be lost (italics, boldface, indentations), the file can be read by most word-processing programs, including the one within your e-mail application. Many graphics files also can be converted from one operating system to another by the e-mail program. Also consult the "Help" files of your software; they often include information about attaching and translating various kinds of files.

Technological problems often have technological solutions. If formatting is critical or your files are complex, consider adding translation software to your computer system. For as little as $100, software can help you cross platforms (DOS, Windows, Macintosh, Unix) or programs without losing anything in your original files.

For examples of effectively written e-mail, see Figures 4.7 through 4.9. Notice that e-mail can be cordial without being unprofessional.

Using an E-Mail Address

With most software you will find a **header** at the top of each e-mail that contains the sender's address. Everyone who has a computer account on the Net has an e-mail address at the top of each piece of e-mail. An e-mail address contains three parts: first the e-mail user's name, then the name of the host computer the person uses, and finally the zone for the type of organization or institution to which the host belongs. This last part of the e-mail address is called the **Internet domain** for the computer system through which mail flows. For example, *dkehler@mail.sdsu.edu* refers to Dorothy Kehler's mailbox, located at San Diego State University, an educational and research institution (note the zone suffix *edu*).

The address *lpolowski@aol.com* refers to Leslie Polowski's mailbox on America Online, a commercial provider. Note that e-mail addresses are usually listed in lowercase letters and an underscore, as in *josh_reynolds@trans.gov*, which means Josh Reynolds's e-mail address is at the Department of Transportation, a government agency.

Be careful not to omit any part of an e-mail address and to punctuate it completely and accurately. Even a small error will stop your message from reaching the recipient.

In addition to your e-mail address you will also want to include your fax and voice mail numbers. Together these addresses constitute your signature file, as in the following example.

>>>>Marvin Cooper/Senior Sales Rep/RTS Technologies<<<<
Voice: (708) 555-1970 FAX: (708) 555-1980
mcooper@rts.com

Guidelines for Using E-Mail

Using e-mail technology does not mean you can forget about preparing and organizing your messages carefully with your reader's needs in mind. Below are several guidelines for writing effective e-mail.

Format

E-mail looks different from a printed memo or letter, as the examples in Figures 4.6 through 4.9 show. E-mail is formatted differently from regular mail. Readers will expect you to follow all the conventions associated with a courteous and helpful e-mail message.

1. Do not send messages in all capital letters. This is the equivalent of shouting at your readers. Moreover, your message will be harder to read, and if the receiver decides to make a hard copy, it will not look like professional correspondence.
2. Always include a return mail address as part of the signature, saving your reader the trouble of looking it up.

FIGURE 4.6 An example of interoffice e-mail.

```
Message4/24 From: peter_zacharias@craftworks.com
Date: Friday 24 July 1998 11:38:35-0400
To: marge_parish@craftworks.com
cc:
Subject: Status of Hinson Order

At last the Hinson-Davis Company received its order, and
they are very pleased with our service.  In fact, Victor
Arana, their district manager, made a point of calling me
first thing this morning to say the order came in at 8:00
a.m. and by 8:15 it was operable on their docks.

Things could not have gone smoother.  Congrats to all.  In
my last update (24 July) to you, we had anticipated some
delays, but luckily we avoided them.

I am going to send Arana a thank you letter today to keep
up the goodwill.
```

3. Use clear and specific descriptions for your subject line. Avoid vague one-word subjects such as "Software" or "Conference." Be more precise—"Advantages of New Software" or "Conference Is Set." One e-mail included only the word "Bill" for a subject line, which left readers wondering if the e-mail was about a person, an unpaid account, or a notice just sent.
4. Keep your subject line short. Get the essential point of the subject into the first three words; otherwise the recipients may not see what you want them to see until they open the message itself. Most people see only about 30 characters of a header when scanning a screen of e-mail messages in their in-box. Remember that the average screen has only 80 characters to work with, and most e-mailers use 60–80 percent of that for displaying the name and address of the sender, and the time and date of the message.

FIGURE 4.7 An example of an ongoing e-mail.

```
Message4/24 From: peter_zacharias@craftworks.com
Date: Friday 24 July 1998 11:55:26-0400
To: lee_thornton@craftworks.com
cc:
Subject: Status of Hinson Order

Lee, IMHO, you did a great job expediting Hinson's order this
a.m.  Getting the order there by 7:38 took a lot of hard work,
and I appreciate your effort.

Hinson's previous order was three days late because it had to be
specially designed at the factory.  I am glad we could deliver
the order exactly according to the specs.

I received a warm phone call this a.m. from Mr. Hinson himself
thanking us.

It might be wise to pay him a visit next week when you circle
back to Dayton while he still recalls our (and your) good work.
Who knows--Hinson might place yet another order with us.  TIA.
```

Observing Netiquette

Respect your e-mail audience as much as you do the audience for your memos, letters, or reports. The following guidelines will help you to observe "netiquette," or the etiquette of the Net. Netiquette is appropriate and courteous behavior for e-mail users. It means being courteous and professional and projecting the best image of yourself and your company. By following netiquette, you can avoid embarrassing yourself or angering your readers.

1. Send your message to the reader's correct address. Sending your e-mail to the wrong address wastes time. As with your traditional, nonelectronic mail, there is no guarantee that mail will get through if the address is wrong. An incorrectly addressed e-mail will be returned to you—99 percent of the

FIGURE 4.8 E-mail sent to a distribution list of co-workers.

```
Message8/18 From: melinda_bell@netech.com
Date: Tue 11 May 1999 13:18:33-0400
Subject: Collaboration on annual report
To: allen_cranston@netech.com;peter_maxwell@netech.com;
       margaret_habermas@netech.com
X-VMS-To: IN% "allen_cranston@netech.com;peter_maxwell@netech.com;
       margaret_habermas@netech.com"
From: NAME: Melinda Bell
         FAX: (603)555-2162
         Voice: (603)555-1505
 FUNC:
TEL: <BELL, MELINDA AT NORTHEAST TECHNOLOGIES COM>
To: IN% "allen_cranston@netech.com;peter_maxwell@netech.com;
       margaret_habermas@netech.com"

To follow up on our conversation yesterday regarding working
together on this year's annual report, I'm glad our schedules are
flexible. Let's meet next Tues the 18th--at 10:30 a.m. in the
upstairs conference room.

Don't forget we have to first draft a two- to three-page overview
that explains Northeast's strategic goals and objectives for
fiscal year 2000.  Not an easy assignment, but we can do it,
gang.

It would be a big help if Allen would bring copies of the reports
for the last three years.  Would Peter call Ms. Jhandez for a
copy of the speech she gave last month to the Powell Chamber of
Commerce?  BTW, if memory serves me correctly, she did a first-
rate job summarizing Northeast's accomplishments for 1998.  We
will certainly want to quote some of her remarks.  Margaret,
thanks for doing stats on the last quarter's outlays for us
before Tuesday.
```

time—by a "gatekeeper." Maintain a current e-mail address file. With so many e-mail users, your e-mail address file will grow and change monthly. Go straight to the Net itself to find current addresses. Also have on hand Meckler Media's *Office Internet Yellow Pages,* edited by George Newby (JDG, 1997 ed.), a convenient, up-to-date directory.

2. Do not send unsolicited e-mail without first verifying whether the individual or group wants to receive such mail. Unsolicited sales messages can cost you an order instead of winning you a contract. E-mail users do not like to receive junk mail (unwanted solicitations are often referred to on the Internet as "spam") and may well refuse to read any further communications from you or your company if you violate this rule. To observe Netiquette, read the introductions to listservers and home pages (see pp. 309–314) of companies or individuals on the Net before sending them e-mail.
3. Determine whether an address is for one individual or an entire group, as in a listserver.
4. Learn as much as you can about the recipient of your e-mail, and let your relationship with that person shape the conversational style of your message.
5. Find out how much information your reader needs. Don't waste his or her time with chatter. Readers of e-mail want only relevant, usable information, not "hype."
6. Always tell your readers what you would like them to do—verify information; send additional data; respond in writing. Maybe you do not need a response at all. If so, tell them.
7. When responding to a request, don't just repeat an e-mail message or question from a sender and give a curt one- or two-word reply. Courtesy dictates that you reply with a message of your own, not simply tack a comment to the original message and send both back.
8. Let the sender know you received his or her e-mail message even if you cannot answer it until later.
9. Answer e-mail the same day it arrives, whenever possible. Your readers expect a much quicker reply to e-mail than for conventional mail. Gone are the days when you can blame your own procrastination on slow postal service.
10. Refrain from sending the same messages over and over again. If you know your e-mail was received, don't annoy your recipient. Just as important, don't send the same message by fax, phone, and e-mail. Such duplication signals not only your own anxiety but also your lack of trust in your recipient.
11. Check your e-mail every day. Don't let it pile up in your mailbox. Not only do you risk offending the individuals who sent you mail, if your mailbox is overcrowded with unread messages, you also run the risk of blocking other messages. Eventually, your mail system may even delete unread messages.
12. Respect the cultural communication traditions of your ESL audience (review pp. 163–170). Smileys (or emoticons), as seen in Figure 4.10, are culturally determined. E-mail coming from other parts of the world is likely to use different symbols with other meanings.

Observing the rules above, you'll get high marks for your netiquette.

FIGURE 4.9 An e-mail response to a customer.

```
From: waller@bnet.tcb.com
To: luceyc@lance.netdoor.com
Subject: Internet addresses for upgrades, etc.
Date: 10 Sept. 1997 09:10:13-0400

Hello Lucy,

Thanks for your e-mail asking about how to obtain print drivers
and other components and software.  To find out about the
availability of drivers, upgrades, bios, specs, fixes, and csd's
(corrective service disks) via the Internet, check out the
following addresses:

        *http://www.pc.bnet.com
        *ftp://ps.omaha.bnet.com
        *ftp://software.jensen.bnet.com

These addresses will be able to tell you about what we have in
stock and how we can get it to you asap.

You might also be interested in the following address for
feedback, FAQ's and tips.  Visit us at
        news://msnews.gen.com/bnet.public internet mail

I hope these leads help.  If you run into any trouble, please e-
mail me back.  I'm at the keyboard all day today until 7:00 p.m.
CST.

Cheers,

Larry Waller
```

FIGURE 4.10 Emoticons and other "body language" symbols on the Internet.

:) = smile
:D = smile/laughing/big grin
:* = kiss
;) = wink
:X = my lips are sealed
:P = sticking out tongue
{ } = hug
:(= frown
:'(= crying
0:) = angel
:-@ = scream
%-) = confused
>:< = anger

Style

Keep in mind that the recipient of your e-mail will be reading it on a computer screen rather than on a piece of standard business stationery. The smaller number of lines on a screen requires that you keep your message succinct and to the point. Follow these guidelines to write concise and clear e-mail.

1. Greet your reader with some type of "hello" instead of with just a header—Greetings, Hello, Hi, Howdy, Good Day, or something similar. These salutations would be too informal for your letters (see pp. 147–149).
2. Writing skills are also important even when you send e-mail. Informal does not mean sloppy. Although e-mail is informal, you should still be concerned about spelling, proper punctuation, and effective word choice. Proofread your e-mail before you press the "send" key. Use the e-mail system's spell-checker.
3. Keep your e-mail concise. Edit it carefully to see how many words you can cut or how many long phrases you can reduce. In your attempt to be concise, though, do not turn your e-mail into telegrams—short, clipped commands. Write in full sentences and do not leave words out. You do not want to appear as if you begrudge the reader your time or consideration. A telegraphic message—"Report immediately; need for meeting"—makes you sound discourteous and demanding.
4. Put the most important part of your message first. Busy readers appreciate seeing key details right away. Also, do not rule out the possibility that e-mail users, because of the vast amount of mail they receive, may only read the first part of your message.

5. Where appropriate, divide your message into an introduction, body, and conclusion.
6. Watch the length of your paragraphs. Seeing the screen filled with one long, unbroken paragraph can intimidate the reader. Keep your paragraphs to three or four lines and always double-space between paragraphs. Space between paragraphs is as important for readers of e-mail as for readers of long reports. The principles of document design (see Chapter 12) pertain to e-mail as well.
7. Your e-mail commands for underlining, boldface, and italics may be different from those on your word processing program. For example, to emphasize a point with italics in e-mail you may have to put an asterisk (*) before and after the word or phrase.
8. Slang and jargon are acceptable, **if your reader understands them and they are proper for the context.** But avoid unfamiliar abbreviations and slang or jargon. Not everyone in your office or company may be aware of the local meanings of a catchword or phrase. Some commonly used cyberspace abbreviations found in e-mail are in Figure 4.11. Computer users have also created a method of shorthand to express feelings and show actions or "body language" (some are shown in Figure 4.10). These symbols are most commonly used in chatting on the Internet, on message boards, or in friendly electronic mail.
9. Always end politely. Let your recipient know that you are concluding your message and that you have appreciated his or her help, or would welcome it.

FIGURE 4.11 Some commonly used e-mail abbreviations.

AFK	Away from keyboard
BAK	Back at keyboard
BRB	Be right back
BTW	By the way
FYI	For your information
IMHO	In my humble opinion
IOW	In other words
LOL	Laughing out loud
OTOH	On the other hand
TIA	Thanks in advance
WB	Welcome back
WRT	With respect to
WTG	Way to go

Confidentiality and Ethics
Privacy and respecting a sender's rights are major problems in today's computerized world. E-mail is available to everyone on the Net. Because e-mail is not completely private and restricted, avoid using it for confidential or sensitive messages. Since the Net is not secure, your message could be intercepted and read by anyone in your company, or anywhere else. For this reason, some firms use encryption, scrambling of computerized information to prevent unauthorized access. Keep in mind, though, that information sent to newsgroups is regarded as public, whereas e-mail is regarded as private. Observing the following rules will help protect confidential and sensitive information.

1. Be careful about what you write. Never attack your company, your boss, a co-worker, the competition, or a customer. Your message could be read and stored to be used later against you or your employer.
2. Send only the information needed to answer a reader's questions or concerns. If you elaborate, you defeat the conciseness of e-mail as well as risk divulging something confidential or sensitive.
3. Send nothing through e-mail that you wouldn't want to see posted on your company bulletin board. Do not spread gossip. You never know who will read (and reveal) your message.
4. When you attach a document, clearly inform readers what can or cannot be copied. (See the Tech Note on p. 122.)
5. Never forward (repost) something that is copyrighted without first securing permission from the copyright holder(s).
6. Tell people if you plan to publish something they will send you, and always get their permission before doing so.
7. Do not change the wording of a message that you expected simply to read and forward.
8. Do not refer readers to another document and ask them to copy it without the writer's permission.
9. Avoid **flaming,** or using strong, angry language. Flaming is verbally attacking, insulting, or mocking your e-mail readers. Figure 4.12 contains the type of e-mail you should avoid. Firing off a flaming message will hurt you and your company and may lead to charges of libel against you and/or your employer.

The last point is especially important. Business e-mail is your company's property, not yours. Because your employer can be sued for libel, many firms demand that their workers follow strict guidelines. Violating these rules may be cause for dismissal.

E-Mail: A Review of Its Uses

As a quick review of the chapter, Table 4.1 (p. 134) briefly summarizes when you should use e-mail rather than send a printed message such as a memo or letter.

FIGURE 4.12 A flaming e-mail.

```
Date: Fri, 06 Mar 1998 09:59:15-0400
To: smith@eagle.com
From: sammy@trinet.rad.com
Subject: Upgrades

Hey guys and gals,

Are you awake out there? I have been trying for over three
blooming hours to get someone --ANYONE!!!!!--at your place to get
back to me asap about the upgrades. But no luck. WTG. What will
it take?  A note from Bill Gates himself at Microsoft, your arch
competitor, to let you know we are going to jump ship.

Once more, I STRESS that your sales team--Rob T., Janet S., and
Tim A.--have not given the upgrades we were promised.  I don't
like to say you lied but what else would you call it, huh?

To put it bluntly, you obviously don't want to stay in business.
If you fold, we won't cry.

TTFN
:P
Sammy
```

✓ Revision Checklist

Memos

- ❑ Used appropriate and consistent format.
- ❑ Announced purpose of memo early and clearly.
- ❑ Organized memo according to reader's need for information—main ideas up front; supplied clear conclusion.
- ❑ Made style and tone of memo suitable for audience.
- ❑ Included bullets, lists, underscoring where necessary to reflect logic and organization of memo and for ease of reading.
- ❑ Refrained from overloading reader with unnecessary details.

Faxes

- ❑ Verified reader's fax number.
- ❑ Sent cover sheet with number of pages faxed and phone number to call in the event of transmission trouble.
- ❑ Excluded anything confidential or sensitive if reader's fax machine is not secure.
- ❑ Followed guidelines for effective correspondence (see p. 118).
- ❑ Returned any calls regarding transmission difficulties with fax promptly.

E-Mail

- ❑ Avoided sending unsolicited mail.
- ❑ Sent to reader's correct address.
- ❑ Formatted e-mail with acceptable margins and spacing.
- ❑ Observed netiquette, especially by avoiding flaming.
- ❑ Wrote a message rather than returning sender's message with short reply.
- ❑ Kept paragraphs short but used full—not telegraphic—sentences.
- ❑ Avoided unfamiliar abbreviations or terms that would cause reader trouble.
- ❑ Received permission to repeat or incorporate another person's e-mail.
- ❑ Safeguarded employer's confidentiality and security by excluding sensitive or privileged information.
- ❑ Included enough information for reader's purpose.
- ❑ Honored reader by observing proper courtesy.
- ❑ Begin with friendly greeting; ended politely.

Exercises

1. Write a memo to your employer saying that you will be out of town two days next week and three the following week for **one** of the following reasons: (a) to inspect some land your firm is thinking of buying, (b) to investigate some claims, (c) to look at some new office space for a branch your firm is thinking of opening in a city five hundred miles away, (d) to attend a conference sponsored by a professional society, or (e) to pay calls on customers. In your memo, be very specific about dates, places, times, and reasons.

2. Write a memo to two or three of your co-workers on the same subject you chose for Exercise 1.

3. Send a memo to your public relations department informing it that you are completing a degree or work for a certificate, and indicate how the information could be useful for its publicity campaign.

TABLE 4.1 The Uses of E-Mail versus Memos or Letters

	E-Mail	*Memo*	*Letter*
Brief Messages	X	X	
Informal	X	X	
Formal			X
Legal Record		X	X
Relaxed Tone	X	X	
Confidential Material		X	X
Multiple Pages			X
Reports		X	X
In-house Messages	X	X	
Proofreading	X	X	X

4. Write a memo to the payroll department notifying it that there is a mistake in your last paycheck. Explain exactly what the error is and give precise figures.

5. You are a manager of a local art museum. Write a memo to the Chamber of Commerce in which you put the following information into proper memo format.

   Old hours: Mon.–Fri. 9–5; closed Sat. except during July and August when you are open 9–12
   New hours: Mon.–Th. 8:30–4:30; Fri.–Sat. 9–9
   Old rates: Adults $3.00; senior citizens $1.00; children under 12 free
   New rates: Adults $4.50; senior citizens $1.00; children under 12 free but must be accompanied by an adult
   Added features: Paintings by Thora Horne, local artist; sculpture from West Indies in display area all summer; guided tours available with a party of six or more; lounge areas will offer patrons sandwiches and soft drinks during May, June, July, and August

6. Select some change (in policy, schedule, or personnel assignment) you encountered in a job you held in the last two or three years and write an appropriate memo describing that change. Assume you are your former employer explaining the change to employees.

7. How would any of the memos in Exercises 1–6 have to be rewritten to make them suitable as e-mail messages? Rewrite one of them.

8. Bring five or six examples of company e-mail to class. As a group evaluate them for style, tone, and layout.

9. Send a fax to a company or organization requesting information about the products or services it offers. Include an appropriate cover sheet with your fax.

10. Write a piece of e-mail with one of the following messages, observing the guidelines discussed in this chapter.

a. You have just made a big sale and you want to inform your boss.

b. You have just lost a big sale and you have to inform your boss.

c. You write to a co-worker about a union meeting.

d. You notify a company to cancel your subscription to one of their publications because you find it to be dated and no longer useful in your profession.

e. You request help from a listserver about research for a major report you are preparing for your employer.

f. You advise your district manager to discontinue marketing one of the company's products because of poor customer acceptance.

g. You send a short article (about two hundred words) to your company newsletter about some accomplishment your office, department, or section achieved in the last month.

h. You write to a friend studying finance at a German university about the biggest financial news in your town or neighborhood in the last month.

11. Rewrite the following e-mail to make it more suitable.

```
Hi--

This new territory is a pain. Lots of stops; no sales.
Ughhhh. People out here resistant to change. Could get
hit by a boulder and still no change. Giant companies
ought to be up on charges. Will sub. reports asap as
long as you care rec.

The long and short of it is that market is down. No
news=bad news.
```

12. As a collaborative venture, join with three or four classmates to prepare one or more of the e-mail messages for Exercise 10. Send each other drafts of your messages for revision. Submit the final copy of the group's effort.

13. Assume you have received permission to repost in an e-mail all or part of the article on microwaves (pp. 31–32) or virtual reality and law enforcement. Prepare an e-mail message to a listserver or usenet group containing part of the article you have chosen.

14. Send your teacher an e-mail message about the project you are now working on for his or her class, outlining your progress and describing any difficulties you are having.

15. You have just missed work or a class meeting. E-mail your employer or your teacher explaining the reason and telling how you intend to make up the work.

CHAPTER 5

Letter Writing: Some Basics

Letters are among the most common kind of writing you will do on your job. Although you may receive helpful criticism as you prepare your letters (as Randy Taylor did in Figure 3.2), more often than not you'll be solely responsible for them. Your signature on your letter tells readers that you are accountable for everything in it. Because letters are so important to your career, Chapters 5, 6, and 7 are devoted exclusively to effective letter writing. This chapter introduces the entire process and provides some guidelines, problem-solving techniques, definitions, and revision strategies common to all letter writing.

The Importance of Letters

Even in this age of electronic communication, letters are still vital. They can be sent over fax machines and via e-mail, as well as through such conventional routes as the U.S. Postal Service. In terms of materials and time, the average business letter can cost between $12 and $14 to compose, keyboard, proofread, transmit, store, and retrieve. Even with the speed of telecommunications, companies need qualified employees to research, draft, revise, edit, proofread, and transmit letters. Many businesses offer employee seminars on how to write clear and appropriate letters, a skill that can lead to promotions and raises.

Why are letters so important? They are both a personal and a professional means of communication and occupy an essential place in the business world for the following reasons.

1. **Letters represent your company's public image and your competence as well.** Effectively written letters can create goodwill; poorly written letters can anger customers and cost your company business.
2. **Letters are more personal than a report, yet more formal than memos or e-mail.** There are occasions when a memo or e-mail message would be inappropriately casual.
3. **Letters constitute a permanent legal record of an agreement.** They state, modify, or respond to a commitment and become part of a company's records.

When sent to a customer, a signed letter constitutes a legally binding contract. Be extra careful about what you put in a letter about prices, warrantees, guarantees, equipment specifications, delivery dates, or other promises. Your reader can hold you and your company to such written commitments. Double-check your facts. Letters are more legally binding than e-mail, which is easy to delete.

4. **Letters follow up on telephone calls and other types of oral communications.** They provide documentation and clarification of oral agreements and prevent misunderstandings about what was said or agreed upon and by whom.
5. **Letters provide a wide range of information.** They give instructions, announce or amend policies, describe changes in a product, service, or procedure, or report events or the results of a study or test.
6. **Letters can prompt action.** They can help a company collect money from overdue accounts, alter a city ordinance, institute a policy, call a meeting, or waive a requirement.
7. **Letters sell.** They can promote a product, a service, an idea, even the writer's own skills.
8. **Letters are efficient for targeted mass mailings.** It is easier for a company to buy a mailing list of addresses than it is to obtain hard-to-find e-mail addresses.

Letters accomplish these goals by following certain conventions, or customary practices. These conventions include the ways readers expect letters to look and to sound. Effective letters (1) announce their purpose clearly, (2) follow an appropriate format, (3) address the reader courteously, and (4) use more formal language.

This chapter will show you how to incorporate these conventions into the drafts, revisions, and final versions of your letters.

The Process of Writing a Letter

Though far shorter than a proposal, report, or manual, a one- or two-page business letter still requires planning and research. All the techniques of the writing process discussed in Chapter 2 apply to letters as well. An effective letter may require several drafts, revisions, and edits before the final copy is satisfactory. Any letter you just sit down and dash off is unlikely to portray you and your company in the best light.

Analyze Your Audience

Before you even start to draft your letter, you need to ask and successfully answer five questions.

1. Who is my audience? Do I know them or are they strangers? Are they familiar with my company and its products/services, or is everything I am writing about new to them?

2. Will my audience be favorably disposed to what I am going to say or will they be disappointed or angry about my news?
3. What kinds of information will my audience expect me to supply—schedules, costs, model numbers, descriptions, measurements, copies of reports?
4. How will my audience use the information I am sending?
5. What impression do I want my letter to make on readers? Do I want to sound courteous and friendly? Informed and efficient? Firm and decisive?

Do Necessary Research

Once you have a strong sense of your audience's need for information, you are in a better position to know what to research, the next stage in the process of creating successful letters. Your research could be as simple as refreshing your memory about one of your company's products or services, looking through your company's files to study previous correspondence with the customer or client, conferring with a co-worker about the client's special needs, or seeing whether your client has a home page on the Internet.

Draft and Revise

Armed with information about your audience and their purpose, you can get your thoughts down first by drafting and then revising and editing to arrive at the appropriate language and tone. Many writers wrongly believe that the hardest part of writing a letter is putting their ideas into the right language. In truth, the hardest part of writing a letter is knowing what must be said to whom, why, and where in the letter. (Chapter 6, on the various types of business letters, will give you helpful information on organizing and preparing your correspondence.) Once you have successfully resolved these points, you can focus more intently on language during the later revising and editing stages. Pages 160–163 suggest guidelines for making the language of your letters clear, concise, and contemporary.

Printing and Proofreading Letters

The first thing a reader will notice about your letter is how it looks—the way the words and paragraphs are arranged on the page. Your audience will be influenced by the way your letter looks even before they read its contents. A neat, professional-looking letter implies to the reader that the work, service, or skill that you promise to deliver will be done in the same way. Strikeovers (crossing out one letter by typing another letter on top of it), blotches of liquid corrector smeared across the page, or handwritten changes inserted to fix errors all look shabby and immediately detract from your message. Word processors of course can help you produce professional-looking letters with ease.

Printing Your Letter

The way a letter is typed or printed significantly affects the visual impression it makes. You can avoid crowded or lopsided letters if you take a few minutes to estimate the length of your message before you print it. Don't start a brief letter at the top of the page and then leave the bottom three-fourths of the page blank. Readers will feel as if you left them hanging. Plan to start a shorter letter near the center of the page. Also avoid cramming a long letter onto one page; sometimes you will have to use a second sheet. A letter that is squeezed onto one page deprives readers of necessary and pleasant white space.

Guidelines on Printing Your Letter

Here are a few general hints on printing your letters:

- **Leave generous margins of approximately 1 1/2 inches all around your type area.** Set 1 1/2-inch margins as the default on your word processor. Leave more white space at the top than at the bottom, and watch right-hand margins in particular, since it is easy to exceed that limit. Shorter letters may require wider margins than longer letters, but don't exceed a margin of 1 1/2 inches on the right-hand side.
- **Always use a dark ribbon or fresh toner cartridge for your printer.** A fuzzy, faint, or messy script hurts your chances of convincing your reader.
- **Use a letter-quality laser printer.** Letters from a dot matrix printer generally do not look as crisp, fresh, or professional as those printed on a laser or laser-jet printer.
- **Choose a type font that is pleasant to read and inviting to the eye.** Crowding too many letters on a line makes your letter harder to read. You will not win any points from your readers if your letter looks cramped. Avoid script or other fancy type fonts.
- **Use only high-quality paper.** Twenty-pound bond paper says your message is important and substantial; thin, light-weight paper signals that your message is flimsy.

Proofreading Your Letter

Proofread everything that has your name on it, even if you did not keyboard it. You cannot blame a keyboarder for spelling mistakes or other blunders. As the writer, you are responsible for the final product. Proofread your letter for errors of fact, miscalculations, and misrepresentations; contradictions between what you say in your letter and what is stated in a brochure or other company document are unethical. Pay special attention to the accuracy of prices, dates, and serial numbers.

Typographical errors can also be costly and embarrassing. If you want to tell a steady customer, "The order will be hard to fill," but you keyboard, "The order will be hard to bill," confusion will result. Poor keyboarding and proofreading can also lead to omitted letters ("the ill arrived" for "the bill arrived"), transpositions ("het" for "the," "nad" for "and"), or omitted words ("the market value of the was high").

TECH NOTE

A word processor's spell-check program is useful, but be careful not to rely too heavily on such a program. Mistakes can still creep in. A spell-check will not differentiate between *their* and *there, its* and *it's,* or *effect* and *affect;* the wrong word spelled correctly is still the wrong word. And after you have run a spell-checker, you still need to proofread your letter carefully to make sure each correction was made (none overlooked) before you print a final copy.

Proofreading Methods

You can't be too careful when you proofread. Proofreading is reading in slow motion. Here are seven ways to proofread effectively.

1. Read your letter word for word from the bottom to the top.
2. Read your letter from start to finish aloud. Pronounce each word carefully to make yourself more aware of typographical errors or omitted words. Look at every letter of every word. Don't skim.
3. Place your finger under each word as you silently read the letter.
4. Double-check the spelling of all names. Errors here are sure to cause problems. Also watch for inconsistencies (*Phillip* in one place, *Philip* in another; *Anderson* for *Andersen*).
5. Have a friend read the letter. Four eyes are better than two.
6. Have your friend read aloud a hard copy of your letter while you follow the copy on your computer screen.
7. Never proofread when you are tired, and avoid proofreading large amounts of material at one sitting.

Letter Formats

Letter format refers to the way in which you type or print a letter—where you indent and where you place certain kinds of information. A number of letter formats exist. Two of the most frequently used business letter formats are the full-block format and the semiblock format. Gaining popularity is the Administrative Management Society (AMS) simplified style. Before you choose a letter format, find out whether your employer has a preference.

Full-Block Format

The full-block format is the easiest to use because all information in the letter is flush against the left-hand margin, with space between paragraphs. You will not have to worry about indenting paragraphs or aligning dates with signatures. Figure 5.1 shows a full-block letter on letterhead stationery (specially printed giving a company's name, business and World Wide Web addresses, fax and telephone numbers, and sometimes the names of executives, and the company logo). If you don't use letterhead, the writer's address is placed flush with the left-hand margin, directly above the date.

Semiblock Format

The semiblock style (Figure 5.2) positions the writer's address (if it is not imprinted on a letterhead), date, complimentary close, and the signature at the right-hand side of the letter. The date aligns with the complimentary close, and notations of any enclosures with the letter flush left below the signature. Paragraphs in the semiblock style can be flush against the left-hand margin or indented.

Simplified Format

Like the full-block format, the simplified letter begins every line at the left-hand margin, including any numbered items such as those in Figure 5.3. But unlike the block format, the simplified style omits the salutation and complimentary close. In place of the salutation it includes a subject line (without using the word *subject*), typed or printed in all capital letters, three spaces down from the inside address. The first paragraph of the letter follows three spaces after the subject line. The writer's name and title are typed in capital letters on the same line four lines after the last paragraph of the letter. Because the simplified style saves keyboarding time (and thus company money), many offices now use it.

Indented Paragraph Format

A fourth format—indented paragraph—is illustrated in Figure 5.10. This style is identical to the semiblock except each paragraph is indented five spaces. Although this format is far less popular today than it once was, it offers some visual benefits, as you'll see on page 165.

FIGURE 5.1 Full-block letter format.

NIRA Nevada Insurance Research Agency
7500 South Maplewood Drive, Las Vegas, Nevada 89152-0026
(702) 555-9876 Fax (702) 555-9876 http://www.NIRA.org

Bradley Fuller, CPCU
Chairperson

Carmen Tredeau, CPCU
President

Theodore Kendrick
Vice President, Public Affairs

Shelly Lampier-Hawn, CPCU
Vice President, Research

Dora Salinas-Sanchez, CPCU
Vice President, Actuary

April 4, 1999

Ms. Molly Georgopolous, C.P.A.
Business Manager
Meyers, Inc.
3400 South Madison Rd.
Reno, NV 89554-3212

Dear Ms. Georgopolous:

All typing/printing lined up against the left-hand margin

As I promised in our telephone conversation this afternoon, I am enclosing a study of the Nevada financial responsibility law. I hope that it will help you prepare your report.

I wish to emphasize again that probably 95 percent of all individuals who are involved in an accident do obtain reimbursement for hospital and doctor bills and for damages to their automobiles. If individuals have insurance, they can receive reimbursement from their own carrier. If they do not have insurance and the other driver is uninsured and judged to be at fault, the State Bureau of Motor Vehicles will revoke that party's driver's license and license plates until all costs for injuries and damages are paid.

Please call me again if I can help you.

Sincerely yours,

Carmen Tredeau, President

CT/IMB

Encl.

FIGURE 5.2 Semiblock letter format.

Writer's address and date are indented.

7239 East Daphne Parkway,
Mobile, AL 36608-1012

January 31, 1998

Mr. Travis Boykin, Manager
Scandia Gifts
703 Hardy St.
Hattiesburg, MS 39401-4633

Dear Mr. Boykin,

I would appreciate knowing if you currently stock the Crescent pattern of model 5678 and how much you charge per model number. I would also like to know if you have special prices per box order.

The name of your store is listed in the Annual Catalog as the closest distributor of Copenhagen products in my area. Would you please give me directions to your shop from Mobile and the hours you are open?

I look forward to hearing from you.

Complimentary close and writer's name are indented.

Sincerely yours,

Arthur T. McCormack

FIGURE 5.3 AMS simplified letter format.

Office Property Management Associates
2400 South Lincoln Highway
Livingston, NJ 07040-9990
(201)555-3740 Fax (201)555-6565

April 19, 1999

Mr. W.T. Albritton
Albritton and Sharp Accounting Services
Suite 400
Suburban Office Complex
Livingston, NJ 07038-2389

IMPROVED SERVICES AT SUBURBAN OFFICE COMPLEX

At our April meeting, Office Property Management Associates discussed a number of requests you and other tenants made at the Suburban Office Complex. I am happy to inform you that the following improvements in services will go into effect at the suburban office complex within 45 days.

1. Effective June 6, you will have an on-site manager, Judy Fiorelli, who will be happy to answer any questions you may have about the Complex and will help you solve any problems.
2. The parking lot on the southwest side will be resurfaced during the week of June 13-20. During this time, would you and your staff please park your vehicles in the north or the east lots.
3. A new outdoor security system will be installed June 17. Work on this system should not inconvenience you.

I welcome your comments on these changes or suggestions for additional ones. Please feel free to call or write me.

Gladys T. Mullins-Osborne

GLADYS T. MULLINS-OSBORNE, VICE PRESIDENT

tc

Continuing Pages

There are several ways to indicate subsequent pages if your letter runs to a second (or additional) page. Here are three conventions.

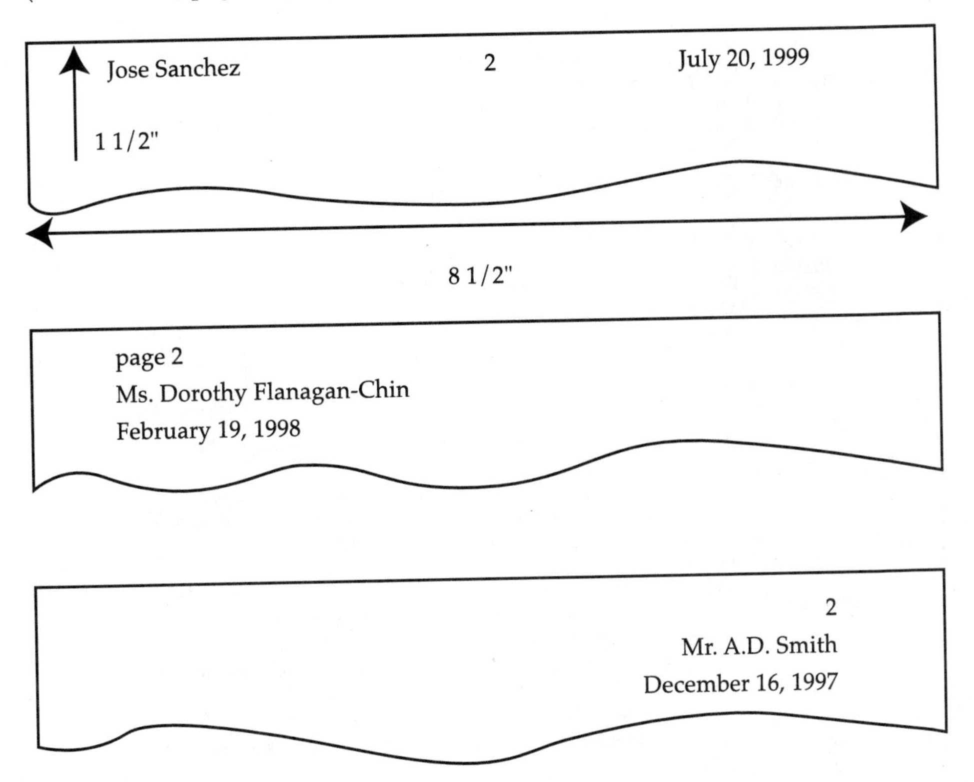

Parts of a Letter

A letter contains many parts, each of which contributes to your overall message. These parts and their placement in your letter form the basic conventions of effective letter writing. Readers will look for certain information in key places. It is your responsibility as a letter writer to meet your reader's expectations, and in doing so, to create a favorable impression.

In the following sections, those parts of a letter marked with an asterisk should appear in every letter you will write. Figure 5.4 is a sample letter containing all the parts discussed below. Note where each part is placed in the letter.

*Date Line

Where you place the month, day, and year depends on the format you are using. I the full-block or the simplified style, the date line is flush with the left-hand m gin. The semiblock style places the date at the center point, centered under a two pany letterhead, or flush with the right-hand margin. The date line appe lines below the writer's address or letterhead.

FIGURE 5.4 A sample letter, full-block format, with all parts labeled.

Letterhead	M &M Madison and Moore, Inc. Professional Architects 7900 South Manheim Road, Crystal Springs, NE 71003-0092 Phone 402-555-2300 FAX 402-555-2310 http://www.MMI.com
Date line	December 12, 2000
Inside address	Ms. Paula Jordan Systems Consultant Broadacres Development Corp. 12 East River Street Detroit, MI 48001-0422
Salutation	Dear Ms. Jordan:
Subject line	SUBJECT: Request for alternate duplex plans, No. 32134
Body of letter	Thank you for your letter of December 6, 2000. I have discussed your request with the officials in our planning department and have learned that the forms we used are no longer available. In searching through my files, however, I have come across the enclosed catalog from a California firm that might be helpful to you. This firm, California Designers, offers plans very similar to the ones you are interested in, as you can tell from the design checked on page 23 of their catalog. I hope this will help your project and I wish you success in your venture.
Complimentary close	Sincerely yours,
Company name	MADISON AND MOORE, INC.
Signature	
Writer's name and title	William Newhouse Office Manager
Keyboarder's identification	WN/kpl
Enclosure	Encl. Catalog
Copy to	cc: Planning Department

Spell out the name of the month in full—"September" or "March" rather than "Sept." or "Mar." The date line is usually keyboarded this way: November 15, 2000. Yet many international firms prefer to date correspondence with the day first, followed by month and year (15 November 2000), with no commas separating the day, month, and year.

*Inside Address

The inside address, which is the same address that appears on the envelope, is always placed against the left-hand margin, two lines below the date line. It contains the name, title (if any), company, street address, city, state, and ZIP code of the person to whom you are writing. Single-space the inside address but do not use any punctuation at the ends of the lines.

Dr. Mary Petro
Director of Research
Midwest Laboratories
1700 Oak Drive
Rapid City, SD 56213-3406

Always try to write to a specific person rather than just "Sales Manager" or "President." To find out the person's name, check previous correspondence, e-mail lists, the company's or individual's home page on the Internet, or call or fax the company. Abbreviate courtesy titles (Mr., Dr., Ms.); however, do not abbreviate military titles (Captain, Sergeant), academic ranks (Professor, Assistant Professor), religious designations (Reverend, Father, Sister), or civic or political titles (Senator, Councilperson). When writing to an elected official, use an honorific (Honorable) and her or his exact title.

The Honorable Barbara Karnes-Bolton
Representative, 10th Congressional District

The Honorable J. T. Morales
Mayor of Freemont

The initials M.D., Ph.D., or D.P.H. should not be added after a name if you use Dr. as a title. Use either *Janice Howell, M.D., or Dr. Janice Howell.* When writing to a woman, use Ms. unless she expressly asked to be called Mrs. or Miss.

The last line of the inside address contains the city, state, and ZIP code. Table 5.1 lists the official U.S. Postal Service abbreviations—two capital letters without periods—for the states and territories and the accepted abbreviations for Canadian provinces.

Salutation

The greeting part of your letter, or the salutation, is placed flush against the left-hand margin in both the full-block and semiblock formats. Begin with *Dear,* a convention showing respect for your reader, and then follow with a courtesy title, the reader's last name, and a colon (Dear Mr. Brown:). Use a comma only for an informal letter, but if you are on a first-name basis with

TABLE 5.1 U.S. Postal Service Abbreviations and Canadian Province Abbreviations

U.S. state/ territory	*Abbreviation*	*U.S. state/ territory*	*Abbreviation*
Alabama	AL	Montana	MT
Alaska	AK	Nebraska	NE
Arizona	AZ	Nevada	NV
American Samoa	AS	New Hampshire	NH
Arkansas	AR	New Jersey	NJ
California	CA	New Mexico	NM
Colorado	CO	New York	NY
Connecticut	CT	North Carolina	NC
Delaware	DE	North Dakota	ND
District of Columbia	DC	Ohio	OH
Florida	FL	Oklahoma	OK
Georgia	GA	Oregon	OR
Guam	GU	Pennsylvania	PA
Hawaii	HI	Puerto Rico	PR
Idaho	ID	Rhode Island	RI
Illinois	IL	South Carolina	SC
Indiana	IN	South Dakota	SD
Iowa	IA	Tennessee	TN
Kansas	KS	Texas	TX
Kentucky	KY	Utah	UT
Louisiana	LA	Vermont	VT
Maine	ME	Virginia	VA
Maryland	MD	Virgin Islands	VI
Massachusetts	MA	Washington	WA
Michigan	MI	West Virginia	WV
Minnesota	MN	Wisconsin	WI
Mississippi	MS	Wyoming	WY
Missouri	MO		
Canadian province	***Abbreviation***	***Canadian province***	***Abbreviation***
Alberta	AB	Nova Scotia	NS
British Columbia	BC	Ontario	ON
Labrador	LB	Prince Edward Island	PE
Manitoba	MB	Quebec	PQ
New Brunswick	NB	Saskatchewan	SK
Newfoundland	NF	Yukon Territory	YT
Northwest Territories	NT		

your reader and using his or her last name would be awkward, by all means write "Dear Bill:" or "Dear Sue:".

Sometimes you may not be sure of the sex of the reader. There are women named Stacy, Robin, and Lee, and men named Leslie, Kim, and Kelly. If you aren't certain, you can use the reader's full name: "Dear Terry Banks:". If you know the person's title, you might write "Dear Credit Manager Banks:". Or you might use the simplified letter format (see Figure 5.3), which omits the salutation altogether. Never use the sexist "Dear Sir:" and "Dear Madam:".

If you are writing to a large group of readers, you might use the occupational designation: "Dear Pilot," "Dear Homeowners." When writing to a company, use the company name—"Dear Apple," "Dear Saperstein Textiles"—not "Dear Gentlemen." Similarly, avoid the stilted "Ladies and Gentlemen" or "Dear Sir/Madam." (For a discussion of sexist language and how to avoid it, see pp. 61–65.)

Also avoid salutations such as "Hello," "Greetings" (which sounds as if it came from a draft board or is a holiday message), and "Good Morning" (it may be late afternoon when your reader opens your letter). And never begin your letter "To whom it may concern," which is trite and rudely impersonal.

Subject Line

The subject line can provide a concise summary of your topic (like a title for your letter) or it can list account numbers, order notations, or referral numbers so that the reader can at once check the files and see the status of your account or policy. In both the full-block and semiblock formats, the subject line, preceded by the word SUBJECT in capital letters, is double-spaced below the salutation, flush with the left-hand margin.

> Dear Ms. Salazar:
>
> SUBJECT: Repair of model 7342

Alternatively, it can be moved to the right-hand side of the letter, on the same line as the salutation.

> Dear Ms. Salazar: SUBJECT: Repair of model 7342

Review page 141 and Figure 5.3 for style and placement of the subject line in the simplified style letter.

*Body of the Letter

The body of the letter contains your message. In the full-block format, paragraphs are never indented; in the semiblock and simplified formats, paragraphs may or may not be indented five spaces. Whichever style you choose, single-space within paragraphs, but double-space between paragraphs.

Some of your letters will be only a few lines long, and others may extend to three or more paragraphs. Keep your sentences short and try to hold your paragraphs to under six or seven lines. Your letters will look more inviting and professional when your paragraphs are balanced in length and not crowded onto the page.

In organizing the body of your letter, follow this plan:

- As a rule, begin your letter with a statement of purpose. In your first paragraph tell readers why you are writing and why your letter is important to them. Depending on your message, begin with your most persuasive point. Also let your readers know, at the start, that you value your relationship with them. Be concerned and courteous.
- Put the most significant point of each paragraph first to make it easier for the reader to find. Never bury important ideas in the middle or end of your paragraph.
- In a second (or subsequent) paragraph, develop your message with factual support. Consider using lists to break your message up and to help readers recall details or groups of items. (See the correspondence in Figure 5.3.) Readers may benefit from your using headings, too. Also, where appropriate, boldface key words or sentences, but don't overdo the visual effects.
- In your last paragraph, bring readers to a true sense of conclusion. Tell them what you have done for them, what they should do for you, what will happen next, when they will hear from you again, or any combination of these messages. Don't leave readers hanging. Above all, your last paragraph should encourage readers to want to continue their association with you and your company.

Complimentary Close

Except in a letter using AMS simplified style (in which the complimentary close is omitted), the complimentary close always appears two spaces below the body of the letter, flush with the left-hand margin in the full-block format and at the center point, aligned with the date, for the semiblock format. Capitalize only the first letter of the complimentary close, and follow the entire close with a comma.

For most business correspondence, use one of these standard closes:

Sincerely,
Sincerely yours,
Respectfully,
Yours sincerely,

If you and your reader know each other well, you might try

Cordially,
Best wishes,
Warmest regards,
Regards,

But avoid flowery closes such as

Forever yours,
Faithfully yours,
Devotedly yours,
Admiringly yours,

These belong in a romance novel, not a business letter.

*Signature

Type your name four spaces below the complimentary close (or in the AMS simplified style below the last line of the body), either on the left side (full-block or

AMS simplified format) or at the center point (semiblock format). Allow four spaces so that your name, when you write it, will not look squeezed in. Always sign your name in ink, just as it is typed. Your name not only identifies you but also verifies that the contents of your letters have your approval. An unsigned letter indicates carelessness or, worse, indifference toward your reader.

If you are a woman, you have the option of indicating how you want your letters addressed. You can sign your letter "Julie Macklin," which signals to the recipient to address you as Ms. Julie Macklin. But if you prefer, you can sign your letter "(Mrs.) Julie Macklin." Married women should always sign their own name, never "(Mrs.) Harold Macklin."

Some firms prefer using their company name along with the employee's name in the signature section. If so, type the company name in capital letters two spaces below the complimentary close, and then sign your name. Add your title underneath your typed name. Here is an example:

Sincerely yours,

THE FINELLI COMPANY

Robert Stravopoulos
Cover Coordinator

Reference Initials

When a letter is keyboarded for you by someone else, your initials and the keyboarder's initials are placed two spaces below your typed name. Your initials appear in capital letters, followed by the keyboarder's initials in lower-case letters. The notation WBT/vgh or WBT:vgh, for example, means that Winnie B. Thompson's letter was keyboarded by Victor G. Higgins. The company thus has a record of who wrote the letter and who keyboarded it. Do not list any initials if you keyboarded your own letter.

Enclosure(s) Line

The enclosure line is placed two spaces beneath the reference initials (or your name if you typed your own letter). This line informs the reader that additional materials (such as brochures, diagrams, forms, contract(s), a proposal) accompany your letter. Examples include:

Enclosure

Enclosures (2)

Encl.: 1998 Sales Report

Copy Line

The abbreviation *cc:* informs your reader that a copy of your letter has been sent to other readers.

cc: Service Dept.

cc: Janice Algood
    Ivor Vas

Letters are copied and sent to third parties for a variety of reasons. You may be required to send copies of your correspondence to your boss, to a specific co-worker who needs to be apprised of an ongoing situation, or to another department in your firm. Professional courtesy dictates that you tell the reader if others will receive a copy of your letter.

Addressing an Envelope

When sending a letter, use a standard 9 1/2" × 4 1/8" white envelope or, as most firms do when they mail statements, an envelope measuring 6 1/2" × 3 5/8" with a window (a transparent cellophane opening) through which the customer's inside address appears. Because most mail is sorted by high-speed electronic scanning equipment, the U.S. Postal Service has strict regulations concerning envelope size. In particular, avoid odd-shaped envelopes and small, invitation-size ones. Preprinted return envelopes or labels carry bar codes that indicate the type of correspondence at the top of the envelope and the address at the bottom.

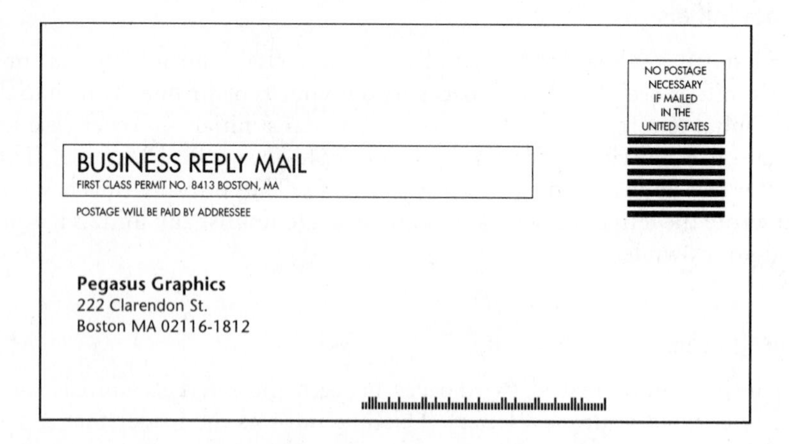

The recipient's address should be single-spaced and centered on the envelope as in Figure 5.5. The postal service recommends using all capitals and no punctuation on the envelope. Always use a ZIP code, even if a letter is being mailed to someone in your city. Every mail delivery area in the United States now has a nine-digit ZIP code. The first five digits direct mail to a particular geographic location (a neighborhood or area in New York City or San Francisco, for example),

FIGURE 5.5 A properly addressed envelope and a preprinted return envelope.

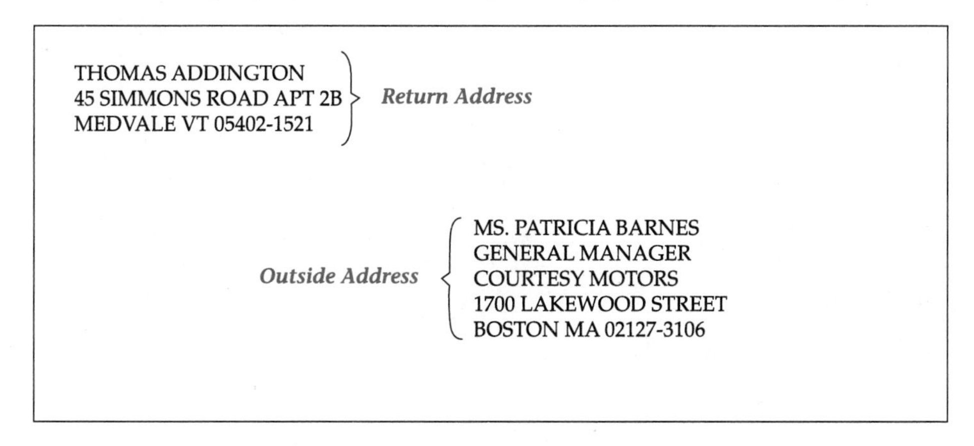

while the last four digits provide a narrower geographic breakdown (a unit within a company, a floor of a large office building, a college post office box).

TECH NOTE

The U.S. Postal Service has a Web site—http://www.usps.gov—to help you locate your reader's ZIP code. All you do is keyboard in your reader's street address, city, and state, and his or her ZIP code automatically appears.

Sometimes special mailing directions are also required. In such cases, one of the following designations is added to the envelope.

- **Attention.** When an attention line is necessary, always put it first, as in Figure 5.6. The attention line is particularly helpful when you have been dealing regularly with one section, department, or individual in a large company—such as the credit officer, parts warehouse, or statistics office.
- **Hold for Arrival.** Individuals may be away on business or on vacation, and you want the letter to reach them on their return.
- **Personal** or **Confidential.** No one but the person named should read the letter.
- **Please Forward.** Note that the U.S. Postal Service forwards mail for only six months.

All such special instructions, except the attention line, are placed at the top left, two spaces below the return address.

FIGURE 5.6 An envelope with an attention line.

KATHY KOOPERMAN
769 EAST 45TH STREET
BALTIMORE MD 21224-6025

HOLD FOR ARRIVAL

ATTENTION MS FAYE GLADSTONE
THE PLACEMENT OFFICE
EAST CENTRAL COMMUNITY COLLEGE
BALTIMORE MD 21228-0710

Making a Good Impression on Your Reader

You have just learned about the formatting and printing requirements your letter must fulfill. Now we will turn to the content of your letter—what you say and how you say it. Writing letters means communicating to influence your readers, not to alienate or antagonize them. Keep in mind that writers of effective letters are like successful diplomats in that they represent both their company and themselves. The image you want readers to have of you and your company is projected through your letter. You want readers to see you as courteous, credible, and professional.

To write an effective letter, first put yourself in the reader's position. What kinds of letters do you like to receive? You would at once rule out letters that are vague, impersonal, sarcastic, pushy, or condescending. You want letters addressed to you to be polite, businesslike, and considerate of your needs and requests. If you have questions, you want them answered honestly, courteously, and fully. You do not want someone to waste your time with a long, puffy letter when a few well-chosen sentences would suffice.

What do you as a writer have to do to send such effective letters? Adopt the **you attitude**; in other words, signal to readers that they and their needs are of utmost importance. Incorporating this "you attitude" means that you should be able to answer "Yes" to these two questions:

1. Will my readers receive a positive image of me?
2. Have I chosen words that convey both my respect for the readers and my concern for their questions and comments?

The first question deals with your overall view of readers. Do your letters paint them as clever or stupid, practical managers or spendthrifts? The second question deals with specific language and tone conveying your view of the reader. Words can burn or soothe. Choose them carefully. As you revise your letters, you will become more aware of and concerned about the ways a reader will respond to you and your message.

Figures 5.7 and 5.8 contain two versions of the same letter. Which one would you rather receive? Why?

Achieving the "You Attitude": Four Guidelines

As you draft and revise your work, pay special attention to the following four guidelines for making a good impression on your reader.

1. Never forget that your reader is a real person. Avoid writing cold, impersonal letters that sound as if they were form letters or instructions on voice mail. Let the readers know that you are writing to them as individuals. Neglecting this rule, a large clinic once sent its customers this statement: "Your bill is overdue. If you pay it by the 15th of this month, no one except the computer will know that it was late." Similarly, the letter below violates every rule of personal and personable communications.

> It has come to our attention that policy number 342q-765r has been delinquent in payment and is in arrears for the sum of $302.35. To keep the policy in force for the duration of its life, a minimum payment of $50.00 must reach this office by the last day of the month. Failure to submit payment will result in the cancellation of the aforementioned policy.

This example displays no sense of one human being writing to another, of a customer with a name, personal history, or specific needs. The letter uses cold and stilted language ("delinquent in payment," "in arrears for," "aforementioned policy"). Revised, this letter contains the necessary personal (and human) touch.

> We have not yet received your payment for your insurance policy (#342q-765r). By sending us your check for $50.00 within the next three weeks, you will keep your policy in force and can continue to enjoy the financial benefits and emotional security it offers you.

The benefits to an individual reader are stressed, and that reader is addressed directly as a valued customer.

Don't be afraid of using "you" in your letters. Readers will feel more friendly toward you and your message. (Of course, no amount of "you's" will help if they appear in a condescending context such as the letter in Figure 5.7.) In fact, you might even use the reader's name or the name of his or her company within your letter to create goodwill and to show your interest.

2. Keep the reader in the forefront of your letter. Make sure that the reader's needs control the tone, message, and organization of your letter. This is the essence of the "you attitude." No one likes people who talk about themselves all the time. What is true about conversation is equally true of letters. Stress the

FIGURE 5.7 A letter lacking the "you attitude."

County Building, Room 200
Ventura Missouri 56780-0101

March 4, 1999

Mr. Ted Ladner
451 West Hawthorne Lane
Morris, MO 64507-3005

Dear Mr. Ladner:

You have written to the wrong office here at the County Building. There is no way we can attempt to verify the kinds of details you are demanding from Brown County.

Simply put, by carefully examining the 1998 tax bill you said you received, you should have realized that it is the Tax Collector's Office, not the Tax Assessor's, that will have to handle the problem you claim exists.

In short, call or write the Tax Collector of Brown County.

Thank you!

Tracey Kowalski

Tracey Kowalski

FIGURE 5.8 A you-centered revision of Figure 5.7.

Brown County
Office of the Tax Assessor

County Building, Room 200
Ventura Missouri 56780-0101

March 4, 1999

Mr. Ted Ladner
451 West Hawthorne Lane
Morris, MO 64507-3005

Dear Mr. Ladner:

Thank you for writing about the difficulties you encountered with your 1998 tax bill. I wish I could help you, but it is the Tax Collector's Office that issues your annual property tax bill. Our office does not prepare individual homeowners' bills.

If you will kindly direct your questions to Paulette Sutton at the Brown County Tax Collector's Office, Room 100, County Building, Ventura, Missouri 56780-0100, I am sure that she will be able to assist you. Should you wish to call her, the number is 458-3455, extension 212.

Respectfully,

Tracey Kowalski

Tracey Kowalski

"you," not the "I" or the "we." Again, try to find out about your readers. Here is a paragraph from a letter that forgets about the reader.

Draft

```
I think that our rug shampooer is the best on the mar-
ket. Our firm has invested a lot of time and money to
ensure that it is the most economical and efficient
shampooer available today. We have found that our cus-
tomers are very satisfied with the results of our ma-
chine. We have sold thousands of these shampooers, and
we are proud of our accomplishment. We hope that we can
sell you one of our fantastic machines.
```

This example talks the reader into boredom by spending all its time on the machine, the company, and the sales success. Readers are interested in how *they* can benefit from the machine, not in how much profit the company makes from selling it. To win the readers' confidence, the writer needs to show how and why they will find the product useful, economical, and worthwhile at home or at work. Here is a reader-centered revision.

Revision

```
Our rug shampooer would make cleaning your Happy Rest
Motel rooms easier for you. It is equipped with a heavy-
duty motor that will handle your 200 rooms with ease.
Moreover, that motor will give frequently used areas,
such as the lobby or hallways, a fresh and clean look
you can be proud of.
```

Note how in this revision the writer shifted attention away from bragging about how "we have sold thousands" to the specific benefits the reader will gain by purchasing the product.

3. Be courteous and tactful. However serious the problem or the degree of your anger at the time, refrain from turning your letter into a punch through the mail. Don't inflame your letter or e-mail readers (see pp. 131–132). Capture the reader's goodwill, and the rewards will be greater for you. The following words can create a bad taste in the reader's mouth.

it's defective
I demand
I insist
we reject
that's no excuse for
totally unacceptable
unprofessional (job, attitude, etc.)
your failure
you contend
you allege
you should have known
your outlandish claim

Use words that emphasize the "you attitude," and avoid offensive language. Compare the discourteous sentences on the left with the courteous revisions on the right.

Discourteous	Courteous Revision
We must discontinue your service unless payment is received by the date shown.	Please send us your payment by November 4 so that your service will not be interrupted.
Your claim that our product was defective on delivery is outlandish.	We are sorry to learn that you were dissatisfied with the condition of our product when it reached you.
The rotten coil you installed caused all my trouble.	The trouble may be caused by a malfunctioning coil.
You are sorely mistaken about the contract.	We are sorry to learn about the difficulty you experienced over the service terms in our contract.
The new printer you sold me is third-rate and you charged first-rate prices.	Since the printer is still under warranty, I hope that you can make the repairs easily and quickly.
Obviously your company is wrong. I wonder if all the people at Acme are as inept as you.	I would appreciate receiving a more detailed explanation from your home office about this matter.
Needless to say, you have misread your warranty agreement.	Clause 17 in your warranty agreement does not cover the problem you have called to our attention.
It goes without saying that your suggestion is not worth considering.	It was thoughtful of you to send me your suggestion, but unfortunately we cannot implement it right now.

The last two discourteous examples above begin with phrases that frequently set readers on edge. It's best to avoid using *needless to say* or *it goes without saying* since they can quickly set up a hostile barrier between a reader and writer.

4. Be neither boastful nor meek. These two strategies—one based on pride and the other on humility—often lead inexperienced letter writers into trouble. On the one hand, they believe that a forceful statement will make a good impression on the reader. On the other hand, they assume that a cautious and humble approach will be the least offensive way to earn the reader's respect. Both paths are wrong.

Aggressive letters, filled with boasts, rarely appeal to readers. Letters should radiate confidence without sounding as if the writer had written a letter of self-recommendation. Letters should let the facts speak directly and pleasantly for themselves. The sentences on the left boast; those on the right express confidence with grace.

Boastful	Graceful Revision
You will find me the most diplomatic employee you ever hired.	Much of my previous work has been in answering and adjusting customer complaints.
The Sun and Sea unqualifyingly promises the nicest rooms on the Coast.	Each room at the Sun and Sea has its own private bath and bar refrigerator.

Boastful	Graceful Revision
I have performed that procedure so many times I can do it in my sleep.	I have performed all kinds of IV therapy as part of standard care.
The Debit Card offers you incomparable customer convenience.	The Debit Card gives you a 50 percent discount on a safe-deposit box.

At the other extreme, some writers stress only their own inadequacy. Their attitude as projected in their letters is "I am the most unworthy person who ever lived, and I would be eternally grateful if you even let my letter sit on your desk, let alone open it." Readers will dismiss such writers as pitiful, unqualified weaklings. Note how the meek sentences on the left are rewritten more positively on the right.

Meek	Positive Revision
I know that you have a busy schedule and do not always have time to respond, but I would be appreciative if you could send me your brochure on how to apply Brakelite.	Please send me your brochure on how to apply Brakelite.
I know the season is almost over, but could you possibly let me know something about rates for the rest of the summer.	I am interested in renting a cabin in late August (24–31) and would like to know about your rates for that week.
I will be grateful for whatever employment opportunities you could kindly give me.	I well welcome the opportunity to discuss my qualifications with you.

Using the Most Effective Language in Your Letters

How you say something in a letter is just as crucial to your success as *what* you say. An effective letter requires you to pay attention, especially during the revision stage, to your words and their tone. Three simple suggestions can help. Your letters should be (1) **clear,** (2) **concise,** and (3) **contemporary.** Regard these principles of letter writing as the three Cs.

1. Be clear. Clarity obviously is the most important quality of a business letter. If your message cannot be understood easily, you have wasted your reader's time. Confusion costs time and money. Plan what you are going to say—what your objective is—by taking a few minutes to jot down some questions you want answered or some answers to questions asked of you. Doing this will actually save you time. Review pages 42–43 in Chapter 2.

Choose precise details appropriate for your audience. In choosing exact words, answer the reader's five fundamental questions—*who? what? why? where?* and *how?* Supply concrete words, facts, details, numbers. On the left are some examples of vague sentences that will puzzle a reader because necessary details are missing. In the revisions on the right, exact words have replaced unclear ones.

Vague	Clear Revision
Please send me some copies of your recent brochure I can use at work.	Please send me 4 copies of your brochure on the new salt substitute to share with my fellow dietitians.
You can expect an appraisal in the next few weeks.	You will receive an estimate on the installation of a new 50,000 BTU air-conditioning unit no later than July 12.
One of our New York stores carries that product.	Our store at 856 East Fifth Avenue sells the entire line of Texworld gloves.
The fee for that service is nominal.	The fee for caulking the five windows on the first floor will be $50.00.

2. Be concise. "Get to the point" is one of the most frequent commands in the business world. A concise letter does not ramble; instead, it is easy to read and to act on. As you draft and then revise your letter, ask yourself these two questions: (1) What is the main message I want to tell my reader? (2) Does every sentence and paragraph stick to the main point? The secret to efficient correspondence is to get to the main point at once, as in the following examples.

Your order will be delivered by July 26, as you requested.

I am happy to confirm the figures we discussed via e-mail last Wednesday.

I request an extension of two weeks in paying my note.

Please accept our apologies for the damaged Movak shipped to you last week.

Here is the report you asked our accountant to prepare. It does contain the new figures on the Manchester store you wanted.

Many letters writers get off to a deadly slow start by repeating, often word for word, the contents of the letter to which they are responding.

First Draft

I have your letter of March 23 before me in which you ask if our office knows of any all-electric duplexes for rent less than five years old and that would be appropriate for senior citizens. You also ask if these duplexes are close to shopping and medical facilities.

Revised

Thank you for your letter of March 23. Our office does rent all-electric duplexes suitable for senior citizens. We have two units, each renting for $375 a month, that are four blocks from the Conyers Clinic and two blocks from the Edgewater Mall.

Another way to write a concise letter is to include only material that is absolutely relevant. For example, in a letter complaining about inadequate or faulty telephone service, mentioning color preferences for cellular telephones would be inappropriate. In a request for information on transferring credits from one college to another, do not ask about intramural sports.

Finally, make sure that your letter is not wordy (review the pertinent sections of Chapter 2, pp. 59–61). By taking a few minutes to revise your letters before they are keyboarded, you can write shorter, more useful letters.

3. Be contemporary. Being contemporary does not mean you should use slang expressions ("I had a tire ripped off"; "That rejection was a bummer") or informal language that is inappropriate ("Doing business with Bindex is a hassle"). Nor should you go to the other extreme and become too stiff and formal. Sound friendly and natural. Write to your reader as if you were carrying on a professional conversation with him or her. Business letters should be upbeat, simple, and to the point. A business letter needs to be readable and believable; it should not be old-fashioned and flowery.

Often individuals are afraid to write naturally because they fear that they will not sound important. They resort to using phrases that remind them (and the reader) of "legalese"—language that smells of contracts, deeds, and starched collars. The following list of words and phrases on the left contains musty expressions that have crept into letters for years; the list on the right contains contemporary equivalents.

Musty Expression	Contemporary Revision
aforementioned	previously mentioned
ascertain	find out
at this present writing	now
I am in receipt of	I have
attached herewith	enclosed
at your earliest possible date	soon
we beg to advise	we believe, think
I am cognizant of	I know
contents duly noted	we realize
endeavor	try
forthwith	at once
henceforth	after this
hereafter, heretofore, hereby	(drop these three "h's" entirely)
immediate future	soon
in lieu of	instead of
kindly advise	let us know
optimum	best
pursuant	concerning
please be advised that	I am happy (or sorry) to tell you that
please find enclosed	I'm enclosing
pending your reply	until I hear from you
per our conversation	when we spoke
prior to	before

Musty Expression	Contemporary Revision
we regret to inform you that	we are sorry that
remittance	payment
remuneration	cost, salary, pay
rest assured that	you can be sure that
your letter arrived and I have same	I have your letter
thanking you in advance	thank you
under separate cover	I'm also sending you
the wherewithal	the way
yours of recent date	your recent letter
your communication	your phone call, your fax, your e-mail

Figure 5.9 shows a letter in stilted language written by Brendan T. Mundell to Patricia Lipinski, an executive whose firm has been overcharged for airplane tickets. Mundell's letter overflows with flowery, old-fashioned expressions. The effect is that Mundell's message—offering an apology, a credit, and a promise to correct the situation—is long-winded and pompous. It even sounds insincere. Note how the revision in Figure 5.10, free of such stilted expressions, is shorter, clearer, and far more personable.

Writing for International Readers

Electronic communications have made the world so compact that, in effect, we live and work in a global village. Companies today depend on international trade to stay in business. Many American business are multinational corporations with sales branches, manufacturing plants, and customers throughout the world. And, in fact, many businesses located in the United States are themselves branches of international firms. Each year a larger share of America's gross national product (GPN) depends on foreign markets. Some American firms estimate that as much as 70 percent of their business is done outside the United States. Every country is affected by every other, connected by vast interlinked communication networks like the Internet.

As a result, don't presume that you will be writing only to native American English speakers. As part of your job, you will very likely write to readers for whom English is not their first (or native) language; these individuals constitute a large and important **English as a Second Language (ESL) audience,** and writing for them requires you to broaden your sense of audience analysis. An ESL audience might reside in a foreign country or in the United States.

Expect your ESL audience to have varying degrees of proficiency in English. Some ESL readers will have an excellent command of American (or British) English; others will have only basic literacy in English. Keep in mind, too, your reader may not speak any English at all but who will rely on an English grammar book and a foreign language dictionary to translate your work. Your ESL audience may be a customer, a co-worker, or a specialist in your field who needs written information from you to get a job done right.

FIGURE 5.9 A letter written in stilted, old-fashioned language.

NORTHERN AIRWAYS

July 12, 1998

Ms. Patricia Lipinski
Vice President
Lindsay Electronics
4500 South Mahoney Drive
Buffalo, NY 14214-4514

Dear Ms. Lipinski:

Please be advised that I am in receipt of yours of July 6th. I would like to take this opportunity to say that we are cognizant of our commitment to good corporate customers like Lindsay and extend our deepest regret and disappointment for the problems your firm has experienced with Northern.

Remittance is beyond any doubt due your firm, and I hasten to rectify the situation with regard to our error. Forthwith we are adjusting your account #7530, crediting it with the $706.82 you were surcharged erroneously. Moreover, inasmuch as Lindsay is a valuable procurer of our services, we are also enhancing your Travel-Pass account with another 3,000 miles.

I would also like to bring to your attention that in an endeavor to correct such billing errors in the immediate future, I have routed copies of your communication to the manager of the billing department, A.T. Padua. I am confident he will take necessary action at his earliest possible convenience to ascertain the situation and make the necessary adjustments in our procedures.

Once again, I want to take the liberty to assure you that Lindsay Electronics is one of our most valued clients. Rest assured that we will take every step imaginable not to jeopardize our long-standing relationship with you. I hope that all the afore-mentioned problems have now been satisfactorily resolved for you. Thanking you, I am

Sincerely yours,

Brendan T. Mundell

Brendan T. Mundell
General Manager

3000 Airline Highway Tyler, ME 04462-3000 (207)555-6300 FAX (207)555-6320
E-MAIL: cserve morair.com

FIGURE 5.10 A revised version of the stilted letter in Figure 5.9.

NORTHERN AIRWAYS

July 12, 1998

Ms. Patricia Lipinski
Vice President
Lindsay Electronics
4500 South Mahoney Drive
Buffalo, NY 14214-4514

Dear Ms. Lipinski:

Thank you for your letter of July 6. I am sorry to learn about the billing problems your employees encountered while traveling on Northern earlier this month. We care very much when we have inconvenienced a good customer like Lindsay Electronics. Please accept my apology.

Lindsay is unquestionably entitled to compensation for our billing error. I am crediting your account #7530 with $706.82, the amount you were overcharged. Furthermore, in appreciation of Lindsay's business, I am also crediting your Travel-Pass account with a bonus 3,000 miles.

To help us avoid similar incidents, I have sent a copy of your letter to the manager of our billing department, A.T. Padua. I know he will want to learn about Lindsay's experience and will use your comments constructively in an effort to revise our billing procedures.

As a Frequent Flyer account member, Lindsay is one of our most important customers, and we will continue to work hard to deserve your support. Please call me if I can help you in the future.

Sincerely yours,

Brendan T. Mundell

Brendan T. Mundell
General Manager

3000 Airline Highway Tyler, ME 04462-3000 (207)555-6300 FAX (207)555-6320
E-MAIL: cserve morair.com

You can expect to write a variety of documents to these readers—e-mail, news releases and brochures, home pages and sales promotions, product descriptions, proposals, and even operating instructions. If you find the set of directions accompanying your computer or a software package confusing, imagine how much more intimidating such a document would be to an ESL reader. **You cannot write too clearly or carefully for these readers.** Because so much technical and scientific literature is written in English, your work must be clearly understood by ESL readers in order for them to translate it into their native language and share your message with fellow specialists. Your company will profit from having its documents written clearly. In fact, the easier and more understandable you make your written communications to ESL audiences, the better your chances of doing business with them. Educate yourself to be sensitive to your audience's multicultural and multilingual background.

Guidelines for Communicating with ESL Readers

To communicate with ESL customers and co-workers, you will have to write "international English," a language that is easily understood in a world that international trade has condensed to a single global marketplace. It would be impossible to give you information about how best to communicate with every ESL audience; there are at least 100 major languages representing diverse ethnic and cultural communities around the globe. But the two most crucial points to keep in mind are these: (1) you need to be aware of cultural differences between you and your reader, and (2) the conventions of writing—the words, sentences, even the type of information you offer—can and do change from one culture to another.

Following the eight guidelines below will help you to communicate more successfully with ESL audiences and will significantly reduce the chances of their misunderstanding you.

Use Common, Easily Understood Vocabulary

Write basic, simplified English. Choose words that are widely understood as opposed to those that are not used or understood by many speakers. Consult a helpful dictionary of basic English such as *The New York Times Everyday Dictionary*, edited by Thomas M. Paikeday (New York Times Books, 1982). Avoid such low-frequency words as the following by substituting simpler synonyms, like those in parentheses: *refrain* (*stop*); *forestall* (*prevent*); *exude* (*discharge*); *exultant* (*happy*).

Avoid Ambiguity

Words that have double meanings force ESL readers to wonder which one you mean. For example, "We fired the engine" would baffle your ESL reader because he or she would not automatically be aware of the multiple meanings of *fire*. Unfamiliar with the context in which *fire* means "start up," an ESL reader might think you're referring to "setting on fire, or inflaming," which is not what you intend. Such interpretation is likely because most bilingual dictionaries used by your ESL reader would list these two meanings. Or because *fire* also means "dismiss" or "let go," an ESL reader might even suspect the engine was replaced by another model. Be especially careful, too, of using synonyms just to vary your word choice. For example, do not write *quick* in one sentence and then, referring to the same action, describe it as *rapid.* Your ESL reader will assume you have two different things in mind instead of just one.

Be Careful About Technical Vocabulary

While an ESL reader may be more familiar with technical terms than with other English words, make sure the technical word or phrase you include is widely known and not just a word or meaning used only at your plant or office. If you write something about a *trackball* and your audience does not know that this object has replaced a *mouse,* your point will be missed. Also steer clear of technical terms in fields other than the one with which your ESL reader is familiar.

Avoid Idiomatic Expressions

Idioms are the most difficult part of a language for the non-native audience to master. As with the example of *fire* above, the following colorful idiomatic expressions will confuse and may even startle an ESL reader: "I'm all ears"; "sleep on it"; "throw cold water on it"; "burn the midnight oil"; "hit the nail on the head"; "he landed in hot water"; "easy come, easy go"; "you bet your life"; "get a handle on it"; "cut off your nose to spite your face." The meaning of these and similar phrases is not literal but figurative, a reflection of *our* culture, *not* necessarily your reader's. An ESL reader will approach these phrases as combinations of the separate meanings of the individual words, not as a collective unit of meaning. By using idioms, you risk confusing and offending your audience.

Imagine the horror an ESL reader—a potential customer in Asia or Africa, for example—might experience if you wrote about a sale concluded at a branch office this way: "Last week we made a killing in our office." Omit the idiomatic expression and substitute a clear, unambiguous translation easily understood in international English. "We made a big sale last week." Or in place of "You hit the nail on the head," write "You have clearly understood what must be done." For "Sleep on it," you might say, "Please take a week or two to make your decision."

Delete Sports (or Gambling) Metaphors

These metaphors, which are often rooted in American popular culture, do not translate word for word for non-native speakers and so again can interfere with your communication with them. In all likelihood, your audience has no equivalent in its culture for American sporting games and events. Here are a few examples to avoid: "out in left field," "a ballpark figure," "we struck out," "we fumbled the ball," "go out for a long pass," "the bases are loaded," "drop the ball," "out of bounds," "down for the count," "made a pass," "top of the ninth." Use a basic English dictionary, and your common sense, to find nonfigurative translations for these and similar expressions so your ESL readers can understand them or discover their meaning in a bilingual dictionary.

Watch Units of Measure

Adapt your references to units of measurement, money, and time to your reader's culture. It is too easy to fall into the cultural trap of assuming that your reader measures distances in miles and feet (instead of kilometers and meters), buys gallons of gasoline (instead of liters), and spends dollars (rather than pesos, marks, rupees, or yen). It might even be insulting to tell an ESL reader he or she can "save bucks" by buying from your company. Just as not everyone in the world uses 110/220 wiring, keep in mind that not everyone sees the world marketplace solely in terms of the United States economy. Adapt your message to the reader's practices. Moreover, in your e-mail, faxes, and other telecommunications, don't assume that because it is late (1:00 A.M.) where you are, it is also late for your reader. Be aware of seasonal differences, too. While New York is in the middle of the winter, Australia and Chile are enjoying summer. Be respectful of your readers' cultural (and physical) environment. Thanksgiving is celebrated in the United States in November, but in Canada the holiday is observed in October, and elsewhere around the globe it may not be a holiday at all.

Avoid Culture-Bound Descriptions of Place and Space

For example, when you tell an ESL reader in Singapore about the Sunbelt, or a potential client in Africa about the Big Easy or the Big Apple, will he or she know what you mean? When you write from California to an ESL reader in India about the eastern seaboard, meaning the East Coast of the United States, the directional reference will not mean the same thing to your audience as it does to you.

Keep Your Sentences Simple and Easy to Understand

Short, direct sentences will cause an ESL reader the least amount of trouble. World languages, especially those in Asia, divide information into sentence units far differently from the way English does. A good rule of thumb is that the shorter and less complicated your sentences, the easier they will be for an ESL reader to process. Long (more than fifteen words) and complex (multiclause) sentences can be so difficult for ESL readers to unravel that they may skip over them or guess at your message. Always try to avoid the passive voice; it is one of the most difficult sentence patterns for an ESL reader to comprehend. Stick to the common subject-verb-object pattern as often as possible.

Respecting the Cultural Traditions of Your ESL Readers

Using simple words and concise sentences certainly will help you to write more effectively and clearly to an ESL audience. But to avoid even greater troubles, you also must be concerned with respecting the cultural traditions, customs, and preferences of your readers. Cultures differ widely in the way they send and receive information. What is acceptable in one culture may be offensive in another.

For example, in Japan you would impress a potential client by bowing rather than shaking hands. But before you bow, learn the protocol involved in bowing—who bows first, for how long, and how deeply. Similarly, Japanese businesspeople are more accustomed to negotiate sitting side by side rather than the more aggressive practice of Americans, who prefer to sit face to face across from each other.

Respecting the cultural practices of your readers can have far more impact on them than touting the price and quality of your product or service.

Here are some culturally conditioned elements of communication you may need to adjust when writing to ESL readers:

- how you address the reader in your salutation
- the beginning and conclusion of your letter
- the type and amount of information you give
- the overall tone you use
- the format and design of your document

TECH NOTE

Visit your reader's country on the Internet. Once you locate the country's home page (often prepared by its Division of Tourism), you can click your way into a deeper knowledge of and respect for your reader's customs. The country's home page should contain information about the geography, language, religion, education, government, and economy. You should also find a map and other statistical information.

Check to see if your reader's company is also listed on the Internet, and explore it.

Finally, many general sources of information, like encyclopedias, guide books, and catalogs, are also available on-line for cross-cultural research.

Sources of Information on Your Reader's Culture

Before you dash off a letter to a prospective ESL client, find out as much as you can about his or her country's customs, especially the ways in which business is conducted and what behaviors are regarded as courteous and discourteous. The sources below provide excellent help.

- Consult the U.S. State Department's *Background Notes* on your reader's country, a pamphlet that includes essential information on the country's government, economy, history, and people; the State Department also prepares *Post Reports,* primarily for embassy staffs, on the food, currency, holidays, religious customs, and many other customs of the country.
- Interview an ESL student or co-worker raised in that culture.
- Call the country's embassy or consulate for specific information, pertinent publications, and spellings/pronunciations.

Each of these sources will help you to understand proper communication etiquette for crossing cultural boundaries.

Case Study: Writing to a Foreign Client

Let's assume that you have to write a sales letter to an Asian business executive. You will have to employ a very different strategy in writing to an Asian executive as opposed to an American executive. For an American reader, the best strategy would be to take a direct approach—fast, hard-hitting, to the point, and stressing your product's strengths versus the opposition's weaknesses. A sales letter to an American reader would be polite but direct.

But such a strategy would be counterproductive in a sales letter to an Asian reader. Business in East Asia is associated with religion and friendship, and is wrapped up in a great many social courtesies. The Asian way of doing business, including writing and receiving letters, is far more subtle, indirect, and complimentary than in America. The American style of directness and forcefulness would be perceived as rude or unfair in, say, Japan, China, or Korea. A hard-sell letter to an Asian reader would be a sign of arrogance, and arrogance suggests inequality for such a reader.

Courtesy for an Asian reader, however, would be of a paramount importance, more persuasive than a thorough description of a product or service. A sales letter to an Asian reader, therefore, should establish a friendship, a relationship in which trust is established first and business details are dealt with later. It is standard in Japanese companies, for example, to have executives meet three or four times just to socialize before beginning business negotiations.

To better understand the differences between communicating with an American reader and an Asian one, study the two versions of the sales letter in Figures 5.11 and 5.12. These letters, written by Susan DiFusco for Starbrook Electronics, sell the same product, but Figure 5.11 is addressed to an American executive while Figure 5.12 adapts the same message for a businessperson in Seoul. Not only do the two letters differ in content, but also in the way each is printed.

The full-block style of the letter to the American reader signals a no-nonsense, all-business approach. Everything is lined up in neat, orderly fashion. For the Korean reader, however, DiFusco wisely chose the more varied pattern of indenting her paragraphs. For Asian readers the visual effect suggests a much more relaxed and friendly, yet duly respectful communication.

Pay special attention to how the American letter in Figure 5.11 starts off politely but much more directly, an opening the Korean reader would regard as blunt and discourteous. The sales letter written to the Korean audience (Figure 5.12) starts not with business talk but with a compliment to the reader and his company, praising them for trustworthiness and wishing them much prosperity in the future. The successful writer addressing a Korean administrator would not get to "the bottom line" right away but would use the introductory paragraph to show his or her respect for the company and the reader—the equivalent of the Japanese business executives socializing first before any mention of business is made.

FIGURE 5.11 Sales letter to a native English speaker in an American firm.

Starbrook Electronics
Perry, TX 75432-3456
phone (713)555-2121 fax(713)555-3172
http://www.starbrook.org

December 2, 1998

Mr. Ellis Fanner
Administrator
Morgan General Hospital
Morgan, OR 97342-0091

Dear Mr. Fanner:

How many times have the physicians who work at Morgan General asked when you would be getting MRI (magnetic resonance imaging) equipment? Having the latest, state-of-the-art imaging equipment is important to maintain your reputation as a leading health care provider in the Morgan River Valley.

As the world leader in designing and manufacturing MRI equipment, Starbrook can offer you the latest technology available. This diagnostic technology will save your patients time and improve the care you give them. With the MRI capability of our Imaging 500, Morgan General can deliver more accurate and timely diagnoses. Within two hours you can determine whether a patient has had a stroke rather than having to wait a day or longer with more conventional X-ray or scanner technology. Thanks to our Imaging 500 model, Morgan General can also improve diagnoses for orthopedic and cardiac problems. Our Imaging 500 delivers much more extensive internal imaging than any other of our competitors' equipment.

By obtaining the Imaging 500, Morgan General will surpass all other health care providers in the Morgan River Valley. No other hospital within a hundred mile radius has one. By acting now, you, too, will receive Starbrook's unsurpassed guarantee of service and clarity. You are guaranteed one year's free maintenance by our team of experts.

Continued

FIGURE 5.11 (Continued)

And we will even give you free upgrades to make sure your Imaging 500 stays state of the art. Since software updates change so often and so radically, no other MRI dealer dares make such an offer. We deliver what we promise. Ask any of our recent satisfied customers–Tennessee General, New York Uptown General, or Nevada Statewide HMO.

We are hosting a demonstration for hospital administrators on the 4th of January in Portland and would like to see you there. Don't hesitate to call me to arrange for your free showing.

Sincerely,

Susan DiFusco
Assistant Manager

Compare the second and third paragraphs of the letters in Figures 5.11 and 5.12. While the letter to the American reader launches an aggressive campaign to get the reader's business, the letter to the Korean reader avoids the hard sell of American business tactics. Susan DiFusco knew that she must not promote too strenuously for her Korean reader. The more she boasted about Starbrook's work, the less likely it was that she would make a sale. She recognized from discussions with other Asian businesspeople over the years that she had to supply key information—such as the application and the advantages of her product—without overwhelming or pressuring her audience. Yet DiFusco subtly reassures Mr. Kim that her company is honorable and worthy to be recommended to his friends—Starbrook is well known for quality work.

Observe, too, how the letter to the American executive (Figure 5.11) undermines the competition by stating how much better Starbrook's offer is. For most Asian readers, on the other hand, it would be considered impolite to claim that your product is better than another company's or that your firm is currently doing business with other firms in the reader's country. Asian audiences prefer to avoid anything that hints of impoliteness or assertiveness.

Finally, contrast the conclusions of the two letters. In the letter to an American audience, DiFusco strongly urges her potential customer to get in touch with her. Such a "call to action" is customary in a sales letter to an American firm. But in her concluding paragraph to Mr. Kim, DiFusco adopts a more reserved and personal tone. Her use of such appropriate phrases as "kindly let me know" and "it

FIGURE 5.12 Sales letter to an ESL reader in a foreign firm.

Starbrook Electronics
Perry, TX 75432-3456
phone (713)555-2121 fax(713)555-3172
http://www.starbrook.org

December 2, 1998

Mr. Kim Sun-Lim
Administrator, Tangki Hospital
210-214 Jenji Road
Seoul, Korea

Dear Mr. Kim:

In your beautiful language I say "Hang wenur paddamnya nensa trenemyada" on behalf of my firm Starbrook Electronics. I am honored to introduce myself to you through this letter.

Please let us know how we might be of service to you. One of the ways we may be able to serve you is by informing you about our new Imaging 500, MRI (magnetic resonance imaging) equipment. This new model can offer you and your patients many advantages. It is far better than conventional X-ray or even scanner models. Your physicians can offer quicker diagnoses for patients with strokes, heart attacks, or orthopedic injuries. MRI pictures will give you clearer and deeper pictures than any X-ray can.

It would be an honor to provide Tangki Hospital with one of our Imaging 500s. If you select the Imaging 500, we will be happy to give you all maintenance and software updates free for one year. Such service will provide the best in health care for the many people who come to you for help. Our firm is well known for its quality service.

Kindly let me know if I might send you information about the Imaging 500. It would be a privilege to meet you and to give you and your staff a demonstration of our Imaging 500.

Sincerely yours,

Susan DiFusco

Susan DiFusco
Assistant Manager

would be a privilege" express the friendly sentiments of respect and esteem that would especially appeal to her reader.

Obviously, some American readers and some Asian readers might respond differently to Figures 5.11 and 5.12 than has been suggested here. Nor is it only in addressing an ESL audience that a writer must consider the readers' communication patterns, use protocols, and cultural (or subcultural) traditions. The same guidelines apply whether you are writing to ESL readers or to people like yourself. Your message and vocabulary should always be clear, understandable, and appropriate for the intended audience.

✓ Revision Checklist

Audience Analysis and Research

- ❑ Made sure reader's name and job title are right.
- ❑ Found out something about my audience—interests, background; well informed or unfamiliar with topic; former clients, new ones.
- ❑ Determined whether audience will be friendly, hostile, or neutral about my message.
- ❑ Did sufficient research—in print and on-line sources—to give audience what they need.
- ❑ Acknowledged previous correspondence.
- ❑ Spent sufficient time drafting and revising letter before printing final copy.

Content/Organization

- ❑ Clearly understood my purpose in writing to reader.
- ❑ Put most important point first in my letter.
- ❑ Started each paragraph with the central idea of that paragraph.
- ❑ Answered all the reader's questions and concerns.
- ❑ Omitted anything offensive, irrelevant, or repetitious.
- ❑ Used numbers and/or boldfacing to make it easier for reader to find and remember main ideas.
- ❑ Stated clearly what I want reader to do.
- ❑ Used last paragraph to summarize and encourage reader to continue cordial relations with me and my company.

Style—Words, Tone, Sentences, Paragraphs

- ❑ Emphasized the "you attitude" by seeing things from reader's perspective.
- ❑ Was not too casual or colloquial.

- ❑ Chose words that are clear, precise, and friendly.
- ❑ Cut anything sounding flowery or stuffy, especially legalese.
- ❑ Ensured that my sentences are readable, clear, and not too long (less than 15–20 words).
- ❑ Wrote paragraphs that are easy to read and flow together.

Writing to ESL Reader(s)

- ❑ Did appropriate research about reader's culture, especially accepted ways of communicating, both in print and on-line sources.
- ❑ Adopted a respectful, not condescending, tone.
- ❑ Avoided anything offensive to my reader, especially references to politics, religion, or cultural taboos.
- ❑ Used language that my reader would understand.
- ❑ Tested my sentences for length and active voice.
- ❑ Made sure nothing in my letter might be misinterpreted by my ESL reader.
- ❑ Selected the right format, salutation, and complimentary close for my reader.

Format/Appearance

- ❑ Followed one letter format (full-block, semiblock, simplified) consistently.
- ❑ Left wide enough margins to make my letter look attractive and well proportioned.
- ❑ Included all the necessary parts of a letter.
- ❑ Made sure that my letter looks neat and professional—toner cartridge is fresh and type font is not crowded or showy.
- ❑ Printed my letter on company letterhead or quality bond paper.
- ❑ Proofread my letter carefully and made sure each correction was made before final copy was printed.
- ❑ Eliminated any grammatical and spelling errors.
- ❑ Signed my letter legibly in blue or black ink.

Exercises

1. What kinds of letters do you receive addressed to you at home? At your work? Write a letter to the sender/company about one of these letters, evaluating how successful it was in getting you to act.

2. Find two business letters and bring them to class. Be prepared to identify the various parts of a letter discussed in this chapter.

3. Find a form letter that is addressed to "Dear Customer," "Postal Patron," or "Dear Resident," and rewrite it to make it more personal.

4. Correct the following inside addresses:
   a. Dr. Ann Clark, M.D.
1730 East Jefferson
Jackson, MI. 46759
   b. To: Tommy Jones
Secretary to Mrs. Franks
Donlevey labs
Cleveland, O. 45362
   c. Debbie Hinkle
432 Parkway
N. Y. C. 10054
   d. Mr. Charles Howe, Acme Pro.
P.O. Box 675
1234 S. e. Boulevard
Gainesville, Flor. 32601
   e. Alex Goings, man.
Pittfield Industries
Longview, TEXAS 76450
   f. ATTENTION: G. Yancy (Mrs.)
Police Academy
1329 Tucker
N. O., La. 3410-70122
   g. David and Mahenny
Lawyers
Dobbs Build.
L.A. 94756
   h. CONFIDENTIAL
Jordon Foods, INC.
Miller Str.
Lincoln, Neb. 2103

5. Write appropriate inside addresses and salutations to (a) a woman who has not specified her marital status; (b) an officer in the armed forces; (c) a professor at your school; (d) an assistant manager at your local bank; (e) a member of the clergy; (f) your congressperson.

6. Rewrite the following sentences to make them more personal.
   a. It becomes incumbent upon this office to cancel order #2394.
   b. Management has suggested the curtailment of parking privileges.
   c. ALL USERS OF HYDROPLEX: Desist from ordering replacement valves during the period of Dec. 19–29.
   d. The request for a new catalog has been honored; it will be shipped to same address soon.
   e. Perseverance and attention to detail have made this writer important to company in-house work.
   f. The Director of Nurses hereby notifies staff that a general meeting will be held Monday afternoon at 3:00 P.M. sharp. Attendance is mandatory.
   g. Reports will be filed by appropriate personnel no later than the scheduled plans allow.

7. The following sentences from letters are discourteous, boastful, excessively humble, vague, or lacking the "you attitude." Rewrite them to correct these mistakes.
   a. Something is obviously wrong in your head office. They have once more sent me the wrong model number. Can they ever get things straight?
   b. My instructor wants me to do a term paper on safety regulations at a small factory. Since you are the manager of a small factory, send me all the information I need at once. My grade depends heavily on all this.
   c. It is apparent that you are in business to rip off the public.

d. I was wondering if you could possibly see your way into sending me the local chapter president's name and address, if you have the time, that is.
e. I have waited for my confirmation for two weeks now. Do you expect me to wait forever or can I get some action?
f. Although I have never attempted to catalog books before, and really do not know my way around the library, I would very much like to be considered at some later date convenient to you for a part-time afternoon position.
g. It goes without saying that we cannot honor your request.
h. May I take just a moment of your valuable time to point out that our hours for the next three weeks will change and we trust and pray that no one in your agency will be terribly inconvenienced by this.
i. Your application has been received and will be kept on file for six months. If we are interested in you, we will notify you. If you do not hear from us, please do not write us again. The soaring costs of correspondence and the large number of applicants make the burden of answering pointless letters extremely heavy.
j. My past performance as a medical technologist has left nothing to be desired.
k. Credit means a lot to some people. But obviously you do not care about yours. If you did, you would have sent us the $249.95 you rightfully owe us three months ago. What's wrong with you?

8. The following letter, filled with musty expressions and in old-fashioned language, buries key ideas. Rewrite and reorganize it to make it shorter, clearer and more reader-centered.

Dear Ms. Granedi:

This is in response to your firm's letter of recent date inquiring about the types of additional services that may be available to business customers of the First National Bank of Bentonville. The question of a possible time frame for the implementation of said services was also raised in the aforementioned letter. Pursuant to these queries, the following answers, this office trusts, will prove helpful.

Please be advised that the Board of Directors at First National Bank has a continuing reputation for servicing the needs of the Bentonville community, especially the business community. For the last fifty years—half of a century—First National Bank has provided the funds necessary for the growth, success, and expansion of many local firms, yours included. This financial support has bestowed many opportunities on a multitude of business owners, residents of Bentonville, and even residents of surrounding local communities.

The Board is at this present writing currently deliberating, with its characteristic caution, over a variety of options suggested to us by our patrons, including your firm. These options, if the Board decides to act upon them, would enhance the business opportunities for financial transactions at First National Bank. Among the two options receiving attention by the Board at this point in time are the creation of a branch office in the rapidly growing north side of Bentonville. This area has many customers who rely on the services of First National Bank. The Board may also place a business loan department in the new branch.

If this office of the First National Bank of Bentonville might be of further helpful assistance, please advise. Remember banking with First National Bank is a community privilege.

Soundly yours,

M. T. Watkins
Public Relations Director

9. Either individually or with a small group, write a business letter to one of the following individuals and submit an appropriate envelope with your letter.

- **a.** your mayor, asking for an appointment and explaining why you need one
- **b.** your college president, stressing the need for more parking spaces or for additional computer terminals in a library
- **c.** the local water department, asking for information about fluoride supplements
- **d.** an editor of a weekly magazine, asking permission to reprint an article in a school newspaper
- **e.** the author of an article you have read recently, telling why you agree or disagree with the views presented
- **f.** a disc jockey at a local radio station, asking for more songs by a certain group
- **g.** a computer dealer, asking about software packages and explaining your company's special needs

10. Rewrite the following letters, making them appropriate for an ESL reader. As you revise these letters, pay attention to the words you use as well as the sentence constructions you employ. Be sure to consider the reader's cultural traditions.

a. Dear Chum,

Our stateside boss hit the ceiling earlier today when she learned that our sales quota for this quarter fell precipitously short. Ouch! Were I in her spot, I would have exploded too. Numerous missives to her underlings warned them to get off the dime and on the stick, but they were oblivious to such. These are the breaks in our business, right?

Let's hope that next quarter's sales take a turn for the best. If they are as disasterous, we all may be in hot water. Until then, we will have to watch our p's and q's around here.

Cheers,

b. Dear Mr. Wong,

It's not every day that you have the chance to get in on the ground floor of a deal so good you can actually taste it. But Off-Wall Street Mutual can make the difference in your financial future. Give me a moment to convince you.

By becoming a member of our international investing group you can just about ensure your success. We know all the ins and the outs of long-term investing and can save you a bundle. Our analysts are the hot shots of the business and always look long and hard for the most propitious business deals. The stocks we select with your interests in mind are as safe as a bank and not nearly so costly for you. We can save you money by investing your money. We are penny pinchers with our client's initial investments, but we are King Midas when it comes to transforming those investments into pure gold.

I am enclosing a brochure for you to study, and I really hope you will examine it carefully. You would be foolish to let a deal like Off-Wall Street Mutual pass you by. Go for it.

Hurriedly,

c. Dear Mr. Bafaloukos,

My firm is taking a survey of businesses in your part of the world to see if there is any likelihood of getting you on board our international computer network and so I thought I would drop you a line to see if you might like to take the chance. In today's uncertain world, business events can change overnight and without the proper scoop you could be left out in the cold. We can alleviate that mess.

Not only do we interface with major exchanges all around the globe but we make sure that we get the facts to you pronto. We do not sit on our hands here at Intertel. Check out the enclosed data sheet on who and how we serve and I have no doubts that you will e-mail or ring us up to find out about joining up.

One last point: can you really risk going out on a limb without first knowing that you have all the facts at your fingertips about worldwide business events? Intertel is there to save you.

Fondly,

11. Interview a student at your school or a co-worker who was born and raised in a foreign country about the proper etiquette in writing a business letter to someone from his or her country. Collaborate with that student to write a letter (a sales letter or a letter asking for information) to an executive from that country.

12. In a letter to your instructor, describe the kinds of adaptations you had to make for the ESL reader you wrote to in Exercise 11.

13. As a collaborative project, team up with three other students in your class to write separate letters tailored to executives in each of the following three countries:
    a. Buenos Aires, Argentina
    b. Tokyo, Japan
    c. Munich, Germany

    Assume you are selling the same product or service to each reader but you will have to adapt your communication to each of the cultures represented by these readers.

    Turn in the three letters and explain to your instructor in an accompanying memo how you met the needs of these diverse cultural audiences in terms of style, tone, level of content, format, and sales tactics. Describe the research tools you used to find out about your reader's particular culture (and communication protocols) and how you benefited from using these sources.

CHAPTER 6

Types of Business Letters

Receiving and answering business correspondence is vital to the success of a company and its employees. Businesses take their letter writing very seriously and employers pay a very high price, as we saw in Chapter 1, for employees to write effective letters. Letter writing is a prized skill in the world of work. In fact, the higher up the corporate ladder you climb, the more you will be expected to write letters—and write them well. For this reason, we cannot overemphasize how important it is to your career to be an effective letter writer.

Why Letters Are Important

- **Letters reflect a company's reputation.** A firm's corporate image is on the line when it sends out a letter. The way a letter looks, what it says, and how it conveys that message determine how the company will be judged in the marketplace—as clear, honest, and professional, or as sloppy, questionable, and unreliable.
- **Letters are formal documents intended primarily for readers outside your organization.** Letters go out regularly to established customers, prospective clients, community and government officials—individuals who can make sure your firm or agency stays in business or does not. Letters are the most formal type of correspondence, while memos and internal e-mail, written for individuals within the company, are the least formal. A formal business letter says that a company is putting its best written work on display. Letters say that the occasion, as well as the message, deserves formal attention.

Preparing Letters

One of the biggest mistakes you can make is to think that writing a letter, even a short one, does not require preparation. Don't think you can just sit down for a few minutes at your keyboard and fire off an acceptable, well-crafted letter. Like other types of occupational writing—proposals, reports, instructions—writing an effective letter requires you to follow, though perhaps in abbreviated fashion, the writing process discussed in Chapter 2. This is true whether you print your letter and drop it in a mailbox or you upload it to the Internet via e-mail.

Accordingly, you need to identify your audience and have a clear sense of your purpose and theirs; do appropriate research; select the best strategy for communicating with your audience; and draft, revise, and edit your letter.

Although it is more likely that you will collaborate on longer documents, you still can expect to work with others on certain letters. For instance, you may be asked to confer with specialists in other divisions of your company to answer a complaint about one of your company's products. You may be required to meet with individuals in your company's law or marketing departments when drafting a sales letter.

TECH NOTE

The success of your letters depends on your willingness to revise them. Never settle for sending your readers a draft. Take time to check your facts—names, dates, costs, places, terms of an agreement or warranty. Be sure to double-check your math; a mistake here could be costly for you and your employer. Run a spell-check but also proofread the final printed copy of your letter to make sure it is reader-ready. If you failed to delete an earlier sentence or even an entire paragraph from a previous draft and printed your final copy that way, your letter would confuse and possibly offend your readers.

There may be times when your boss will ask to edit a letter (see pp. 83–85) before you send it.

Types of Letters

This chapter discusses five common types of business correspondence that you will be expected to write on the job:

1. order letters
2. inquiry letters
3. special request letters
4. sales letters
5. customer relations letters

These five types of letters will involve a variety of formats and writing strategies and techniques. Various letters can be classified as **positive, neutral,** or **negative,** depending on their message and the anticipated reactions of your audience. Order, inquiry, and special request letters are examples of neutral letters. They carry neither good nor bad news; they simply inform, responding to routine correspondence. Neutral letters request information about a product or service, place an order, or respond to some action or question. Sales letters promoting a product carry good news, according to the companies that spend millions of dollars a year preparing these documents. Customer relations letters can be positive (responding favorably to a writer's request or complaint) or negative (refusing a request; saying no to an adjustment; denying credit; seeking payment; critiquing poor performance; or announcing a shipping delay).

TECH NOTE

Keep in mind that your letters must make a good visual, as well as verbal, impression on your readers. Take advantage of the "print preview" function of your word-processing software to see how your letter will look on the page before you print. If necessary, readjust margins and add white space. But don't insert too much white space because your letter will look inflated and insincere. And never print a letter in all capital letters; not only will you make it harder to read but your letter will look unprofessional, like an old-fashioned telegram. Choose a print font that is appropriate for your message and audience. Your readers will find a serif font much more open and inviting than a sans serif one. Among the most cordial serif fonts are Goudy and Palatino, available on many word-processing software packages. Depending on the length and details of your letter, you might want to use bullets to itemize a series of points, boldface type to emphasize a key point or word, or underscoring and color for highlighting. But use these devices sparingly. Overdoing color, boldfacing, underscoring, and italicizing makes a letter busy-looking and conveys an image of you as over-anxious, even self-conscious, not professional and self-controlled.

Order Letters

Order letters are straightforward notices informing a seller that you want to purchase a product or service. To make sure that you receive exactly what you want, your letter must be clear, precise, and accurate. Double-check the seller's brochure, catalog, manual, or Web site before you send your order letter.

FIGURE 6.1 An order letter.

DAVIS CONSTRUCTION COMPANY
1200 South Devon
Millersville, Pennsylvania 17321
(712)555-1000 FAX (712)555-4221

April 6, 1998

E. F. Johnson and Associates
820 Frontage Road
P.O. Box 1007
St. Louis, MO 64211-1007

ORDERING MATERIALS

Please send us by first-class mail the following materials listed in your 1998 catalog entitled *Personnel Forms for Monitoring Selection, Administration, and Evaluation* advertised on your Web site.

Catalog Number	**Title**	**Number of Copies**	**Total Cost**
P-4	Personnel Interviewer's Survey	25	$20.00
P-5	Health Information Questionnaire	100	48.00
EA-4	Personnel Application Forms	25	24.00
ATS-6	Health History	1	3.00
			$95.00

Thank you for sending them to the Personnel Department in care of my attention within the next ten days. Please e-mail (youngblood@dcc.com) or telephone me if there will be any delay.

A check for one half the amount ($47.50) is enclosed, and the balance will be paid when the forms arrive.

Roberta Youngblood

ROBERTA YOUNGBLOOD, PERSONNEL OFFICER

Encl.: Check 2467 for $47.50

Order letters address the following five points.

1. **Description of the product or service.** Specify the name, model or stock number, quantity, color, weight, height, width, size, or any special features that separate one model from another (e.g., chrome as opposed to copper handles). Make your letter easy to read by itemizing when you order more than one product. Listing the products or materials in tabular form will set them apart and allow the seller an opportunity to check off each item as it is being prepared for shipment.

2. **Price of the product or service.** Indicate precisely the price per unit, per carload, or per carton, and then multiply that price by the number you are ordering. For example, ask for "twelve units @ $5 a unit." Do not put down the cost of one item ($5) for the dozen you are requesting. You will receive only one.

3. **Shipping instructions.** Do you want the product sent by an express carrier or priority mail? Specify any special handling instructions—Do not fold; Use hand stamp; Refrigerate or Pack in Dry Ice; Ship to the Production Department.

4. **Date needed.** Is there a rush date? Is there a date after which you do not want the order filled at all? Include this information in your letter.

5. **Method of payment.** Businesses with good credit standing are sent a bill. Individuals, however, may be required to pay at the time the order is placed. If so, are you enclosing a check or money order? Is the product to arrive COD (cash on delivery)? If you use a charge card, specify which card, and include your account number. Will you pay by e-cash (see pp. 631–652)? Will you be paying in installments? State how much you are including and when and how you intend to pay the balance.

Figure 6.1 shows a sample order letter written in the AMS format for letters.

Inquiry Letters

An inquiry letter asks for information about a product, service, publication, or procedure. Businesses frequently exchange such letters. As a customer, you too have occasion to ask in a letter, over e-mail, or through a fax for catalogs, names of stores in your town selling a special line of products, the price, size, and color of a particular object, or delivery arrangements. Businesses are eager to receive such inquiries and will answer then swiftly because they promise a future sale.

Figure 6.2 illustrates a letter of inquiry. Addressed to a real estate office managing a large number of apartment complexes, Michael Ortega's letter follows the three basic rules for an effective inquiry letter by

- stating exactly what information the writer wants
- indicating clearly why the writer must have this information
- specifying when the writer must have the information

FIGURE 6.2 A letter of inquiry.

Michael Ortega
403 South Main Street Kingsport, TN 37721-0217

March 2, 1999

Mr. Fred Stonehill
Property Manager
Acme Property Corporation
Main and Broadway
Roanoke, VA 24015-1100

Dear Mr. Stonehill,

States precise request

Explains need for information

Please let me know if you will have any two-bedroom furnished apartments available for rent during the months of June, July, and August. I am willing to pay up to $425 a month plus utilities. My wife, one-year-old son, and I will be moving to Roanoke for the summer so that I can take classes at Virginia Western Community College.

If possible, we would like to have an apartment that is within two or three miles of the college. We do not have any pets.

Specifies exact date when a reply is needed

I would appreciate hearing from you within the next two weeks. My e-mail address is mortega@aol.com or you can call me at home (606-555-8957) any evening from 6-10 P.M.

Offers to confer with reader

If you have any suitable vacancies, we would be happy to drive to Roanoke to look at them and give you a deposit to hold an apartment. Thanks for your help.

Sincerely yours,

Michael Ortega

Michael Ortega

Whenever you request information, be sure to supply appropriate stock and model numbers, pertinent page numbers, or exact descriptions. You might even clip and mail the advertisement describing the product you want. Vague or general letters delay a response to you. Had Michael Ortega written the following letter to Acme, he would not have helped his family move: "Please send me some information on housing in Roanoke. My family and I plan to move there soon." Such a letter does not indicate whether he wants to rent or buy, whether he is interested in a large or small apartment, furnished or unfurnished, where he would like to be located, or the rent he is able to pay. Similarly, a letter asking a firm to "send me all the information you have on microwave ovens" might bring back a detailed service manual or a firm but polite request that the writer be more specific when the writer only wanted prices of the top-selling models.

Special Request Letters

Special request letters make a special demand, not a routine inquiry. Among other things, these letters can ask a company for information that you as a student will use in a paper, an individual for a copy of an article or a speech, or an agency for facts that your company needs to prepare a proposal or sell a product. The person or company being asked for help stands to gain no financial reward for supplying this information; the only reward is the goodwill a response creates.

Make your request clear and easy to answer. Supply readers with an addressed, postage-paid envelope, an e-mail address, and fax and telephone numbers in case they have questions. But don't ask a company to fax a long document to you. It is discourteous to ask someone else to pay the fax charges for something you need.

Saying "please" and "thank you" will help you get the information you want. Also, do not expect your reader to write the paper or proposal for you. Asking for information is quite different from asking readers to organize and write it for you. Follow these seven points when asking for information in a special request letter.

1. State who you are and why you are writing.
2. Indicate clearly your reason for requesting the information.
3. Give the reader precisely and succinctly the questions you want answered. List, number, and separate the questions.
4. Specify exactly when you need the information. Allow sufficient time—at least three weeks.
5. Offer to forward a copy of your report, paper, or survey in gratitude for the help you were given.
6. If you want to reprint or publish the materials you ask for, indicate that you will secure whatever permissions are necessary. State that you will keep the information confidential if that is appropriate.
7. Thank the reader for helping.

Figure 6.3 gives an example of a letter that follows these guidelines.

Sales Letters: Some Preliminary Guidelines

A sales letter is written to persuade the reader to buy a product, try a service, support some cause, or participate in some activity. A sales letter can also serve as a method of introducing yourself to potential customers. No matter what profession you have chosen, knowing how to write a sales letter is an invaluable skill. There will always be times when you have to sell a product, a service, an idea, a point of view, or yourself!

You have undoubtedly received numerous sales letters from military recruiters, large, international companies, local merchants, charitable organizations, and campus groups. Home pages on the Internet constitute a special type of sales announcement, discussed in Chapter 8.

FIGURE 6.3 A special request letter.

234 Springdale Street
Rochester, NY 14618-0422
February 3, 2000

Ms. Victoria Lohrbach-Vitelli
Research Director
Creative Marketing Association
198 Madison Avenue
New York, NY 10016-0092

Dear Ms. Lohrbach-Vitelli:

I am a sophomore at Monroe College in Rochester, and I am preparing a term paper on the topic "Current Marketing Practices." As part of my research, I am writing an overview of marketing practices for the past ten years. Two of your publications would be of great help to me. Would you please send me the following pamphlets:

1. A History of Marketing in the United States (CMA 15)
2. Creative Marketing on the Internet (CMA 27)

I would appreciate receiving these materials by February 28 and will be pleased to send you a copy of my paper in late May. Creative Marketing Association will, of course, be fully cited in my bibliography.

Thank you for your assistance. I look forward to hearing from you. Should you have any questions or need to speak to me about my request, you can reach me at jkawatsu@golf.edu or at (716)555-4329.

Sincerely yours,

Julie Kawatsu

Julie Kawatsu

Because of the great volume of sales letters in the business world, the ones you write face a lot of competition. To write an effective sales letter that stands out and does its job, you have to do the following:

1. Identify and limit your audience. In deciding whom you want to reach, you will have to determine how many people are in your audience. Sometimes a sales letter is written to just one person (see Figure 6.4) or to hundreds of readers (see Figure 6.5). (However, Cory Soufas's sales letter in Figure 6.5 could also be addressed personally to individual readers if Workwell Software purchased a targeted audience mailing list.) Sales appeals on the Internet have the potential of reaching millions of readers. Do you want to write a letter to all the nurses in your state district nurses' association or only to the pediatric nurses at a particular hospital? Are you writing to all homeowners who purchased aluminum siding from your company in the last three years or just the customers in the last three months? Decide before you start writing.

2. Use reader psychology. If you think like a reader, you will find that the central question to ask yourself is: "What are we trying to do for you, our customer?" Ask yourself that question before you begin writing and you will be using effective reader psychology. The "you attitude" is essential to your sales letter. To persuade your readers to buy, support, or join, everything in your letter should be directed toward putting that goal in terms of helping and satisfying the reader. Appeal to your readers' emotions, health, security, comfort, or pocketbooks by focusing on the right issues (for instance, many companies make a point of informing buyers that their product research involves no animal testing). Consider what issues or attitudes are important to your audience, and shape your sales letters accordingly.

TECH NOTE

Companies can obtain mass mailing lists of target customers in a variety of ways. Many businesses use a grand opening or special sales event as an opportunity to have potential customers sign up for a free gift: the sign-up list then becomes a marketing list of motivated customers' names, addresses, and phone numbers. Firms use their home pages on the Internet not only to advertise but also to solicit information about a target audience—asking Net surfers to answer a brief questionnaire, fill out a form, or sign up for a tour or free gift. Once the potential customer clicks on "send," the company has information to include in a directory. Internet directories are also available from which your company can establish a client base. Finally, direct mail and marketing companies will, for a fee, supply you and your company with specific demographic data about a presorted, target audience, including age, income, occupation, and recent types of purchases. Some of these specialized firms will even supply you with a database of such information on CD-ROM. Always find out where such lists have been generated so you don't duplicate your company's other efforts or send sales messages to individuals who are definitely not part of your target audience.

3. Don't boast or be a bore. A sales letter is not written to provide elaborate explanations about a product. That kind of information comes after the sale. So save such detailed documentation for instruction booklets and warranties. Further, do not turn your sales letter into a glowing commendation of your company, or yourself. Either error—boasting about company merits or burying readers in technical details—causes you to lose sight of your audience's needs.

4. Use words that appeal to the readers' senses. To succeed, your sales letters must employ the most effective language possible. Use concrete, specific words instead of abstract, vague ones. Select words that appeal to the reader's senses. You will have a greater chance of selling the reader if he or she can hear, see, taste, or touch your product. As you draft and revise, find verbs that are colorful, that put the reader in the picture, so to speak. Try to make your *readers* the subject of most of your sentences. That way they can visualize themselves buying or using your product or service.

5. Be ethical. Be as objective with your evidence as you are with your language. Do everything you can to gain your readers' confidence. Avoid untruths, insincere flattery, exaggerations, false comparisons, and unsupported generalizations. Honesty is the best way to make a sale. If readers suspect you are playing false with them, they lose confidence in you and your company. If you do your homework about your product or your service, you will be able to give readers the honest, essential evidence they deserve and demand. Remember that your key selling strategy is winning the reader over to your point of view. (You might want to review pp. 23–27 in Chapter 1.)

The Four A's of Sales Letters

Successful sales letters follow a time-honored and workable plan; that is, each sales letter follows what can be called the "four A's":

1. gets the reader's *attention*
2. highlights the product's *appeal*
3. shows the customer the product's *application*
4. ends with a specific request for *action*

These four goals can be achieved in less than four or five paragraphs. Look again at Figure 6.5, a one-page letter in which these four parts are labeled for you. Holding a sales letter down to one page or less will keep the reader's attention. Television commercials and magazine ads provide useful models of this fourfold approach to the customer. The next time you see one of these variations of the "sales letter," try to identify the four A's.

Getting the Reader's Attention

Your opening sentence is crucial. That first sentence is bait on a hook. If you lose readers here, you will have lost them forever. A typical television commercial has thirty to sixty seconds to sell viewers; your first sentence has about two to five seconds to catch the readers' attention and prompt them to read on. It must show readers how their problems could be solved, their profits increased, or their pleasures enriched. The reader's attitude will be, "What's in this for me?" Tell them right away that you *can* increase their profits or happiness or decrease their problems or troubles.

Avoid an opening that is flat, vague, or lengthy. "I have great news for you" tells the readers nothing. They are more likely to toss your letter away if it does not contain something personally relevant in the first sentence. Select something that will appeal to their wallets, their emotions, or their chances to look better in the eyes of others. A letter for a pesticide that begins, "The cockroach could become the next endangered species if a California manufacturer has its way," offers some interesting, specific news. Keep your opening short, one or two sentences at most.

The following six techniques are a few of the many interest-grabbing ways to begin a sales letter. Each technique requires you to adapt it to your product or service.

1. Ask a question. Mention something that readers are vitally concerned about that is also relevant to your product or service. Look, for example, at the opening questions in Figures 6.4 and 6.5. Avoid such general questions as "Are you happy?" or "Would you like to make money?" Use more specific questions with concrete language. For example, an ad for a home-study training course in crime investigation and identification asks readers, "Are you promotable? Are you ready to step into a bigger job?" And an ad for Air Force Reserve Nursing asks nurses, "Are you looking for something 30,000 feet out of the ordinary?" These are intriguing questions relevant to the offers they introduce. Similarly, a sales letter beginning with, "Could you use $100?" zeroes in on one particular desire of the reader. Since your main goal is to persuade readers to continue reading your letter, ask a question that they will want to see answered.

2. Employ a "how to" statement. This is one of the most frequently used openers in a sales letter. Note that the opening sentence of Cory Soufas's sales letter in Figure 6.5 combines both a "how to" and a question approach. The reason for its success is simple—the letter promises to tell readers something practical and profitable. Here are some effective "how to" statements: "We can show you how to increase your plant growth up to 91%." "Here is how to save $500 on your next vacation." "This is how to provide nourishing lunches for less than eighty cents a person."

FIGURE 6.4 A sales letter soliciting a financial contribution.

ELMWOOD VOLUNTEER FIRE DEPARTMENT
Elmwood, Idaho 87549

To report a fire 911 **To volunteer 555-7878**

January 31, 1998

Mr. Alex B. Sutton
1453 North Prentiss Drive
Elmwood, ID 87549

Dear Mr. Sutton:

Do you want to know how to better protect your home from life-threatening fire? Support Elmwood Volunteer Fire Department. The safety and security of our homes depend on the proven ability of our Volunteer Fire Department.

Thanks to your support in the past, the Elmwood VFD purchased a new Powers V-10 pumping engine last year that offers state-of-the-art capabilities in firefighting. Most large city fire departments own this engine to protect their citizens. All of our volunteers have been trained to operate the Powers and have proven themselves many times in the past year. These volunteers work to guard you, and their invaluable service costs you, the taxpayer, nothing.

In 1997 alone the Elmwood volunteer firefighters logged over 1,000 hours in responding to 36 emergency calls, three of them within a few blocks of your home on Prentiss Drive. Insurance adjusters have estimated that without the Elmwood VFD more than 100 lives and more than $10 million in property might have been lost. One of your neighbors, Ms. Sarah Capsky, told the *Elmwood News*: "Our volunteer fire department saved my house, my three kids, and our two dogs. They deserve everyone's respect and gratitude."

To continue to protect you and your home, the Elmwood VFD needs your support. We cannot make it on our tax allocation from the county alone. In fact, tax dollars covered only the down payment for the Powers V-10. The balance of $145,000–nearly $110,000–must come from all of us in Elmwood.

Won't you take a minute to fill out the enclosed, postage-paid pledge card and return it by February 28? You can even authorize a monthly contribution from your checking account. At the end of our pledge drive I will send you a full financial report. Please help us protect Ms. Capsky's neighbors as if your life depended on it.

Respectfully,

Alice Sano

Alice Sano
Captain, Elmwood VFD

FIGURE 6.5 A sales letter sent to a business reader.

Workwell Software
3700 Stewart Ave., Chicago IL 60637-2210
Phone: (312)555-3720 Fax: (312)555-7601 Voice: (312)555-3232
Web: http://www.workwell.com

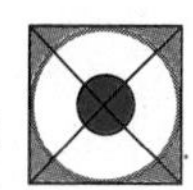

Dear Office Manager:

Arouses reader's interest with a question

Do you know how much money your company loses from one of the greatest dangers in the workplace-repetitive strain injury? Each year employers spend millions of dollars on employee insurance claims and the resulting decreased productivity because of back pains, fatigue, eye strain, and carpal tunnel syndrome.

Calls attention to the product's appeal

Workwell can solve your problems with its easy to use Exercise Program Software. This program will automatically monitor the time an employee spends at his or her computer and also measure keyboard activity. After each hour (or the specified number of keystrokes), the software will take your employees through a series of exercises that will help prevent carpal tunnel syndrome and strains and give you the assurance that you have taken necessary precautions to reduce a workplace hazard.

Shows specific application of the product

Workwell's Exercise Program Software is easy to use and will not interfere with busy schedules. Developed by a leading orthopedic surgeon, Dr. Anna Chang, each of the 27 exercises is demonstrated on screen with audio instructions. The entire program takes less than 3 minutes and can be performed at the employee's workstation. The software is available for Windows or Mac.

You can protect your employees for a fraction of the money you will spend on claims. For $1499.00, you can provide a networked version of this valuable software to all of your employees. And if you place your order within the next week, Workwell will supply you with free upgrades and maintenance for a year.

Ends with a call to action

To make sure that your employees are at the peak of their efficiency in a safe work environment, please call us at 1-800-555-WELL or contact us at http://www.workwell.com to order your Workwell Software today.

Thank you,

Cory Soufas

Cory Soufas
Manager, Sales

3. Compliment your reader. Appeal to the reader's ego. But remember that readers are not naive; they will be suspicious of false praise.

4. Offer a free gift. Often you can lure readers further into your letter by telling them they can save money or get the second product free or at half price. A local realtor tempts customers to see lots for sale with, "Enclosed is a coupon worth $25 in gas for your drive to Deer Run Trails Estates."

5. Introduce a comparison. Compare your product or service with conventional or standard products or procedures. For instance, a jacket firm told police officers that if they purchased a particular jacket, they were really getting three coats in one, because this product had a lining for winter and a covering used for greater visibility at night.

6. Announce a change. Link your sales offer to a current event that will directly affect your prospective customer. When the sales tax on cars was about to be increased by 3 percent in one state, an automobile dealer sent sales letters to potential buyers, alerting them to the implications of delaying their purchase: "The sales tax on new cars will jump a WHOPPING 3 percent effective the first of next month. You may not think that 3 percent will mean that much money, but on a new 2000 model that increase could cost you an extra $900." The sales letter continued: "Couldn't you use that money for something else, say, those extras you've always wanted, like a CD player or an extended warranty?"

Calling Attention to the Product's Appeal

Once you have aroused your reader's attention, introduce your product or service. Make it so attractive, so necessary, and so profitable that the reader will want to buy or use the product or service. Don't lose the momentum you have gained with your introduction by boring the reader with petty details, flat descriptions, elaborate inventories, or trivial boasts. Appeal to the reader's intellect, emotions, or both, while introducing the product. In Figure 6.5 Workwell Software's mass mailing letter to office managers appeals to their desire for greater employer productivity and improved safety. In Figure 6.4 the captain of a volunteer fire department urges citizens to help pay for a piece of fire equipment that might save their lives and property. Here is an emotional appeal by the Gulf Stream Fruit Company:

> Can you, when you bite into an orange, tell where it was grown? If it tastes better than any you have ever eaten . . . full of rich, golden flavor, brimming with juice, sparkling with sunshine . . . then you know it was grown here in our famous Indian River Valley where we have handpicked it, at the very peak of its flavor, just for your order.[1]

From a leading question, the sales letter moves to a vivid description of the product, the name of the supplier, and the customer's ability to recognize how special both the product and he or she are to the Gulf Stream Fruit Company. Using another kind of appeal, the manufacturers of Tree Saw show how their product can decrease customers' fears and increase their safety:

[1]"Gifts from Gulf Stream," Gulf Stream Fruit Company, Ft. Lauderdale, Fla. Reprinted by permission.

> The greatest invention since the ladder, and a lot safer. With the Tree Saw, you can cut branches thirty feet above the ground. No need to call in a tree surgeon. And you don't have to risk your life on a shaky ladder. Just toss the beanbag weight of the flexible Tree Saw over a branch. Pull on each end of the control rope to make a clean cut. Spring pruning is quick and safe with this perfect tool.[2]

Showing the Customer the Product's Application

The third part of your sales letter lets the readers know how and why the product is worthwhile for them. Here is the place where you give evidence of the value of what you are selling. You have to be careful, though, that you do not overwhelm readers with facts, statistics, detailed mechanical descriptions, or elaborate arguments. The emphasis is still on the reader's use of the product and not on the company that manufactures or sells it. Shifting the focus from your company to your prospective customer is essential to any sale.

Supplying the Right Evidence

What evidence best convinces readers about a product's or service's appeal?

- **descriptions** of the product or service that emphasize state-of-the-art design and construction, efficiency, convenience, usefulness, and economy.
- **special features** or changes in the product or service that make it more attractive—a greenhouse manufacturer stressed that in addition to using its structure just for growing plants, customers would also find it a "perfect sun room enclosure for year-round 'outdoor' activities, gardening, or leisure health spa."
- **testimonials,** or endorsements, from previous customers as well as from specialists. Rather than saying hundreds of people are satisfied with your product, get two or three of those happy customers to allow you to quote them in your letter. A large nursing home published residents' compliments about the food and care as testimony that it was a pleasant place to live. In Figure 6.4 Captain Sano persuasively uses Ms. Capsky's endorsement in the third and fifth paragraphs of her sales letter, and Cory Soufas in Figure 6.5 cites "a leading orthopedic surgeon" by name.
- **guarantees, warranties, services, or special considerations** that will make the customer's life easier or happier—loaner cars, free home deliveries, ten-day trial period, thirty free hours of Internet access.

Do I Mention Costs?

You may be obligated to mention costs in your letter. But postpone discussion of them until the reader has been shown how appealing and valuable the product is. Of course, if price is a key selling point mention it early in the letter. Readers will react more favorably to costs after they have seen the reasons why the product or service is useful.

As a general rule, do not bluntly state the cost. Relate prices, charges, or fees to the benefits provided by the services or products to which they apply. Cus-

---

[2]Reprinted courtesy Green Mountain Products, Inc., Norwalk, Conn.

tomers then see how much they are getting for their money. An electric blanket ad tells customers that it costs only four cents a night to be warm and comfortable. That sounds more inviting than just listing the cost of the blanket as $40. A dealer who installs steel shutters does not tell readers the exact price of the product, but does indicate that they will save money by buying it: "Virtually maintenance free, your Reel Shutters also offer substantial savings in energy costs by reducing your loss through radiation by as much as 65% . . . and that lowers your utility bills by 35%." One charitable organization appeals to readers' generosity by asking them to contribute $200 a year, which amounts to less than sixty cents a day, or less than the price of a cup of coffee.

Ending with a Specific Request for Action

This last section of your letter is vital. If the reader ignores your request for action, your letter has been written in vain. Tell readers exactly what you want them to do, by when, and make it easy for them to

- fill out a postcard ordering your product
- send for a brochure
- come into your store
- take a test drive
- participate in a meeting
- fill out a pledge card (as in Figure 6.4)
- respond via the Internet (as in Figure 6.5)
- sign an order blank
- fill out an enclosed stamped envelope for their convenience

As with price, link the benefits the customers will receive to their responses. "Respond and be rewarded" is the basic message of the last section of your letter. Note that in Figure 6.5 the call to action is made in the next-to-last paragraph, followed by a brief reminder of the significance of the action, and urges the reader to act immediately in order to take advantage of the free upgrades and maintenance. In Figure 6.4 the call to residents to contribute to the Elmwood Fire Department is in the last sentence.

Customer Relations Letters

Much business correspondence deals explicitly with establishing and maintaining friendly working relations. Such correspondence, known as **customer relations letters,** sends readers good news or bad news, acceptances or refusals. Good news tells customers that

- you have the product or service they want at a reasonable price
- you agree with them about a problem they brought to your attention
- you are solving it exactly the way they want
- you are approving their loan
- you are grateful to them for their business

Thank-you letters, congratulations letters, or adjustment letters saying "Yes" are examples of good news messages.

Bad news messages inform readers that

- you do not like the work or equipment they have sold you
- you do not have the equipment or service they want or you cannot provide it at the price they want to pay
- you are rejecting a proposal they offered
- you are denying someone further use of a facility
- you cannot refund their purchase price or perform a service again as they requested
- you are raising their rent or not renewing their lease
- you want them to pay what they owe you now

Bad news messages often come to readers through complaint letters, adjustment letters that say "No," or collection letters.

Diplomacy and Reader Psychology

Regardless of the type of news—good or bad—you have to convey to readers, you need to be a diplomatic and persuasive writer. Customer relations letters show how you and your company regard the people with whom you do business. These letters should reveal your sensitivity to their needs. After all, you want to keep them as customers and you never know when you may work with them again. Writing effective customer relations letters requires skill in human relations and reader psychology. The first lesson to learn is that you cannot look at your letter only from your (the writer's) perspective. You have to see the letter from the reader's perspective and anticipate the reader's needs and reactions. Ask yourself how you would feel if you received the same letter. What would be your view of the writer and the writer's company?

The Customers Always Write

As you read this section, keep in mind the two basic principles captured in the words "the customers always write."

1. Customers will write about how they would like to be or have been treated—to thank, to complain, to request an explanation.
2. Customers have certain rights that you must respect in your correspondence with them. They deserve a prompt and courteous reply, whether or not they are correct. If you refuse their request, they deserve to know why; if they owe you money, you should give them an opportunity to explain and a chance, up to a point, to set up a payment schedule.

Planning Your Customer Relations Letters

Whether you are sending good news or bad, you have to determine what to say and how to say it. As with your other written work, do some preliminary planning. Outline for a few minutes to find your ideas. Your outline does not have to be formal or even neatly written—a few scribbles sometimes will be enough to get you started. By outlining, you will save (not lose) time, for you can identify your main points, exclude unnecessary or unimportant ones, and avoid the risk of forgetting something essential.

Fortified by your outline, you will feel more confident as you draft your letter. In the process of drafting and revising that letter, consider whether your reader will bristle at or accept the words you use. Your choice of words will determine the success or failure of your letter. You might want to review the discussions on tone and the "you attitude" (pp. 154–160).

Being Direct or Indirect

Your message, tone, and knowledge of your reader are essential ingredients in a successful customer relations letter. But success involves something more than knowing how to use the right words to get across your main ideas. It is also crucial to know where and how to start, and especially, where to present your main point. Not every customer relations letter starts by giving the reader the writer's main point, judgment, conclusion, or reaction. *Where you place your main idea is determined by the type of letter you are writing.* Good news messages require one tactic; bad news, another.

Good News Message

If you are writing a good news letter, use the direct approach. Start your letter with the welcome, pleasant news that the reader wants to hear. Don't postpone this opportunity to put your reader in the right frame of mind. Then provide any relevant supporting details, explanations, or commentary. Being direct is advantageous when you have good news to convey.

Bad News Message

If you have bad news to report, do *not* open your letter with it. Be indirect. Prepare your reader for the bad news. Do everything you can to keep the tension level down. If you throw the bad news at your reader right away, you run the risk of jeopardizing the goodwill you want to create and sustain. Again, put yourself in the reader's place. Consider how you would react to a letter that began, "We regret to inform you that . . . ," "Your order cannot be filled," or "Your application for a loan has been denied." Having been denied, disappointed, or even offended in a first sentence or paragraph, the reader is not likely to give you his or her attentive cooperation thereafter.

Notice how A. J. Griffin's bad news letter in Figure 6.6 curtly starts off with the bad news of a rent increase. Receiving such a letter, the owner of Flowers by Dan could certainly not be blamed for looking for a new place of business. Or if he did pay the increase, Griffin's letter would hardly ensure that Mr. Sobol would remain a happy tenant at River Road. Griffin was too direct when he should have been diplomatically indirect. He did not consider his reader's reaction; all he was concerned about was delivering his message.

Compare the curt version of Griffin's letter in Figure 6.6 with his revised message in Figure 6.7. In the revised version, Griffin begins tactfully with a few pleasant, positive words designed to put his reader in a good frame of mind about the

FIGURE 6.6 An ineffective bad news letter.

River Road Mall

December 1, 1999

Mr. Daniel Sobol
Flowers by Dan
Lower Level
River Road Mall

Dear Mr. Sobol:

This is to inform you of a rent increase. Starting next month your new rent will be $2500.00, resulting in a 15 percent increase.

Please make sure your January rent check includes this increase.

Sincerely yours,

A J Griffin

A. J. Griffin
Manager

Canton, Ohio
(216)555-6700
e-mail:
ajg@rrdmall.com

FIGURE 6.7 A diplomatic revision of the bad news letter in Figure 6.6.

River Road Mall

December 1, 1999

Mr. Daniel Sobol
Flowers by Dan
Lower Level
River Road Mall

Dear Mr. Sobol:

It has been a pleasure to have you as a tenant at the Mall for the past two years, and we look forward to serving you in the future.

Over the last two years we have experienced a dramatic increase in costs at River Road Mall for security, maintenance, pest control, utilities, insurance, and taxes. Last year we tried to absorb these increases and so did not have to raise your rent. Unfortunately, we find we cannot do it again for 2000 and so regretfully we must increase your rent by 15 percent, to $2,500, effective January 1, 2000.

We do not like to raise rents, and we know that you do not like it either. But we also know that you do not want us to compromise on the quality of service that you and your customers expect and deserve from River Road Mall.

Please let us know how we can assist you in the future. We wish you a very successful and profitable 2000. If you have any questions, please call or visit my office.

Cordially,

A J Griffin

A. J. Griffin
Manager

Canton, Ohio
(216)555-6700
e-mail:
ajg@rrdmall.com

management of River Road Mall. Then Griffin gives some background information that the owner of Flowers by Dan can relate to. A businessperson himself, Mr. Sobol doubtless has experienced some recent increases in his own costs. Griffin makes one more attempt to encourage Sobol to recall his good feelings about the Mall—last year they did not raise rents—before introducing the bad news of a rent increase. And even after giving the bad news, Griffin softens the blow by saying that the Mall knows it is bad news. Griffin's tactic here is to defuse some of the anger that Sobol will inevitably feel. Griffin then ends on a positive, upbeat note: a prosperous future for Flowers by Dan.

Follow-Up Letters

A follow-up letter is sent by a company after a sale to thank the customer for buying a product or using a service and to encourage the customer to buy more products and use more services in the future. A follow-up letter is, therefore, a combination thank-you note and sales letter. The letter in Figure 6.8 is sent to customers soon after they have purchased an appliance and offers them the option of a continued maintenance policy. The letter in Figure 6.9 shows how an income tax preparation service attempts to obtain repeat business. Both of these letters follow helpful guidelines.

1. They begin with a brief and sincere expression of gratitude for having served the customer.
2. They discuss the benefits (advantages) already known to the customer. Then, they transfer the company's dedication to the customer from the product or service already sold to a new or continuing sales area.
3. They end with a specific request for future business.

Occasionally, a follow-up letter is sent to a good customer who, for some reason, has stopped doing business with the company. Perhaps the customer has closed an account of long standing, no longer comes to the store, discontinued a subscription, or fails to send in an order for a product or service. Such a follow-up letter should try to find out why the customer has stopped doing business and to persuade that customer to resume business dealings. Study the letter in Figure 6.10, in which Jim Margolis first politely inquires whether Mr. Janeck has experienced a problem and then urges him to come back to the store.

Complaint Letters

Each of us, either as consumers or businesspeople, at some time has been frustrated by a defective product, inadequate service, or incorrect billing. Usually our first response is to write a letter dripping with juicy insults. But a hate letter rarely gets results and can in fact hurt the writer and create an unfavorable image of the company being represented. A complaint letter is a delicate one to write.

Establishing the Right Tone

A complaint letter is written for more reasons than just blowing off steam. You want some specific action taken. By adopting the right tone, you increase your

FIGURE 6.8 A follow-up letter to sell a maintenance agreement.

100 Walden Parkway
Denver, Colorado 80203-4296
(303) 555-9681
e-mail cmorrow@dac.com

August 9, 1998

Mr. John H. Abbott
3715 Mayview Drive
Cottage Grove, MN 53261-1852

Dear Mr. Abbott:

We are delighted that you have purchased a Dynamic appliance. To help ensure your satisfaction, this appliance is backed by a Dynamic warranty. At the same time, we realize that you bought the appliance to serve you not just for the period covered by the warranty but for many years to come. That's why purchasing a Dynamic Maintenance Agreement at this time is one of the wisest investments you can make.

A Dynamic Maintenance Agreement provides savings benefits many cost-conscious customers want and look for today. It helps extend the life of your appliance through an annual, on-request maintenance check-up. And if you need service, it provides for as many service calls as necessary for repairs due to normal use–at no extra charge to you.

All this coverage is now available at a special introductory price of $55 a year. This price includes the warranty coverage you have remaining.

Please act now by filling out and returning the enclosed form.

Sincerely,

Carole Morrow

Carole Morrow
Sales Representative
Extension 285

Encl.

FIGURE 6.9 A follow-up letter to encourage repeat business.

Taylor Tax Service
Highway 10
North Jennings, TX 06324
phone (888) 555-9681 fax (888) 555-9670

December 1, 1999

Ms. Laurie Pavlovich
345 Jefferson St.
Jennings, TX 78326

Dear Ms. Pavlovich:

Thank you for using our services in February of this year. We were pleased to help you prepare your 1998 federal and state income tax returns. Our goal is to save you every tax dollar to which you are entitled. If you ever have questions about your return, we are open all year long to help you.

We are looking forward to serving you again next year. Some new federal tax laws, which go into effect January 1, will change the types of deductions you can declare. These changes might appreciably increase your refund. Our consultants know these new laws and are ready to explain them to you and apply them to your return.

Another important tax matter influencing your 1999 returns will be any losses you may have suffered because of the hailstorms and tornadoes that hit our area three months ago. Our consultants are specially trained to assist you in filing proper damage claims with your federal and state returns.

To make using our services even easier, we can help you file your tax return electronically, to speed up any refund you are entitled to. Please call us at 555-9681 as soon as you have received all your forms in order to set up an appointment. We are waiting to serve you any day of the week from 9:00 A.M. to 9:00 P.M.

Sincerely yours,

TAYLOR TAX SERVICE

J. P. Sanchez

J. P. Sanchez
Manager

FIGURE 6.10 A follow-up letter to maintain customer goodwill.

BROADWAY CLEANERS

April 6, 2000

Mr. Edward Janeck
34 Brompton Lane
Apartment 143
Baltimore, MD 21227-0102

Dear Mr. Janeck:

Thank you for allowing us to take care of your cleaning needs for more than three years now. It has been our pleasure to see you in the store each week and to clean your shirts, slacks, and coats to your satisfaction. Since you have not come in during the last month, we are concerned that in some way we have disappointed you. We hope not, because you are a valuable customer whose goodwill we do not want to lose.

If there is something wrong, please tell us about it. We welcome any suggestions on how we can serve you better. Our goal is to have a spotless reputation in the eyes of our customers.

The next time you need your garments cleaned, won't you please bring them to us, along with the enclosed coupon worth $10 on your next bill? We look forward to seeing you again–soon.

Cordially,

Jim Margolis

Jim Margolis, Manager

Encl. coupon

Broadway at Davis Drive Baltimore, Maryland 21228-6210 555-1962

chances of getting what you want. Do not call the reader names, hurl insults, or refuse to do business with the company again. Register your complaint courteously and tolerantly. Companies want to be fair to you in order to keep you as a satisfied customer and correct defective products so that other customers will not be inconvenienced. The "you attitude" is especially important here to maintain the reader's goodwill.

An effective complaint letter can be written by an individual consumer or by a company. Figure 6.11 shows Michael Trigg's complaint about a defective fishing reel; Figure 6.12 expresses a restaurant's dissatisfaction with an industrial dishwasher.

Writing an Effective Complaint Letter

To increase your chances of receiving a speedy settlement, follow these five steps in writing your letter of complaint.

1. Begin with a detailed description of the product or service. Give the appropriate model and serial numbers, size(s), quantity, and color. Specify check and invoice numbers. Indicate when and where (specific address) you purchased it, and also how much warranty time remains. Indicate if you are returning the product to the company and how you are sending it—U.S. mail, UPS, through a sales representative, or the like. If you are complaining about a service, give the name of the company, the date of the service, the personnel providing it, and their exact duties.

2. State exactly what is wrong with the product or service. Precise information will enable the reader to understand and act on your complaint.

- How many times did the machine work before it stopped?
- What parts were malfunctioning?
- What parts of a job were not done or were done poorly?
- When did all this happen?

Stating that "the brake shoes were defective" tells very little about how long they were on your car, how effectively they may have been installed, or what condition they were in when they ceased functioning safely. Reach some conclusion, even if you qualify your remarks with words like "apparently," "possibly," or "seemingly" when you describe the difficulty.

3. Briefly describe the inconvenience you have experienced. In this section of the complaint letter, show that your problems were directly caused by the defective product or service. To build your case, give precise details about the time and money you lost. Don't just say you had "numerous difficulties." If you purchased a calculator and it broke down during a mathematics examination, say so (but do not blame the calculator company if you failed the course). Did you have to pay a mechanic to fix your car when it was stalled on the road, did you have to take time away from your other chores to clean up a mess made by a leaky new washing machine, or did you have to buy a new fax machine or modem?

FIGURE 6.11 A complaint letter from a consumer.

17 Westwood
Magnolia, MA 02171

September 15, 1998

Mr. Ralph Montoya
Customer Relations Department
Smith Sports Equipment
P.O. Box 1014
Tulsa, OK 74109-1014

Dear Mr. Montoya:

On August 31, 1998, I purchased a Smith reel, model 191, at the Uni-Mart Store on Marsh Avenue in Magnolia. The reel sold for $54.95 plus tax. Since the reel is not working effectively, I am returning it to you under separate cover by first-class mail.

I had made no more than five casts with the reel when it began to malfunction. The button that releases the spool and allows the line to cast will not spring back into position after casting. In addition, the gears make a grinding noise when I try to retrieve the line. Because of these problems, I was unable to continue my participation in the Gloucester Fishing Tournament.

I am requesting that a new reel be sent to me free of charge in place of the defective one I returned. I would also like to know what was wrong with this defective reel.

I would appreciate your handling my claim within the next two weeks, if at all possible.

Sincerely yours,

Michael Trigg

Michael Trigg

FIGURE 6.12 A complaint letter faxed from a business.

The Loft
Cameron and Dale
Sunnyside, California 91793-4116
213-555-7500

June 17, 1998

Priscilla Dubrow
Customer Relations Department
Superflex Products
San Diego, CA 93141-0808

Dear Ms. Dubrow:

On September 15, 1997, we purchased a Superflex industrial dishwasher, model 3203876, at the Hillcrest store at 3400 Broadway Drive in Sunnyside, for $3000. In the last three weeks, our restaurant has had repeated problems with this machine. Three more months of warranty remain on the dishwasher.

The machine does not complete a full cycle; it stops before the final rinsing and thus leaves the dishes dirty. It appears that the cycle regulators are not working properly because they refuse to shift into the next necessary gear. Attempts to repair the machine by the Hillcrest crew on June 3, 10, and 16 have been unsuccessful.

The Loft has been greatly inconvenienced. Our kitchen team has been forced to sort, clean, and sanitize utensils, dishes, pans, and pots by hand, resulting in additional overtime. Moreover, our expenses for proper detergents have increased.

We want your main office to send another repair crew at once to fix this machine. If your crew is unable to do this, we want a discount worth the amount of the warranty life on this model to be applied to the purchase of a new Superflex dishwasher. This amount would come to $600, or 20 percent of the original purchase price.

So that our business is not further disrupted, we would appreciate your resolving this problem within the next week.

Sincerely yours,

Emily Rashon

Emily Rashon
Manager

Browse our menu, which changes daily, at www.theloft.com

4. Indicate precisely what you want done. Do not simply write that you "want something done," that "adequate measures must be taken," or that "the situation should be corrected." State that you want

- your purchase price refunded
- your model repaired or replaced
- a completely new repair crew provided
- an apology from the company for discourteous treatment

If you are asking for damages, state your request in dollars and cents, and include a copy of any bills documenting your expenses related to the problem. Perhaps you had to rent a car, were forced to pay a janitorial service to clean up, or had to rent equipment at a higher rate because the company did not make its deliveries as promised.

5. Ask for prompt handling of your claim. Ask that an answer be provided to any question you may have (such as finding out where calls came from that you were billed for but did not make). And ask that your claim be handled as quickly as possible. You might even specify a reasonable time by which you want to hear from the writer or need the problem fixed.

Adjustment Letters

Adjustment letters respond to complaint letters by telling customers dissatisfied with a product or service how their claim will be settled. Adjustment letters should reconcile the differences that exist between a customer and a firm and restore the customer's confidence in that firm.

The Importance of Complaint Letters to a Business

Rather than ignoring or quarreling with complaint letters, most companies view answering them as good for business. Many large firms maintain separate claims and adjustment departments just to respond to disappointed customers. By writing to complain about a product or service, the customer alerts your company to a problem that can be remedied to avoid similar complaints in the future. Customers who have taken the time to put complaints in writing obviously want and deserve a reply. If you do not answer the customer's letter politely, you may lose a lot of business—not just the customer's business, but also that of his or her friends, family, and associates, who will all have been told about your discourtesy.

How to (and Not to) Write an Adjustment Letter

An effective adjustment letter requires diplomacy; be prompt, courteous, and decisive. Do not brush the complaint aside in hopes that it will be forgotten. Investigate the complaint quickly and determine its validity by checking previous correspondence, warranty statements, guarantees, and your firm's policies on merchandise and service. In some cases you may even have to send returned damaged merchandise to your firm's laboratory to determine who is at fault.

A noncommittal letter signals to the customer that you have failed to investigate the claim or are stalling for time. Do not resort to vague statements like the following:

- We will do what we can to solve your problems as soon as possible.
- A company policy prohibits our returning your purchase price in full.
- Your request, while legitimate, will take time to process.
- We will act on your request with your best interest in mind.
- While we cannot now determine the extent of an adjustment, we will be back in touch with you.

Customers want to be told that they are right; if they cannot get what they request, they will demand to know why, in the most explicit terms. At the other extreme, do not overdo an apology by agreeing that the company is "completely at fault," that "such shoddy merchandise is inexcusable," or that "it was a careless mistake on our part." An expression of regret need not jeopardize all future business dealings. If you make your company look too bad, you risk losing the customer permanently. When you comply with a request, a begrudging tone will destroy the goodwill created by your refund or replacement.

Adjustment Letters That Tell the Customer "Yes"

If investigation reveals the customer's complaint to be valid, you must write a letter saying "Yes, you are right; we will give you what you asked for." Such a letter is easy to write if you remember a few useful suggestions. As with a good news message, start with the favorable news the customer wants to hear and that will put him or her in a positive frame of mind to read the rest of your letter. Also, let the customer know that you sincerely agree with them—don't sound as if you are reluctantly honoring their request. For example, if your airline lost or misplaced luggage, apologize to the customer before you offer a settlement.

The two examples of adjustment letters saying "Yes" show you how to write this kind of correspondence. The first example, Figure 6.13, says "Yes" to Michael Trigg's letter in Figure 6.11. You might want to reread the Trigg complaint letter to see what problems Ralph Montoya faced when he had to write to Mr. Trigg. The second example of an adjustment letter that says "Yes" is in Figure 6.14. It responds to a customer who has complained about an incorrect billing.

Writing a "Yes" Letter

The following four steps will help you write a "Yes" adjustment letter.

1. Admit immediately that the customer's complaint is justified and apologize. Briefly state that you are sorry and thank the customer for writing to inform you.

2. State precisely what you are going to do to correct the problem. Let the customer know that you will

- cancel a bill
- return a damaged camera in good working order

FIGURE 6.13 An adjustment letter saying "Yes."

Smith Sports Equipment
P.O. Box 1014
Tulsa, Oklahoma 74109-1014
Phone (918)555-0164 Fax (918)555-0170

September 21, 1998

Mr. Michael Trigg
17 Westwood
Magnolia, MA 02171

Dear Mr. Trigg:

Thank you for alerting us in your letter of September 15 to the problems you had with one of our model 191 spincast reels. I am sorry for the inconvenience the reel caused you. A new Smith reel is on its way to you.

We have examined your reel and found the problem. It seems that a retaining pin on the button spring was improperly installed by one of our new soldering machines on the assembly line. We have thoroughly inspected, repaired, and cleaned this soldering machine to eliminate the problem. Our company has been making quality reels since 1955. We hope that your new Smith reel brings you years of pleasure and many good catches, especially next year at the Gloucester Fishing Tournament.

We appreciate your business and look forward to serving you again.

Respectfully,

SMITH SPORTS EQUIPMENT

Ralph Montoya

Ralph Montoya, Manager
Customer Relations Department

FIGURE 6.14 An adjustment letter saying "Yes."

Brunelli Motors

Route 3A Giddings, Kansas 62034-8100 (913)555-1521
Drive a new *Phantom* at our Web site
http://www.brunelli.com

October 6, 1998

Ms. Kathryn Brumfield
34 East Main
Giddings, KS 62034-1123

Dear Ms. Brumfield:

We appreciate your notifying us, in your letter of September 30, about the problem you experienced regarding warranty coverage on your new Phantom Hawk GT. The bills sent to you were incorrect, and I have already canceled them. Please accept my apologies. You should not have been charged for a shroud or for repairs to the damaged fan and hose, since all of these parts, and labor on them, are covered by warranty.

The problem was the result of an error in the way the charges were listed. Our firm has begun using a new computer system to give customers better service, and the mechanic apparently entered the wrong code for your account. I have instructed our mechanics to double-check code numbers before submitting them to the Billing Department. We hope that this policy will help us serve you and our other customers more efficiently.

We value you as a customer of Brunelli Motors. When you are ready for another Phantom, I hope that you will once again visit our dealership.

Sincerely yours,

Susan Chee-Saafir

Susan Chee-Saafir
Service Manager

- repaint a room
- enclose a free pass
- provide a complimentary dinner
- give the customer credit toward another purchase
- upgrade software

Do not postpone the good news the customer wants to hear. The rest of your letter will be much more appreciated and convincing. In Figure 6.13 Michael Trigg is told that he will receive a new reel; in Figure 6.14 Kathryn Brumfield is informed that she will not be charged for parts or service.

3. Tell customers exactly what happened. They deserve an explanation for the inconvenience they suffered. Note that the explanations in Figures 6.13 and 6.14 give only the essential details; they do not bother the reader with side issues or petty remarks about who was to blame. Don't threaten to fire one of your employees because of a customer's problem. Assure customers that the mishap is not typical of your company's operations. While your comments should not shift the blame, your letter should center on the unusual reason or circumstance for the difficulty. Avoid promising, however, that the problem will never recur. Not only is such a guarantee unnecessary, but also keeping it may be beyond your control.

4. End on a friendly, and positive, note. Do not remind customers of the trouble they have gone through. Leave them with a good feeling about your company. Say that you are looking forward to seeing them again, that you will gladly work with them on any future orders, or that you can always be reached for questions.

Adjustment Letters That Tell the Customer "No"

Writing to tell customers "No" is obviously more difficult than agreeing with them. You are faced with the sensitive task of conveying bad news, while at the same time convincing the reader that your position is fair, logical, and consistent. Do not accuse or argue.

What Not to Say

Avoid remarks such as the following that blame, scold, or remind customers of a wrongdoing and are likely to cost you their business.

- You obviously did not read the instruction manual.
- Our records show that you purchased the set after the policy went into effect.
- The company policy plainly states that such refunds are unallowable.
- You were negligent in running the machine.
- You claim that our word processor was poorly constructed.
- Your error, not our merchandise, is to blame.
- You must be mistaken about the merchandise.
- As any intelligent person could tell, the switch had to be "off."
- Your complaint is unjustified.

How to Say "No" Diplomatically

The following five suggestions will help you say "No" diplomatically. Practical applications of these suggestions can be found in Figures 6.15 and 6.16. Contrast the refusal of Michael Trigg's complaint in Figure 6.15 with the favorable response to it in Figure 6.13.

1. Thank customers for writing. Make a friendly start by putting them in a good frame of mind. The letter writers in Figures 6.15 and 6.16 thank the writers for bringing the matter to their attention. As with other bad news letters, never begin with a refusal. You need time to calm and convince customers. Telling them "No" ("We regret to inform you") in the first sentence or two will negatively color their reactions to the rest of the letter. Also, never begin letters with "I was surprised to learn that you found our product defective (*or* our service inefficient)" or "We cannot understand how such a problem occurred. We have been in business for years, and nothing like this has ever happened." Such openings put customers on the defensive. Use the indirect approach discussed on pages 198–201.

2. State the problem so that customers realize that you understand their complaint. You thereby prove that you are not trying to misrepresent or distort what they have told you.

3. Explain what happened with the product or service before you give customers a decision. Provide a factual explanation to show customers that they are being treated fairly. Convince them of the logic and consistency of your point of view. Rather than focusing on the customer's mishandling of merchandise or failure to observe details of a service contract, state the proper ways of handling a piece of equipment or the terms outlined in an agreement. For instance, instead of writing "By reading the instructions on the side of the paint can, you would have avoided the streaking condition that you claim resulted," tell the customer that "Hi-Gloss Paint requires two applications, four hours apart, for a clear and smooth finish." In this way you remind customers of the right way of applying the paint without pointing an accusing finger at them. Note how the explanations in Figures 6.15 and 6.16 emphasize the right way of using the product.

4. Give your decision without hedging. Do not say "Perhaps some type of restitution could be made later," or "Further proof would have been helpful." Indecision will infuriate customers who believe that they have already presented a sound, convincing case. Never apologize for your decision. Avoid using the words *reject, claim,* or *grant. Reject* is too harsh and impersonal. *Claim* implies your distrust of the customer's complaint and suggests that questionable differences of opinion remain. *Grant* signals that you have it in your power to respond favorably but decline to do so; a grant is the kind of favor a ruler might give a subject. Instead, use words that reconcile.

5. Leave the door open for better and continued business. Whenever possible, help customers solve their problem by offering to send them a new product or part, and quote the full sales price. Note how the second-to-last paragraphs of the letters in Figures 6.15 and 6.16 do this diplomatically.

FIGURE 6.15 An adjustment letter saying "No."

Smith Sports Equipment
P.O. Box 1014
Tulsa, Oklahoma 74109-1014
Phone (918)555-0164 Fax (918)555-0170

September 21, 1998

Mr. Michael Trigg
17 Westwood
Magnolia, MA 02171

Dear Mr. Trigg:

Thank you for writing to us on September 15 about the trouble you experienced with our model 191 spincast reel. We are sorry to hear about the difficulties you had with the release button and gears.

We have examined your reel and found the trouble. It seems that a retaining pin on the button spring was pushed into the side of the reel casing, thereby making the gears inoperable. The retaining pin is a vital yet delicate part of your reel. In order to function properly, it has to be pushed gently. Since the pin was not used in this way, we are not able to refund your purchase price.

We will be pleased, however, to repair your reel for $9.98 and return it to you for hours of fishing pleasure. Please let us know your decision.

I look forward to hearing from you.

Respectfully,

SMITH SPORTS EQUIPMENT

Ralph Montoya

Ralph Montoya, Manager
Customer Relations Department

FIGURE 6.16 An adjustment letter saying "No."

Health Air, Inc.
4300 Marshall Drive
Salt Lake City, Utah 84113-1521
(801)555-6028

August 19, 1998

Ms. Denise Southby, Director
Bradley General Hospital
Bradley, IL 60610-4615

Dear Ms. Southby:

Thank you for your letter of August 10 explaining the problems you have encountered with our Puritan Bennett MAII ventilator. We were sorry to learn that you could not get the high-volume PAO_2 alarm circuit to work.

Our ventilator is a high-volume, low-frequency machine that can deliver up to 40 ml. of water pressure. The ventilator runs with a center of gravity attachment on the right side of the diode. The trouble you had with the high oxygen alarm system is due to an overload on your piped-in oxygen. Our laboratory inspection of the ventilator you returned indicates that the high-pressure system had blown a vital adaptor in the machine. Our company cannot be responsible for any overload caused by an oxygen system. We cannot, therefore, send you a replacement ventilator free of charge. Your ventilator is being returned to you by National Express.

We would, however, be glad to send you another model of the adaptor, which would be more compatible with your system, as soon as we receive your order. The price of the adaptor is $600, and our factory representative will be happy to install it for you at no charge. Please let me know your decision.

We welcome the opportunity to assist you in providing quality health care at Bradley General.

Sincerely yours,

R. P. Gifford

R. P. Gifford
Customer Service Department

Refusal of Credit Letters

A special set of bad news letters deals with a company refusing credit to an individual or another company. Writing such a letter requires a great deal of sensitivity. You want to be clear and firm about your decision; at the same time, you do not want to alienate the reader and risk losing his or her business in the future.

How to Say "No"

1. When you refuse someone credit, begin on a positive—not a negative—note. Find something to thank the reader about; starting with something positive will make the bad news easier to take. Compliment the reader's company or previous good credit achievements (if known); certainly express gratitude to the individual for wanting to do business with your company.

2. In a second paragraph provide a clear-cut explanation of why you must refuse the request for credit, but base your explanation on facts, not personal shortcomings or liabilities. Appropriate reasons to cite for a refusal of credit include

a. reminding the reader of a lack of business experience or prior credit
b. pointing out that the individual or company is "overextended" and needs more time to pay off existing obligations
c. calling attention to current unfavorable or unstable financial conditions
d. indicating that an order is too large to process without some prepayment
e. noting that a company lacks the equipment or personnel to do the business for which they are seeking credit

3. Conclude your letter on a positive note. Encourage the reader to reapply when business conditions have improved or when the reader's firm is in a better financial position. Make an attempt to keep the reader as a potential customer, eager to try you again. Figure 6.17 illustrates an effective letter that denies credit, following the organizational plan just discussed.

Writing About Credit to an ESL Reader

Writing a letter denying credit to an ESL reader requires double tact. As you saw on pages 163–170, you have to consider the cultural expectations of such an audience and use easily understood "international English." Compare the inappropriate refusal letter contained in Figure 6.18 with the far more diplomatic, and more acceptably worded, letter in Figure 6.19. The letter in Figure 6.18 is rude, uses words an ESL reader may not understand ("expedited," "herewith"), and does not encourage future business dealings with Consolidated Plastics.

The diplomatic letter in Figure 6.19 follows the guidelines for effective communication with ESL readers and adheres to the suggestions for denying credit. In Figure 6.19 Emma Corson compliments her reader and his firm, expresses an interest in doing business with Mendson SA, and helps her reader to understand how Consolidated's credit policy can even help Mendson in the future.

FIGURE 6.17 An effective letter refusing credit.

West Coast Credit, Inc.
4800 Ridge Road
Los Angeles, CA 91666
Phone (914)555-3500 FAX(914)555-4323

October 19, 1998

Mr. Otto L. King
Sunshine Interiors
8235 Mimosa Highway
Vinedale, CA 92004

Dear Mr. King:

Begins on a positive note

We appreciate your interest in wanting to do business with West Coast Credit. It is always gratifying to our institution to see a store like yours open in an expanding community like Vinedale.

Denies credit but explains why

In reviewing your credit application, we checked into the business history and credit references you supplied. We also called your local credit bureau. While we found nothing negative in your credit history, we did determine that for a business of your size you already have reached a maximum level of indebtedness. For that reason, we believe that it would not be the best time to extend your credit line.

Encourages reader to reapply

We would, however, encourage you to resubmit your application in six to eight months. By that time we hope that the growing market in your area will profitably allow you to take on additional credit lines. In the meantime, we wish you every success.

Cordially,

B. Rimes-Assante

B. Rimes-Assante
Manager

FIGURE 6.18 An inappropriate letter refusing credit to an ESL reader.

CONSOLIDATED PLASTICS

May 25, 1998

Mr. Jan Buwalda
Mendson SA
Hoofdstraat 23
Dokkum, The Netherlands 1234 XK

Dear Mr. Jan Buwalda:

I have received herewith your request and news about your company. Thanks.

Regarding that request to open a credit account with us, it just cannot be done. I don't know how things are expedited in your country, but in America giving credit to a first-time foreign customer is just not standard business practice. As you will understand, your credit rating is unacceptable as far as we are concerned. You will be expected to pay in cash for your first transaction with us. We'll evaluate the situation thereafter.

Let me know how you anticipate proceeding.

Sincerely,

Emma Corson

Emma Corson
Accounts Executive

999 Industrial Blvd. Bambrake, NH 01243
Phone (603)555-7000 FAX (603)555-4321 Voice (603)555-4300

FIGURE 6.19 A diplomatic revision of Figure 6.18, a letter refusing credit to an ESL reader.

CONSOLIDATED PLASTICS

May 25, 1998

Mr. Jan Buwalda
Mendson SA
Hoofdstraat 23
Dokkum, The Netherlands 1234 XK

Dear Mr. Jan Buwalda:

Thank you very much for your letter inquiring about opening a credit account with our firm. It is always a pleasure to hear from potential customers in Holland. I was most interested to learn about Mendson's diverse activities.

We understand and share your company's wish to have an American supplier to work with you. Having Mendson as a customer would be beneficial for Consolidated Plastics, too. Working with you would allow us to enter a new market.

However, I am afraid that we cannot open new accounts on credit. If you would kindly send us your check for the first month's supplies you need, we would rush your shipment to you. This will establish an account with us, and you can charge your second month's supplies on that account.

Please write or fax me if you have any questions. My fax number is at the bottom of the page. I look forward to serving you and Mendson in any way I can.

Cordially,

Emma Corson

Emma Corson
Accounts Executive

999 Industrial Blvd. Bambrake, NH 01243
Phone (603)555-7000 FAX (603)555-4321 Voice (603)555-4300

✓ Revision Checklist

- ❑ Planned what I am going to say to my readers. Did necessary homework and double-checking to answer any questions. Proved to my readers that I am knowledgeable about my topic.
- ❑ Used an appropriate (and consistent) format and page layout for my letters.
- ❑ Followed acceptable company protocol in organization, style, and tone of my letters.
- ❑ Adapted style and length of my message for my readers.
- ❑ Organized and, if necessary, labeled information in my letters in the most effective way for my message and for my readers.
- ❑ Emphasized the "you attitude" with my readers—whether employer, customer/client, or co-worker.
- ❑ Conveyed impression of being courteous, professional, and easy to work with.
- ❑ Used clear and concise language appropriate for my reader.
- ❑ Began my correspondence with reader-effective strategies. If reporting good news, told the reader right away. If reporting bad news, was diplomatically indirect.
- ❑ Followed the four As of effective sales letters. Identified and convinced my target audience.
- ❑ Wrote complaint letters in a calm and courteous tone. Informed the reader what is wrong, why it is wrong, and how I would like problem resolved.
- ❑ Wrote adjustment letters that say "Yes" sincerely and to the point. Made those that say "No" fair. Acknowledged reader's point of view and provided clear explanation for my refusal.
- ❑ Ensured correspondence was timely. Was prompt and reasonable in answering all my correspondence—from both people in my company and customers.

Exercises

1. Write a letter in which you order merchandise from a vendor's or wholesaler's catalog. Specify the quantity, size, stock number, and cost. Also include delivery instructions, the date by which you must receive the merchandise, and the way in which you will pay for it.

2. Write a letter of inquiry to a utility company, a safety or health care agency, or a company in your town, asking for a brochure describing its services

to the community. Be specific about your reasons for requesting this information.

3. In which course(s) are you or will you be writing a paper or report? Write to an agency or company that could supply you with helpful information for the paper and request its aid. Indicate why you are writing, precisely what information you need, and why you need it, and offer to share your paper or report with the company.

4. Examine an ad in a magazine or a TV commercial, and then write a one-page assessment in which you identify the four parts of its sales message.

5. Choose one of the following and write a sales letter addressed to an appropriate audience on why they should
   a. major in the same subject you did
   b. live in your neighborhood
   c. be happy taking a vacation where you did last year
   d. dine at a particular restaurant
   e. shop at a store you have worked for or will work for
   f. have their cars repaired at a specific garage
   g. give their real estate business to a particular agency
   h. visit your home page

6. Find at least two sales letters you, your family, or your firm have received, and in a memo to your instructor or employer, evaluate how well they follow the four parts of a sales letter discussed in this chapter. Attach these sales letters to your evaluation. If your memo is addressed to your boss, indicate how you would improve on your competition's sales letters.

7. Write a sales letter about the effectiveness of sales letters to an individual who has told you that almost everyone throws them away as soon as they arrive.

8. As a collaborative project, rewrite the following sales letter to make it more effective. Add any details you think are relevant.

   Dear Pizza Lovers:

   Allow me to introduce myself. My name is Rudy Moore and I am the new manager of Tasty Pizza Parlor in town. The Parlor is located at the intersection of North Miller Parkway and 95th Street. We are open from 10 a.m. to 11 p.m., except on the weekends, when we are open later.

   I think you will be as happy as I am to learn that Tasty's will now offer free delivery to an extended service area. As a result, you can get your Tasty Pizza hot when you want it.

   Please see your weekly newspapers for our ad. We also are offering customers a coupon. It is a real deal for you.

```
I know you will enjoy Tasty's and I hope to see you. I
am always interested in hearing from you about our serv-
ice and our fine product. We want to take your order
soon. Please come in.
```

9. Send a follow-up letter to one of the individuals below:
   a. a customer who informs you that he or she will no longer do business with your firm because your prices are too high
   b. a family of four who stayed at your motel for two weeks last summer
   c. a church group that used your catering services last month
   d. a customer who exchanged a dress or coat for the purchase price
   e. a customer who purchased a new (or used) car from you and who has not been happy with warranty service
   f. a company that bought software from you nine months ago, alerting them to improvements in the software

10. Write a bad news letter to an appropriate reader about one of the following:
   a. Your company has to discontinue Saturday deliveries because of rising labor and fuel costs.
   b. You are the manager of an insurance company writing to tell one of your customers that because of reckless driving, his or her rates are going to increase.
   c. You have to refuse to send a bonus gift to a customer because that customer sent in an order after the expiration date for qualifying for the gift.
   d. You have discontinued a model that a business customer wants to reorder.
   e. You have to notify residents of a community that a bus route is being discontinued.
   f. You represent the water department and have to tell residents of a community that they cannot water their lawns for the next month because of a serious water shortage in your town.
   g. You cannot send customers a catalog—which your company used to send free of charge—unless they first send $10 for the cost of that catalog.
   h. You cannot repair a particular piece of equipment because the customer still owes your company for three previous service visits.

11. Write a good news letter about the opposite of one of the situations listed in Exercise 10 above.

12. You just found out that a business that applied for credit has missed its last mortgage payment. You have to refuse credit to this local firm that has been in business successfully for eight years. Write a refusal letter without jeopardizing future business dealings.

13. Write a complaint letter about one of the following:
   a. an error in your utility, telephone, or credit card bill
   b. discourteous service you received on an airplane or bus
   c. a frozen food product of poor quality
   d. a shipment that arrives late and damaged

e. an insurance payment to you that is $100 less than it should be
f. a public television station's policy of not showing a particular series
g. junk mail that you are receiving
h. equipment that arrives with missing parts
i. misleading representation by a salesperson
j. incorrect information given at a Web site

14. Write the complaint letter to which the adjustment letter in Figure 6.14 responds.

15. Write the complaint letter to which the adjustment letter in Figure 6.16 responds.

16. Rewrite the following complaint letter to make it more precise and less emotional.

    Dear Sir:

    We recently purchased a machine from your Albany store and paid a great deal of money for it. This machine, according to your home page, is supposedly the best model in your line and has caused us nothing but trouble each time we use it. Really, can't you do any better with your technology?

    We expect you to stand by your products. The warranties you give with them should make you accountable for shoddy workmanship. Let us know at once what you intend to do about our problem. If you cannot or are unwilling to correct the situation, we will take our business elsewhere, and then you will be sorry.

    Sincerely yours,

17. Write an adjustment letter saying "Yes" to the manager of The Loft whose letter is included in Figure 6.12.

18. Write an adjustment letter saying "No" to the customer who received the "Yes" adjustment letter included in Figure 6.14.

19. Rewrite the following ineffective adjustment letter saying "Yes."

    Dear Mr. Smith:

    We are extremely sorry to learn that you found the suit you purchased from us unsatisfactory. The problem obviously stems from the fact that you selected it from the rack marked "Factory Seconds." In all honesty, we have had a lot of problems because of this rack. I guess we should know better than to try to feature inferior merchandise along with the name-brand clothing that we sell. But we originally thought that our customers would accept poorer quality merchandise if it saved them some money. That was our mistake.

Please accept our apologies. If you will bring your "Factory Second" suit to us, we will see what we can do about honoring your request.

Sincerely yours,

20. Rewrite the following ineffective adjustment letter saying "No."

Dear Customer:

Our company is unwilling to give you a new toaster or to refund your purchase price. After examining the toaster you sent to us, we found that the fault was not ours, as you insist, but yours.

Let me explain. Our toaster is made to take a lot of punishment. But being dropped on the floor or poked inside with a knife, as you probably did, exceeds all decent treatment. You must be careful if you expect your appliances to last. Your negligence in this case is so bad that the toaster could not be repaired.

In the future, consider using your appliances according to the guidelines set down in warranty books. That's why they are written.

Since you are now in the market for a new toaster, let me suggest that you purchase our new heavy-duty model, number 67342, called the Counter-Whiz. I am taking the liberty of sending you some information about this model. I do hope you at least go to see one at your local appliance center.

Sincerely,

21. You are the manager of a computer software company, and one of your salespeople has just sold a large order to a new customer whose business you have tried to obtain for years. Unfortunately, the salesperson made a mistake writing out the invoice, undercharging the customer $229. At that price, your company would not break even and so you must write a letter explaining the problem so the customer will not assume all future business dealings with your firm will be offered at such "below market" rates. Decide whether you should ask for the $229 or just "write it off" in the interest of keeping a valuable new customer. Write a letter
    a. to the new customer, asking for the $229 and explaining the problem while still projecting an image of your company as accurate, professional, and very competitive

b. to the new customer, not asking for the $229 but explaining the mistake and emphasizing that your company is both competitive and professional
c. to your boss explaining why you wrote letter **a**
d. to your boss explaining why you wrote letter **b**
e. to the salesperson who made the mistake, asking him or her to take appropriate action with regard to the new customer

CHAPTER 7

How to Get a Job: Résumés, Letters, Applications, and Interviews

Obtaining a job today involves a lot of hard work. Before your name is added to a company's payroll, you will have to do more than simply walk into the personnel office and fill out an application form. Furthermore, finding the *right* job takes time. And finding the right person to fill that job also takes time for the employer.

Steps the Employer Takes to Hire

From the employer's viewpoint, the stages in the search for a valuable employee include the following:

1. deciding on what duties and responsibilities go with the job and determining the qualifications the future employee should possess
2. advertising the job on their home page, in newspapers, and in professional publications
3. reading and evaluating résumés and letters of application
4. having candidates complete application forms
5. requesting further proof of the candidates' skills—letters of recommendation, transcripts
6. interviewing selected candidates
7. offering the job to the best-qualified individual

Sometimes these steps are interchangeable, especially steps 4 and 5, but generally speaking, employers go through a long and detailed process to select employees. Step 3, for example, is among the most important for employers (and the most crucial for job candidates). At this stage employers often classify job seekers into one of three groups: those they definitely want to interview; those they may want to interview; and those they have no interest in.

Steps to Follow to Get Hired

As a job seeker you will have to know how and when to give the employer all the kinds of information the seven steps require. You will also have to follow a certain schedule in your search for a job. The following eight procedures will be required of you:

1. analyzing your strengths and restricting your job search
2. preparing a dossier (placement file)
3. looking in the right places for a job
4. constructing a résumé
5. writing a letter of application
6. filling out a job application
7. going to an interview
8. accepting or declining a job

Your timetable should match that of your prospective employer.

Chapter 7 shows you how to begin your job search and how to prepare appropriate letters that are a part of the job-search process. You will need to write a letter of application, letters requesting others to write recommendations for you, letters thanking employers for interviews, and letters accepting or declining a job offer. In addition to discussing each of these kinds of letters, this chapter shows you how to assemble the supporting data—dossiers, résumés—that employers request. You will also find some practical advice on how to handle yourself at an interview.

The eight steps of your job search are arranged in this chapter in the order in which you are most likely to proceed when you start looking for a job. By reading about these stages in sequence, you will have the benefit of going through a dry run of the employment process itself.

Analyzing Your Strengths and Restricting Your Job Search

Two "Fatal Assumptions"

Individuals who advise students about how to get a job have isolated two "fatal assumptions" that many job seekers make. If you assume either of the following two statements to be true, chances are that you will *not* be very successful in your job search.

1. I should remain loose (vague) about what I want so I will be free to respond to any opportunity.
2. The employer has the upper hand in the whole process.

The first "fatal assumption" will disqualify you for any position for which your major has prepared you. Your first responsibility is to identify your professional qualifications. Employers want to hire individuals with highly developed technical skills and training. Your education and experience should help you to identify and emphasize your marketable skills.

The second "fatal assumption"—assuming that the employer controls the entire job-search process—is equally misleading. To a large extent, *you* can determine whether you are a serious contender for a job by the letters and résumés you write and the self-image you present. Even in today's highly competitive job market, you can secure a suitable job if you keep in mind that the basic purpose of all job correspondence is to sell yourself. Letters and résumés are sales tools to earn you an interview and eventually a job. Be confident and convincing. Believe in yourself and your abilities. Employers almost always have a shortage of good, qualified employees.

Finding the Right Job

Here are some pointers on finding the right job.

1. Make an inventory of your most significant accomplishments in your major and/or on the job. What are your greatest strengths—writing and speaking, working with people, organizing and problem solving, developing software, performing accounting audits?

2. Decide which specialty within your chosen career appeals to you most. If you are in a nursing program, do you want to work in a large teaching hospital, for a home health or hospice agency, or in a physician's office? What kinds of patients do you prefer to care for—geriatric, pediatric, psychiatric?

3. What are the most rewarding prospects of a job in your profession? What most interests you about a position—travel, international contacts, on-the-job training, helping people, being creative?

4. Avoid applying for positions for which you are either overqualified or underqualified. If a position requires ten years of related work experience and you are just starting out, you will only waste the employer's time, and your own, by applying. However, if a job requires a certificate or license and you are in the process of obtaining one, go ahead and apply.

5. Take advantage of career counseling available at your school, through your state employment agency, and from numerous guides and books. One of the most important career guides is *What Color Is Your Parachute? A Practical Manual for Job Hunters and Career Changers,* by Richard Boles. Now in its thirtieth edition, this book has been read by over 6 million people. It will give you sound advice on how to identify your most marketable strengths, package your credentials, and prioritize your job goals.

Changing Careers: Some Guidelines for Success

Because of downsizing, reorganization, and mergers, many experienced professionals are being forced to find a job after being laid off. Other individuals, such as Patrice Cooper Bolger, whose résumé is shown in Figure 7.7, are reentering the job market after years of absence. And still others are seeking a more fulfilling career. A laid-off engineer found a satisfying new position as a private investigator and an insurance salesperson forced to take an early retirement went to work for a hospital as a patient accounts manager. If you are forced to make a career change, follow the guidelines above as well as these below.

1. Identify those activities that brought you success (raises, promotions, recognitions) in your old career or job and relate them to what you can do for a new employer. Doubtless you have skills in working with people and delegating authority, submitting and managing a budget, advising and evaluating personnel, working closely with vendors and customers, and keeping a schedule. Being a team player and an effective communicator are also invaluable assets to any employer (as we saw in Chapter 3).

2. Don't panic and don't think that you have to start all over from scratch. Translate your experiences and achievements in step 1 above into marketable skills that you can use on a new job or in a new career. Show the continuity of your achievements.

3. Seek outplacement counseling, if available, from your former employer; if it is not available, identify placement counselors through such agencies as your state employment agency or the Career Planning and Adult Development Network.

4. Investigate job retraining programs available through federal, state, and local agencies.

Preparing a Dossier

The job placement office or career center at your school will assist you by providing counseling, notifying you of available, relevant jobs, and arranging on-campus interviews. The job placement office may also help you establish your **dossier,** sometimes referred to as your placement file.

Your *dossier,* French for a "bundle of documents," is your personal file stored at the placement office. This file contains information about you that substantiates and supplements the facts listed on your résumé and letter of application. Basically your dossier contains

- solicited letters of recommendation
- unsolicited letters that awarded you a scholarship, praised your work on the job, or honored you for community service
- your résumé, including job experiences
- your academic transcript(s)

Be very selective about unsolicited letters; you do not want to crowd your dossier with less important items that will compete for attention with your academic recommendations. You may ask that your dossier be sent to an employer, or employers may request it themselves if you have listed the placement office address on your résumé.

Who You Should Ask for Letters of Recommendation

The most important part of your dossier consists of your letters of recommendation. Ask the following individuals to write letters describing your work qualifications and habits.

- your present or previous employer (even for a summer job)
- two or three of your professors who know and like your work, have graded your papers, or have supervised you in fieldwork or laboratory activities
- superiors who evaluated your work in the military
- community leaders or officials with whom you have worked on civic projects

Recommendation from these individuals will be regarded as more objective—and more relevant—than a letter from your clergy or a neighbor. Of course, if you are asked specifically for a character reference, by all means ask a member of the clergy.

Always Ask for Permission

Ask permission before you list an individual as a reference. You could jeopardize your chances for a job if a prospective employer called someone you named but did not ask to be a reference and that person responds that he or she did not even know you were looking for a job or, worse yet, reveals that you did not have the courtesy to ask to use his or her name.

When you ask for permission in a letter or in person, stress how much a strong letter of support means to you and find out whether the individual is willing to write such a letter for your dossier. Generalized or weak letters may hurt your chances in your job search, so be specific in your requests. Tell your references what kind of jobs you are applying for and keep them up-to-date about your educational and occupational achievements, as Robert Jackson does in Figure 7.1.

Should You Ask Your Current Boss?

Asking your boss to recommend you for another job can be tricky. Be cautious here. If your current employer knows that your education is preparing you for another profession, or if you are a student working at a part-time job, you should obtain a letter of recommendation to include in your dossier. However, if you are employed and are looking for professional advancement or a better salary elsewhere, you may not want your present employer to know that you are searching for another job. You have the right to ask a prospective employer to respect your confidence (for instance, not to call you at work) until you become a leading candidate. At that point you may be happy to have your current employer consulted for a reference. On the other hand, if you are not on particularly good terms with your current employer and want to find another more suitable position, inform the prospective employer as honestly and professionally as you can with the least damage to yourself. Use your best judgment depending on the circumstances of your search.

FIGURE 7.1 Request for a letter of recommendation.

5432 South Kenneth Avenue
Chicago, IL 60651
March 30, 1999

Mr. Sonny Butler, Manager
Empire Supermarket
4000 West 79th Street
Chicago, IL 60652-4300

Dear Mr. Butler:

I was employed at your store from September 1997 through August 1998. During my employment, I worked part time as a stock clerk and relief cashier, and during the summer months I was a full-time employee in the produce department, helping to fill in while Bill Dirksen and Vivian Rogers were away on their vacations.

I enjoyed my work at Empire, and I learned a great deal about ordering stock, arranging merchandise, and assisting customers.

This May I will receive my A.A. degree from Moraine Valley Community College in retail merchandising. I have already begun preparing for my job search for a position in retail sales. Would you be willing to write a letter of recommendation for me in which you mention what you regard as my greatest strengths as one of your employees? To assist you, I can send you a letter of recommendation form from the Placement Office at Moraine Valley. Your letter would become part of my permanent placement file.

I look forward to hearing from you. I thought you might like to see the enclosed résumé, which shows what I have been doing since I left Empire.

Sincerely yours,

Robert B Jack

Robert B. Jackson

Encl. Résumé

Should You See Your Letters?

You have a legal right to determine whether you want to see your letters or not. If you have read them, the fact is noted on the dossier. Some employers believe that if the candidates see what is written about them, the references may be less frank and may withhold critical information. If you waive your right to see the letters written about you, you must sign an appropriate form, a copy of which is then given to the individual recommending you. Remember, though, that some individuals may refuse to write a letter that they know you will see; they may prefer absolute confidentiality. Before you make any decision about seeing your letters, get the advice of your instructors and placement counselor.

When Should You Establish Your Dossier?

Do not wait until you begin applying for jobs to compile your dossier. Most placement offices recommend that candidates set up their dossiers at least three to six months before they begin looking for jobs. With that lead time, you can be sure that your letters of recommendation are on file and that you have benefited from the placement officer's services. In advising you, the placement counselor will ask you to complete a confidential questionnaire about your geographic preferences, salary expectations, and the types of positions for which you are qualified. With this information on hand, the placement office will be better prepared to notify you of appropriate openings.

Some placement offices charge a small fee for their services, while others provide their services free of charge.

TECH NOTE

Many college and university placement offices now offer students and alumni the opportunity to file electronic dossier. When an employer or graduate requests a dossier, the placement office can send it electronically, thus streamlining the job search. For example, the University of Chicago utilizes the ProNet electronic dossier service. Also, Boston's Emerson College has an on-line Career Resource Library.

Looking in the Right Places for a Job

One way to search for a job is simply to send out a batch of letters to companies you want to work for. But how do you know what jobs, if any, these companies have available, what qualifications they are looking for, and what application procedures and deadlines they want you to follow? You can avoid these uncertainties by knowing where to look for a job and knowing what a specific job entails. Such information will make your search easier and, in all likelihood, more successful. Consult the following resources for a wealth of job-related information.

1. The Internet. Information about numerous employment opportunities exists in cyberspace. As many as 200,000 jobs are posted on the Net. The Net gives employers a vehicle for finding and recruiting qualified applicants in many fields—technical, sales, marketing, legal, and more. Companies can post jobs and openings and describe precisely what they are looking for in far more detail than in a classified ad.

You can learn about jobs on the Net in several ways. If you have a particular company in mind, you can call up its home page to see whether position openings are posted there. Or you can try clicking on such large categories as "Employment Opportunities" or "Jobs" and follow leads through various Web sites. You can also check into the many on-line job services that list positions and sometimes give advice as well. The largest of these on-line services are

- JobTrak (http://www.jobtrak.com)
- Online Career Center (http://www.occ.com)
- Helpwanted.com (http://helpwanted.com)

Similar services are devoted to specific types of jobs or employers; their Web sites are listed below in other sources to help you find a job. Once you see a position advertised, you can e-mail your letter of application and résumé directly to the prospective employer's Web site. Of course, not every company advertises its openings on-line, but enough do to make an Internet job search worthwhile.

2. Newspapers. Look at local newspapers as well as large city papers with a wide circulation, such as *The New York Times,* the *Chicago Tribune,* the *Los Angeles Times,* the *New Orleans Times-Picayune,* and the *Cleveland Plain Dealer.* The Sunday editions usually advertise positions available all over the country. Check every possibly relevant category in the classified section (for example, "Computer Programmers" as well as "Programmers"). The *National Business Employment Weekly,* published by the *Wall Street Journal,* lists jobs in many different areas, including technical and managerial positions. Many newspapers are now circulated on the Net. One invaluable on-line reference is *American Employment Weekly* (http://branch.com/aew/aew.html), which offers the classified ads from the Sunday editions of more than 100 major U.S. newspapers.

3. Professional and trade journals in your major. Identify the most respected periodicals in your field and search their ads. The *American Journal of Nursing,* for example, carries notices of openings arranged by geographic location in each of its monthly issues; and each issue of *Food Technology* features a section called "Professional Placement," listing jobs all over the country. Consult the *Encyclopedia of Associations* for a list of journals and newsletters in your profession. Check the Internet, too, to see whether one or more of the professional journals in your field might also be available on-line.

4. Your college placement office. Counselors keep an up-to-date file of available positions and can also tell you when a firm's recruiter will be on campus to conduct interviews. They can also help you locate summer and part-time work, both on and off campus.

5. Personal contacts. Let your professors, friends, neighbors, relatives, and even your clergy know you are looking for a job. They may hear of something and can notify you. Better yet, they may recommend you for the position—with a phone call, a visit to their own company's personnel department, or a letter. This sort of networking definitely pays off; referrals are among the most successful ways to land a job. John D. Erdlen and Donald H. Sweet, experts on the job search, cite the following as a primary rule of job hunting: "Don't do anything yourself you can get someone with influence to do for you."

6. Local, state, and federal employment offices. These agencies maintain a current file of positions and offer some counseling (free of charge). The U.S. government is one of the biggest employers in the country. If you are interested in working for the federal government, visit a Civil Service Commission Office, a Federal Job Information Center, or another government agency. Also check *Federal Jobs Digest* (http://www.jobsfed.com), a privately published sourcebook available in most libraries that lists more than 30,000 civil service jobs each month. On-line sources of federal jobs include America's Job Bank (http://www. ajb.dni.us), which lists jobs in the federal government and the military, and USA Jobs (http://www.usajobs.opm.gov), which also posts federal government jobs.

7. The personnel department of a company or agency you would like to work for. Often you will be able to fill out an application even if there is no current opening. But do not call employers asking about openings: a visit shows a more serious interest. Also, the personnel officers are more likely to remember you if a job does develop. New openings often arise unexpectedly in business and industry and a visit may have put you in the right place at the right time.

8. Your local Chamber of Commerce. Although not a placement center, the Chamber of Commerce can give you the names and addresses of employers likely to hire individuals with your qualifications, as well as information about these companies to use in a letter of application or at an interview. Many companies planning to relocate or expand first notify a Chamber of Commerce to assess the potential labor force.

9. A résumé database service. Several on-line services will put your résumé in a database and make it available to prospective employers, who scan the database regularly to find suitable job candidates. The Tech Note on page 253 lists the addresses of some of the more popular databases. Figure 7.2 describes a résumé database service offered by one professional organization—the Association for Com-

FIGURE 7.2 A description of one résumé database service.

Employers Benefit from ACM Résumé Database

The ACM Résumé Database is composed of résumés of ACM members–high-caliber, information-technology professionals who can bring expertise to your company. All résumés are up-to-date and can be searched according to criteria you provide, in a short period of time and at low rates. Single searches as well as annual subscriptions to the database can be requested from the database administrator, Resume-Link, at 614-529-0429. ACM Institutional members get a 10% discount off the cost of the search.

And for ACM members! Our database also is now searchable for internships and co-op positions, as well as for full-time and consultant placements. Use this free career development service and submit your résumé online at http://www.Resume-Link.com/.

puting Machinery (ACM)—for its members. Check to see if a professional society to which you belong (or might join) offers a similar service.

10. A video résumé. In some parts of the country, job hunters have made twenty-second "video résumés" that are aired on local television stations. In a video résumé, the job seeker must make a sales pitch quickly and effectively, emphasizing the one or two strengths—like the amount of sales dollars generated or the ability to speak several languages—that will prompt a potential employer to call the station for the candidate's phone number. Inquire at your local state employment service about whether any local television stations offer this public service.

11. Recruiters at a professional employment agency. Some agencies list two kinds of jobs—those that are found for the applicant free of charge (because the employer pays the fee) and those for which the applicant pays a stiff fee—usually a percentage of your first year's salary. If you do use an agency, be sure to ask who pays the fee for the service. Because employment agencies often find out about jobs through channels already available to you, exhaust all the services listed here before you rely on an agency.

Preparing a Résumé

The résumé, sometimes called a **data sheet** or **curriculum vitae,** may be the most important document you prepare in your job search. It deserves your attention.

What Is a Résumé?

A résumé is a factual and concise summary of your qualifications for a job. It is not your life history or your emotional autobiography, nor is it a transcript of your college work. A résumé highlights your proven accomplishments and abilities. It is a record of results, showing a prospective employer that you have what it takes (in education and experience) to do the job. The résumé is a short (preferably one-page, never longer than two) outline accompanying your letter of application. Never send a résumé alone, but do bring one with you to an interview. And you should certainly have copies available for recruiters if you schedule campus interviews.

What Employers Look for in a Résumé

Employers may receive as many as 300 or 400 résumés for one position and often spend as little as thirty to sixty seconds on each. After a quick glance a prospective employer can reject your résumé. All employers are looking for employees who are intelligent, cooperative, and responsible. They will look for résumés that show evidence of the following most desirable qualifications a candidate can have:

- necessary experience or education
- ability to communicate well
- computer literacy
- capacity to organize effectively
- ability to learn material quickly
- qualities of a team player

Your goal is to prepare a résumé that shows you possess these sought-after job skills.

The Process of Writing Your Résumé

As with other types of writing, preparing your résumé requires that you work through a process. You cannot produce a polished résumé in an hour. But having a clear idea about the type of job you want and your qualifications for it should get you off to a good start.

It might be to your advantage to prepare several versions of your résumé and then adapt each one you send out to the specific job skills a prospective employer is looking for. Following the process below will help you prepare any résumé.

Getting Started—Ask Key Questions

Begin with the prewriting strategies discussed in Chapter 2 to identify your strengths and achievements. Ask yourself important questions about what you are most proud of from your school and job work.

1. What classes did you excel in?
2. What papers or reports earned you your highest grades?
3. What computer skills have you mastered—computer languages, software knowledge, navigating the Net?

4. On the job, what technical skills have you acquired?
5. Have you won any awards or scholarships, or received a raise or other promotion?
6. Do you work well with people? What skills do you possess as a member of a team? Can you organize complicated tasks or solve problems quickly?
7. Do you have any hobbies or other interests that would show your professional or occupational talents and ability?

Write down everything you think is a skill or achievement—from jobs (full- or part-time, summer), school (course work, workshops, labs, extracurricular activities, sports), the military, or community work. Circle or underline any awards or honors as well as any major responsibilities. Don't be worried if you produce a rather lengthy list of seemingly unrelated activities. From this brainstormed list (or clustered grouping) you will have a working inventory of accomplishments from which you can begin to pull pertinent material.

Scrutinize Your Accomplishments

The next step is to be highly critical of your working inventory. Cross off any repetitions, eliminate unimportant or irrelevant items, try to sort items into related categories, and add whatever information you think is valuable.

As you criticize and supplement your list, examine your accomplishments the way a prospective employer might. The chief questions an employer will ask about you are

- What can this person do for our company?
- How does this person compare with other job applicants?

Employers are results-oriented; they read résumés looking for clear proof of your marketable skills. You have to convince your reader (or group of readers, since résumés often circulate among a hiring committee) that you have succeeded in the past and will do so in the future.

Pay special attention to your four or five most significant, job-worthy strengths and work especially hard on listing them concisely. While not everything you have done relates directly to a particular job, indicate how your achievements could be relevant to the employer's overall needs. For example, handling money responsibly or supervising staff in a grocery store points to your ability to perform the same duties in another business context.

Use "Selling Clauses"

The next step in preparing a winning résumé is to translate the items in your list of accomplishments above into appropriate résumé entries. There is a big difference between résumés and other types of work-related documents. Résumés generally are written in short sentences (clauses really) that omit the subject "I." Instead of complete sentences, résumés use action-packed *selling clauses* that convince prospective employers that you are the right person for the job. Use the action verbs in Table 7.1.

TABLE 7.1 Action Verbs to Use in Your Résumé

accommodated	conducted	implemented	reconciled
achieved	coordinated	improved	reduced
administered	customized	increased	researched
analyzed	dealt in	informed	scheduled
arranged	determined	initiated	searched
assembled	developed	installed	selected
attended	directed	instituted	served
awarded	drafted	instructed	settled
built	earned	maintained	sold
calculated	established	managed	supervised
coached	estimated	navigated	taught
collected	evaluated	negotiated	tracked
communicated	expedited	operated	trained
compiled	figured	organized	tutored
completed	guided	oversaw	verified
composed	handled	performed	weighed
computed	headed	prepared	wrote

How Much Should You Include on a Résumé?

How much should you include on your résumé? Both experienced candidates and recent graduates with limited experience ask this question. The dangers involve including too much or putting in too little.

How Long Should My Résumé Be?

Some employment counselors advise candidates to prepare no more than a one-page résumé. However, depending on your education and job experience, you may want to include a second page. On-line résumés (pp. 251–256) can exceed a page. A good rule of thumb is that if you have more than one degree beyond high school and if you have held more than two full-time professional jobs, you will probably need a two-page résumé to present your experience adequately. You might want to experiment with preparing a two-page chronological résumé, as Anna Cassetti has done (Figure 7.3), or a one-page bullet résumé (explained on pp. 247–249), like Donald Kitto-Klein's (Figure 7.8).

Balancing Education and Experience

If you have years of experience, don't flood your prospective employer with too many details. You cannot possibly include every detail of your job(s) for the last ten or twenty years. Emphasize only those skills and positions most likely to earn you the job. Eliminate your earliest jobs that do not relate to your present employment search. You also may have to combine and condense the skills you have acquired over many years and through many jobs. Figure 7.3 shows the résumé of Anna Cassetti, who had years of job experience before she returned to school. Take a look, too, at Patrice Cooper Bolger's résumé (Figure 7.7) and Donald Kitto-Klein's (Figure 7.8). These individuals also have a great deal of experience to offer prospective employers.

FIGURE 7.3 Résumé from an individual with ten years' job experience.

RÉSUMÉ
ANNA C. CASSETTI

6457 Blackstone Avenue	MacMurray Real Estate
Fort Worth, TX 76321-6733	1700 Ross Boulevard
(817)555-5657	Haltom City, TX 77320-1700
ACassetti@aol.com	(817)555-7211

CAREER OBJECTIVE

Full-time sales position with large real estate office in the Phoenix or Tucson area with opportunities to use proven skills in real estate appraisal and tax counseling.

EXPERIENCE

1998–present *MacMurray Real Estate, Haltom City, Texas*
Real estate agent. Excelled in small suburban office (four salespersons plus broker) with limited listings; sold individually over one million dollars in residential property; appraised both residential and commercial listings.

1992–1997 *Dallman Federal Savings and Loan, Inc., Fort Worth, Texas*
Chief Teller. Responsible for supervising, training, and coordinating activities of six full-time and two part-time tellers. Promoted to Chief Teller, March 1994.

1992 (Sept.–Dec.) *H&R Block, Westover Hills, Texas*
Tax Consultant. Prepared personal and business returns.

1989–1991 *Cruckshank's Hardware Store, Fort Worth, Texas.* Salesperson.

1985–1989 *U.S. Navy*
Honorably discharged with rank of Petty Officer, Third Class. Served as stores manager.

EDUCATION

1991–1997 *Texas Christian University, Fort Worth, Texas*
Awarded B.S. degree in Real Estate Management. Completed thirty-three hours in business and real estate courses with a concentration in real estate finance, appraising, and property management. Also took twelve hours in computer science and data processing. Wrote reports on appraisal procedures as part of supervised training program.

Continued

FIGURE 7.3 (Continued)

Cassetti 2

1990 (Sept.–Dec.)	*H&R Block, Westover Hills, Texas* Earned diploma in Basic Income Tax Preparation after completing intensive ten-week course.
1985–1986	*U.S. Naval Base, San Diego, California* Attended U.S. Navy's Supply Management School. Applied principles of stores management at Newport Naval Base.

PERSONAL

Texas Realtor's License: 756a2737

HOBBIES AND INTERESTS

Chair, Financial Committee, Grace Presbyterian Church, Fort Worth. Advised teenagers in Junior Achievement about business practices and management. Enjoy golfing and hiking.

REFERENCES

My complete dossier is available from the Placement Office, Texas Christian University, Fort Worth, TX 76119-6811.

Many job candidates who have spent most of their lives in school are faced with the other extreme: not having much job experience to put down. The worst thing to do is to write "None" for experience. Any part-time, summer, or other seasonal jobs, as well as volunteer work done at school for a library or science laboratory, shows a prospective employer that you are responsible and knowledgeable about the obligations of being an employee. So, too, do internships or community work. Figure 7.4 shows a résumé from Anthony Jones, a student with very little job experience; Figure 7.5 shows one from Maria Lopez, a student with only a few years of experience.

What Should You Exclude from a Résumé?

Knowing what to exclude from a résumé is as important as knowing what to include. Here are some details best left out of your résumé:

- salary demands or expectations
- preferences for work schedules, days off, or overtime
- comments about fringe benefits
- travel restrictions
- reasons for leaving your last job
- your photograph (unless you are applying for a modeling or acting job)
- comments about your family, spouse, or children
- height, weight, hair, or eye color
- any handicaps

Save questions and statements of preference for your interview. The résumé should be written appropriately to get you that interview.

FIGURE 7.4 Résumé from a student with little job experience.

RÉSUMÉ

ANTHONY H. JONES
73 Allenwood Boulevard
Santa Rosa, California 95401-1074
(707) 555-6390

CAREER OBJECTIVE

Full-time position as a layout artist with a commercial publishing house using my knowledge of state-of-the-art technology

EDUCATION

1997–1999 SANTA ROSA JUNIOR COLLEGE, A.S. degree to be awarded in May 1999
Dean's list in 1998, GPA 3.4
Major: Industrial Graphics Illustration, with specialty in design layout
Related courses included:

Design Principles	Desktop Publishing
Photography	Graphic Software Programs (Illustrator, Photoshop)

APPRENTICESHIP, McAdam Publishers
Major projects included:
Assisting layout editors.
Writing detailed reports on designs, photographs, and artwork used in Living in Sonoma County and Real Estate in Sonoma County magazines.

1994–1997 SANTA ROSA HIGH SCHOOL
Electives included drawing, photography, and industrial arts.
Provided major artwork for student magazine, Thunder.

EXPERIENCE

1992–1999 SALESPERSON (part-time), Buchman's Department Store
Duties included:
Assisting customers in sporting goods and appliance departments.
Coordinating sport shop by displaying merchandise.

PERSONAL

VOLUNTEER, Santa Rosa Humane Society, designed posters for the 1998 fund drive.

REFERENCES

Mr. Albert Fong	Ms. Margaret Feinstein	Dr. Gloria Cernek-Willis
Art Department	Layout Editor	Art Department
Santa Rosa Junior College	McAdam Publishers	Santa Rosa Junior College
Santa Rosa, CA 95401-1099	Santa Rosa, CA 95401-1079	Santa Rosa, CA 95401-1099
(707) 555-6300	(707) 555-8699	(707) 555-6300

Parts of a Résumé

Name, Address, Phone

Center this information at the top of the page under the heading RÉSUMÉ. Capitalize all the letters of your name (do not use a nickname) to make it stand out, but do not capitalize every letter of your address. Include your ZIP code, telephone and fax number, and e-mail address, if you have them, so a prospective employer will know how to reach you for an interview. If you have two addresses, it is wise to list both. List all appropriate phone numbers: where you can be reached

FIGURE 7.5 Résumé from a student with some job experience.

RÉSUMÉ

MARIA H. LOPEZ
1725 Brooke Street
Miami, Florida 32701-2121
(305) 555-3429

Career Objective	Full-time position assisting dentist in providing dental health care and counseling and performing preventative dental treatments; especially in applying my clinical skills in the practice of pedodontics.
Education	
August 1996– May 1998	**Miami Dade Community College, Miami, Florida** Will receive A.S. degree in dental hygiene in May. Have completed nine courses in oral pathology, dental materials and specialties, periodontics, and community dental health. Currently enrolled in clinical dental hygiene program. Experienced with procedures and instruments used with oral prophylaxis techniques. Subject of major project was proper nutrition for preschoolers. Minor area of interest is psychology (twelve hours completed). Received excellent evaluations in business writing course. GPA is 3.3. Plan to take American Dental Assistants' Examination on June 2.
1989–1993	**Miami North High School, Miami, Florida** Took electives in computers, electronics, and public relations.
Work Experience	
April 1994– July 1996	**St. Francis Hospital, Miami Beach, Florida** Full-time unit clerk on the pediatric floor. Duties included ordering supplies, maintaining records, transcribing orders, and greeting and assisting visitors.
June 1993– April 1994	**Murphy Construction Company, Miami, Florida** Secretary-receptionist. Did keyboarding, filing, and mailing in small office (three employees).
Summers 1991–1992	**City of Hialeah, Florida** Water meter reader.
Personal	Health: Excellent; Bilingual: Spanish/English.

Continued

FIGURE 7.5 (Continued)

Lopez 2

Hobbies and Interests	Swimming, reading (especially applied psychology), and tennis. Have done volunteer work for church day-care center.
References	The following individuals have written letters of recommendation for my placement file, available from the placement center, Miami-Dade Community College, Medical Center Campus, Miami, FL 33127-2225.

Sister Mary James
Head Nurse
Pediatric Unit
St. Francis Hospital
10003 Collins Avenue
Miami Beach, FL 33141-4041
(305) 555-5113

Professor Mitchell Pelbourne
Department of Dental Hygiene
Miami-Dade Community College
Medical Center Campus
Miami, FL 33127-2223
(305) 555-3872

Mildred Pecos, D.D.S.
9800 Exchange Avenue
Miami, FL 33167-6028
(305) 555-1039

Mr. Jack Murphy
1203 Francis Street
Miami, FL 33157-6819
(305) 555-6767

or where you receive messages during the day and evening, or home and work numbers.

Career Objective Statement

One of the first things a prospective employer reads is your career objective statement, also called an **employment objective.** This sentence tells the employer what kind of job you are looking for and in what ways you are qualified to hold it. Such a statement involves focused self-evaluation and will influence everything else you include. To write an effective career objective statement, ask yourself four basic questions:

1. What kind of job do I want?
2. What kind of job am I qualified for?
3. What kinds of capabilities do I possess?
4. What kinds of skills do I want to learn?

As part of your objective, indicate the title of the position you are seeking, the skills you have for that position, and what benefits you can promise the employer.

Formulate your career objective statement precisely. Let a prospective employer know that you have carefully defined goals and skills that will benefit his or her company. Avoid trite or vague goals as "looking for professional advancement" or "want to join a progressive company." Compare the vague objectives on the left with the more precise ones on the right:

Unfocused	Focused
Job in sales to use my aggressive skills in expanding markets.	Regional sales representative using my proven skills in marketing and communication to develop and expand a customer base.
Full-time position as staff nurse.	Full-time position as staff nurse on cardiac step-down unit to offer excellent primary care nursing and patient/family teaching.

Do not apply for a position that requires experience you lack or demands skills you do not possess. On the other hand, do not give the impression that you will take anything. Be careful not to make your objective either too broad or too restricted. If your focus is too broad, the employer won't have a clue as to the particular job you are seeking. If your focus is too restricted, you might not be considered for other related openings available at a company.

Depending on your background and the types of jobs you are qualified for, you might formulate two or three different career or employment objectives to use with different versions of your résumé. If your career objective statement is irrelevant or even slightly off the mark for an employer, you are likely to be excluded from the competition.

Credentials

The order of the next two categories—**Education** and **Experience**—can vary. Generally, if you have lots of work experience before, during, or after college, list experience first. Years of experience will impress a prospective employer. Note that for Anna Cassetti (Figure 7.3), Patrice Cooper Bolger (Figure 7.7), and Donald Kitto-Klein (Figure 7.8), experience is their best selling point and so they placed this category ahead of education. However, if you are still in school or are a recent graduate short on job experience, list education first, as Anthony Jones (Figure 7.4) did. Maria Lopez (Figure 7.5) also decided to place her education before her job experience because the job she was applying for required the formal training she received at Miami-Dade Community College.

Education

Begin with your most recent education first, and then list everything significant since high school. Give the names of the schools and the dates you attended. Don't overlook any military schools or major training programs (EMT, court reporter), institutes, workshops, or apprenticeships you have completed. Indicate when you received your latest degree, diploma, or certificate or when you expect to receive it.

Remember, however, that a résumé is not a transcript. Simply listing a series of courses will not set you apart from hundreds of other applicants taking similar courses across the country. Avoid vague titles such as Science 203 or Nursing IV. Instead, concentrate on describing the kinds of skills you learned.

> 30 hours in planning and development courses specializing in transportation, land use, and community facilities and 12 hours in field methods of gathering, interpreting, and describing survey data in reports.

> Completed 28 hours in major courses in business marketing, management, and materials in addition to 12 hours in computer science, including LAN.

Also note any laboratory work, fieldwork, internship, or cooperative educational work. Such extras are important to employers looking for someone with previous practical experience. List your GPA (grade point average) only if it is 3.0 or above and your rank in class only if you are in the top 35 percent. Otherwise, indicate your GPA in just your major or during your last term, if it is above 3.0.

Also list any academic honors you have won (dean's list, department awards, school honors, scholarships, grants, or honorable mentions). Memberships in honor societies in your major also demonstrate that you are professionally accomplished and active.

Experience

Your job history is the key category for many employers. It shows them that you have held jobs before and that you are responsible. Here are some guidelines to follow in listing information about your experience.

1. Begin with your most recent position and work backward in reverse chronological order. List the company or agency name, location (city and state), and your title. Do not mention why you left a job.

2. Give the most attention to your latest, most relevant position. If it happens to be your second most recent position, keep the correct order, but spend more time on it. If you have held many jobs, highlight whatever positions are most relevant for the job you are seeking now. Your summer job as a lifeguard who knew lifesaving techniques may help you in finding a position as a respiratory therapist. Discuss jobs that you held eight or nine years ago only if your experiences then are relevant to your present job search. Avoid stringing out five or six temporary, short-term jobs (each under three months). Combine them into one brief statement or omit them. Remember, space is at a premium on your résumé. If you have been a full-time parent for ten years, indicate the management skills you developed in running a household and any community or civic service, as Patrice Cooper Bolger does in her résumé in Figure 7.7. She skillfully relates her family and community accomplishments to the specific job she seeks.

3. Provide a short description of your duties and achievements using your "selling clauses" (pp. 237–238). Don't just say you "worked for a newspaper"—your prospective employer won't know whether you wrote editorials, sold advertising space, or delivered papers. Perhaps you were an assistant to the advertising editor and were responsible for arranging and verifying copy. Say so. That is impressive and informative. Rather than saying that you were a secretary, indicate that you wrote business letters and contracts, learned various software programs, prepared schedules for part-time help in an office of twenty-five people, or assisted the manager in preparing accounts. See how Donald Kitto-Klein's résumé (Figure 7.8) uses effective selling clauses to describe the skills he has acquired in several positions.

4. In describing your position(s), emphasize any responsibilities that involved handling money, managing other employees, dealing with customer accounts, services, and programs, or writing letters and reports. Prospective employers are interested in your leadership abilities, financial shrewdness (especially if you saved your company money), tact in dealing with the public, and communication skills. They are also favorably impressed by promotions you may have earned.

5. Never exaggerate or lie about your job duties. It will catch up with you. Don't call yourself an assistant buyer when you were a sales clerk. If you were a clerical assistant to an attorney, don't describe yourself as a paralegal. Do not inflate your role with fancy terminology; a receptionist is not "communications consultant" nor is a waiter a "food services manager." On the other hand, don't assume that your work experience was so routine, so ordinary, that it was unimportant and unhelpful in your job search. Show your prospective employer that no matter what you did, you did it well, exceptionally well in fact. For example, rather than just saying you were a nurse's aide, include some selling clauses about working with patients, carrying out various procedures, and assisting the nursing staff. Emphasize your responsibilities and how you carried them out professionally as part of a health care team.

Personal (Optional)

Federal employment laws prohibit discrimination on the basis of sex, race, national origin, religion, or marital status, and you need not include any such information on your résumé. But if you think that any of this personal information might help land you the job, then you are free to include it. If you are applying for a position in a day-care center or as a teacher's aide, the fact that you have children may be important to your employer. Ultimately, common sense dictates that you reveal only personal information that is required or that underscores your qualifications for the job. Here are some personal details you might include on your résumé:

- foreign languages you speak or write
- extensive travel
- certificates or licenses you hold
- memberships in professional associations
- memberships in community groups (Lions, Red Cross, Elks; list any offices you hold—recorder, secretary, fund drive chairperson)

Hobbies or Interests (Optional)

This category is of least value to a prospective employer because usually there is no connection between it and the rest of your résumé. Of course, if a pastime, sport, or hobby has a direct bearing on the kind of job you are applying for or the kinds of subjects you studied in school, then by all means list and briefly describe it. Exclude such obvious interests as reading, meeting new people, or listening to music.

References

You can inform readers of your résumé that you will provide references on request or that they may obtain a copy of your dossier (which contains reference letters) from your placement center; or you can list on your résumé the names, titles, e-mail and street addresses, and telephone numbers of three or four individuals. Prospective employers can then write to you, ask for your dossier, or directly write or call the individuals whose names you have provided. List your references only when they are well known in the community or belong to the same profession in which you are seeking employment. In these cases you profit from your association with a recognizable name or title. If you list your references on your résumé, include no more than three or four names.

In this section of your résumé, you may also indicate that a portfolio of your work is available for review.

Organizing Your Résumé by Chronology or by Function

There are two primary ways to organize your résumé.

Chronologically

The résumés in Figures 7.3 through 7.5 are organized chronologically. Information about the job applicants is listed year by year under two main categories—Education and Experience. This is the traditional way to organize a résumé. It is straightforward and easy to read, and employers find it acceptable. The chronological sequence works especially well when you can show a clear continuity toward progress in your career through your job(s) and in schoolwork or when you want to apply for a similar job with another company. A chronological résumé is appropriate for students with limited experience who want to emphasize recent educational achievements.

By Function or Skill Area

Depending on your experiences and accomplishments, you might organize your résumé according to function or skill areas. According to this plan, you would *not* list your information chronologically within categories labeled "Experience" and "Education." Instead, you would sort your achievements and abilities—whether from course work, jobs, extracurricular activities, or technical skills—into two to four key skill areas, such as "Sales," "Public Relations," "Training," "Management," "Research," "Technical Capabilities," "Counseling," "Group Leadership," "Communications," "Working with People," "Computer Skills," "Problem-Solving Skills," and so forth.

Under each area you would list three to five points illustrating your achievements in that area. Skills or functional résumés are often called **bullet résumés** because they itemize the candidate's main strengths in a bulleted list. Some

FIGURE 7.6 Anna Cassetti's résumé organized by function, or skill areas.

RÉSUMÉ
ANNA C. CASSETTI

Home	*Office*
6457 Blackstone Avenue	MacMurray Real Estate
Fort Worth, TX 76321-6733	1700 Ross Boulevard
(817) 555-5657	Haltom City, TX 77320-1700
acassetti@aol.com	(817) 555-7211

CAREER OBJECTIVE

Full-time sales position with large real estate office in the Phoenix or Tucson area with opportunities to use proven skills in real estate appraisal and tax counseling.

SALES/ FINANCIAL MANAGEMENT SKILLS

- Licensed (Texas) real estate appraiser with extensive knowledge of real estate codes, appraisal procedures, and market conditions
- Achieved notable success in selling residential property in Fort Worth
- Served as a tax consultant with special interest and competence in real estate sales
- Performed general banking procedures as chief teller
- Responsible for maintaining, purchasing, and ordering supplies for ship's store in U.S. Navy

PUBLIC RELATIONS SKILLS

- Have developed positive, professional image in helping clients select appropriate property for their needs and income
- Counseled commercial and individual clients about taxes
- Supervised, trained, and coordinated the activities of six bank tellers
- Established rapport in assisting customers with their banking needs
- Chaired a financial committee at Grace Presbyterian Church, Fort Worth, Texas

BUSINESS COMMUNICATION SKILLS

- Prepared standard real estate appraisals
- Wrote in-depth business reports on appraisal procedures, property management problems, and banking policies affecting real estate transactions
- Proficient in Corpsheet and other spreadsheet programs
- Conducted small group training and sales sessions

Continued

FIGURE 7.6 (Continued)

Cassetti 2

EDUCATION	B.S. in real estate management, 1997 Texas Christian University, Fort Worth, Texas Advanced course work taken in business, finance, and real estate; minor in data processing Diploma, Basic Income Tax Preparation, 1990 H&R Block, Westover Hills, Texas
EMPLOYMENT HISTORY	MacMurray Real Estate, Haltom City, Texas: 1998–present: sales agent Dallman Federal Savings and Loan, Inc., Fort Worth, Texas: 1992–1997; Chief Teller H&R Block, Westover Hills, Texas: 1992 Sept.–Dec; consultant, tax preparer Cruckshank's Hardware Store, Fort Worth, Texas: 1989–1991; salesperson U.S. Navy, San Diego, California (last duty station): 1985–1989; stores manager; honorably discharged with rank of Petty Officer, Third Class
HOBBIES AND INTERESTS	Golfing; hiking; worked with Junior Achievement advising teenagers about business management
REFERENCES	Complete dossier is available from the Placement Office, Texas Christian University, Fort Worth, TX 76119-6811

employers prefer the bullet résumé because they can skim the candidate's list of qualifications in a few seconds.

The most important goal of a functional résumé is to highlight the strengths or skills in which you believe a prospective employer is most interested.

Figure 7.6 shows what Anna Cassetti's chronological resume (Figure 7.3) might look like if she had organized it according to function; Figure 7.7 shows Patrice Bolger's résumé organized by function. Note that the three functional areas Cassetti selected ("Sales/Financial Management," "Public Relations," and "Business Communication") effectively allowed her to capitalize on her diverse experiences.

Who Should Use a Functional Résumé?

The following individuals would probably benefit from organizing their résumés by function instead of by chronology:

- nontraditional students who have diverse job experiences
- individuals who are changing their profession because of downsizing or seeking new professional opportunities
- individuals who have changed jobs frequently over the last five to ten years

FIGURE 7.7 Patrice Cooper Bolger's résumé organized by skill areas.

RÉSUMÉ
PATRICE COOPER BOLGER
1215 Lakeview
Westhampton, Michigan 46532
phone 616-555-4772 e-mail pcbolger@aol.com

EMPLOYMENT OBJECTIVE — **Seek full-time position as public affairs officer in health care, educational, or charitable facility**

SKILLS, RESPONSIBILITIES, EXPERIENCES

Organizational Communication

- **Delivered** 20 presentations to neighborhood and civic groups on educational and civic issues
- **Transcribed** minutes and helped formulate agenda as president for large, local PTA for last 6 1/2 years
- **Possess** excellent keyboard skills and knowledge of **Wordmatch 6.0**
- **Updated** and **maintained** computerized mailing lists for Teens in Trouble and Foster Parents' Association

Money Management

- **Spearheaded** 3 major fundraising drives (total of $120,000 collected)
- **Prepared** and **implemented** large family budget (3 children, 8 foster children) for 15 years
- **Served as financial secretary**, Broad Street United Methodist Church for 4 years
- **Awarded** "Volunteer of the Year" (1996) by Michigan Foster Child Placement Agency for Budget Planning

Administration

- **Organized** volunteers for American Kidney Fund (last 5 years)
- **Established** and **oversaw** neighborhood carpool (17 drivers; more than 50 children) for 7 years
- **Coordinated** after-school tutoring program for Teens in Trouble; president since 1990

EDUCATION

A.A. Metropolitan Community College, 1992
B.S. Mid-Michigan College, expected 1999; major: public administration; minor: psychology. GPA 3.35

WORK EXPERIENCE

Secretary, 1983–1993 (full and part-time):
Merrymount Plastics; Foley and Wasson;
Westhampton Health Dept.; G & K Electric

Assembly Line Worker, 1980–1982
Universal Motors

REFERENCES

Available upon request

- individuals who are entering the civilian marketplace after retiring from the military

Preparing a Skills Résumé
When you prepare a functional or skills résumé, start with your name, address, telephone number, and career objective, just as in a chronological résumé. To find the best two or three functional areas to include, use the prewriting strategies (especially clustering and brainstorming) discussed in Chapter 2. Think of your skills areas as the common denominators that cross job and educational boundaries—the common threads that link your diverse experiences. Note how Donald Kitto-Klein (Figure 7.8) was able to pull together a series of related, marketable skills from the many different jobs he had held over several years. After you discover and suitably revise the information to be included in your categories, only briefly list your educational and work experiences, as Anna Cassetti (Figure 7.6), Patrice Cooper Bolger (Figure 7.7), and Donald Kitto-Klein (Figure 7.8) do.

Advantages of a Functional Résumé
A functional résumé has a number of advantages for some job seekers.

1. It can offer a productive way to fill in gaps in education or employment, since a job seeker is not tied to a strict chronological account. If you were laid off, left school for several years, or moved around, you do not have to worry about accounting for the gaps in time. Note Patrice Cooper Bolger's profitable use of a functional résumé format in Figure 7.7. She was out of school for more than ten years because of family commitments, yet she uses the experiences she acquired during those years to her advantage in her résumé organized by "Skills, Responsibilities, Experiences." She successfully translates her many accomplishments in managing a home and working on charitable and community projects into marketable skills of great interest to a prospective employer, and no gap of ten years interrupts a work experience list.
2. Because a functional résumé clearly emphasizes general skills acquired over long periods of time, individuals who have had many different types of jobs in diverse fields may prefer this format over a chronological résumé, as Donald Kitto-Klein does in Figure 7.8.
3. Unlike a chronological résumé, a functional résumé does not force the job seeker to emphasize his or her most recent experience or educational achievement at the expense of more pertinent earlier accomplishments.

You might want to prepare two different versions of your résumé—one functional and one chronological—to see which sells your talents better. Don't hesitate to seek the advice of your instructor or placement counselor about which one works best for you.

The On-Line Résumé

With many jobs being advertised on-line, more and more prospective employers want applications sent to their Web addresses. Consequently, job seekers can

FIGURE 7.8 Functional résumé by a candidate who has held a variety of jobs.

RÉSUMÉ
DONALD KITTO-KLEIN
kitto@gar.com

56 South Ardmore Way Petersburg, NY 15438 (716) 555-9032	Garland Industries Grand Banks, NY 15532 (716) 555-4800, Ext. 5398

Objective	Seek supervisory position in Computer Maintenance and Service Department to provide excellent service to staff and clients
Computer Languages	MC68000 Assembler, Fortran, Ada
Systems Experience	MACHINES: IBM, RS60000, Macintosh, Sun NETWORKS: Novell, Banyan; set up Internet node
Computer Maintenance/ Service Skills	• Serviced PCs and workstations on a regular basis for 3 1/2 years • Worked extensively on spreadsheets/database software • Modified software billing program • Coordinated maintenance/service activities
People Skills	• Supervised Computer Servicing with three technicians • Worked closely with computer manufacturers and suppliers to minimize hardware down time • Elected to employee benefits committee • Promoted to First-Shift Computer Services Manager
Communication Skills	• Collaboratively wrote safety manual for power company road crews • Taught in-house training sessions on computer maintenance, networking, data security • Devised routing systems to expedite work orders and follow-ups • Coordinated small group meetings in systems analysis
Employment	Garland Industries, Team Leader, Maintenance, 1997–present Business Graphics and Computers Store, Salesperson, 1994–1998 U.S. Army, Specialist, 4/E 1987–1994
Education	B.S., Grand Valley Technical Institute, 1995 U.S. Army schools in computer programs, 1989–1993
References	Available on request.

expect to apply for at least some positions on-line and will have to send their résumés by e-mail. The on-line, or electronic, résumé contains essentially the same information that can be found in the various types of résumés discussed above. Of course, you might take advantage of the multimedia options on the Internet to add voice and picture, though you would have to include a small photograph (a large one would take too long to display).

To compete for jobs on-line, you will have to adapt the conventional résumé for transmission as a hypertext document. Follow the eight guidelines below to prepare an effective on-line résumé; as you read these guidelines refer to Figure 7.9, which illustrates an on-line résumé.

1. Format your résumé properly and consistently—in ACSII text—to be received and read around the globe. Test your formatting by sending your résumé to a friend's e-mail address.
2. Make your résumé readable, functional, and easy to scroll. Use plenty of white space.
3. Do not use italics, underlining, boldfacing, fancy scripts, or logos. Each of these features might interfere with the transmission of your résumé and garble it when a prospective employer clicks on it.

TECH NOTE

Don't worry if you cannot format your résumé in an ASCII text. You might want to take advantage of one of the résumé database services (see pp. 234–235) that prepare and post on-line résumés in the proper format for a small fee. Not only will such a service format your résumé, but it will also classify it according to your area(s) of expertise, making it easy for employers to find you. Below are the addresses of some of the most popular database services:

The Best Résumés on the Net
http://tbrnet.com

Résumé Net
http://www.resumenet.com

At On-line Résumés
http://ol-resume.com

Virtual Résumé
http://virtualresume.com

FIGURE 7.9 An on-line résumé.

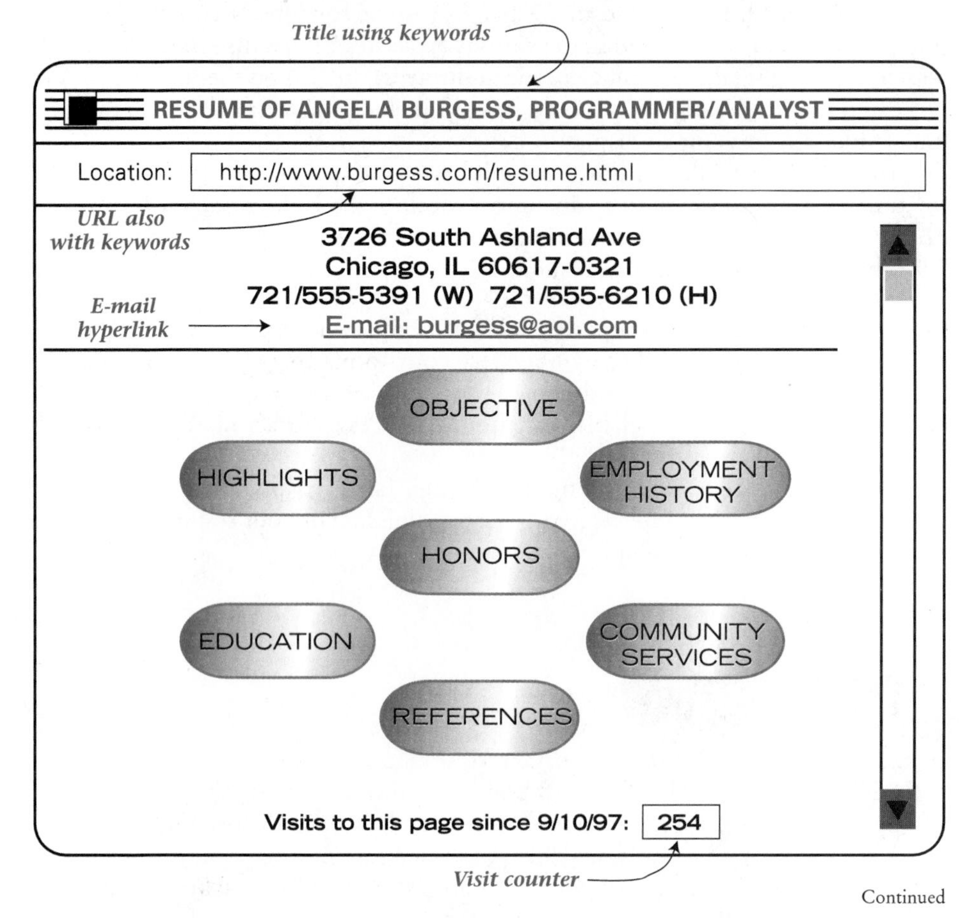

Continued

4. Start with your on-line address to make it easy for employers to reach you. If you use a résumé database service, you could begin with that location and list your e-mail address as well. Angela Burgess gives readers both locations in her on-line résumé in Figure 7.9.
5. Use hyperlinks at the top of the résumé to connect to key categories on the same or a continuing page. Angela Burgess's hyperlinks are OBJECTIVE, HIGHLIGHTS, EXPERIENCE, EDUCATION, COMMUNITY SERVICE, and REFERENCES. Highlighting these categories makes it easy for a prospective employer to jump to the résumé section deemed most important. Keep in mind a prospective employer may be scrolling as many as 200 or 300 résumés a day to compile a short list of candidates to interview.
6. Use keywords as hyperlinks. The electronic résumé emphasizes nouns, whereas the conventional résumés in Figures 7.6 and 7.7, for example, use strong verbs. Nouns function as the keywords by which a résumé is scanned by Web search engines and organized in a database. Prospective

FIGURE 7.9 (Continued)

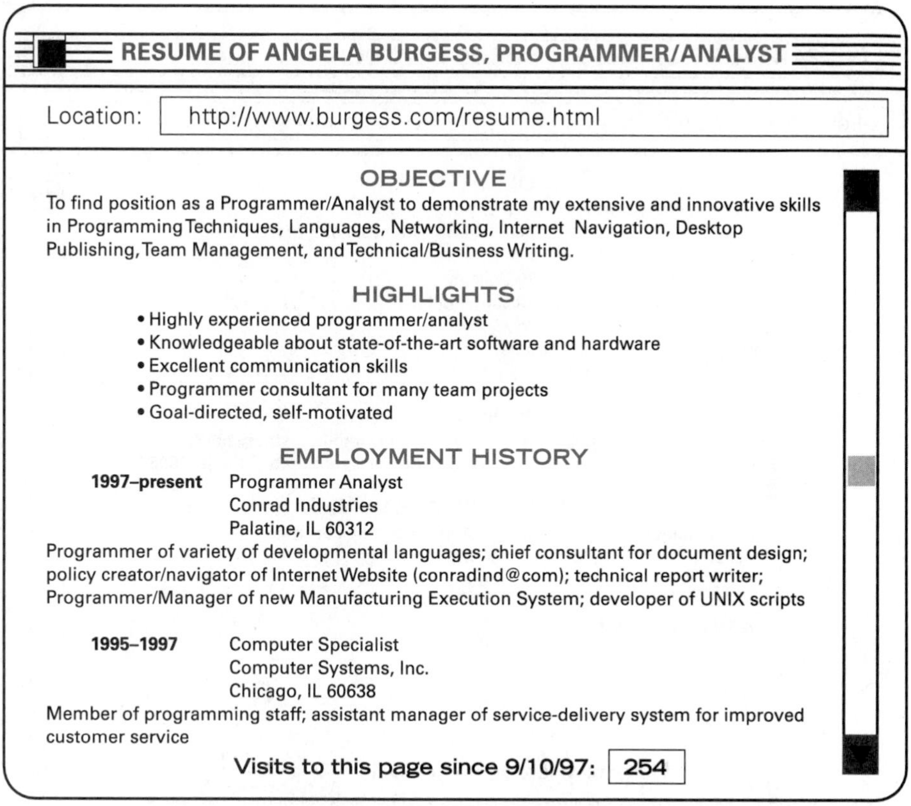

RESUME OF ANGELA BURGESS, PROGRAMMER/ANALYST

Location: http://www.burgess.com/resume.html

OBJECTIVE

To find position as a Programmer/Analyst to demonstrate my extensive and innovative skills in Programming Techniques, Languages, Networking, Internet Navigation, Desktop Publishing, Team Management, and Technical/Business Writing.

HIGHLIGHTS

- Highly experienced programmer/analyst
- Knowledgeable about state-of-the-art software and hardware
- Excellent communication skills
- Programmer consultant for many team projects
- Goal-directed, self-motivated

EMPLOYMENT HISTORY

1997–present Programmer Analyst
Conrad Industries
Palatine, IL 60312

Programmer of variety of developmental languages; chief consultant for document design; policy creator/navigator of Internet Website (conradind@com); technical report writer; Programmer/Manager of new Manufacturing Execution System; developer of UNIX scripts

1995–1997 Computer Specialist
Computer Systems, Inc.
Chicago, IL 60638

Member of programming staff; assistant manager of service-delivery system for improved customer service

Visits to this page since 9/10/97: 254

Continued

employers, for example, search the Web by keywords to find what they want to see in the job seeker's experience, education, or activities. Repeat several times in your résumé the two or three key nouns that most accurately reflect your accomplishments, as Angela Burgess does. The on-line résumé keywords on the right replace the action verbs from conventional résumés on the left.

edited company newsletter	newsletter editor
wrote technical report	technical writer
performed laboratory tests	laboratory technologist
responsible for managing accounts	accounts manager
won two awards	award winner
solved software problems	software specialist

Don't be afraid of using shop talk (or jargon) for your keywords. An employer searching for a specialist will expect the résumé writer to be aware of current terminology, especially in computer programming or networking.

FIGURE 7.9 (Continued)

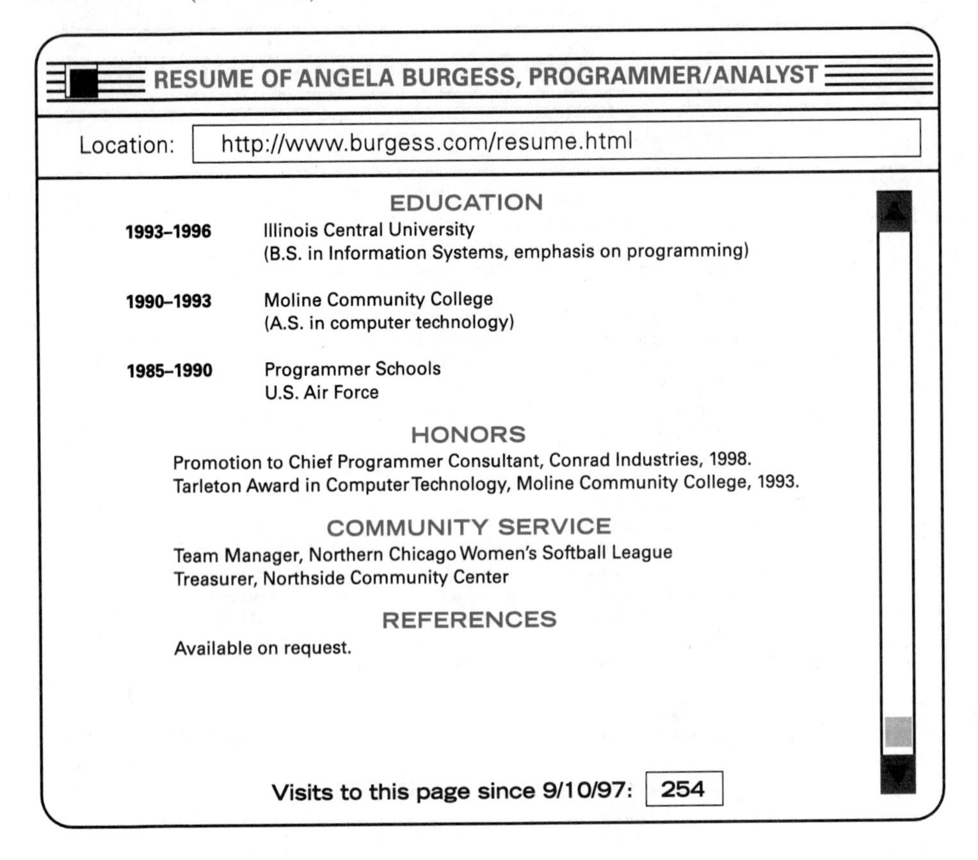
RESUME OF ANGELA BURGESS, PROGRAMMER/ANALYST

Location: http://www.burgess.com/resume.html

EDUCATION

1993–1996	Illinois Central University (B.S. in Information Systems, emphasis on programming)
1990–1993	Moline Community College (A.S. in computer technology)
1985–1990	Programmer Schools U.S. Air Force

HONORS

Promotion to Chief Programmer Consultant, Conrad Industries, 1998.
Tarleton Award in Computer Technology, Moline Community College, 1993.

COMMUNITY SERVICE

Team Manager, Northern Chicago Women's Softball League
Treasurer, Northside Community Center

REFERENCES

Available on request.

Visits to this page since 9/10/97: 254

7. You are not limited to just one page; include an additional page or two, but don't inflate your accomplishments with unnecessary details or boasts.

TECH NOTE

Each Web site contains a "visits counter" recording the number of times Web users have visited a particular site. A visits counter is especially helpful for job seekers because it alerts them to the amount of interest their résumé is generating. The way to increase the number of visits your Web site receives is to include key words and phrases as hyperlinks so that prospective employers will be directed to your résumé a maximum number of times. The more visits your on-line résumé receives, the better your chances of getting an interview and eventually a job.

The Appearance of Your Résumé

The appearance of your résumé is as important as its content. Preparing a résumé involves skills highly valued on the job—neatness, the ability to organize and summarize, and most important, a sense of proportion. If your résumé looks professional, employers will predict that the work you do for them will be done the same way. A résumé can make that good first impression for you; a poorly prepared one assures that you will not get a second chance.

There are a variety of ways you can prepare your résumé. You can put it on disk with your PC, as Patrice Cooper Bolger and Donald Kitto-Klein did (Figures 7.7 and 7.8). You can take it to a printer who for a relatively modest fee will phototypeset your résumé, giving it a highly professional look. This is what Anna Cassetti did (Figure 7.3 and 7.6). Printing offers many advantages—your résumé will look neat, professional, crisp—but you will have to buy at least twenty-five to fifty copies of the same one. There is no way to tailor your résumé if the need arises; you cannot add, delete, or emphasize information as your job search progresses. If you know how to use desktop publishing software, you can produce a professional-looking printed résumé yourself.

Regardless of the method you use, follow these guidelines:

1. **Spacing.** Avoid the twin dangers of crowding too much information on the page or of leaving huge, highly conspicuous chunks of white space at the bottom and sides. A crowded résumé suggests that you cannot summarize; too much blank space points to a lack of achievements. Leave plenty of white space between categories to emphasize certain points and to make reading your résumé easier. Study the sample résumés in this chapter again. Print a number of versions of your résumé to experiment with spacing.

2. **Type.** With your word processor, take advantage of different sizes of type, and use boldface or italics to separate and highlight information. Help employers spot your achievements by using clearly divided sections with headings in boldface type. Don't make the print size so small that it is difficult for a prospective employer to read. Stay as close to 10-point type as possible. Be careful that you do not overuse visual effects. You can also justify your margins (that is, make the right-hand side of your résumé line up just as the left-hand side does).

3. **Paper.** Print or type your résumé on good quality 8 1/2" × 11" white or off-white bond paper (at least 20-pound stock). In fact, recruiters in one study preferred white to colored paper. Never use flimsy computer paper with perforated strips. Do not run your résumé off on a dot matrix printer; use a letter-quality or laser printer with dark type.

4. **Copies.** Your prospective employer will expect a professional-looking original of your résumé so never send a poor copy. Ensure your printer works well. Avoid rushing to the nearest photocopy machine in the library or student union to make a copy. You may end up with a résumé peppered with black dots or smudgy streaks. Try to locate the best-working copier available, and try to have your résumé duplicated on bond paper. It will be worth the extra money.

5. Proofreading. Proofread your résumé to make sure that it is letter-perfect. Prospective employers will be looking for accuracy (a spelling mistake can be ruinous) and consistency. Double-check the spelling of any words or names about which you are unsure. Ask two or three people to check your document for you, too. Because you will be supplying many prospective employers with your résumé, a single error is multiplied by the number of times you send it out.

Writing a Letter of Application

Along with your résumé, you must send your prospective employer a letter of application, one of the most important pieces of correspondence you may ever write. Its goal is to get you an interview and ultimately the job. Letters you write in applying for jobs should be personable, professional, and persuasive—the three Ps. Knowing how the letter of application and résumé work together and how they differ can give you a better idea of how to compose your letter.

How the Application Letter and Résumé Differ

The résumé is a compilation of facts—a record of dates, your important achievements, names, places, addresses, and jobs. You will have your résumé duplicated, and you will send a copy to each prospective employer. As we saw, you may even prepare two or three different résumés depending on your experiences and the job market. Your letter of application, however, is much more personal. You must write a new, original letter to each prospective employer. That is, you might write (or adapt) many different letters. Photocopied letters of application say that you do not care enough to spend the time and energy to answer the employer's help wanted ad personally.

Each letter of application should be tailored to a specific job. It should respond precisely to the kinds of qualifications the employer seeks. The letter of application is a sales letter emphasizing and applying the most relevant details (of education, experience, and talents) on your résumé. In short, the résumé contains the raw material that the letter of application transforms into a finished and highly marketable product—you.

Résumé Facts to Exclude from Your Letter of Application

The letter of application should not simply repeat the details listed in your résumé. In fact, the following details belong *only* on your résumé and should *not* be restated in the letter.

- personal data including license or certificate numbers
- specific names of courses in your major
- your hobbies
- names and addresses of all your references

Duplicating these details in your letter gives no new information that might persuade prospective employers that you are the individual they are seeking.

FIGURE 7.10 A company's home page with information about jobs and company products and services.

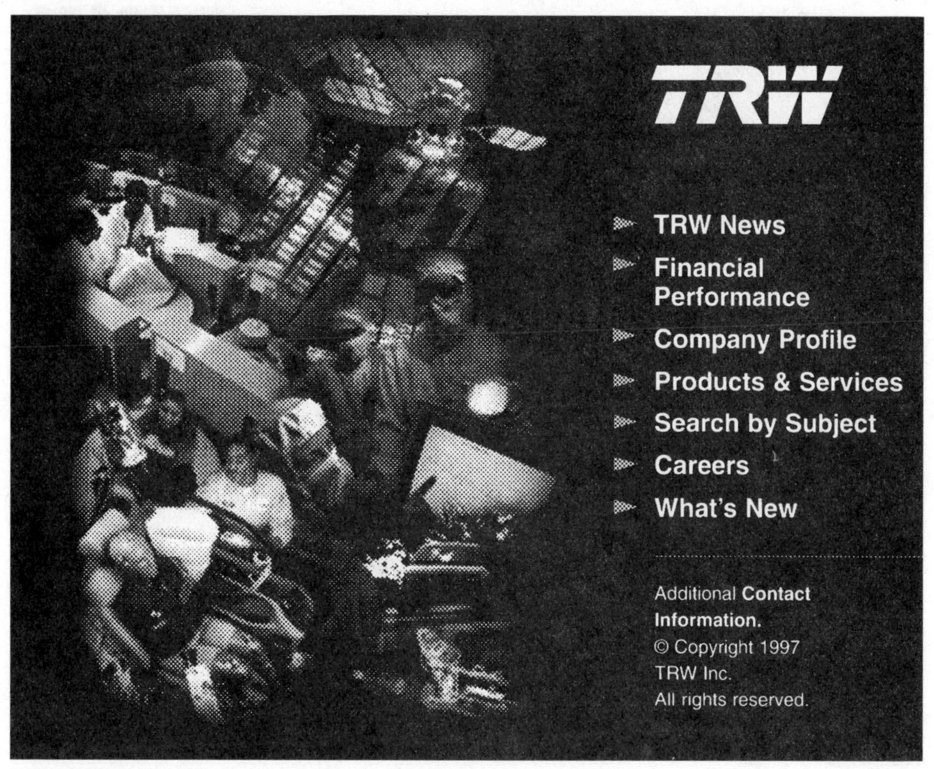

© Copyright 1997, TRW is the name and mark of TRW Inc.

Continued

Finding Information About Your Prospective Employer

One of the best ways to sell yourself to future employers is to demonstrate that you have some knowledge of their company. Recently, a software firm was looking for a computer programmer and was impressed with one candidate who referred to the company's specific products by name in her letter of application. This individual went to the profitable trouble of reading the company's home page on the Web before applying.

Do a little similar homework; investigate the job and the company. It is to your advantage to find out as much as possible about a prospective employer. Some things you need to find out include whether the company is privately or publicly owned; what its chief products or services are; whether it has subsidiaries or is a subsidiary of a larger company in the United States or overseas; who its chief officers are.

There are many ways to find out about a prospective employer.

1. See if the firms you are interested in have a Web site. Most of them will; all larger companies do. Figure 7.10 shows the first page of the home page for TRW, a

FIGURE 7.10 (Continued)

Company Profile

- What is TRW?
- Facts about TRW
- Financial Goal
- Mission and Values Statement
- Industry Segments
  TRW Inc. provides high-technology products and services to the automotive and space & defense markets. The financial results of the company's operations are reported in two core business segments: Automotive and Space & Defense.
  - **Automotive**
    TRW designs, manufactures, and sells five automotive product lines for cars, trucks, buses, and off-highway vehicles **steering and suspension systems**, **engine components**, **occupant restraints**, **electrical and electronics**, and fasteners. These products are distributed directly to vehicle makers and through independent distributors.
  - **Space & Defense**
    Space & Defense consists of **space & electronic systems** and **systems integration**.
  - **Other Businesses**
- **History**
- **Community Relations**
- **TRW Global Locations**

TRW Home Page | Search | Contacts

Continued

large research and manufacturing company. Note the valuable options a job seeker can explore, including TRW News, Company Profile, and Careers.

2. In addition to consulting home pages on the Web, look at the following directories, all of which contain much useful information about businesses you are maybe applying to.

- *Million Dollar Directory* and *Principal International Businesses* are both published by Dun and Bradstreet. They do not have separate Internet addresses. Information on both can be found at the Dun and Bradstreet Website. http://www.dbisna.com
- *Directory of Corporate Affiliations* does not have its own address either, but you can find information on the Knight Ridder Information, Inc. home page. http://www.krinfo.com
- *Standard and Poor's Register of Corporations, Directors and Executives* does not have an individual address but you can find information at the McGraw-Hill Website (S&P is a division of McGraw-Hill) and the Standard & Poor's Compustat page. http://www.mcgraw-hill.com, http://www.compustat.com

FIGURE 7.10 (Continued)

For more information, contact:

TRW Communications
1900 Richmond Road
Cleveland, Ohio 44124

TRW Home Page | Search

Additional Contact Information

3. You can consult two valuable databases as well. One is *The Business Periodicals Index*, a helpful guide to articles written about different companies in various business publications (like the *Wall Street Journal* or *Money* magazine). Also check the Job Search file of NewsBank, which searches "thousands of recent articles in Business NewsBank by company name, location, and industry and [finds] background information on local and regional companies everywhere in the U.S."

4. You might also consult the following sources as part of your search for information about a particular company: *The Thomas Register of Manufacturers; Forbes* magazine (especially the Annual Directory in the May issue); and *Fortune* magazine (lists the Fortune 500 companies).

5. You can obtain information from companies directly by writing an inquiry letter or e-mail requesting brochures, newsletters, company magazines, and annual reports several months before you actually apply. It goes without saying that current employees of a company are an invaluable source of information.

Drafting the Letter of Application

The letter of application can make the difference between getting an interview and being eliminated early from the competition. Keep in mind that employers receive many letters and that you will have to compete for attention. You want your letter to be placed in the "definitely interview" category.

The best application letters are (a) professional, (b) readable, and (c) to the point. Limit your letters to one page. As you prepare your letter, follow these general guidelines:

1. Follow the standard conventions of letter writing discussed in Chapter 5. Type or print your letter on good-quality, 8 1/2" × 11" bond paper. Make sure your printer works perfectly; the print should be dark and distinct. Proofread meticulously; a spelling error will harm your chances.

2. Make sure your letter looks attractive. Use wide margins and don't crowd your page.

3. Send your letter to a specific person. Never address an application letter "To Whom It May Concern," "Dear Sir or Madam," or "Personnel Director." Try to get an individual's name; double-check the company's home page; if you cannot find the human resources director there, try calling the company's switchboard and then verify the spelling of the person's name and his or her title.

4. Don't forget the "you attitude" discussed on pages 154–160. Approach your qualifications in terms of how and why they are valuable to a particular employer. You will need to make appropriate changes in details and emphasis from one letter to another so as to focus on each reader. Remember that employers are not impressed by vain boasts ("I am the most efficient and effective safety engineer"; "I am a natural-born nurse"). One applicant spent so much time on the advantages he would get from the job that he forgot the employer entirely: "I have worked with this kind of equipment before, and this experience will give me the edge in running it. Moreover, I can adjust more quickly to my working environment."

5. Strive for brevity and clarity. Summarize your qualifications convincingly without being long-winded. But don't write telegraphic messages in two or three sentences. Stay away from abbreviations ("nite" for "night," "thru" for "through"). Avoid slang and legalese.

6. Don't be tempted to send out your first draft. Remind yourself to be thorough in following the steps of the writing process. A first or even second draft rarely sells your abilities as well as a third, fourth, or even fifth revision does. Write and rewrite your letter of application until you are convinced it presents you in the best possible light. Your job may depend on it.

The discussion that follows will give you some suggestions on how to prepare the various parts of an application letter successfully.

Your Opening Paragraph

The first paragraph of your application letter is your introduction. It must get your reader's attention by answering three questions:

1. Why are you writing?
2. Where or how did you learn of the company or the job?
3. What is your most important qualification for the job?

Begin your letter by stating directly that you are writing to apply for a job. Don't say that you "want to apply for the job"; such an opening raises the question, "Why don't you, then?" And don't waste the employer's time and your space on the page repeating verbatim the words of the advertisement.

Avoid an unconventional or arrogant opening: "Are you looking for a dynamic, young, and talented photographer?" Do not begin with a question; be more positive and professional.

If you learned of the job through a newspaper or journal, as Anthony Jones notes in his letter (Figure 7.11), make sure that you italicize or underscore the title.

> I am applying for the food-service manager position you advertised in the May 10 edition of the *Los Angeles Times* on the Internet.

Since more and more companies are announcing positions via the Internet, you will impress employers by checking the Net to see if their position is listed online. As one job counselor observed, if a candidate can use this most recent technological tool, he or she can probably do many other related tasks as well.

If you learned of the job from a professor, friend, or employee at the firm, state that fact. Take advantage of a personal contact who is confident you are qualified for the position, as Maria Lopez in Figure 7.12 and Patrice Cooper Bolger in Figure 7.13 do. But first confirm that your contact gives you permission to use his or her name.

The Body of Your Letter

This section of your letter, comprising one or two paragraphs, provides the evidence based on information from your résumé to show you are qualified for the job.

See Yourself as Your Employer Would

Concentrate very carefully on relating your education and job experiences to the employer's needs. But don't be egotistical and begin each sentence with "I." Vary your sentence structure. Focus on how your education or experience meets your employer's needs, as Anthony Jones does by citing Megalith Publications as models he followed, or as Patrice Cooper Bolger does in framing her public speaking abilities as an asset to the Tanselle Agency.

Highlight your qualifications by citing specific accomplishments. Don't simply state what you have done. Tell your reader exactly how your schoolwork and job experience qualify you to function and advance in the job advertised. Here, your homework on the company's history, structure, and goals will pay off.

In her letter of application in Figure 7.13, Patrice Cooper Bolger demonstrates how her wealth of previous experiences relates specifically to Tanselle Mental Health's outreach programs. Note, too, in Figure 7.12 how Maria Lopez uses her knowledge of Dr. Henrady's specialty in pedodontics to her advantage.

Keep your paragraphs readable and short. They should not exceed four or five sentences. Avoid long, complex sentences in the passive voice. Emphasize yourself as a doer with the active voice.

Education

You might want to spend one paragraph on your educational qualifications and one on your job experience; or perhaps your work or civic and community experiences are so rich that you will spend an entire paragraph on them, as Patrice Cooper Bolger does. At any rate, do not neglect education for experience or vice

FIGURE 7.11 Letter of application from Anthony Jones, a recent graduate with little job experience.

73 Allenwood Boulevard
Santa Rosa, CA 95401-1074

May 23, 1999

Ms. Jocelyn Nogasaki
Personnel Manager
Megalith Publishing Company
1001 Heathcliff Row
San Francisco, CA 94123-7707

Dear Ms. Nogasaki:

I am applying for the layout editor position advertised on May 22 in the San Francisco Chronicle. Early next month, I will receive an A.S. degree in industrial graphics illustration from Santa Rosa Junior College.

With a special interest in the publishing industry, I have successfully completed more than forty credit hours in courses directly related to layout design, where I acquired experience using QuarkXPress as well as Illustrator and Photoshop. You might like to know that many of the design patterns of Megalith publications were used as models in my graphic communications and photographic technology classes.

My studies have also led to practical experience at McAdam Publishers, as part of my Santa Rosa apprenticeship program. While working at McAdams, I was responsible for assisting the design department in photo research and in the creation of advertising layouts. Other related experience I have had includes artwork and proofreading for the student magazine, Thunder. As you will note on the enclosed résumé, I have also had experience in displaying merchandise at Buchman's Department Store.

I would appreciate the opportunity to discuss with you my qualifications in industrial graphics. After June 12, I will be available for an interview at any time that is convenient for you.

Sincerely yours,

Anthony H. Jones

Encl. Résumé

FIGURE 7.12 Letter of application from Maria Lopez, a recent graduate with some job experience.

1725 Brooke Street
Miami, FL 32701-2121

May 14, 1998

Dr. Marvin Henrady
Suite 34
Medical/Dental Plaza
839 Causeway Drive
Miami, FL 32706-2468

Dear Dr. Henrady:

Mr. Mitchell Pelbourne, my clinical instructor at Miami-Dade Community College, informs me that you are looking for a dental hygienist to work in your northside office. I am writing to apply for that position. This month I will graduate with an A.S. degree in the dental hygienist program at Miami-Dade Community College, and I will take the American Dental Assistants' Examination in early June.

I have successfully completed all course work and clinical programs in oral hygiene, anatomy, and prophylaxis techniques. During my clinical training, I received intensive practical instruction from a number of local dentists, including Dr. Mildred Pecos. Since your northside office specializes in pedodental care, you might find the subject of my major project—proper nutrition for preschoolers—especially relevant.

I have also had some related job experience in working with children in a health care setting. For a year and a half, I was employed as a ward clerk on the pediatric unit at St. Francis Hospital, and my experience in greeting patients, transcribing orders, and assisting the nursing staff would be valuable to you in running your office. You will find more detailed information about me and my experience in the enclosed résumé.

I would welcome the opportunity to talk with you about the position and my interest in pedodontics. I am available for an interview any time after 2:30 until June 11. After that date, I could come to your office any time at your convenience.

Sincerely yours,

Maria H. Lopez

Maria H. Lopez

Encl. Résumé

FIGURE 7.13 Letter of application from Patrice Cooper Bolger, a recent graduate with community and civic experience.

Patrice Cooper Bolger
1215 Lakeview Avenue
Westhampton, MI 46532
616-555-4772
pcbolger@aol.col

February 10, 1998

Dr. Lindsay Bafaloukos
Tanselle Mental Health Agency
4400 West Gallagher Drive
Tanselle, MI 46932-3106

Dear Dr. Bafaloukos:

At a recent meeting of the County Services Council, a member of your staff, Homer Strickland, informed me that you will be hiring a public affairs coordinator. Because of my extensive experience and commitment to community affairs, I would appreciate your considering me for this opening. I expect to receive my B.S. in public administration from Mid-Michigan College next year.

For the last ten years, I have organized community groups with outreach programs similar to Tanselle's. I have held administrative positions in the PTA, the Foster Parents' Association, and was president of Teens in Trouble, a volunteer group providing assistance to dysfunctional teens. My responsibilities with Teens have included coordinating our activities with various school programs, scheduling tutorials, and representing the organization before government agencies. I have been commended for my organizational and communication skills. My twenty presentations on foster home care and Teens in Trouble demonstrate that I am an effective speaker, a skill your agency would find valuable.

Because of my work at Mid-Michigan and in Teens and Foster Parents, I have the practical experience and education in communication and psychology to promote Tanselle's goals. The enclosed résumé provides details of my experience and education.

I would enjoy discussing my work with Teens and the other organizations with you. I am available for an interview any day after 11:00 a.m. Thank you for reaching me at the phone number or e-mail address at the top of this letter.

Sincerely yours,

Patrice Cooper Bolger

Patrice Cooper Bolger

Encl. Résumé

versa. Refer to your résumé, and do not forget to say that you are including it with your letter.

Recent graduates with little work experience will, of course, spend more time on their education, but even if you have much experience, don't forget to mention your education. Stress your most important educational accomplishments. Employers want to know what skills and expertise your education has given you and how those skills apply to their particular job. For instance, simply saying that you will graduate with a degree in criminal justice does not explain how you, unlike all the other graduates of all the other criminal justice programs, are best qualified for the job. But when you indicate that in thirty-six hours of course work you have specialized in industrial security and that you have twelve course hours in business and communications, that says something specific. Rather than just claiming that you are qualified, give the facts to prove it. If your GPA is relatively high or if you have won an award, mention it. Do not worry about repeating important information in both your letter and résumé.

Job Experience

After you discuss your educational qualifications, turn to your job experience. But if your experience is your most valuable and extensive qualification for the job, put it before a discussion of your education. If you are switching careers or are returning to a career after years away from the work force, start the body of your letter with your work experiences. Then mention your specific educational achievements. The best letters show how the two are related. Employers like to see continuity between a candidates's school and job experience. Provide that link by showing how the jobs you have held have something in common with your major—in terms of responsibility, research, customer relations, community service. Show how your course work in computer science helped you to be a more efficient programmer for your previous employer, or how your summer jobs for the local park district reinforced your skills in providing client services. However, do not dwell on being a nurse's aide three years ago when you have nearly completed a degree program to be a registered nurse. Busing tables in a restaurant last summer was good experience, but do not let that job overshadow your current work as a management trainee for a large hotel chain.

Your Closing Paragraph

Make your closing paragraph short—about two or three sentences—but be sure it fulfills these three important functions.

1. Emphasize once again your major qualifications.
2. Ask for an interview or a phone call.
3. Indicate when you are available for an interview.

End gracefully and professionally. Don't leave the reader with a single weak, vague sentence: "I would like to have an interview at your convenience." That does nothing to sell you. Say that you would appreciate talking with the employer

FIGURE 7.14 Letter of application from Donald Kitto-Klein, who has strong work experience.

Donald Kitto-Klein
56 South Ardmore Way
Petersburg, NY 15438
(716) 555-9032 kitto@gar.com

November 4, 1999

Ms. Akki Shibuto
Vice President, Operations
Patterson Corporation
Sun Valley, CA 94356

Dear Ms. Shibuto:

I am applying for the position of Director of Computer Services advertised on your home page last week. My knowledge of computers and my proven ability to work with people in a large organization such as Patterson's qualify me to be a productive member of your management team.

For the last three years, I have been responsible for all phases of computer maintenance and service at Tazwell Industries. As a regular part of my duties I have serviced and repaired 15 to 20 PCs a month as well as supervised a mainframe. I have successfully coordinated the activities of my department with Tazwell's other offices and worked closely with vendors and manufacturers. The training programs and in-house conferences I have conducted repeatedly receive high praise from Tazwell's management. Because I have worked so effectively with management and staff, I was promoted to repair team leader.

My educational achievements include a degree in computer technology from Grand Valley and certificates from U.S. Army schools in computers. The enclosed résumé will give you information about these and my other accomplishments.

I would like to put my ability to motivate personnel and to manage computer services to work for you and the Patterson Corporation. I am available for an interview at your convenience. I look forward to hearing from you.

Sincerely yours,

Donald Kitto-Klein

Encl. résumé

further to discuss your qualifications. If the employer's office is far from your city, you might ask for a phone call instead. Then mention your chief talent. You might also express your willingness to relocate if the job required it.

After indicating your interest in the job, give the times you are available for an interview and tell specifically where you can be reached. You might even repeat your phone numbers or e-mail address. If you are going to a professional meeting that the employer might also attend, or if you are visiting the employer's city soon, say so.

The following samples show how *not* to close your letter, and why not.

Pushy: I would like to set up an interview with you. Please phone me to arrange a convenient time. (That's the employer's prerogative, not yours.)

Too Informal: I do not live far from your office. Let's meet for coffee sometime next week. (Say instead that since you live nearby, you will be available for an interview.)

Too Humble: I know that you are busy, but I would really like to have an interview. (Say you would like to discuss your qualifications further.)

Introduces New Subject: I would like to discuss other qualifications you have in mind for the job. (How do you know what the interviewer might have in mind?)

Note that the closing paragraphs in Figures 7.11 through 7.14 avoid these errors.

Filling Out a Job Application

At some point in your job search, you will be asked to complete a prospective employer's application form. A recruiter may hand you a job application form at a campus interview, or you may be e-mailed a form in response to your letter of application. Most often, though, you will be given an application to complete when you are at the employer's office. The job application form, your letter of application, and your résumé are the three key written documents employers use to screen applicants.

Application forms can vary tremendously. But they all ask you to give information about your education, any military service, present and previous employment, references, general state of health, and reasons for wanting to work for the company or agency. Since these topics overlap with those on your résumé, bring the résumé with you to the employer's office to make sure you don't omit something important. Some forms even require you to attach your résumé. But *under no circumstances* should you attach a résumé to a blank form instead of filling the form out. Employers want their own forms completed by job seekers.

Some forms ask applicants to give reasons for leaving previous jobs and also require them to write a "personal essay" stating why the company should hire them. Both of these requirements will require tact and thought. If you were fired from a past job, it is not to your advantage to simply state that fact. Indicate further relevant information, such as that your company was downsized and you were laid off, or that your company merged and your department was eliminated. More frequently, though, your reasons for leaving a job will be financial, educational, or geographic. You may have received a better offer, decided to return to school, or planned to relocate.

Going to an Interview

An interview can be challenging, threatening, friendly, or chatty; sometimes it is all of these. By the time you arrive at the interview stage, you are far along in your job search. Basically, there are two kinds of interviews. One is a *screening interview,* to which numerous applicants have been invited so that a company can narrow down the candidates. Campus interviews are an example of screening interviews. The other kind of interview is known as a *line interview;* the employer invites only a few select applicants to the company's office for a tour and detailed conversation.

Preparing for an Interview

Interviews can last half an hour or extend to two or three days. Most often, though, an interview will last approximately one hour. It has been estimated that the applicant will do about 80 to 90 percent of the talking. Since you will be asked to speak at length, make the following preparations before your interview.

1. Do your homework about the employer—history, types of products or services provided, number of employees, location of main, branch, and overseas offices, contributions to industry or the community. Consult such relevant sources as those listed on pages 259–261 as well the company's home page on the Web.
2. Review the technical skills most relevant for the job. You might want to reread sections of a textbook, study some recent journal articles, or talk to a professor or an employee you know from the company.
3. Prepare a brief (one- or two-minute) review of your qualifications to deliver orally should you be asked about yourself.
4. Be able to elaborate on and supplement what is on your résumé. Your interviewer will have a copy on the desk, so you can be sure that its contents will be the subject of many questions. Any extra details or information that bring your résumé up to date ("I received my degree last week"; "I'll get the results of my state board examinations in one week") will be appreciated.

Questions to Expect at an Interview

You can expect questions about your education, job experience, and ambitions. An interviewer will also ask you about courses, schools, technical skills, and job goals. Through these questions an interviewer attempts to discover your good points as well as your bad ones. A common interviewer strategy is to postpone questions about your bad points until near the end of the interview. Once a relaxed atmosphere has been established, the interviewer thinks that you may be less reluctant to talk about your weaknesses. The following fifteen questions are typical of those you can expect from interviewers.

1. Tell us something about yourself. (Here's where the prepared one-minute oral presentation of yourself comes in handy.)
2. Why do you want to work for us? (Recall any job goals you have and apply them specifically to the job under discussion.)
3. What qualifications do you have for the job? (Mention educational achievements in addition to relevant work experience, especially computer skills.)
4. What could you possibly offer us that other candidates do not have? (Say "enthusiasm," being a team player, and problem-solving abilities in addition to educational achievements.)
5. Why did you attend this school? (Be honest—location, costs, programs.)
6. Why did you major in "X"? (Do not simply say financial benefits; concentrate on both practical and professional benefits. Be able to state career objectives.)
7. Why did you get a grade of "C" in a course? (Do not hurt your chances of being hired by saying that you could have done better if you tried; that response shows a lack of motivation most employers find unacceptable. Explain what the trouble was and mention that you corrected it in a course in which you earned a B or an A.)
8. What extracurricular activities did you participate in while in high school or college? (Indicate any duties or responsibilities you had—handling money, writing memos, coordinating events; if you were not able to participate in such activities, tell the interviewer that a part-time job, community or church activities, or commuting a long way to school each day prevented your participating. Such answers sound better than saying that you did not like sports or fraternities or clubs in school.)
9. Did you learn as much as you wanted from your course work? (This is a loaded question. Indicate that you learned a great deal but now look forward to the opportunity to gain more practical skill, to put into practice the principles you have learned; say that you will never be through learning about your major.)
10. Why was your summer job important? (Highlight skills you learned, people you helped, employers you pleased.)
11. What is your greatest strength? (Being a team player, cooperation, willingness to learn, ability to grasp difficult concepts easily, managing time or money, taking criticism easily, and profiting from criticism are all appropriate answers.)
12. What is your greatest shortcoming? (Be honest here and mention it, but then turn to ways in which you are improving. You obviously don't want to say something deadly like, "I can never seem to finish what I start" or "I hate being criticized." You should neither dwell on your weaknesses nor keep silent about them. Saying "None" to this kind of question is as inadvisable as rattling off a list of faults.)
13. How much did you earn at your last job and what salary would you expect from us? (Some job counselors wrongly advise interviewees to lie about their

past salaries in order to get a larger one from the future employer. But if the prospective employer checks your last salary and finds that you have lied, you lose. It is better to round off your last salary to the nearest thousand.)

14. Why did you leave your last job? (Usually you will have educational reasons:—"I returned to school full-time"—or geographical reasons:—"I moved from Jackson to Springfield." *Never attack your previous employer.* This only makes you look bad.)
15. Is there anything else you want to discuss? (Here is your opportunity to end the interview with more information about yourself. You might take time to reiterate your strengths, to correct an earlier answer, or to express your desire to work for the company.)

Of course, you will have a chance throughout the interview to ask questions, too. Do not forget important points about the job: responsibilities, opportunities for further training, security, and chances for promotion. You will also want to ask about salary (but do not dwell on it), fringe benefits, schedules, vacations, and bonuses. If you spend time on these subjects, especially during the first part of the interview, you tell the interviewer that you are more interested in the rewards of the job than in the duties and challenges it offers. Do not go to the interview with dollar signs flashing in your eyes.

Some questions an interviewer may not legally ask you. Questions about your age, marital status, ethnic background, race, or any physical handicaps violate equal opportunity employment laws. Even so, some employers may disguise their interest in these subjects by asking you indirect questions about them. A question such as "Will your husband care if you have to work overtime?" or "How many children do you have?" could probe into your personal life. Confronted with such questions, it is best to answer them positively ("My home life will not interfere with my job," "My family understands that overtime may be required") rather than bristling defensively, "It's none of your business if I have a husband."

Interview Do's and Don'ts

Keep in mind some other interview "do's" and "don'ts."

1. Be on time. If you are unavoidably delayed, telephone to apologize and set up another interview. Go to the interview alone.
2. Dress appropriately for the occasion.
3. Speak slowly and distinctly; do not nervously hurry to finish your sentences, and never interrupt or finish an interviewer's sentences.
4. Do not smoke, even if the interviewer offers you a cigarette. Also refrain from chewing gum, fidgeting, or tapping your foot against the floor or a chair.
5. Maintain eye contact with the interviewer; do not sheepishly stare at the floor or the desk. Body language is equally important. For instance, don't fold your arms—that's a signal indicating you are closed to the interviewer's suggestions and comments.
6. And one last point: When the interview is over, thank the interviewer for considering you for the job.

FIGURE 7.15 A follow-up letter.

2739 East Street
Latrobe, PA 17042-0312
September 20, 1999

Mr. Jack Fukurai
Personnel Manager
Transatlantic Steel Company
1334 Ridge Road N.E.
Pittsburgh, PA 17122-3107

Dear Mr. Fukurai:

I enjoyed talking with you last Wednesday and learning more about the security officer position available at Transatlantic Steel. It was especially helpful to take a tour of the plant's north gate section to see the challenges it presents for the security officer stationed there.

As you noted at my interview, my training in surveillance electronics has prepared me to operate the sophisticated equipment Transatlantic has installed at the north gate. I was grateful to Ms. Turner for taking time to demonstrate this equipment.

I am looking forward to receiving the brochure about Transatlantic's employee services. Would it also be possible for you to include a copy of the newsletter introducing the new security equipment to the employees?

Thank you for considering me for the position and for the hospitality you showed me. I look forward to hearing from you. After my visit last week, I know that Transatlantic Steel would be an excellent place to work.

Sincerely yours,

Marcia Le Borde

Marcia Le Borde

FIGURE 7.16 Letter accepting a job.

73 Park St.
Evansville, WI 53536-1016
June 29, 1998

Ms. Melinda Haas, Manager
Weise's Department Store
Janesville Mall
Janesville, WI 53545-1014

Dear Ms. Haas:

I am pleased to accept the position of assistant controller that you offered me in your letter of June 22. Starting on July 18 will be no problem for me. I look forward to helping Ms. Meyers in the business office. In the next few months I know that I will learn a great deal about Weise's.

As you requested, I will make an appointment for early next week with the Personnel Department to discuss travel policies, salary payment schedules, and insurance coverage.

I am eager to start working for Weise's.

Cordially,

Kevin Dubinski

Kevin Dubinski

The Follow-up Letter

Within a week after the interview, it is a wise strategy to send a follow-up letter thanking the interviewer for his or her time and interest in you. The letter will keep your name fresh in the interviewer's mind. Do not forget that this individual interviewed other candidates, too—some of them probably on the same day as you.

In your follow-up letter, you can reemphasize your qualifications for the job by showing how they apply to conditions described by the interviewer; you might also ask for some further information to show your interest in the job and the employer. You could even refer to a detail, such as a tour or film that was part of the interview. A sample follow-up letter appears in Figure 7.15.

Accepting or Declining a Job Offer

Even if you have verbally agreed to take a job, you still have to respond formally in writing. Your letter will make your acceptance official and will probably be included in your permanent personnel file. Accepting a job is easy. Make this communication with your new employer a model of clarity and diplomacy. Respond to the offer as soon as possible (certainly within two weeks). Often a time limit is specified.

A sample acceptance letter appears in Figure 7.16. In the first sentence tell the employer that you are accepting the job, and refer to the date of the letter offering you the position. Indicate when you can begin working. Then mention any pleasant associations from your interview or any specific challenges you are anticipating. That should take no more than a paragraph. Do not fill your letter with praise for the employer or the job.

In a second paragraph express your plans to fulfill any further requirements for the job—going to the personnel office, taking a physical examination, having a copy of a certificate or license forwarded, sending a final transcript of your college work. A final one-sentence paragraph might state that you look forward to starting your new job.

Refusing a job requires tact. You are obligated to inform an employer why you are not taking the job. Since the employer has spent time interviewing you, respond with courtesy and candor. For an example of a refusal letter, see Figure 7.17.

Do not bluntly begin with the refusal. Instead, prepare the reader for bad news by starting with a complimentary remark about the job, the interview, or the company. Then move to your refusal and supply an honest but not elaborate explanation of why you are not taking the job. Many students cite educational opportunities, work schedules, geographic preference, health reasons, or better, more relevant professional opportunities. End on a friendly note, because you may be interested in working for the company in the future and do not want to arouse any bad feelings.

✓ Revision Checklist

- ❑ Restricted the types of job(s) for which I am qualified.
- ❑ Prepared a dossier at school placement office with letters from professors, employers, and community officials.
- ❑ Identified places where relevant jobs are advertised.
- ❑ Notified instructors, friends, relatives, clergy, and individuals who work for the companies I want to join that I am in the job market.
- ❑ Checked with Chamber of Commerce, state employment office, and relevant government agencies.
- ❑ Researched the companies I am interested in—on the Internet, through printed sources, and in interviews with current employees I know.
- ❑ Inventoried strengths carefully to prepare résumé.
- ❑ Eliminated weak, irrelevant, repetitious, and dated material from inventory.

FIGURE 7.17 Letter refusing a job.

345 Melba Lane
Bellingham, WA 98225-4912
March 8, 2000

Ms. Gail Buckholtz-Adderley
Assistant Editor
The Everett News
Everett, WA 98421-1016

Dear Ms. Buckholtz-Adderley:

I enjoyed meeting you and the staff photographers at my recent interview for the photography position at the **News**. Your plans for the special weekend supplements are exciting, and I know that I would have enjoyed my assignments greatly.

However, because I have decided to continue my education part time at Western Washington University in Bellingham, I have accepted a position with the **Bellingham American**. Not having to commute to Everett will give me more time for my studies and also for my freelance work.

Thank you for your generous offer and for the time you and the staff spent explaining your plans to me. I wish you much success with the supplements.

Sincerely yours,

George Alexander

George Alexander

- ❑ Wrote a focused and persuasive career objective statement.
- ❑ Determined the most beneficial format of résumé to use—chronological or functional or both.
- ❑ Prepared an electronic résumé to send on-line if employer so directs.
- ❑ Made résumé attractive and easy to read with logical and persuasive headings and hyperlinks.
- ❑ Made sure résumé contains neither too much nor too little information.
- ❑ Proofread résumé to ensure everything is correct, consistent, and accurate.
- ❑ Wrote letter of application that shows how my specific skills and background apply to and meet an employer's exact needs.
- ❑ Prepared short oral presentation about myself and my accomplishments for an interview.
- ❑ Sent prospective employer a follow-up letter within a few days after interview to show interest in position.
- ❑ Sent prospective employer a polite acceptance or rejection letter, depending on situation.

Exercises

1. Make a brainstormed list of your marketable job skills. To do this, first concentrate on the specialized kinds of skills you learned in your major or on your job (for example, giving injections, fingerprinting, preparing specialized menus, keeping a ledger book, operating a computer, learning a computer language). List as many of these skills as you can think of; then organize them into three or four separate categories that reflect your major abilities.

2. Translate the list of skills from Exercise 1 into a number of "selling clauses" (see pp. 237–238), each introduced with a strong action verb.

3. Using at least four different sources, including the Internet, compile a list of ten employers for whom you would like to work. Get their names, addresses, phone numbers, and the names of the managers or personnel officers. Then select one company and write a profile about it—location, services, kinds of products or services offered, number of employees working for it, clients served, types of schedules used, and other pertinent facts.

4. Write an e-mail message to your instructor describing the resources you used to find three to four jobs in your area that you could apply for upon graduation.

5. Obtain some personal evaluation forms from your placement office. Write a sample letter to a former or current teacher and employer, asking for a recommendation. Tell these individuals what kinds of jobs you will be looking for and politely mention how a strong letter would help you in your job search.

Make sure that you bring them up-to-date about your educational progress and any employment you have had since you worked for them.

6. Which of the following would belong on your résumé? Which would not belong? Why?
   a. student I.D. number
   b. Social Security number
   c. the ZIP codes of your references' addresses
   d. a list of all your English courses in college
   e. section numbers of the courses in your major
   f. statement that you are recently divorced
   g. subscriptions to journals in your field
   h. the titles of any stories or poems you published in a high school literary magazine or newspaper
   i. your GPA
   j. foreign languages you studied
   k. years you attended college
   l. the date you were discharged from the service
   m. names of the neighbors you are using as references
   n. your religion
   o. job titles you held
   p. your summer job washing dishes
   q. your telephone number
   r. the reason you changed schools
   s. your current status with the National Guard
   t. the name of your favorite professor in college
   u. your volunteer work for the Red Cross
   v. hours a week you spend reading science fiction
   w. the title of your last term paper in your major
   x. the name of the agency or business where you worked last

7. Indicate what is wrong with the following career objective statements and rewrite them to make them more precise and professional.
   a. Job in a dentist's office.
   b. Position with a safety emphasis.
   c. Desire growth position in a large department store.
   d. Am looking for entry position in health sciences with an emphasis on caring for older people.
   e. Position in sales with fast promotion rate.
   f. Want a job working with semiconductor circuits.
   g. I would like a position in fashion, especially one working with modern fashion.
   h. Desire a good-paying job, hours: 8–4:30, with double pay for overtime. Would like to stay in the Omaha area.
   i. Insurance work.
   j. Computer operator in large office.
   k. Personal secretary.
   l. Job with preschoolers.

m. Full-time position with hospitality chain.
n. I want a career in nursing.
o. Police work, particularly in suburb of large city.
p. A job that lets me be me.
q. Desire fun job selling cosmetics.
r. Any position for a qualified dietitian.
s. Although I have not made up my mind about which area of forestry I shall go into, I am looking for a job that offers me training and rewards based upon my potential.

8. As part of a team or on your own, revise the following poor résumé to make it more precise and persuasive. Include additional details where necessary and exclude any details that would hurt the job seeker's chances. Also correct any inconsistencies.

RÉSUMÉ OF

Powell T. Harrison
8604 So. Kirkpatrick St.
Ardville, Ohio
345 37 8760
614 234 4587
harrison@gem.com

PERSONAL	Confidential
CAREER OBJECTIVE	Seek good paying position with progressive Sunbelt company.
EDUCATION	
1991–1994	Will receive degree from Central Tech. Institute in Arch. St. Earned high average last semester. Took necessary courses for major; interested in systems, plans, and design development.
1987–1991	Attended Ardville High School, Ardville, OH; took all courses required. Served on several student committees.
EXPERIENCE	None, except for numerous part-time jobs and student apprenticeship in the Ardville area. As part of student app. worked with local firm for two months.
HOBBIES	Surfing the Net, playing Nintendo video games. Member of Junior Achievement.
REFERENCES	Please write for names and addresses.

---

9. Determine what is wrong with the following sentences in a letter of application. Rewrite them to eliminate any mistakes, to focus on the "you attitude," or to make them more precise.
   a. Even though I have very little actual job experience, I can make up for it in enthusiasm.
   b. My qualifications will prove that I am the best person for your job.
   c. I would enjoy working with your other employees.
   d. This letter is my application for any job you now have open or expect to fill in the near future.
   e. Next month, my family and I will be moving to Detroit, and I must get a job in the area. Will you have anything open?
   f. If you are interested in me, then I hope that we make some type of arrangements to interview each other soon.
   g. I have not included a résumé since all pertinent information about me is in this letter.
   h. My GPA is only 2.5, but I did make two Bs in my last term.
   i. I hope to take state boards soon.
   j. Your company, or so I have heard through the grapevine, has excellent fringe benefits. That is what I care about most, so I am applying for any position which you may advertise.
   k. I am writing to ask you to kindly consider whether I would be a qualified person for the position you announced in the newspaper.
   l. I have made plans to further my education.
   m. My résumé speaks for itself.
   n. I could not possibly accept a position which required weekend work, and night work is out, too.
   o. In my own estimation, I am a go-getter—an eager beaver, so to speak.
   p. My last employer was dead wrong when he let me go. I think he regrets it now.
   q. When you want to arrange an interview time, give me a call. I am home every afternoon after four.

10. Explain why the following letter of application is ineffective. Rewrite it to make it more precise and appropriate.

Apartment 32
Jeggler Drive
Talcott, Arizona

Monday

Grandt Corporation
Production Supervisor
Capital City, Arizona

Dear Sir:

I am writing to ask you if your company will consider me for the position you announced in the newspaper yesterday. I believe that with my education (I have an associate degree) and experience (I have worked four years as a freight supervisor), I could fill your job.

My schoolwork was done at two junior colleges, and I took more than enough courses in business management and modern technology. In fact, here is a list of some of my courses: Supervision, Materials Management, Work Experience in Management, Business Machines, Safety Tactics, Introduction to Packaging, Art Design, Modern Business Principles, and Small Business Management. In addition, I have worked as a loading dock supervisor for the last two years, and before that I worked in the military in the Quartermaster Corps.

Please let me know if you are interested in me. I would like to have an interview with you at the earliest possible date, since there are some other firms also interested in me, too.

Eagerly yours,

George D. Milhous

---

11. From the Sunday edition of your local newspaper, or from one of the other sources discussed on pages 232–235, find notices for two or three jobs you believe you are qualified to fill and then write a letter of application for one of them.

12. Write a chronologically organized résumé to accompany the letter you wrote for Exercise 11 above.

13. Write a functional résumé for your application letter in Exercise 11.

14. Bring the two résumés you prepared for Exercises 12 and 13 to class to be critiqued by a collaborative writing team. After your résumés are reviewed, revise them.

15. Write an appropriate job application letter to accompany Anna Cassetti's résumés in Figures 7.3 and 7.6.

16. Write a letter to a local business inquiring about summer employment. Indicate that you can work only for one summer and that you will be returning to school by September 1.

Writing Promotional Literature: News Releases, Brochures, Newsletters, and Home Pages

News releases, brochures, newsletters, and home pages are important marketing tools found in all businesses and organizations. These four types of documents—sometimes called promotional literature—disseminate important information about your company. They are vital to the success of any company or agency because they promote its products and services and inform and congratulate its work force. Without marketing, a company will fail. The fact that your company or agency designs state-of-the-art equipment or offers essential services will mean very little if customers know nothing about them.

You may not be a marketing or public relations major, but you are more familiar with these marketing tools than you may think. Brochures and newsletters are found everywhere—you get them in the mail, see electronic versions on the Internet, pick them up at local businesses or through your memberships in organizations. You may never have seen an actual news release, but you certainly have read stories in your local newspaper or in a professional publication that are based on them. Which brochures or newsletters caught your attention recently and why? Which ones did you glance at for only a second or two before discarding? Has a brochure ever helped you to understand a subject you had some interest in, whether it related to your career, your specific job responsibilities, your health, home, car, or family? What stories in a newsletter have you enjoyed reading and which ones did you find irrelevant and skip? Questions like these go to the heart of what makes a successful news release, brochure, newsletter, or home page.

Why You Need to Know How to Write Promotional Pieces

The goal of this chapter is to introduce you to some of the strategies necessary to write and design successful news releases, brochures, newsletters, and home pages.

You might think that you will never be asked to write these documents. After all, don't large companies have their own public relations and promotion departments? Don't they hire high-priced consultants to write these materials? While many companies do have these resources, you may work for a smaller firm that does not, or you may be asked to assist an organization to get its message across to the members. Even in a large company, you may be called on to help prepare a brochure or home page, or to write an article about your department for a newsletter. Knowing how to write and design these documents is among the most marketable skills you can list on your résumé.

As you read this chapter, keep in mind that creating an effective news release, brochure, newsletter, or home page calls for many skills—verbal and visual—necessary in the world of work, including:

- analyzing the needs of a specific audience
- researching a topic of interest to your audience
- collaborating successfully with co-workers
- writing clearly and concisely
- choosing the most appropriate visuals
- correlating visuals with text
- designing an attractive document

Audiences for Promotional Pieces

To write an effective promotional piece, start with some very careful audience analysis (see Chapter 1). Determine *who* you are communicating with, *why*, and *about what*. Though audiences may vary for the four types of documents, the principles of finding out what your audience needs to know and how best to provide that information are the same.

Here are four important questions to ask yourself as you prepare news releases, brochures, newsletters, or home pages:

1. Is the subject of interest to my reader?
2. Does my discussion answer basic questions my readers may have about the topic?
3. Is my work written with my readers in mind—not too detailed or technical for consumers?
4. Have I made information easy for my readers to find?

Audiences for News Releases

Your most significant readers for a news release will be the editors who will decide whether to print your message and the potential customers who will read it. Do not send your releases out at random to every editor of any publication that might look promising. Study your message and decide who would be most interested in it. After you determine if your release is intended for a technical, consumer, or business audience (or a combination of these), you have successfully identified your target audience. Then examine relevant publications and other media outlets to see which editors might be most interested in your release.

If your news release appeals to business readers, for example, send it to the business or financial page editor. If your message is targeted for nonspecialist consumers, don't send it to the editor-in-chief of a local newspaper; find the names of the editors of the most pertinent sections of the publication.

Audiences for Brochures

Your audience for a brochure can be consumers interested in your product or service, company employees, or specialists in your field. Will you be writing for consumers whose reading level is at about the ninth-grade level or will you be writing to technical specialists and related professionals? Will your purpose be to convince a customer to order a product, try a service, or attend a conference; explain a policy; or encourage readers to change a behavior? Brochures sell a reader by proving that the writer sees things from the reader's perspective, so the focus on the audience is paramount. Demonstrate that you know your readers' problems, business goals, even their schedules, if necessary. To win your readers, identify a key benefit or solution. Then showcase that benefit visually and in writing—point out ways it can save readers time or money; improve their communications, lifestyle, or health; or contribute to a cleaner, safer environment.

Audiences for Newsletters

Your audience may be the employees and managers of your firm; customers—past, current, or potential—your company wants to reach; a segment of the consumer population that requires information or guidance; members of an organization to which you belong; or some combination of these groups. Newsletters usually target people who have an ongoing readership with the source (such as members in a professional association; stockholders in a company). Look at previous issues of the newsletter to see what specific audience needs are addressed. Whoever your audience is, they will be reading your newsletter for information that they cannot get elsewhere as quickly or as easily.

Audiences for Home Pages

Your potential Web audience is vast—millions of people all around the globe. How much of this potential you realize, how many people visit your home page, and how many go to your other pages depend on how well you do your homework, how you include hyperlinks, and how attractively you present your material. In general, this audience has an extremely short attention span and unless you immediately catch an individual's attention, he or she will likely go to another site—maybe that of a competitor.

News Releases

A news release, sometimes called a press or media release, is a short announcement (usually a page or two) about your company or agency's specific products, services, or personnel. News releases are written for the media to gain your company free publicity; they bring your company's name and message to the attention of the public—potential and current customers. Releases are sent by mail or (increasingly) by fax machines programmed with a battery of appropriate numbers to editors of trade publications (specialized, technical magazines, newsletters, and journals) as well as to local newspapers; they can also be sent to news directors of television and radio stations. Figure 8.1 is a sample news release that publicizes a safety announcement; the labels show the key parts of a release.

The first goal in writing a news release—and a crucial one—is to convince an editor to print it. That is not always an easy job. Editors of trade publications polled about the number of releases they actually printed revealed that the figure was less than 20 percent.

FIGURE 8.1 Sample news release.

News Release

US Army Corps
of Engineers
Vicksburg District

Contact person

Patty K. Elliott
Media Services Coordinator
2101 N. Frontage Road
Vicksburg, MS 39180-5191

Release No: LMK- 94-118
Phone: (601) 631-5053
FAX: (601) 631-5225

For Immediate Release

August 17, 1994

Slug

CORPS TEST: MERCURY, DIOXIN
NOT A CONCERN FOR DREDGING

Lead

VICKSBURG....After conducting a series of tests, the U.S. Army Corps of Engineers has determined that mercury and dioxin are not problems in the maintenance dredging of the Ouachita River.

Supporting information

The tests were begun in early July because area residents feared that annual maintenance dredging would increase levels of these compounds. The Corps believes that these tests should restore public confidence in the safety of the dredging activity.

The Corps took weekly measurements of mercury both upstream and downstream of the dredging site; dioxin measurements were also taken at five locations in the Crossett area, four below Coffee Creek and one above. Tests were conducted both before and after dredging.

The tests showed that the mercury and critical dioxin levels were well below the EPA criteria for protection of human health, and that no levels of TCDD (an especially toxic form of dioxin) were detectable. Moreover, the tests revealed that levels of dioxin and mercury were consistent with those in other rivers in the county.

Even more promising, the Corps tests agreed with similar recent dioxin tests in the region that confirmed dioxin levels have declined significantly.

Signs end of the release

#

TECH NOTE

In addition to faxing your news release to an editor, you can send it electronically to a hot line, Web site, or computer site. Numerous businesses and organizations use their home page on the Internet to disseminate their news releases. A news release can be retrieved from the Internet. Note how NASA advises readers on how to obtain its on-line releases.

> NASA press releases and other information are available automatically by sending an Internet electronic mail message to domo@hq.nasa.gov. In the body of the message (not the subject line) users should type the words "subscribe press-release" (no quotes). The system will reply with a confirmation via E-mail of each subscription. A second automatic message will include additional information on the service. NASA releases also are available via CompuServe using the command GO NASA.

The Internet now contains news releases in abundance. You may want to surf it to compare the ways they are written. News releases about computer software and hardware—sales and service—are often sent to on-line information services through Newbytes, a newsgroup system that charges subscribers a fee to format and provide first-rate commercially important news. Large firms also subscribe to PR Newswire, a service that will, for a fee, distribute news releases across the nation. Your firm pays a fee each time someone requests one of its releases through PR Newswire.

To write a successful release, therefore, you probably have to sell two key audiences—an editor to print it and a potential customer or client to pay attention to it. Your company's image can be enhanced or tarnished by the release you write. Sending out ineffective or unclear releases is a waste of everyone's time.

Subjects Appropriate for News Releases

News releases can be written about a variety of newsworthy subjects. Here is a short list of topics about which you might write a news release.

1. **New products or services**
   - a new model or line of merchandise or a significantly improved piece of equipment
   - new or expanded services
   - the entrance of your company (service or product line) into a new and productive market
   - the application of the latest technology in creating a product, as when a company announces that its services are available on the Internet

New product releases usually emphasize benefits to users and announce costs and availability dates. The MCI release in Figure 8.2 reports on an important new service.

FIGURE 8.2 A news release announcing a new service.

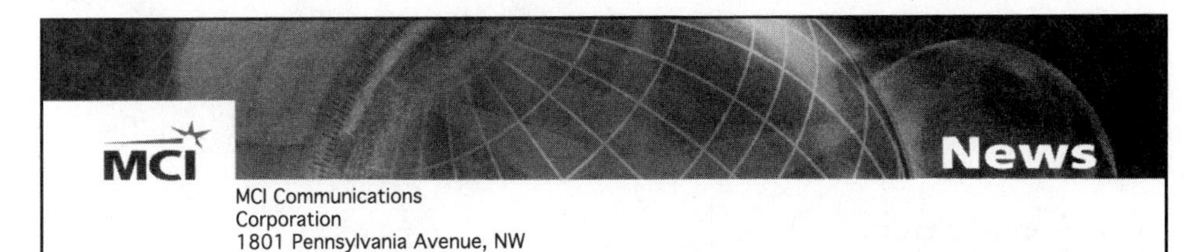

MCI Communications
Corporation
1801 Pennsylvania Avenue, NW
Washington, DC 20006

Contact:

Mark Pettit
MCI News Bureau
1-800-644-NEWS
newsmci@mcimail.com

Robyn Freedman
Ketchum Public Relations
404-877-1800

For Release:

IMMEDIATE

MCI INTRODUCES 1-800 MUSIC NOW (sm)

Leverages Intelligent Infrastructure to Sell Recorded Music

NEW YORK (Nov. 8, 1995) -- MCI Communications Corporation (NASDAQ: MCIC) today announced 1-800 MUSIC NOW, which turns every touch tone telephone in the country into a music store allowing consumers to sample and buy from more than 5,000 music titles, 24 hours a day. This introduction marks MCI's entrance into the $12 billion music distribution business, which is experiencing unprecedented growth rates of 12.5% annually.

1-800 MUSIC NOW allows sampling and sales in 14 different formats, from pop to country to children's. Customers may purchase CDs or cassettes, and all purchases are billed to a major credit card. Merchandise will be delivered within days, or even overnight.

The introduction of this service marks not only the first brand of its kind, but the establishment of a new category for music retailing...

"1-800 MUSIC NOW is another example of MCI's strategy to use our core strengths and existing assets against new opportunities, with little incremental investment," said Timothy F. Price, MCI Telecommunications Corporation president. "Through our existing intelligent network, anyone in America can take advantage of 1-800 MUSIC NOW. A traditional retailer would have to spend years and billions of dollars to achieve the same national presence."

# # #

2. **New policies and/or procedures**

   - acquisition of new technologies to lower production costs
   - changes in production to improve safety

In this type of release, you explain how the new procedure differs from the old one, what benefits are gained, and when and why the change will be introduced.

3. **Personnel changes**

   - appointments
   - promotions
   - employee recognition for earning a certificate or winning an award
   - hiring additional staff
   - restructuring a department

Explain the reasons and stress the benefits to the customer of such personnel decisions. Even retirements or transfers can merit a news release if such changes reflect positively on your company. Figure 8.3 shows a news release about a Native American contracting firm being honored by the Small Business Administration.

4. **New construction**

   - plant or office openings
   - new satellite, branch, or overseas offices
   - major remodeling

Again, emphasize why and how such changes will make doing business with your company more convenient and more profitable. If new jobs are created, local media will *always* follow up on your release.

5. **Financial and business news**

   - stock reports
   - quarterly earnings and dividends
   - sales figures
   - company reorganizations
   - new acquisitions

The news release in Figure 8.4 announces a newsworthy financial event at the Upjohn Company.

6. **Special events**

   - training seminars
   - demonstrations of new models
   - trial runs of machinery
   - charity benefits
   - visits of national speakers
   - dedications
   - tours of key facilities previously closed to the public

FIGURE 8.3 A news release about an honor received.

Idaho National Engineering Laboratory

INFORMATION

Lockheed Martin Idaho Technologies

INEL SUBCONTRACTOR WINS NATIONAL RECOGNITION FROM THE SBA FOR ITS WORK ON IDAHO CONTRACT

May 10, 1996

For more information, contact Scott Hallman at (208) 526-4448; e-mail: sh@inel.gov

WASHINGTON, D.C., May 8, 1996 – Environmental Management Inc., a Native American-owned firm with corporate headquarters in Idaho Falls, has been named the 1996 Small Business Subcontractor of the Year in Region 10 by the United States Small Business Administration for its work at the Idaho National Engineering Laboratory.

The award was presented to Jerry Walkingstick, President of EMI, at recent ceremonies in Washington, D.C.

EMI became eligible for the award after being nominated by Parsons Environmental Services, Lockheed Martin Idaho Technologies' remedial design and remedial action subcontractor at the INEL (Idaho National Engineering Laboratory). EMI won and executed a contract for environmental restoration work for Parsons in 1995. The job entailed excavation and removal of contaminated soil, using unique EMI-developed methods that prevented contaminated dust from escaping.

In conjunction with the project, EMI trained and mentored employees of another Native American firm, Sorrell Builders. Sorrell is affiliated with the Shoshone-Bannock Tribes, whose reservation borders the INEL.

All contract work was completed ahead of schedule and with no accidents or injuries, earning EMI its valued status in the INEL's top priority area, safety.

EMI won the Region 10 Award in a Small Business Administration competition against 30 other nominees from the Pacific Northwest.

"EMI is very deserving of this award," said John Denson, President of Lockheed Martin Idaho Technologies. "No accomplishment is more valuable than keeping your promises, and EMI has managed to complete its commitments thus far with a fine record of doing what it said it would. We congratulate them on this well deserved honor."

# # #

P.O. Box 1625 Idaho Falls, Idaho 83415-3695 (208) 526-0445

FIGURE 8.4 A news release reporting a corporate acquisition.

ON-CALL

Below is the story you requested. To retrieve additional stories immediately by fax, please call (800) 758-5804. Company News on Call is available on the World Wide Web at http://www.prnewswire.com.

PR NEWSWIRE

Story 14351
6/1/95

UPJOHN TO ACQUIRE PARKE-DAVIS VETERINARY BUSINESS

KALAMAZOO, Mich., June 1/PRNewswire/—The Upjohn Company (NYSE: UPJ) today announced the purchase of the Parke-Davis veterinary business from Warner-Lambert Company (NYSE: WLA) and its affiliates. The acquisition will include all of Parke-Davis' animal health products sold in Germany, the U.K., Ireland, Austria and Switzerland. According to Peter A. Croden, vice president, Upjohn Worldwide Animal Health, the acquisition is part of Upjohn's global growth strategy and provides for expansion in important European animal health markets.

"We are committed to growing our business in Europe, and the purchase of Parke-Davis veterinary business gives us a complete product range and a stronger business presence in strategic animal health markets, particularly in Germany and the U.K.," Croden said. "The two businesses complement each other to give us key products for both large and small animals."

The acquisition includes more than 50 animal health products including antibiotics and steroids for use in companion animals as well as some large animal products. In addition, Parke-Davis veterinary business employees in Germany will have the opportunity to continue with Upjohn.

The acquisition was effective May 31.

Warner-Lambert Company is devoted to discovering, developing and marketing pharmaceutical, consumer health care and confectionery products, with headquarters in Morris Plains, New Jersey.

The Upjohn Company is a worldwide, research-based provider of human health care products, animal health products and specialty chemicals. Headquartered in Kalamazoo, Mich., the company has been dedicated to improving health and nutrition for more than a century.

CONTACT: Maury Ewalt of Upjohn, 616-385-6658.

# # #

For information about PR Newswire call:
800-832-5522

Information and news releases furnished by the members of PR Newswire, who are responsible for their fact and content.

Supply dates and times, and explain what kinds of benefits—financial, technological, educational—result from readers participating in or attending such events.

All of the topics above focus on a company's achievements—the positive contributions a firm makes to its customers, clients, and community. Your news releases should project an image of your company as professional and progressive, customer-centered, quality-focused, and eager to improve products, services, and customer relations.

Not all the news you may be asked to announce is pleasant. At times you will have to report events that are difficult—product recalls, work stoppages or strikes, layoffs, limited availability or unavailability of parts, fires, store or plant closings, higher prices. Even when you have to write about such unpleasant topics, portray your company honestly in the most professional and conscientious light.

Topics That Do Not Warrant a News Release

Not every event or change at your company warrants a news release. The chief reason why editors refuse to publish a release (or why Internet users refuse to read it) is that they do not consider it newsworthy. Most news releases about events that have already happened are considered old news and wind up in the wastebasket. Below are some topics that hold little or no interest for editors; avoid writing news releases about them.

1. **Well-known products or services.** News about these subjects is as interesting as yesterday's newspaper. Just because your firm makes a slight modification in a product (such as a change in color or a minor additional feature) does not mean it warrants a news release.
2. **Products or services still in the planning stages.** Not only would such news be premature, but releasing it ahead of time might meet with your boss's severe disapproval, not to mention your competition's delight. (In fact, always check with your superior before you prepare a news release on a topic.)
3. **Controversial products or services.** Refrain from writing a news release on these subjects unless your boss instructs you to do so.
4. **A history of your department, division, or company.** Editors are experts at separating real news from company hype. Save remarks about your firm's history for the annual report to stockholders.
5. **An obviously padded tribute to your boss or to your company.** Again editors will see through a flowery collection of quotations.

Generating News

How do you ethically generate news for a release? That is a large topic discussed in marketing courses. Many sales are generated by ordinary business activity. But there are many ways to highlight your company's initiative. One authority, Seth Godwins, suggests running a contest or conducting a survey. Then you report on the results. Other public relations experts recommend that you offer a free seminar on some aspect of your business and invite the public, or establish a speaker's bureau through which your company's workers can address community groups on relevant topics. You can then send out releases about these newsworthy activities.

Format of a News Release

As you can see from the examples in Figures 8.1 through 8.4, news releases follow a conventional format that editors will expect you to adhere to. Reflecting the logical divisions of a news release, this format lists information in convenient and customary places. Refer to the labeled news release in Figure 8.1 as you read the guidelines below.

1. Never send a cover letter with your news release unless you are faxing it.
2. Print your news release on a special news release form and always supply a name with an address (Internet, e-mail, and street), phone and fax number, of a contact person an editor can call with any questions. Some releases list two names, as Figure 8.3 does, to reflect a collaborative effort. (The name(s) of the contact person(s) may sometimes follow the news release.)
3. Print your release on only one side of the page. Make sure that your print is dark and easy to read. Your release must look professional—leave ample margins.

TECH NOTE

Do not use fancy or script typefaces. Do not print your news release in all capital letters. Both of these poor choices are hard to read and so probably won't be read.

4. About eight lines before the news release begins, provide a **slug** (or headline) summarizing the subject of your release for an editor whose job is made easier by such helpful descriptors. Boldface your title so it stands out. Some news releases even supply a running title after the slug. The extra space before the slug begins gives editors room for annotations or instructions to a keyboarder.
5. Always supply a precise date for the release, not simply "Today" or "Thursday." A release dated "For Immediate Release" means your story is ready to be printed when the editor receives it. You can also "embargo" the release, asking that editors not publish it until a specific time—for instance, only after a prize has been publicly awarded to someone.
6. If your news release runs to a second or additional page(s), type the word "more" at the bottom of the first page and subsequent pages to indicate that another page(s) will follow.

7. When you finish your release, type "30" or "#" five or six lines below the end of text. This is journalistic shorthand signaling that your news release has concluded and no more copy follows. Some news releases simply put "end" a few lines after the close.
8. Indicate at the end of your news release whether you are sending any accompanying documents with your news release—statistical information, lists of names and addresses, specifications, photographs. When you include a photograph, identify the source as well as any people or objects in it, in pencil on the back. That way there's no danger that the photo will be discarded, miscaptioned, or run with another story.

Organization and Style: Writing a Successful News Release

As you can see from studying Figures 8.1 through 8.4, news releases are short and crisp. They are usually only a page, or at most, a couple of pages long, although releases on the Internet sometimes exceed two to three pages by including specifications. Don't forget, though, that a news release may have started out as a much longer draft, but the successful writer has shortened the final copy by including only the most relevant details. If you waste words, you will waste an editor's time.

The cardinal rule in writing a news release is to put the most important piece of news first. In fact, everything in a news release should be arranged in descending order of importance so that your first paragraph is the most significant. Think of your release as an inverted pyramid with the top containing the most significant facts. The reason the most important details come first is simple: you want to catch the editor's and reader's eye. But, also, because editors have limited space, they may cut your release to fit their requirements. If they do, they will start cutting at the end of your release and work their way toward the top. Editors routinely complain that writers bury the most important information in the middle or toward the end of the release or, even worse, never really say it outright. Make the editor's job easy by putting the most important points first.

The following sections contain guidelines for organizing and writing the different sections of your news release.

The Slug, or Headline

One of the most important parts of your release is your headline, or **slug.** It should announce a specific subject for readers, and draw them into it. Write a slug that entices or grabs your reader, such as "Warning: Breathing May Be Hazardous to Your Health." Do not write simply "E-Mail"; instead, tell readers exactly what you are going to say about e-mail—"New E-Mail System Has Automatic Address Memory Built In."

The Lead

The first (and most significant) sentence of a news release is called the **lead.** It introduces your topic, sets the tone, and captures the reader's attention. For that reason it is also called the **hook.** The best leads easily answer an editor's or client's/customer's basic questions—*who? what? when? where? why?* and *how?* (or *how much?*)—the five Ws. Note how the leads in Figures 8.1 through 8.4 arouse interest. Not every lead may answer all these questions, but the more of them you do answer, the better your chances are of capturing your audience's attention. Below are some effective leads that answer the crucial questions:

who when

- Massey Labs announced today that the FDA has
  what
  approved the marketing of its vaccine—
  why
  Viobal—to retard recurrent lesions of herpes simplex.
  when what
- At its monthly meeting on February 15, the Board of
  how
  Directors of Pinellas Credit Union unanimously elected
  who why
  Dora Sanchez to succeed Howard Rush as president.
  who what where
- National University's spring enrollment for all campuses
  how much when why
  is up 3.5 percent over a year ago, reflecting a strong

  carry over from last fall's near record totals.
  what
- Maryville Engineering, Inc., founded and directed by
  who what
  Carmelita Stinn, has been awarded a contract by
  why
  Aerodynamics, Inc., to develop an accoustical system to
  where
  measure and monitor stress levels at the Knoxville
  when
  aircraft plant, D. B. Hall announced today.

Misleading leads that deviate substantially from the guidelines given above annoy editors. Resist the temptation to start off with a question or with folksy humors.

Ineffective question: Does the thought of hypothermia send chills up your spine?
Ineffective humor: If you thought 1998 was great, you ain't seen nothin' yet!

The Body of Your Release

What kinds of information should you give readers in the second and subsequent paragraphs of your release? If the five Ws are answered in your lead, the following paragraphs fill in the necessary supporting details. Regard your lead as a summary. The body of your release then amplifies the *why?* and may also get into the *so what?*

Avoid filling your news release with unnecessary technical details—scientific formulas, intricate specifications. Instead relate your product or service to your targeted audience's needs by emphasizing benefits to them. Focus on special features such as how a new product or service surpasses an earlier one in efficiency and cost. But be careful that you do not load your news release with hype. Editors expect objective, straightforward, and clear reporting.

Use quotations selectively. Too many quotations from management will also arouse suspicion; the implication is that you are apple-polishing. Use a quotation to report vital facts, or to express an idea dramatically, not turn your release into an interview with or a tribute to your employer. See how Scott Hallman effectively uses a quotation to end his release in Figure 8.3.

Style and Tone

The style and tone of your news release should reflect your objectivity as well as take into account your audience's background. Expert publicists who prepare countless releases advise writers to keep their style simple and to the point. Keep in mind the following suggestions on preparing an effectively written news release.

1. Write easy to read paragraphs of three to five sentences, and keep your sentences short (under seventeen words is ideal). Avoid long, convoluted sentences; and avoid fillers like "It has been noted that. . . ." These sentences slow your reader down.
2. Raise your reader's interest level with graphic, understandable words and examples. (MCI does this well in Figure 8.2.) Avoid unfamiliar jargon; it's deadly. For example, a news release for a hospital counseling center lost potential seminar participants by describing it in jargon: "Milford Hospital is pleased to offer an adventure intensive counseling workshop that incorporates trust sequencing; high element activities; and quantifiable decision modules."
3. The tone of your release should be upbeat, easygoing, and direct; stress the human side. Don't become so scientific you lose readers or so biased they mistrust you. Emotional exaggeration is out of place in a news release: "You will wonder how you ever did without the Fourier scanner." Stay away from adjectives dripping with a pushy sell—"incomparable," "fantastic," "incredible."

Brochures

Brochures, or pamphlets, can generate sales, promote business or agency goals, enhance corporate image, and educate readers. Much business can result when potential customers read your brochure, whether you make it available on the Internet, through the mail, or at your office or agency. Brochures are also often sent to customers as a follow-up after they call or visit a firm.

The Consumer Information Center of the U.S. government designs and distributes thousands of brochures on a wide variety of topics, sending them all over the country from its clearinghouse in Pueblo, Colorado, or making them available on the Internet. Doubtless you have received numerous brochures in the last few months, and you have probably discarded many of them as "junk mail." When these "rejects" failed to catch and keep your attention, the company that produced them lost a potential customer and a great deal of money designing and printing an unread brochure. In the world of work, successful brochures overcome a lot of resistance. This section of Chapter 8 shows you some strategies for writing and designing effective ones.

Figures 8.5 and 8.6 show two highly effective brochures; in addition to studying these examples, pay close attention to the brochures you receive at home and at work to see how and why they succeed or fail.

Brochures Can Be Collaborative Ventures

Brochures are often a team effort. Like other business communications, preparing a brochure goes through several stages: writing copy, designing the graphics and selecting the illustrations, creating a mock-up, proofreading, securing management's approval, and, finally, printing and distributing.

Since the process may involve many individuals, you can expect to write for a hierarchy of readers. First you might interview your supervisor about what he or she wants to see in a brochure and why. The boss may have only a general idea that you and others will be expected to turn into reality. Depending on the topic, you might then initially work closely with specialists (engineers, physicians, marketing consultants) to ensure that your brochure is accurate yet clear enough for general readers. In the early stages, you can expect to confer often with all these readers.

Expect these in-house consultants to add, change, delete, and question your work. In the world of business, feedback is essential to the success of a project. When you are in doubt about a technical matter, do not be afraid to ask for clarification or assistance. No one ever lost a reader because of the desire to explain simply and clearly. You may also collaborate with people in your graphics department to determine the most appropriate layout and visuals. Your company may be spending a lot of money on brochures and it expects an effective product.

The Parts of a Brochure

As Figures 8.5 and 8.6 show, brochures are often organized around a central theme, question, or problem. Then they unfold like a story or scenario detailing

FIGURE 8.5 A promotional company brochure.

Cardiac/Pulmonary Rehabilitation

CENTER FOR OUTPATIENT REHABILITATION

INSTITUTE FOR WELLNESS & SPORTS MEDICINE

Cardiac/Pulmonary Rehabilitation

The Center for Outpatient Rehabilitation located at the Institute for Wellness and Sports Medicine (IWSM) currently offers outpatient programs for cardiac and pulmonary rehabilitation.

The framework of outpatient rehabilitation is a multiple intervention, team approach designed to assist the primary physician in offering optimal management of cardiac/respiratory disorders for his/her patients. A team of qualified specialists works with the referring physician to manage and alleviate symptoms associated with the patient's disorder.

Maintenance Program

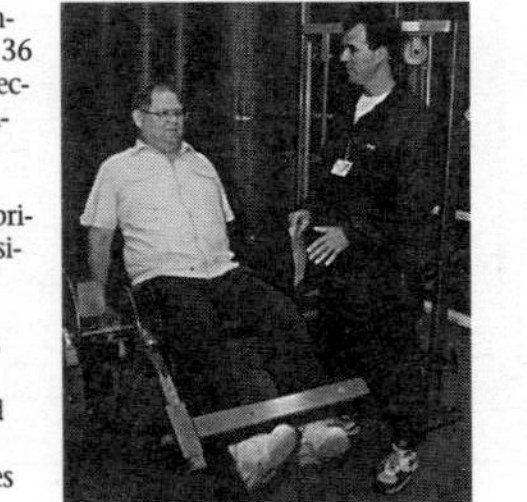

Upon completion of 36 sessions, recommendations are requested from the primary physician for advancement into closely supervised exercise three times weekly.

CENTER FOR OUTPATIENT REHABILITATION

Post Office Box 16796
Hattiesburg, MS 39404

(601) 268-5010

METHODIST HOSPITAL
A commitment to quality » for life.

Reprinted with permission of Methodist Hospital of Hattiesburg.

FIGURE 8.6 An informational government brochure.

Electronic Access to the Consumer Information Center

The electronic world of cyberspace can seem like a library with randomly shelved books and incomplete index files; there's a tremendous amount of information available, but finding what you need can be difficult, if not impossible.

The Consumer Information Center (CIC) brings useful information on a variety of important subjects together in one place, simplifying the search process. And while CIC is famous for its Pueblo, Colorado, mailing address, it can also be reached through its Bulletin Board System and Internet site, including access via World Wide Web (WWW), Gopher, anonymous FTP, Telnet and a Listserv. Instructions for each are detailed inside.

By accessing CIC electronically, you'll find the current Consumer Information Catalog, which lists more than 200 general interest publications from the federal government. You'll also find the full text of listed publications and a library of consumer news features. You can learn about applying for federal benefits, buying a home, raising your children, managing money, eating right and staying healthy, planning a trip and much more. Best of all, it's free.

If you have questions, comments, or just need help, send an e-mail message to:

catalog.pueblo@gsa.gov

BULLETIN BOARD SYSTEM

```
♦ MAIN MENU ♦                          BBS:    202-208-7679
Consumer Information Center BBS        Voice:  202-501-1794
                                       E-mail: catalog.pueblo@gsa.gov
                                       http://www.gsa.gov/staff/pa/cic/cic.htm
MENUS
  F File Menu        B Bulletin Menu        M Message Menu
OPTIONS
  C Calendar of Events      W Who's Online         G Goodbye/Logoff
  A Special Announcements   C Comments to CIC      H Help Level
  N Newsletter              Q Questionnaires       P Page Sysop
  Y Your Settings           V Verify a User
  O CD-ROM Online           I Initial Welcome
  • Press [!] for quick & easy downloading of popular files •
Main Menu Command Bob >>
```

CIC BBS Main Menu

The Consumer Information Center Bulletin Board provides the public with the full text of hundreds of publications. To connect to the BBS, follow the instructions below.

- Set your communications software to 8-bit, no parity, and 1 stop bit (8N1) with ANSI/BBS terminal emulation
- Dial 202/208-7679
- Follow the on-screen instructions
- Baud rates of up to 28.8K accepted

```
♦ FILE MENU ♦                          BBS:    202-208-7679
Consumer Information Center BBS        Voice:  202-501-1794
                                       E-mail: catalog.pueblo@gsa.gov
                                       http://www.gsa.gov/staff/pa/cic/cic.htm
OPTIONS
  L List File Areas    Y Your Settings    V View ZIP File   O CD-Rom Online
  D Download File(s)   P Personal Stats   I Info on a File
  U Upload File(s)     G Goodbye/Logoff   E Edit Mark List
  N New Files Since    ? Command Help     T Transfer Info
  S Search Files       H Help Level       R Read Text File
MENUS
  M Message Menu     Q Quit to Main Menu
  • Press [!] for quick & easy downloading of popular files •
File Menu Command Bob >>
```

CIC BBS File Menu

INTERNET

The Internet is a global network allowing millions of people around the world access to information on every subject imaginable. The Consumer Information Center has hundreds of federal publications that can be viewed online or downloaded through the Internet. If you have an Internet connection or an Internet service provider, follow the directions below.

- **WORLD WIDE WEB**

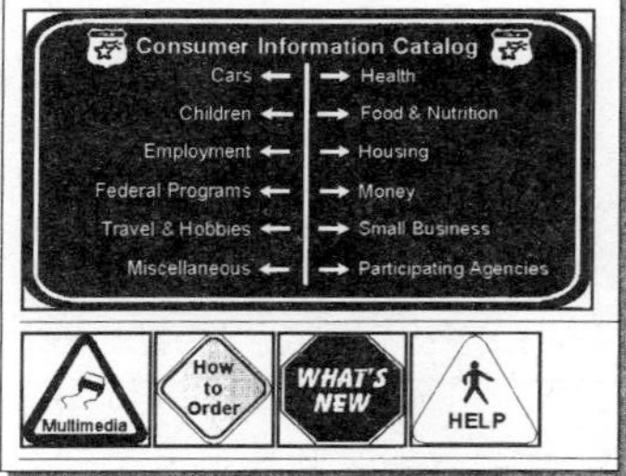

CIC WWW Home Page

The most graphical Internet connection is also the easiest to use. The World Wide Web (WWW) allows users to view formatted text, hear sound recordings, and see movies and pictures, all on the same page. For those users who are using non-graphical browsers, the CIC web site has a text only page available. The Consumer Information Center's WWW pages can be reached by opening the following URL:

http://www.gsa.gov/staff/pa/cic/cic.htm

• GOPHER

Since some computers are incapable of displaying complex pictures or movies, the directory and file structure of Gopher allows users to bypass graphics and navigate through the Internet quickly. To reach the Consumer Information Catalog through Gopher, follow these instructions:

- Connect to gopher.gsa.gov
- Select STAFF OFFICES, then PUBLIC AFFAIRS, then CONSUMER INFORMATION CENTER
- The list of Catalog categories appears
- Select your area of interest

Gopher Menu:

- Instructions
- Order Information
- Catalog in .txt format
- Catalog in .zip format
- Cars
- Children
- Employment
- Federal Programs
- Food & Nutrition
- Health
- Housing
- Money
- Small Business
- Travel & Hobbies
- Miscellaneous
- News for Consumers
- Federal Information

CIC Gopher Menu

• ANONYMOUS FTP

When an Internet user has the location of a file they would like to obtain, and where to find it, they can use File Transfer Protocol (FTP). FTP is a command-level text system that is the most basic Internet interface. File descriptions are in the "index" file of each directory. The Consumer Information Center can be accessed using FTP as follows:

- Point to ftp.gsa.gov
- Login with username "anonymous"
- Use your email address as the password
- Change the directory to: /pub/cic
- A list of Catalog categories appears as subdirectories

```
../ (parent directory)
-rw-rw-r--    cic   1847 Apr  5 15:38 lectronc.txt
-rw-rw-r--    cic   8286 Apr  5 15:36 lorder.txt
drwxrwxr-x    cic    512 Apr  6 13:05 acrobat
drwxrwxr-x    cic    512 Apr  6 07:57 cars
drwxrwxr-x    cic   1024 Apr  4 16:46 children
drwxrwxr-x    cic    512 Apr  4 16:48 employ
drwxrwxr-x    cic   1024 Apr 11 16:25 fd&nut
drwxrwxr-x    cic   1024 Apr  4 16:53 fed_prog
drwxrwxr-x    cic    512 Apr  6 08:03 fedinfo
drwxrwxr-x    cic   2048 Apr  4 16:57 health
drwxrwxr-x    cic   1024 Mar  1 15:18 housing
drwxrwxr-x    cic   1024 Apr 11 16:26 misc
```

CIC FTP Directory

• TELNET

Telnet provides the means to connect to and interact with another computer using standard Internet protocols. Through FedWorld, a popular Telnet-capable system, you can Telnet to the Consumer Information Center BBS:

- Telnet to fedworld.gov
- From the main menu, type "/go gateway"
- Select "D - Connect to Government Sys/Database"
- Type "6" to access the CIC
- Follow the on-screen instructions

• LISTSERV

Internet users can also obtain information from the Internet through a Listserv. Users can retrieve documents as email by simply sending computer commands through email messages to special Internet computers called Listservs. Information on accessing the CIC electronically can be retrieved by:

- Send an email message to:
- cic.info@pueblo.gsa.gov
- with no subject and the words **SEND INFO** in the body of the message

CIC ADDRESSES & NUMBERS

E-MAIL

catalog.pueblo@gsa.gov

BBS

(202) 208-7679

WWW

http://www.gsa.gov/staff/pa/cic/cic.htm

GOPHER

gopher.gsa.gov

FTP

ftp.gsa.gov

MAIL

To get a printed copy of the Consumer Information Catalog, mail your name and address to:

Consumer Information Catalog
Pueblo, Colorado 81009

or call: (719) 948-4000

CIC Form 40 May 1995

the various parts of that theme or solutions to that problem. By thinking of your brochure as telling a story or solving a problem, you will organize it better for readers.

TECH NOTE

Graphics are vital to the success of your brochure. Given the opportunities available through scanners and publishing software, do not settle for anything less than topnotch graphics. Avoid amateurish line drawings, faded sketches, or overused computer clip art. Include relevant action photographs to motivate readers, and arrange them in a logical order. However, don't include a photo of an employee or your office just for looks. Be sure to obtain signed releases from all people you photograph.

1. Identify the exact subject of and reason for your brochure. If you can summarize in fifteen words or fewer why you are writing your brochure, you are already well on your way to creating a possible front cover.
2. Develop a cover that informs your intended readers of that topic and assures them that reading your brochure is in their best interest. Readers will throw your brochure away if it does not provide information they need or want.
3. Write a title that does these two things effectively—that is, that identifies the topic and focuses on reader needs and desires. Here are some brochure titles that work:

   - Fire: Your Guide to Survival
   - We Can Save You Money on Long Distance Service! (Office Depot)
   - Could Your Place Use More Space? ReSpace Your Place with . . . Closet Maid
   - Can I Qualify for Medicaid?
   - Headache Control: Strategies for Relief (Advil)
   - What You Should Know About Wheel Alignment Today (Hunter)

4. Inside your brochure, break your topic (message) into steps or parts. Readers feel more comfortable with (and are favorably disposed to) a message when they can see it divided into sections and unfolding as a sequence.
5. Introduce each subtopic by a heading for your target audience. Your company may expect you or your team to prepare a tentative list of subtopics (or subheadings) for your brochure during the early drafting stage.

TECH NOTE

The success of your brochure depends as much on how it looks as on what it says. If you can use a desktop publishing software program, experiment with several designs. Always put your address into columns rather than lines that extend across two or more pages (folds) of a brochure. Use a print size and font that is crisp and direct. Avoid small, 8-point type (the size in newspaper ads) and never print in all capital letters; both formats make your brochure difficult to read. If your budget allows, use more than one color. Confer with an art director to be sure your color choices and typefaces are compatible. For best results, use dark color inks on light color papers.

The format of brochures requires that they be well organized into information sections. Below are a few examples of main topics that brochure writers have broken into smaller "digestible" units for targeted audiences. The Soil Conservation Service of the U.S. Department of Agriculture prepared a brochure for a carefully focused audience—recreation area planners—to convince them to examine soil surveys. After a short introduction, the brochure divides into these four subsections:

- Why Soil Data Are Needed
- Selecting Recreation Areas
- Maintaining Recreation Areas
- How to Obtain a Soil Survey

By carefully addressing the targeted audience's major concerns, the writers of this brochure are better able to achieve their goal of encouraging recreation planners to order and consult soil surveys.

A brochure advertising the Miracle Ear hearing aid was organized according to the key questions potential consumers usually ask about the product, a convenient and logical scheme. Clearly the brochure staff did a careful job of analyzing their audience's needs. On the cover of the brochure readers immediately see the title "What Everyone Should Know About Hearing Loss." Inside are sections devoted to answering these major questions:

- What causes hearing loss?
- What are the symptoms?
- How can it be helped?
- Do hearing aids really work?
- Are hearing aids right for me?
- What can I expect from my hearing aids?
- Will others be able to tell that I wear a hearing aid?

Perhaps the most important information in a brochure is in the last section. There you should invite readers to take some action. Some brochures omit or fail to convey clearly what a reader is supposed to do after looking at the brochure.

- Do you want the reader to order a product or service, or make a call to get further information?
- Have you made it clear how to contact you (voice mail, e-mail, Web site)?
- Did you supply your company's address, phone, and fax number?
- Did you list the phone numbers and addresses of several branch offices so the reader can choose the most convenient?
- Have you made it easy for the reader to place an order, by including a postage paid tear-out application or order form?

Selecting the Right Content for Your Brochure

Armed with knowledge of your audience's needs and background, you will be able to create appropriate copy, the written message. Don't overwhelm readers with too much information; select just the right amount. Some brochures go on and on for pages and lose readers in the process. Make every word count. Ask yourself if you have included all the essential information for your readers. Are any details unnecessary or too technical? You want readers to feel comfortable with the facts and explanations you provide. You certainly don't want to include inaccurate or misleading information.

On the other hand, a skimpy brochure with just two or three paragraphs can be as inappropriate as one that runs for several columns or pages beyond the reader's interest.

Using the Most Appropriate Style

Writing brochures differs from writing letters or reports. Brochures are far more informal, intentionally designed to be a fast read. Long, unbroken paragraphs discourage readers who will assume at first glance that your message is too difficult or time-consuming. Besides looking intimidating, a series of long paragraphs costs more to print. A series of shorter paragraphs, on the other hand, tells readers that your brochure is easy to handle. So keep paragraphs short, between 100 and 125 words. Your first paragraph should be even shorter: it should grab readers' attention in under twenty words. In your first sentence, use the fewest words possible. Many brochures even use sentence fragments to get the reader's attention. An insurance company brochure advising individuals about what to do in case of fire begins.

> Fire is a killer. But while flames may terrify people, smoke and poisonous gases are just as lethal.

Many brochures even use sentence fragments to get the reader's attention. For example, an Olympic Paints brochure on preventing graying opens with several very short paragraphs and some very short sentences.

> If you're tired of the way water sealers let your wood turn gray, we have some good news for you.

> Introducing all-new Olympic Natural Look Protector Plus.™ It waterproofs.
> Plus it's guaranteed to prevent graying for a minimum of two years.
> This new, unique formula contains Natural Wood Enhancers that keep new wood looking new, and restore the natural beauty to weathered wood.
> If your wood's already weathered gray, just clean it first with Olympic Deck Cleaner. It's fast and easy.
> Then let Natural Look Protector Plus restore and protect your wood's natural beauty.

The Olympic Paint opening packs a lot of power and reader benefits into each sentence.

Be User-Friendly

When writing brochures, keep your language practical and direct, but always be user-friendly. Your tone should be cooperative and caring. Never talk down to or lecture readers. Focus on what is most important to them. Imagine that you are talking face to face with a reader who has come to you with a problem or particular business need. Select words consumers will easily understand.

Keep Explanations Simple

Keep explanations short, simple, and sensible. If you have to explain a technical concept, do it through an example or an analogy (an extended comparison). In a brochure for a particular brand of vitamin C, a pharmaceutical company compared their antioxidant vitamins to a mop sweeping up dangerous free radicals. In marketing Enviracare, an air filter that removes indoor pollutants, a Honeywell brochure explained "How does a HEPA filter work?" to consumers by employing the following inventive analogy:

> Basically, as particles pour through the densely packed glass fibers of the paper media [filtering materials], they literally run into one of the fibers and stick to it by mutual attraction. On a large scale, it would be like trying to blow a grain of sand through a stack of hay.

The brochure vividly clarified a highly technical concept readers needed to understand by comparing the pollutants to sand and the filter to a haystack. Describing things "on a larger scale" helped convince consumers to buy the product.

Testing Your Brochure

Test your brochure before you have thousands of copies printed. Ask co-workers and potential customers to read it and give you their reaction. See if they understand it—why you wrote the brochure and how the information will benefit the target audience. Ask your test audience to pay special attention to the sections into which you organized information and the first sentence under each of your subtopics. Finally, invite their comments on the overall layout and the visuals and graphics you used.

Depending on what your readers tell you, you may have to do some further revising and editing. But remember, it is a lot less costly to change a brochure before it is printed than after—don't forget the bottom line.

Newsletters

Newsletters are published at regular intervals (weekly, monthly, bimonthly, quarterly) in almost every business, organization, and agency. In fact, the CD-ROM Global Access to the *Encyclopedia of Associations* lists more than 87,000 organizations worldwide, each one producing its own newsletter.

Why Newsletters Are Published

Companies, large and small, publish newsletters to inform employees about corporate news (meetings, personnel, policies) as well as to recognize individuals for achievements on and off the job. Newsletters are also written to encourage customers to do further business with the firm, as the *Lanier On-Line* newsletter in Figure 8.7 does, or to report financial information for investors. All government agencies issue some type of newsletter explaining and justifying their work and/or research. Hospitals and other health care providers also send out newsletters—to former patients, to community members—to announce continuing medical advances and to showcase acts of caring. A newsletter such as the *Penn State Sports Medicine Newsletter* functions to educate readers about sports procedures and equipment, nutrition and health, and to assist them in their decision making.

Some newsletters are distributed gratis; others are commercial publishing ventures that charge readers a subscription fee, as the *Kiplinger Washington Letter* and *Lanier On-Line* do, for example.

TECH NOTE

Consider putting your newsletter on-line on your Web site or on an electronic bulletin board where readers around the world, surfing the Net, can see it. Publishing your work this way not only expands its circulation and potential impact but also adds new contacts to your network. The feedback you receive from an audience you may not have envisioned engenders new ideas and pushes you to improve your publications.

How Newsletters Relate to News Releases and Brochures

A newsletter shares much in common with news releases and brochures. All three are promotional tools tailored to a target audience. A newsletter, like a news release, is concerned with recent accomplishments and carefully summarized information. Stories in a newsletter may even be based, in part, on information in a news release. Like a brochure, a newsletter must be eye-catching, formatted with attention to layout, visuals, and proportion.

Newsletters—like brochures—are frequently the result of collaboration. You may not be asked to prepare a newsletter by yourself, but you may play a major role in preparing one or in contributing to a regular feature. If so, you could have to depend on fellow employees for information about articles, or they may write stories for you to edit for your newsletter. Finally, as with a news release or brochure, you will work closely with staff in a graphics or layout department and will have to get your newsletter approved by management.

Parts of a Newsletter

Newsletters usually include stories on several topics and resemble a small newspaper. A newsletter is composed of many different parts or subsections, as the Lanier example in Figure 8.7 shows.

TECH NOTE

When you design your newsletters, make your headlines stand out by putting them in dark, bold type. Using type of the same size and font for each headline provides logical consistency that readers will appreciate. If you are producing a print version of your newsletter, don't use dark paper stock. Set apart special features in boxed inserts, such as with this Tech Note. Fit visuals to the space and sense requirements of your story (for example, do not force a horizontal photo into a vertical format).

Some newsletters are highly structured with regular sections readers expect to find in each issue. Regular columns in the Lanier newsletter include an executive message in "Commentary," technology updates in "What's New?" and audience response in "Reader Question."

Depending on your company's needs and activities, you may be responsible for gathering information for a particular column or section. If so, maintain a current file about these (and related) topics so you always have information on hand for forthcoming issues. Always have a clear idea of your **turnaround time,** the time necessary to gather, write, and review information before it goes to publication.

Below are some sections or topics common to company newsletters:

Dates to Remember
Upcoming Events
From the Director's Desk
Promotions
Letters
Awards and Honors
Updates/Follow-Ups
Conferences/Classes
Policy Changes

FIGURE 8.7 A two-page company newsletter.

News and views for today's information management professional. Volume 3 • Number 3 • 1995 • $2.00 Per Issue

Conference participants are online to a "living network" of ideas and information

INPUT '95 attracts 350+ participants from throughout the country

The Internet isn't the only place you'll find people networking to share information on the latest technology. For three days recently in Atlanta, hundreds of Lanier customers became a "living network", exchanging ideas and insights during INPUT '95.

INPUT — an acronym for "INterchange of Products, Users and Technology" — is an organization dedicated to information sharing between Lanier and its system users, as well as among the users themselves. Each year, the annual INPUT Conference provides a more formal and concentrated exchange of ideas.

See INPUT '95 -continued on page 4

Lanier's INPUT '95 Conference offered participants a first-hand look at new technolgy and systems.

Is your organization online with the Internet?

We'd like to know how you're using (or plan to use) the Internet

The Internet is rapidly becoming the greatest communications system the world has ever known.

Currently, it's estimated 25 to 30 million people use the Internet with one million of them using it *every day*. By the year 2000, it's estimated one billion people worldwide will actively use the Internet, providing them with the unique opportunity to interact with an audience on a global scale.

For business, the implications are immense. The Internet can provide business users with detailed, substantive information which they choose to see. By allowing direct, immediate feedback via electronic mail, the Internet also enables businesses to take full advantage of the Internet as an information-exchanging medium.

It's basically easy, fast and limitless,

See Internet-continued on page 3

IN THIS ISSUE...

READ & ROUTE

- ❑ ______
- ❑ ______
- ❑ ______
- ❑ ______
- ❑ ______

FREE Gift Offer inside! (Page2)

Continued

FIGURE 8.7 (Continued)

ON-LINE VOL 3. NO 3. 1995

C O M M E N T A R Y

Only *proven* technology can take you into the digital future

by C. Lance Herrin
Executive Vice President, U.S. Operations

"We live in interesting times." I'm sure you've heard this saying before. Still, it comes to mind whenever I look at the wonders digital technology is bringing to our business world.

With more and more digital products and systems on the market, the purch- ...n-making process is ... well, *interesting*. ... entire

choosing an existing product off the shelf and offering it to you.

In a sense, you might say Lanier acts as *your* evaluation committee — we identify, select and recommend only those products and systems *proven* to satisfy your needs. Moreover, as good as we've become in sourcing, developing ... innovative techn... looking fo...

What's new?

MedWord® for Windows™ brings the power of Windows to medical transcription

In current and future networked environments, probably the most important feature of any single system is its ability to communicate and share information with other systems. That's why Lanier's MedWord has evolved from a proprietary medical transcription system to an open system architecture utilizing industry stand... as WordPerfect®, MicroSoft® W... Novell® and Btrieve®.

Lanier's new M...

VOL 3. NO 3. 1995 ON-LINE

Quick tips for making your fax transmissions more cost-efficient

You can save money while increasing the efficiency of your facsimile usage

Every now and then, it can pay you to take a look at how your office utilizes its facsimile systems. Reviewing fax usage patterns with employees can help reduce or eliminate inefficient, costly procedures.

We've prepared a few tips to help you save money while increasing the efficiency of fax usage at your office:

- **Fax or... ...ments.** If your do... ...s long, ...
- **Share fax machines between offices.** If you're able to share a fax machine between different offices or departments, you might reduce costs for supplies, maintenance and other operating expenses. This approach works best when routine fax volume is low enough so all users can have quick access to the fax machine when it's needed.
- **Consider the alternatives.** If the fax equipment and supplies ... concern, you might c... faxes h...

Internet-from page 1

and many businesses are currently using or plan to use the Internet in some form.

That's why we'd like to know how familiar *your* business or organization is with the Internet. Your responses to our Reader Survey will help us understand how Lanier can better serve you by using the Internet as yet another channel of communication between us.

So, please take a moment to complete the Reader Survey in this issue of ON-LINE. We appreciate your input, and you'll receive a FREE ... for your timely response.

...EE *gift, just* ... *this issue*

Meetings
Membership Drive
Trends
What to do if . . .
Suggestions
Retirements
Did You Know?
New Products/Services
Births and Marriages
Frequently Asked Questions
Recent Research
Highlights
Publications
Staff Profiles
Department in the Spotlight

While not all of these sections/topics will appear in every issue of your newsletter, the headings do give an idea about the range of information you may be expected to gather, verify, and print.

In addition to these customary regular columns of a newsletter, you may include stories, or special features, or relevant and timely events, procedures, or accomplishments. Here are some sample headlines of newsletter articles:

"An Interview with the CEO"
"Employee Honored for Community Service"
"Membership Drive a Tremendous Success"
"A Nutritional Approach to Chronic Fatigue Syndrome"

FIGURE 8.8 A sample newsletter article.

October 1 systemwide E-Mail launch date set

Training, utilization plans now in progress

The launch date has been set. It's now T-minus three months and counting until Union Planters completely changes the way it communicates internally, according to Jack Moore, President of Union Planters Corporation.

"Out with the old and in with E-Mail." Moore said during the quarterly bank presidents' meeting at the UP Administrative Center last month. "The flow of paper and the cost of hard copy have passed the critical stage. A lot of what I see coming across my desk belongs on E-Mail. It's a great tool, and it's going to have a significant impact on the communication process at Union Planters.

"Our system has great E-Mail capability and full utilization will make us much more productive and cost efficient," Moore added.

In this spirit, several corporate departments are now using or making plans to utilize E-Mail...

A few E-Mail sessions will be conducted this month.

"But we will concentrate heavily on E-Mail training in August and September so that as many employees as possible will have knowledge of E-Mail before October 1," said DeVaux.

Source: **UP Beat**, July 1995, Union Planters Administrative Center, 7130 Goodlet Farms Pkwy. (AZE), Cordova, TN 38018; 901/580-6064; Holly H. Fava, Communications Specialist.

"Free Services to All Shareware Subscribers"
"Day Care Services Expand to Meet Flex-time Schedules"
"We Will Merge with Scriptnet Software Nov. 1st"
"New Software Makes Downloading Easier"
"Windows '98 Installed"
"Cyber-Publicists Launch E-Mail Campaign for Local Charity"

These titles reflect the diversity of newsletter contents and approaches. To write such stories, you'll obviously have to do some research. That may include interviewing co-workers, managers, specialists, and possibly even members of the community. You may also research your subject using electronic or print references. While it is unlikely that your article will carry footnotes or references, consider the possibility of listing additional sources of information to help your readers.

Choosing an Appropriate Writing Style for a Newsletter

Your stories, whether they are about people, events, equipment, or procedures, should be highly readable. Your primary goal is to make information accessible, easy to read, and easy to find. Keep in mind that your audience does not have the time to dwell on individual stories. Since readers are not looking to your newsletter for exhaustive and detailed analyses, your stories will, for the most part, be short—200 to 400 words. Figure 8.8 contains a company newsletter article of typical length and style.

Home Pages

The Internet may be the most productive marketing tool ever invented. In Chapter 9 you will learn about the Internet as an information source in greater detail, but in this section of Chapter 8 we will look at the World Wide Web as a marketing tool and what your role might be in writing for it.

The Web connects sets of information—called pages—at millions of sites around the world. Some pages are quite long, while others are just one computer screen long, as in Figure 8.9. Any individual or company can place its own pages on the Web. The first page of a set of information is called a **home page.** The home page is where visitors to a site will go first, and if it catches their attention, then they are likely to click onto the other pages to look in more depth at the products and information presented there.

Benefits of a Home Page

Home pages have to arouse readers' interest, show how a product or service can appeal to on-line shoppers, demonstrate the application of the product or service, and include a call to action. A Web site offers marketing possibilities unheard of before. For example, thanks to Web sites, a company can

FIGURE 8.9 A sample home page.

ASSOCIATION OF

BRITISH INSURERS

Welcome to the WWW site of the Association of British Insurers. The ABI is the trade association for insurance companies and represents virtually the whole of the UK insurance company market.

This site contains five sections. To navigate around the site, select a section from the list below, or use the button bar that appears at the bottom of each page. The grey button indicates the current section.

About the ABI
An introduction to the ABI, its work, and the services it offers.

Members
The ABI has around 440 members covering 95% of the UK insurance market.

Hot Topics
News, press releases, and other items of current interest.

Consumer Information
Information and advice about the insurance industry, provided for the consumer.

Industry Briefing
Statistics and other information, provided for the ABI membership.

Home page	About the ABI	Members	Hot Topics	Consumer Information	Industry Briefing

Association of British Insurers
Gresham St, London EC2V 7HQ Tel: 44+(0) 171 600 3333 Fax: 44+(0) 171 696 8999
Site design: Centre for Computing in the Social Sciences
©1996 Association of British Insurers
Site last updated: 4th July 1997
http://www.abi.org.uk

- attract new and satisfy repeat customers the world over, since the site is public, international, and available at all times on demand
- keep track of the number of customers who visit a particular page
- provide additional pages of in-depth technical information about a product or service for those who are interested
- answer any FAQs (frequently asked questions) customers might have
- expedite a potential customer's talking with satisfied clients who have used the product or service before
- supply sound and graphic support
- accommodate individual customers by tracking previous purchases (such as the size, color, and style of clothing a customer ordered before)
- improve research by keeping track of what the competition is doing at their Web sites, exploring their products and services
- interface customers' orders with credit card debits or receive payments through e-cash

TECH NOTE

There are several things you can do to plan a successful Web site.

1. Look at good sites already on the Web. Use your Web browser's links to "Cool Sites" or "Top 100 Sites" to see what works.
2. Pay attention to the various ways a home page can be designed and organized; you will see a lot of variety in the "Top" sites but notice what they have in common and how their information content is organized and presented.
3. Print out copies of pages of organizations that are similar to yours—know your competition.
4. Read magazines dedicated to the Web for in-depth analyses of page designs. For example, you can check out these magazines dedicated to the Internet: *Internet World* (http://www.iw.com) and *IEEE Internet Computing* (http://computer.org/internet/).

The Web provides a crucial medium from which to present your product, one that is constantly expanding the ways to present information: color, graphics, animation, sound, video, 2-D, 3-D, interactive, immersive—all are available and more are coming. But this does not mean all you have learned is irrelevant; it is, in many ways, even more relevant. The abundance of choices makes it easy to be carried away with the glitz, to lose your focus, and to miss an opportunity.

Your home page has a **major focus** and a **subsidiary focus.** The major focus is to catch your visitor's attention. If you fail to do this, then everything else is a waste of time; your visitors will click on their mouse and be gone. Your subsidiary focus is to sell your product or introduce your organization. This may seem backward but unless you get readers to your home page you cannot promote the product. So your ideal strategy will be to engage your visitor, wherever possible, using your product as the mechanism.

When designing a Web page, do not overload your visitor with information; overload is a turn-off. Make sure there is enough "white space" to give the user breathing space. Do not squeeze all the features of your product onto your home page; leave plenty of unused space—it is easier on the eyes and welcomes the visitor with an unhurried, more relaxed feeling, such as the home page design in Figure 8.9. Use a good font size for text and clearly discernible graphics; anything that is hard to distinguish is easy to ignore.

Employ color wisely; it is an excellent tool for engaging a visitor's attention and interest but can easily be overused, both in variety and extent. An extreme example of the overuse of color would be when nearly everything on the page is a different color. Color can be startling—for example, a splash of bright colors to attract attention—but this splash should be in only a limited area of the page with the actual information easily readable. Wherever possible, and particularly with text, have a sharp contrast, say, between textual characters and their background. Low contrast makes for poor readability.

Make your home page informative. It should at least tell a visitor what your product is or what services are offered. The information can be indirect by providing links to other pages that actually spell out details of the product that the visitor requires. In fact, these links are good in that they draw the visitor further into your set of Web pages where you can promote your product. It is very important for your home page to indicate at least the type of information available within your set of pages. Not every topic needs a direct link but visitors should get a sense of what is available and how to connect to it.

You can link your home page to your other pages by highlighted words (hypertext), by icons (small stylized images), by thumbnail images (shrunken versions of larger pictures), by full-size images, or by a combination of these. (The links in Figure 8.9 are hypertext: **About the ABI, Members,** and so on.) Link indicators should succinctly inform visitors about what will be on the page they are moving to. Present a list of the options available. A retailer of kitchen appliances, for example, might show a set of pictures that represent various appliance groups; clicking on the picture of a refrigerator moves the visitor directly to the page that provides details about all the refrigerators available. Make it easy for visitors to find what they are looking for. In addition to a main list, you can provide key words across the bottom of your home page (as done in Figure 8.9) to let visitors move quickly to the page that interests them.

This raises the issue of time. If it takes too long for your page to download to a computer screen, visitors may get bored and leave. Limit the number of images to what can be displayed in a lively way. Use backgrounds with caution; if they contain images, they can drastically increase display time. One effective compromise is to use a single-color background that is "painted" and thus takes much less time to download. Remember that most people's computers are three or four years old and may be much slower than the one you use.

One other aspect of your page needs consideration. Do you want visitors to come to your page(s), find what they want, and then leave? Or would you like them to make return visits? If the latter, then add some extra enticements. Use sneak previews or other devices to draw visitors back to your pages. Note how The Loft, in Figure 6.12, invited customers to check the Web site for the daily specials. Always encourage visitors to bookmark your pages to make it easy for them to get back to you. Some other ways to motivate them to return are

- offering insider news updates specific to your business area
- providing reviews of new articles, books, and journals in your area
- giving preproduction information on items or services

This Tech Note was written by Dr. Cliff Burgess, a computer science consultant and teacher with extensive experience in creating such sites.

How Do On-Line Visitors Find You?

You may have a wonderful home page, but how can you be sure the world will find it among the millions of other Web pages in cyberspace? When looking for something on the Web, users generally use a search engine such as Alta Vista, Yahoo!, or Lycos (see pp. 355–356). It is therefore important that you contact all the search engine sites and alert them to your pages. They may find you without this notification, but it is smarter not to leave it to chance. When a search engine finds your page as a result of a search for a user, the engine will display just the first few lines of text that it finds on your page. Obviously, it is worth giving some thought to the first text that appears on your page—those words are what will turn potential visitors into actual visitors.

Another way to publicize your pages is to search for pages with content that is associated with your product. Contact the page owners and see if they will provide links to your pages in return for your providing links to theirs.

Then there are on-line directories, like telephone books. To be listed in these you need to complete an on-line form and within a few weeks your pages will be listed in the directory. While directories may charge for this service, a free avenue for getting known is via newsgroups and e-mail mailing lists. Newsgroups and mailing lists exist for a huge range of topics, so contact all the groups and lists that might have an interest in your pages and give them a sample of what you have available, along with your page address.

✓ Revision Checklist

News Releases

- ❑ Identified a truly newsworthy subject.
- ❑ Found appropriate publications to send release to.
- ❑ Followed accepted format for news release.
- ❑ Created headline (slug) that tells topic and hints at reader benefits.
- ❑ Wrote clear, crisp, attention-grabbing lead sentence (hook).
- ❑ Made sure key ideas are not buried in the middle of the release, put near the end, or left out entirely.
- ❑ Deleted irrelevant or unnecessary details.
- ❑ Used clear, direct, and readable style.
- ❑ Checked length of paragraphs.
- ❑ Received boss's approval to send out.

Brochures

- ❑ Organized around key idea, question, or problem.
- ❑ Divided subject of brochure clearly into subtopics with distinct, helpful headings.
- ❑ Created effective cover with informative and catchy title.
- ❑ Stressed audience's needs and benefits.
- ❑ Wrote in appropriate language and tone for targeted audience.
- ❑ Included right amount of detail.
- ❑ Used attractive brochure design with enough white space.
- ❑ Chose colors that work well with typeface.
- ❑ Clarified message and supported text with high-quality visuals, all properly identified and credited.
- ❑ Told readers what to do when they finish.

Newsletters

- ❑ Examined earlier issues to get sense of design and target audience.
- ❑ Gathered and verified information for all sections of newsletter in time for publication.
- ❑ Identified relevant topics for articles.
- ❑ Used crisp, lively, and inviting style that makes readers take interest.

- ❑ Created catchy and informative headlines.
- ❑ Verified all facts and details.
- ❑ Designed layout to be pleasing, not crowded.
- ❑ Used clear photos with proper identification and credits.

Home Pages

- ❑ Used plenty of white space so visitor is not overloaded with text.
- ❑ Chose color carefully so it is easy on the eyes and all text is legible.
- ❑ Gave essential information about product or service.
- ❑ Selected and designed appropriate hyperlinks.
- ❑ Made it easy for visitors to navigate and locate what they need quickly.
- ❑ Ensured that images and backgrounds would not slow visitor down.
- ❑ Opened with key text for search engines to find.

Exercises

1. Find a news story about a company event or personnel appointment printed in your local newspaper. Using the information you find and adding original details to your own, write an appropriate news release. Attach the original newspaper clipping to your news release.

2. Rewrite the poor leads below to make them answer an editor's five questions (*who? what? when? where? why?*).
   a. Several new appointments were made at the plant this week.
   b. A really wonderful new product will hit the market later this month.
   c. Our schools are very fortunate in securing the services of a top-flight computer whiz.
   d. If you have ever been in the dark about screen savers, here is your answer.
   e. Jones Industries regrettably is recalling its defective software packages.
   f. An epidemic has hit the Bridgeview Community.

3. Below are two poorly organized and sloppily written news releases. Reorganize and rewrite them according to the guidelines presented in this chapter.
   a. Friday   Alan Bowerstock

   Metropolitan State University is a four-year urban institution of higher education offering majors in many fields. Located two miles west of Taylorsville Industrial Park, MSU currently boasts more than 7,000 undergraduate students.

Among the many student services currently available at MSU is the Division of Career Placement; this division is located in the Student Services Building, Room 301, just across the hall from the Department for Greek Life.

In the last year, the Division has assisted more than 2,000 MSU students to find part-time jobs. Full-time jobs, too.

The goal of the Division is to help students earn money for their college expenses. The Division also wants to assist local businesses in contacting MSU qualified undergraduates.

The MSU family is well represented on the home page. Every department from the University has been encouraged to report on its activities. The Division of Career Placement is also on the home page.

Counselors at the Division can help MSU students prepare a four- or five-line ad about their qualifications. The Division will run these ads in their home page. The university will also run ads looking for job candidates from the state employment agency and the greater Taylorsville area.

b. For General Information Frank Day

J.T. Bushart, CEO of Bonnetti and Blount Construction for the last three years, asserted today that the company is devoted to progress and change. Bushart came to B&B Engineering after several years working for Capitol City Engineering.

Bonnetti and Blount is a leading firm of contractors and has worked for both national and international corporations.

The firm specializes in construction projects that require special expertise because of their challenges in difficult terrains.

B&B has just been awarded a contract to work on the 10 million dollar renovation of two major Fairfax dams. The firm anticipates hiring more than 200 new workers. These new employees will work on the dams that present dangerous conditions to residents.

```
When completed, the two new dams will further assist
residents of Fairfax and Hamilton Counties receive
all the necessary irrigation and hydroelectric energy
they need.

Engineering shetches and blueprints are in the works.
```

4. As a collaborative writing activity, prepare a brochure for an organization that you and your team are members of (or would like to join) encouraging prospective members to join. As part of your team effort, decide on the appropriate cover, text, photographs, and layout.

5. As a group activity, contact a local charity, a campus organization, or a small local business in your city that does not have a promotional brochure and volunteer to prepare one for them. As part of your collaborative writing activity, determine the responsibilities for (a) interviewing the group, charity, or business, (b) writing copy, (c) choosing visuals, and (d) designing the cover and overall layout. Prepare the brochure.

6. Bring an example of an ineffective brochure to class. Again, as a collaborative team, write the following documents for the company or agency that has distributed the brochure.
   - **a.** a letter diplomatically pointing out the weaknesses of the brochure and why it would hurt, not help, the agency or company
   - **b.** a short (one-page) proposal (see pp. 560–566) offering to prepare a new brochure for the company or agency and why and how you believe it would increase sales
   - **c.** the new brochure itself incorporating all the benefits you outlined in **b** and omitting or improving all the weak points you cited in **a**

7. Assume that the large company you work for has asked you to write a story (about 400 words) for the next issue of the company newsletter, circulation 5,000. Write the story on one of the following topics.
   - **a.** new ergonomic furniture
   - **b.** new security lighting
   - **c.** mandatory classes on computer privacy and security
   - **d.** aerobic classes offered at the lunch hour and before and after work
   - **e.** an exchange or internship program whereby your company will invite students from Asian and African universities to work at your facility for the summer
   - **f.** a training seminar scheduled for next month
   - **g.** extended hours and services of a company-sponsored day-care facility
   - **h.** remodeling at a particular site and the temporary provisions planned for providing services to other workers at your office or plant

8. Bring to class a few examples of carefully designed home pages and be prepared to explain why they are effective, visually and verbally.

9. Find two or three examples of what you regard as poorly designed and written home pages from the World Wide Web. Choose one and write a memo to your instructor pointing out its weaknesses and suggesting ways to revise and improve the page.

10. As a collaborative activity with two or three other students, design a home page about one of the following.
    a. the business or technical writing class you are taking this term
    b. a student group to which you belong
    c. a vacation area your group has visited
    d. a college or community association, club, or team

PART III

Gathering and Summarizing Information

CHAPTER 9

Doing Research: Finding and Using Print, On-Line, and Internet Information Sources

A personnel director asked an applicant during a job interview to name the titles of two or three major journals in the applicant's field. When the applicant could not come up with even one, the applicant's chances for employment at the company became slim. The question was typical, fair, and relevant. The interviewer knew that an applicant's success in the job depends on the quality of information supplied to the employer. Employers expect carefully researched answers; they will not be satisfied with guesses.

The Importance of Research

Research, or the careful investigation of material found in books, magazines, pamphlets, films, the Internet, and other sources (including resource people), is a vital part of every occupation. Companies expect their employees to be able to research a problem, find information, and reach informed conclusions for any number of assignments, such as writing proposals and analytical reports.

You might conduct research by e-mailing someone in another department for information or by scrolling through a company's home page on the World Wide Web. You might engage in research by downloading a section of a reference work or by preparing statistical or marketing surveys. You need to be informed about the latest developments in your field, and you need to be able to communicate your findings accurately and concisely.

Doing research is a practical skill like swimming, running, or keyboarding. Once learned, it is easy to perform. Knowing how to do research in your field brings lifelong benefits. The information in Chapter 9 can save you from suffering the embarrassment experienced by the job applicant described above.

The Process of Doing Research

Just as there is a process for writing—brainstorming, drafting, revising, editing—there is a process involved in doing research. In research, you go through the steps of finding, assessing, and incorporating information into your written work. Basically, that process includes using the following strategies, which sometimes overlap (especially steps 2, 3, and 4 below) since doing research is a recursive activity.

1. Identify a significant topic. Discovering an important subject is crucial to your success. Restating the obvious—mercury in drinking water is dangerous both to human health and to the ecosystem; computers can save a business time and money—unimaginatively duplicates what is well known without providing new interpretations or solutions. Check with your boss or your instructor to make sure that the subject of your research is both timely and significant. Usually, that assignment will involve investigating some problem—its importance, its impact, and how to resolve, reverse, or contain it.

Once you have identified a significant topic, you are well on your way to establishing the **purpose** for your research. Look at the reports on telecommuting (pp. 398–413) and e-cash (pp. 631–652) to see how their authors identified and restricted a major subject to research.

2. Limit the scope of your topic. You can realistically devote only so much time to researching it. For example, you would have trouble writing a restricted report on lasers—certainly an important subject, but one that is too broad and too complex. Because thousands of articles and books have been written on lasers, it would take years to gather information about every aspect of laser technology, certainly more time than you have in a course or in the busy, deadline-bound world of work. To restrict your topic, you need to ask several questions about lasers and determine what your audience needs to know about them. You want to focus on a problem that applies to your audience and their job responsibilities. Your questions could lead you in several ways—through space or environment (the risks posed by lasers used in security) or through reference to key personnel or users (physicians using lasers for gallbladder surgery) or some specific combination of time, space, and personnel. You would then be in a better position to address a limited problem related to lasers in a particular context—e.g., medical, security, or office technology use. In the world of work, you may be given a problem to investigate but have to break it down into different parts to research it thoroughly.

3. Next, identify the location of materials for your research. That is, find the **resources** that will enable you to carry out your research. Identifying resources involves more than going to your college or public library and searching through a few books or journals. To do thorough research, you cannot stop with a handful of books or articles picked up on one visit to a library. Resources often involve a diversity of media—print, electronic, audiovisual, Internet—located in a variety of places. You may start your research in your college library but then continue it at home or at your office with your computer; or you may have to read company reports on file or join a discussion group on the Internet. Doing quality research means exploring a range of opinions and not settling for quick, superficial answers. You will have to locate the most pertinent information to prove you have conscientiously investigated your topic.

4. Know how to use the sources you find. This stage is devoted to the **methods** and **research tools** you use to uncover information—the steps you follow to identify, gather, and record the information you need. Based on the restricted topic (problem) you are researching, you will have to determine what tools you will need to do your research—such as indexes, abstracts, and electronic databases. Determine any special preparations you have to make. For example, if you are having a specialized search done, you may have to work closely with a librarian who will ask you for key terms to conduct that search. You may also have to use special audiovisual equipment or software. If you are conducting an Internet search (see pp. 335–338), you will need to employ various search techniques and search engines.

5. Next, familiarize yourself with these research materials, especially the way they are organized. The quickest way is to look at the parts of a work that readily alert readers to the topics it covers. Prefaces and introductions typically spell out an author's or reference work's purpose and scope. Also examine indexes to find out about their coverage. Home pages on the Internet are equally important in directing you to further, possibly major, information. If you are looking at articles in journals, examine abstracts or summaries and note any key terms. Pay special attention to endnotes, bibliographies, or works cited lists to see whether they refer you to other relevant sources. Additionally, don't neglect dialoguing or collaborating with co-workers and other resource people.

6. Know how to evaluate sources. Doing research means more than just being on a search mission to uncover facts. Don't be a human copying machine, naively repeating everything that you read. Be prepared to ask the right kinds of questions about the information you uncover.

- What are the most significant sources I need for my purpose?
- Is this the most recent opinion?
- What other studies have been done?
- Is this information complete?
- Is it biased?
- Does this article, report, or study raise further questions that I must consider?
- What's missing in this interpretation or in this presentation of data?
- What conclusions must I cite and perhaps modify or adapt for my purpose and my audience?

Never be afraid to question and seek further information. As your research progresses, you will find that the shape and quality of your report or paper will improve. Don't be discouraged if you seem to go down a few blind alleys or deadend streets; those so-called false leads may in fact result in a better informed and more useful report.

7. **Always document where and from whom you received information.** If you are using information from a source, including a co-worker or another department in your company, you are obligated to **document** it. This is not a matter of simple courtesy; it is an ethical and legal necessity. Accordingly, during your research record the author, title, and date (or volume number for journals or publisher for a book) of the source. Don't think that because so much may have been written on your topic or that because you found your source on the Internet that you can regard any information as "public domain" and hence do not have to give the author full credit. As you do your research, pay attention to and respect any copyright notice that appears in print or over the Internet. Chapter 10 will show you how to document your sources correctly and consistently.

Each of these seven steps above involves a major ingredient of successful research—you will need to establish your **purpose,** restrict your **scope,** identify your **resources,** develop appropriate research **methods,** familiarize yourself with a work's **organization,** subject it to your critical **evaluation,** and provide careful and thorough **documentation.**

The Library and the Internet

Doing research in the "Information Age" means taking advantage of all the resources of contemporary technology. To do effective, thorough research, you will have to use the resources of your library as well as the Internet, which has been called "the library without walls" since it potentially offers users any and every kind of information. Accordingly, the rest of this chapter is divided into (1) a survey of the research materials available at a library and (2) beginning on page 347, the types of resources you will find on the Internet. The last half of this chapter, therefore, is devoted to the Net as the virtual library. The chapter concludes with suggestions for taking research notes—both bibliographic and informational.

The Library and Its Services

You may think of your library as a single building with a single function. Yet that building is divided into many sections, each offering many services. When you walk into a library, probably the first section you see is the circulation desk—in many ways, the business center of the library. From there you can move to any one of the following parts of the library to use materials. (The page numbers after each area refer to the page numbers of this chapter where you will find a description of the particular library service and the materials in it.)

- the on-line catalog (pp. 325–329)
- periodical holdings and indexes (pp. 329–332)
- computer resources (pp. 332–338)
- reference books (pp. 338–342)

- government documents (pp. 342–345)
- the popular press (pp. 346–347)
- audiovisual materials (p. 347)

Keep in mind that every library is different and that your library may combine these services and offer many more.

The On-Line Catalog

This is often the starting place to look for research materials. The on-line catalog is your guide to your library; it will tell you what materials your library owns and where you can find them. The on-line catalog is not restricted to books; it lists the microform materials, computer software, CD-ROMs, encyclopedias, periodical indexes, audio recordings and visuals, and the magazines and journals to which your library subscribes. But while the on-line catalog may list specific journal or magazine titles, it will not give you information about the contents of individual articles. For that kind of information you have to consult abstracts or the periodical itself.

Usually, terminals for an on-line catalog can be found near the entrance to your library; they may also be located on different floors of your library. You may even be able to access the on-line catalog from your dorm room, house, or office. Some on-line catalogs include CD-ROM networks and databases such as **First-Search** or **CARL UnCover** (see p. 336).

An on-line catalog links users via a computer system to the automated database containing information about materials in the library. Entering information you seek, you will see bibliographic information displayed on the screen, and you can print or download it onto your disk. With access to that database, you can find information about titles, authors, and subject areas as well as many other things not available to library patrons a decade ago, including

- the status of a book—noncirculating, reserve, available, or checked out (and when it is due)
- the location of a book or other item if your college has several libraries (e.g., business library, fine arts library, regional library) or is part of a library consortium
- type of document (you can choose to retrieve only certain works—like films or dissertations)
- works published by or sponsored by a certain organization (only material released, say, by the American Cancer Society or the National Institute of Technology)

Ways to Search the On-Line Catalog

Perhaps the most significant advantage of an on-line catalog is that it makes your library research easier, giving you more extensive and helpful access to the library's holdings. You can search a computerized catalog by a variety of methods.

```
NOW SEARCHING: ALL LIBRARY COLLECTIONS
LOOK FOR THE FIRST WORD IN THE:
       Subject
       Title
       Author
       All of the Above
       Call Number

LOOK FOR ANY WORD IN THE:
       Subject
       Title
       Author
       All of the Above
       Expert Search
```

By simply pressing the corresponding option for subject, title, and author, or a combination of these fields, you can search the library's holdings all at once. If you know an author's name, you can retrieve a title that she or he has written; similarly, if you know the title of a book or audiovisual, you can call it up on the screen and get relevant publication information.

Subject Searches

If you don't have specific authors or titles in mind, then it's best to do a subject search. By entering two or three key words (see "Key Word Searches" below) and pressing the subject option, you can call up books and other relevant materials that contain the key subject terms you are investigating. Start with broad terms that will generate large general categories. Often, looking at these longer lists will give you ideas about focusing your topic; then you can limit the scope of your investigation.

Key Word Searches

You can also search for key words appearing in the title of works included in the on-line catalog. If you already know the title of a work or the author's name, you can save time by making those choices and entering the specific key word you want. You will get exactly what you ask for. If you enter "Bill Gates" as the key words in an author search, your list will consist of only items written by Bill Gates. The same concept applies to the title search; you will get a limited return.

Researchers generally prefer the broader scope of materials produced by the subject search because it covers more data.

What if I Have Incomplete Names or Titles?

Even if you don't know an author's full name (assuming you know just the surname), the on-line catalog can help. For example, if you know that the author's last name is Ponders, and you key in that name, the on-line catalog will display all works by authors named Ponders—whether they are by Anthony, Mary, S. A., Trey, or Zack. You then could select which work by which Ponders you need.

Similarly, if you know only the first part of a title—*The Growth of the Internet* for the full title *The Growth of the Internet and Reasons for Regulation*—you could enter that incomplete title, and the on-line catalog will list all the titles that begin with those five words on the screen; you can then select the one you are searching for.

Call Numbers

Author and title searches will give you a **call number** that indicates where the book or document is located in the library. Call numbers or letters are usually found in the upper left-hand side of the screen, although this practice can vary from system to system.

Case Study: Women in Business

Let's say you want to find relevant documents in your library on **women in business,** a very general topic. Entering these key words for a subject search, you find a list of pertinent subtopics, one of which is **women-owned business enterprises.** Figure 9.1 shows subheadings for this narrower topic. Note that among these subtopics are areas for investigation, such as job stress or family life, as well as various research tools, bibliographies, and periodicals. Assume that you choose number 2, **African American women-owned businesses.** You would then see your screen fill up with the items that the library holds on this topic. The number of items, of course, will vary from one library to another. Some of these items from a college library are shown in Figure 9.2. Note that the list includes books, periodicals, government documents, and even a video. Also keep in mind that a subject search gives you titles that may include a broader discussion but still cover the area of your investigation.

Choosing one of these items, you then see a full display of bibliographic information—author, title, and publisher for each—as in Figure 9.3. Keep in mind that an on-line catalog gives you the same publication information as does a traditional card catalog; it's just computerized.

Narrowing and Expanding Your Search

Thanks to key word searching on an on-line catalog, you can narrow or expand your topic, as necessary. Suppose you are researching the use of lasers in printing. An on-line catalog subject search finds only documents with *both* words (**lasers** *and* **printing**). To expand your search you might search for **lasers** *or* **printing** separately. On the other hand, to narrow your search, you can specify exactly what

FIGURE 9.1 Subtopics on "women in business" from an on-line catalog.

Your Search: SW=WOMEN BUSINESS 52 headings in list

	Subject Heading	No. of Titles
1.	Women owned businesses	32
2.	Afro-American women owned businesses	17
3.	Women owned businesses-attitudes	8
4.	Women owned businesses-biography	13
5.	Women owned businesses-case studies	21
6.	Women owned businesses-discrimination	6
7.	Women owned businesses-family life	9
8.	Women owned businesses-finance	5
9.	Women owned businesses-government contracts	3
10.	Women owned businesses-job stresses	11
11.	Women owned businesses-psychology	2
12.	Women owned businesses-United States-Bibliography	7
13.	Women owned businesses-United States-Directories	3
14.	Women owned businesses-United States-Handbooks	1
15.	Women owned businesses-United States-Periodicals	5
16.	Women owned businesses-United States-Proceedings	2
17.	Women owned businesses-United States-Statistics	1

FIGURE 9.2 Items in the library on "African American women-owned businesses."

Government Documents *Books*

No.	Title	Author	Date
1.	How to Survive in Spite	Robinson, Jean	1996
2.	Accept No Limitations: A Black Woman	Kimborough, M.	1991
3.	Stresses, Beliefs, and Coping	Harry, Lois	1994
4.	Work, Sister, Word: How Black Women	Shields, Cydney	1994
5.	Black Woman's Career Guide	Nivens, Beatrice	1982
6.	African-American Firsts: Famous Black	Potter, Joan	1994
7.	In the Black Plus-Call Us	Parker, Ophelia	1990
8.	A Study of Success	Lownes, Millicent	1981
9.	Doing Business with HUD: Minority Women	HUD	1991
10.	Minority and Women Business Programs	ABA	1990
11.	Minority and Women-Owned Business	Walker, Marissa	1993
12.	Black Female Executives	Ducksworth, Pam	1987
13.	History of Women Managers in America	Walton, Ind.	1997
14.	Black Studies (1998)		1998

Pamphlet *Film* *Periodical*

FIGURE 9.3 On-line catalog brief display of highlighted item.

```
HD6057
.837

AUTHOR       Harry, Lois
TITLE        Stressors, Beliefs, and Coping Behaviors of Black
             Women Entrepreneurs
PUBLISHER    Garland
DATE         1994
SUBJECTS     Women in business-United States
             Women in business-African-American
             Women-owned businesses
             Women and Stress-Work Related
             Women Executives-Psychology
```

you want *excluded* from your search—**lasers not military.** This allows you to narrow a broad topic (such as **lasers**) by focusing only on nonmilitary aspects. (See pp. 356–357 for some advice on key word searching.)

Periodical Holdings and Indexes

Your research will not be confined exclusively to books. Don't neglect the wealth of information contained in periodicals. A **periodical** is a magazine or journal that is published at established, frequent intervals—weekly, bimonthly (once every two months), quarterly. The word *magazine* refers to periodicals—*National Parks, Redbook,* or *TIME*—that appeal to a diverse audience and that treat popular themes in nontechnical language. The word *journal* characterizes technical or scholarly periodicals, such as the *American Journal of Nursing* or *Chemical Engineering,* whose audiences consult them for professional or scientific information. Some periodicals are published in print and on-line.

Advantages of Using Periodicals

Periodicals have certain advantages over books when it comes to research. Because they take less time to produce than books, periodicals can give you more recent information on a topic. Further, periodicals are not as restricted as books. A book usually discusses a subject from one point of view, but a periodical may contain ten different articles, each covering a separate topic and each offering a different perspective. This is not to imply that you should ignore books and focus solely on magazines or journals. Use both, but be aware of the differences.

Using Indexes

To find appropriate articles in magazines and journals, you need to use indexes. An **index** is a listing by subject, and sometimes also by author, of articles that have appeared within a specified period of time. An index tells you what articles have been published on your subject, where they were published, who wrote them, and whether they contain any special information, such as bibliographies, illustrations, diagrams, or maps.

Indexes are invaluable. You could never successfully thumb through all the periodicals in the library to determine which ones contain an article you could use. Furthermore, your library might not subscribe to all the journals and magazines that contain articles related to your subject. The indexes let you know what is available beyond your library's holdings.

General Indexes

You may already be familiar with some general indexes such as *The Readers' Guide to Periodical Literature* or *The Magazine Index.* These indexes survey periodicals of interest to the general, not technical, reader. *The Readers' Guide,* for example, indexes information from more than two hundred periodicals, mostly popular magazines like *Newsweek, Sports Illustrated,* and *American Health. The Magazine Index* overlaps *The Readers' Guide* and surpasses its scope by surveying more than 500 magazines. To learn about individual periodicals, consult *Ulrich's International Periodical Directory,* which contains information on over 1,700 publications.

Specialized Indexes

You should not rely only on general indexes to do research in your major or for a report on the job because they do not list technical or specialized journals. To find information about these sources, you will need to use a specialized index. These indexes are guides to the literature in particular professions or subject areas—business, computer graphics, finance, food service, nursing, public safety, respiratory therapy, and many others. Some print versions of indexes cover sixty or more years of work (*Applied Science & Technology Index* or *Index Medicus*); other indexes began as on-line references (see the discussion below) more than fifteen years ago. Almost every profession has its own index. Here, for example, are some leading specialized indexes, with the year they began in parentheses:

ACM Guide to Computing Literature (1960)
Applied Science & Technology Index (1958)
Bibliography and Index of Geology (1969)
Biological and Agricultural Index (1964)
Business Periodicals Index (1958)
Computer Literature Index (1980)
Cumulative Index to Nursing and Allied Health (1965)
Current Law Index (1984)
Ecology Abstracts (1975)

Engineering Index (1934)
Index to Legal Periodicals (1926)
Social Sciences Index (1974)

Keep in mind that indexes are available in many forms—in print, through on-line searches, and over the Internet. Many individual indexes like those listed above are included in at least one on-line database (see p. 333).

To give you some idea about what an index includes and how it is organized, let's take a closer look at two: the *Applied Science & Technology Index* and the *Business Periodicals Index.*

Applied Science & Technology Index

If you need to locate information about a technical subject, you might start with the *Applied Science & Technology Index.* This index is a guide to more than 300 specialized journals in a wide range of scientific and technical fields—such as atmospheric sciences, computer technology and applications, food and food industry, geology, metallurgy, oceanography, plastics, textile industry and fabrics, and transportation.

Organized alphabetically by subject, *Applied Science* subdivides these subjects to assist users further. Figure 9.4 reprints a section of the index, showing how useful these subclassifications can be. Information published about concrete is categorized according to aggregate, air entrainment, durability, failure, specifications, and so forth. The subclassifications in the *Applied Science & Technology Index* might give you an idea for a research paper and then assist you in initially gathering relevant information about that topic.

FIGURE 9.4 An excerpt from *Applied Science & Technology Index.*

Concrete-*cont.*

Permeability

Effects of curing conditions and age on chloride permeability of fly ash mortar. A. Alhozaimy and others. *ACI Mater J* v93 p87-95 Ja/F '96

Plastics mixtures

See Polymer concrete

Silica fume additives

Corrosion of reinforcing steel in concrete: effects of materials, mix composition, and cracking. T. Lorentz and C. French. bibl il diag *ACI Mater J* v92 p181-90 Mr/Ap '95; Discussion. v93 p102-4 Ja/F '96

Effects of high temperature and pressure on strength and elasticity of lignite fly ash and silica fume concrete. S. Ghosh and K.W. Nasser. bibl il *ACI Mater J* v93 p51-60 Ja/F '96

Standards

Why certify shotcrete nozzlemen? L. Balck, Jr. *Concr Int* v18 p68-9 F '96

Strength

See also
High strength concrete
Pullout tests

Effects of high temperature and pressure on strength and elasticity of lignite fly ash and silica fume concrete. S. Ghosh and K.W. Nasser. bibl il *ACI Mater J* v93 p51-60 Ja/F '96

Stresses

Investigations on determining thermal stress in massive concrete structures. T. Kawaguchi and S. Nakane, il diags *ACI Mater J* v93 p96-101 Ja/F '96

Microplane model for concrete. I: stress-strain boundaries and finite strain. Z. P. Bazant and others. bibl diags *J Eng Mech* v122 p245-54 Mr '96

Microplane model for concrete. II: data delocalization and verification. Z. P. Bazant and others. bibl diags *J Eng Mech* v122 p255-62 Mr '96

Applied Science and Technology Index, May 1996, pages 73–74.

FIGURE 9.5 An excerpt from *Business Periodicals Index.*

Credit cards
See also
Affinity credit cards
Corporate purchasing cards
Credit card authorization
Credit card industry
Point-of-sale systems
Know your customers: consumers speak [survey of retail payment systems] graphs tabs *Chain Store Age* v72 p11A-14A Ja '96 supp Survey of Retail Payment Systems

Competition
Credit card war hots up [Asia] A. Pawlyna. il *Asian Bus* v32 p46-8 F '96
White paper report: corporate purchasing cards: purchasing takes cards to new level. S. Avery. il *Purchasing* v120p54-5+ F 15 '96

Default
Memo to credit card risk managers. S. Allard. *U S Banker* v106 p 80 F '96

Fees
Antitrust and payment technologies. D. W. Carlton and A. S. Frnkel. bibl graphs tabs *Fed Reserve Bank St Louis Rev* v77 p41-54 N/D '95
Commentary. N. Economides. bibl diags *Fed Reserve Bank St Louis Rev* v77 p60-3 N/D '95
Commentary. J. J. McAndrews. bibl *Fed Reserve Bank St Louis Rev* v77 p55-9 N/D '95
Credit and debit fees hold steady since Price Chopper complaint. il *Chain Store Age* v72 p208 Ja '96

Credit cards-*cont.*

Fraudulent Use
See Credit card fraud

Interest
Consumer behavior and the stickiness of credit-card interest rates. P. S. Calem and L. J. Mester. bibl graph tab *Am Econ Rev* v85 p1327-36 D '95

Marketing
Age differentials [survey of retail payment systems] il *Chain Store Age* v72 p17A Ja '96 supp Survey of Retail Payment Systems
Good food cheap? Pick a card! R. S. Teitelbaum. il *Fortune* v133 p133 Mr 18 '96
Sears revives credit card business. P. A. Murphy. il *Stores* v78 p57 F '96
White paper report: corporate purchasing cards: purchasing takes cards to new level. S. Avery. il *Purchasing* v120p54-5+ F 15 '96

Security Measures
The Internet: marketers deal with buyer perceptions through better security. J. Morris-Lee. *Direct Mark* v58 p38-40 Ja '96

Asia
Visa's Pyne leaves Asian operation at critical period. F. Warner. *Asian Wall St J Wkly* v18 p5-6 Ja 29 '96

Foreign Business
Order approving establishment of a branch. J. J. Johnson *Fed Reserve Bull* v82 p104-7 Ja '96

Business Periodicals Index

The *Business Periodicals Index* surveys more than 300 business publications in accounting, advertising, communications, computer technology and applications, finance, industrial relations, management, occupational health and safety, personnel, real estate, telecommunications, and transportation. Like the *Applied Science & Technology Index,* it is organized alphabetically by subject and offers many useful subheadings. Note how the excerpt from the *Business Periodicals Index,* above, arranges and classifies information about credit cards and how such classification helps users find information.

Do not limit your research to just one index, no matter how valuable and thorough it may be. Since indexes are computerized and their information is downloaded along with other indexes onto centralized databases, always seek to broaden your search. In all likelihood, your research will take you into on-line searches (see pp. 335–338) and into the Internet itself.

Computer Searches in Your Library Research

A computerized search is a speedy alternative to a manual search through an index or abstract because you can go through ten to fifteen years of an index in just a few minutes. Like a paper index, a computer search can also help you locate items that you may otherwise miss and can also give you access to materials your library does not have.

Variety of Databases

The procedure of conducting a computerized search is basically simple. Information found in various indexes and abstracts is stored in on-line databases. A **database** is an electronic index service that contains a file of information from which the computer then retrieves the key citations (and sometimes abstracts) you need. Several thousand databases have been compiled. In fact, almost every professional discipline has its own database—aerospace (*Aerospace Database*), criminal justice (*Criminal Justice Periodical Index*), nursing (*Cumulative Index to Nursing & Allied Health Literature*), chemistry (*Chemistry Abstracts*), business (*Business Periodicals Index; Trade and Industry Index*), psychology (*PsychLit*), engineering (*COMPENDEX*), and so forth.

Searching a Database

To search a database, first narrow your topic and then find a key word or phrase (plus suitable alternatives) that best summarize your topic. (See pp. 356–357 for some guidelines for selecting key words.) For example, to find out specific information about such broad topics as child abuse, artificial intelligence, or virtual reality, you would need to restrict your focus. You can also restrict your topic chronologically, say, limiting it to materials published after 1995. Or perhaps you are interested in finding films, reports, or conference proceedings as well as articles. You will then enter this key phrase or word, along with any limitations you are imposing, into the computer to search the database(s) for materials about your topic.

When the search is complete, you will see on your screen the number of entries (articles, books, reports, and more) that are filed under this key word or phrase. With another command, the computer will print a list of bibliographic citations, thus saving you the trouble of having to write them out by hand.

Now we will take a look at the two ways you can search electronic databases. One of these uses a **compact disk,** and the other employs an **on-line search**.

Compact Disks

This type of computerized reference work is known as a **CD-ROM** ("compact disk, read-only memory"), meaning that information about the periodicals, encyclopedias, books, and other reference works is stored on a self-contained disk. A disk can store an incredible amount of information. Updata, a CD-ROM distributor, states in a recent catalog that "a CD-ROM can hold up to 600 megabytes (mb) of data, the equivalent of about 1600 floppy disks or a quarter of a million single-space typewritten pages. What a space saver!" Your library has the individual disks, which are usually updated each month. A CD-ROM is also a closed system; that is, you are restricted to information placed on the disk.

The following are some widely used indexes on CD-ROM.

Info-Trac

Published monthly, a single Info-Trac disk can store bibliographic information about more than one million articles published in over 2,000 periodicals in business, law, health, technology, psychology, and education. Info-Trac includes general magazines (*Consumer Reports*), business journals (*Harvard Business Review*), and even major newspapers, some of which are given in full text.

Like printed indexes, Info-Trac is organized into subject categories with subheadings. If your library subscribes to the General BusinessFile of Info-Trac, you gain access to more than bibliographic information. General BusinessFile is an excellent source database for research on all aspects of business, management, industries, and companies. All of the information contained in the Business Index, Company ProFile, and Investext databases is integrated. Article citations from business and management publications are linked with company directory information, the full text of newswire reports, and company and industry investment reports.

To search a topic on Info-Trac, you enter a key phrase or phrases into the computer for it to examine the database file and identify all the appropriate citations listed on the compact disk. It can do this in less than a minute. With another simple command, Info-Trac prints a list of complete bibliographic citations of these materials. In some instances, you can also obtain full texts of selected articles.

DISCLOSURE

On this CD-ROM you will find valuable information from the U.S. Securities and Exchange Commission, including extracts from profits and sales reports, market shares, and complete company records as well as names of directors and officers with their salaries. You can search by company name, type of business, geographic area, stock exchange, or Fortune 500 number.

Legal/Trac™

This CD-ROM index includes citations from over 800 legal publications.

Lotus One Source™

Information in this database is indexed from company reports on stocks, research studies, and corporate profiles.

COMPENDEX Plus

COMPENDEX Plus abstracts and indexes journals in technology and engineering, a merging of COMPENDEX (the Computerized Engineering Index, which contains more than 2 million abstracts dating from 1969 to the present) and Ei Engineering Meetings (a database containing papers from engineering conferences dating from 1982 to the present).

On-Line Searches

TECH NOTE

Be aware that your library's on-line search capabilities may require you to conduct a mediated search; that is, only the library's staff may access the databases, and you will have to supply the search strings to be used. Since not all on-line searches are available at every library, check first with your librarian to see which ones you can use and which ones require a librarian's help, or "mediation."

An on-line search works like this. A library subscribes to a particular computerized information retrieval service that has numerous databases stored in its host mainframe computer. Via its computer terminal, your library is able to dial into these databases. There are thousands of different databases, and while no one retrieval service offers access to all of them, these services do supply their customers with vast amounts of information. Many of the databases correspond to printed indexes (such as the *Business Periodicals Index*), while others are available only through on-line searches—that is, there is no print equivalent for them.

An on-line search can offer you bibliographic information (a list of articles and other references for a preliminary working bibliography), copies of abstracts of articles included in certain databases, and, in some cases, copies of the publications themselves. Note that most databases do not go back more than about fifteen years. For earlier publications, you will need to consult print indexes.

Many on-line search services are now making their databases available on disks as well. Keep in mind that in some cases, the same information is included in print, on CD-ROM, and on-line. It depends on the software. For example, DIALOG (see below) and Silver Platter (a CD-ROM) are comparable. Both types of software provide access to similar material. You might want to consult K. Y. Marcaccio's *Computer-Readable Databases: A Directory and Data Sourcebook,* 8th ed. (Detroit: Gale, 1995), and Gale's *Information Industry Directory* for information on the variety of databases available to researchers.

Types of Information an On-Line Search Can Retrieve

Searching an on-line database, you can receive

- **bibliographic citations**—author, title, publication, volume and issue number for journal, publisher, date, and page numbers
- **full text of most articles**

- **factual information**—with selected databases you can get everything from stock quotations to new patents, news of public affairs, and product information, names and addresses of company executives

Helpful On-Line Databases

Below are four on-line services you may be likely to use; note that there is some overlap among these different services and that some (like DIALOG and Wilson) are also available via the Internet.

1. FirstSearch. Among the most popular and widely used databases, FirstSearch is found in many libraries as well as on the Web. FirstSearch contains many of the sources that are also available on CD-ROM.

As its "Welcome" screen announces, FirstSearch will help you research your topic in "books, articles, theses, films, computer software, and other types of material. . . ." FirstSearch accesses more than forty databases in a variety of subject areas, including business and economics; conferences and proceedings; consumer affairs and people; engineering and technology; general and reference sources; and medicine and health sciences. Below are just the first ten databases available on FirstSearch.

	Database	Contents
1	WorldCat	Books and other materials in libraries worldwide.
2	Article1st	Index of articles from nearly 12,500 journals.
3	Content1st	Table of contents of nearly 12,500 journals.
4	FastDoc	Index of articles with text on-line or by e-mail.
5	NetFirst	NEW: OCLC database of Internet resources.
6	A&H Search	Arts & Humanities Search. A citation index.
7	AGRICOLA	Materials relating to aspects of agriculture.
8	AIDS/Cancer	NEW: AIDS and Cancer Research.
9	ArtIndex	Leading publications in the world of the arts.
10	BasicBIOSIS	A wide range of bioscience topics.

2. CARL UnCover. This database indexes more than 20,000 periodicals in business, science, and current events. Over 400 current citations are added each day. You can also access "20 commercial databases and over 420 individual library catalogs that are part of the CARL system." It is possible to search CARL UnCover by key word, topic, author, or through a browse feature that allows you to see the table of contents for any issue of a periodical that is available on-line. As with other searches, CARL UnCover provides a full bibliographic citation, and in some cases, a summary of an article from the list of items found on your topic.

An especially helpful feature of Carl UnCover is its on-line periodical delivery service. For a fee you can order any article, book review, or letter to the editor from the database and have it faxed or e-mailed to you.

3. DIALOG (Knight-Ridder Information, Inc.). This on-line service accesses more than 425 databases (with more than 200 million items) in almost every discipline—from technology and science to public affairs and the arts. Using DIALOG, you can quickly (within a microsecond) search ten years' worth of *Financial Times, Business Week, Fortune, Chemical Week*, and many other publications for information relating directly to your topic. Since DIALOG can be expensive (up to $25 to $30 for a ten-minute search), discuss your search first with a librarian who can advise you about the best strategies to follow.

TECH NOTE

Here are just a few of the many databases you can search with DIALOG.

ABI/Inform (citations and abstracts to articles published in over 800 journals in business—accounting, and taxes, finance, marketing. Two highlights of this source are an early beginning (1987) and its 150-word abstracts.)
AGELINE
AIDSLINE (AIDS research)
CHEMPLANT PLUS/CHEMSEARCH (information about 23,000 chemical plants)
Commerce Business Daily
COMPENDEX (2.8 million citations from over 4,500 journals, reports, conference proceedings in engineering and technology)
FOOD SCIENCE TECHNOLOGY ABSTRACTS
FBR Asian Company Profiles (descriptions of and financial information about Asian companies)
ISMEC: Mechanical Engineering Abstracts
MEDLINE (medicine)
PAPERS (daily editions of more than fifty U. S. newspapers)
TOXLINE (medicine)
WATERNET (abstracts on issues relating to water from drinking water to sludge disposal)
WORLD TEXTILES (indexes and abstracts from textile literature throughout the world)

Some of these databases (for example, ABI/Inform and MEDLINE) are found on CD-ROM as well as on the Internet.

4. WILSONLINE. This database offers subscribers computerized access to the bibliographic information contained in many of the indexes published by the H. W. Wilson Company as well as other sources, twenty-five databases in all. These databases hold up to ten years of information and are updated twice each week. Some of the indexes included on WILSONLINE are

Applied Science and Technology Index
Biological and Agricultural Index

Book Review Digest
Publishers's Directory
Wilson Business Abstracts

Like most databases, WILSONLINE is not cumulative; it does not go back forty or fifty years. Coverage on WILSONLINE began in 1986. As more information and years are added, individual databases will become larger.

Using Computerized Searches

Here is some practical advice on running a computer literature search.

1. **Prepare for your search.** Do not just walk into the library and tell the librarian you are interested in taxes, credit cards, drug enforcement, or other broad topics. Rather than asking for all the information on acid rain, first decide on what specific aspect of acid rain you need to research in detail—its effect on crops, its prevention, its danger to the ozone layer, its political effects on neighboring countries, and so on.

2. **Conduct a preliminary search.** That is, find out approximately how much is written on the topic. This overview will give you some idea of whether you will be asking for ten entries or a hundred. Take a look at relevant subject headings in FirstSearch (see p. 336) or in a print index for the last two or three years to get a rough estimate.

3. **Refine your search.** Experiment with the most appropriate key words for carrying on your search. Often you will have to try two or three different key phrases or words, or combine or modify them. Keep in mind that you are not limited to subject terms found in print indexes. You can search any one of these databases by subject, author, and title. The more precise you make your search with key words, the more effective it will be.

4. **Be realistic.** Realize that databases do not scan an unlimited period of time. In some cases, databases cover only a specified time.

5. **Be prepared to pay.** Libraries often charge for a computer search. Patrons are usually charged per citation retrieval and per minute for time spent connected via telephone lines to the database. It may not be uncommon to spend $30 to $40 for a search on DIALOG, for example. You can set a limit on how much you are willing to spend before the search starts.

6. **Recognize the limitations of your library.** If your library's access to databases in your field is limited, you may need to use another library that subscribes to additional databases and certainly try other sources such as the Internet.

Reference Books

Encyclopedias, dictionaries, abstracts, manuals, and almanacs of all kinds are reference books—sources of useful facts and basic information that are housed in the library's reference room. Reference books, marked **Ref.** or **R** in an on-line catalog, usually may not be checked out of the library. When you look for reference books, always make sure that you use the most up-to-date ones. Ask your librarian for help in identifying them.

Encyclopedias

The word *encyclopedia* comes from a Greek phrase meaning "general education." A general encyclopedia can get you started in your research by giving you relevant background information, explaining key terms, offering quick summaries, and supplying a list of further readings. Three useful, general encyclopedias are *Collier's Encyclopedia, Encyclopedia Americana,* and *New Encyclopaedia Britannica.*

Also consult specialized encyclopedias, which include far more technical information. Two helpful specialized encyclopedias are *The Encyclopedia of Associations,* which gives information (addresses, phone and fax numbers, descriptions of purpose and publications) on over 30,000 groups, and the *Encyclopedia of Careers and Vocational Guidance,* which provides overviews of the requirements and responsibilities for numerous technical and managerial positions.

Some specialized encyclopedias include

Encyclopedia of Banking and Finance
Encyclopedia of Chemical Technology
Encyclopedia of Computer Science and Engineering
Encyclopedia of Consumer Brands
Encyclopedia Food Technology
Goodheart-Wilcox Automotive Encyclopedia
McGraw-Hill Encyclopedia of Environmental Science and Engineering
McGraw-Hill Encyclopedia of Science and Technology

TECH NOTE

Note that, with rare exceptions like the *Encyclopaedia Britannica,* CD-ROM encyclopedias do not give the full text of entries. Instead, multimedia encyclopedias like *Encarta* and *Grolier's* provide direct links from the CD-ROM articles to related sites on the Internet. Do not be mislead into thinking that by consulting CD-ROM encyclopedias you are receiving full coverage of the topics you are researching.

However valuable encyclopedias are, do not attempt to do all your research using only these reference guides. Relying on encyclopedias alone reveals your lack of ability to find and use other sources.

Dictionaries

Dictionaries, in addition to giving the meanings of words, indicate spelling, pronunciation, etymology (word history), usage, and sometimes biographical information. The following unabridged (comprehensive) dictionaries are excellent references:

The American Heritage Dictionary of the English Language
Funk and Wagnalls New Standard Dictionary of the English Language
Random House Dictionary of the English Language
Webster's Third New International Dictionary of the English Language

Specialized, or field, dictionaries define the words used in the literature of a profession and also characterize the scope and importance of that profession.

Dictionary of Architecture and Construction
Dictionary of Computing
Dictionary of Earth Sciences
Stedman's Medical Dictionary
McGraw-Hill Dictionary of Scientific and Technical Terms

Abstracts

In addition to providing bibliographic data, an abstract gives a short summary of the content and scope of a book, article, or report. By condensing this information, an abstract can save users hours of time by letting them know whether the work is relevant to their topic. Many abstracts are available through the on-line databases discussed above.

Figure 9.6 contains an abstract that appears in *Communication Abstracts* of an article about videophones. In Chapter 11 you learn how to write an abstract yourself.

FIGURE 9.6 An abstract of a journal article from *Communication Abstracts.*

COMMUNICATION TECHNOLOGY. ELDERLY. HOME HEALTH CARE. TELECOMMUNICATIONS TECHNOLOGY. VIDEOPHONE.

The authors introduced a videophone system of full-color motion pictures, using Integrated Systems for Digital Networks, into home health care services and evaluated its effects. Twenty households including the disabled elderly were enrolled into the project for a three-month period. Communication and social cognition independence after the trial, as examined by means of the Functional Independent Measure, were statistically improved as compared with those before the trial. The videophone also improved activity of daily living (ADL), instrumental ADL, family health, and accessibility to medical consultations. The advantages of a wider application of this telecommunications technology to home health care services are discussed as well.

Use abstracts with caution, however, for they are not a substitute for the work they summarize. A few sentences highlighting the content of a book or article obviously omit much. When in doubt about what is omitted, read the original work to uncover the details, the rationale, and the dimensions of the whole problem. Never quote from an abstract; always cite material from the original work.

The following titles give you an idea of the many types of abstracts:

Biological Abstracts
Chemical Abstracts
Communication Abstracts
Computer Abstracts
Energy Information Abstracts
Forestry Abstracts
Metals Abstracts
Oceanic Abstracts
Petroleum Abstracts
Pollution Abstracts
Psychological Abstracts
Transportation Research Abstracts
Work Related Abstracts
World Textile Abstracts

Manuals and Almanacs

Manuals supply explanations of procedures and authoritative overviews of practical and professional issues you will encounter on the job. Consult more than one manual to compare their discussions of the same topic.

In business, among the most useful references are *Moody's Manuals,* which provide a wealth of information on the history of companies, descriptions of products and services, and basic financial details (such as stocks, earnings, and mergers). Depending on the type of business you're interested in, you may consult one or more of the eight Moody manuals, such as *Industrial Manual, Moody's Bank and Finance Manual,* or *Moody's Transportation Manual.*

Along with *Moody's Manuals*, investigate several company directories that give helpful information on specific companies, including

Million Dollar Directory
Standard and Poor's Register of Corporations and Executives
Try Us: National Minority Directory
World Business Directory

And don't forget to check a company's home page on the Internet.

Almanacs contain carefully organized statistical information—charts, tables, graphs, price indexes, federal and state budgets—as well as descriptions of events by year or by region. Almanacs are published on a variety of specialized subjects, as the following titles indicate:

Almanac of Science and Technology: What's New and What's Known
Business Week Almanac
Everybody's Business: An Almanac
Information Please Almanac, Atlas and Yearbook

Weather Almanac
World Almanac and Book of Facts

Government Documents

The U.S. government is the country's biggest publisher. Through its diverse agencies and departments, the U.S. government engages vigorously in conducting research and in publishing its findings. This published material, collectively referred to as **government documents,** includes journal articles, pamphlets, research reports, transcripts of government hearings, speeches, statistical reports, films, maps, and books.

These materials can have immense practical value for your research. Government reports, for example, discuss virtually every imaginable topic: from care of the aged, computer programs, flood insurance, and farming techniques, to fire precautions, housing costs, outdoor recreation, transportation, and urban development.

Three indexes to government documents are especially helpful in guiding you through this vast store of information.

1. *The Monthly Catalog of U.S. Government Publications (MOCAT).* Published since 1895, this index lists government documents published during that month. The catalog is arranged by agencies that publish or sponsor works (the Departments of Agriculture, Commerce, Interior, State, and so forth). Each entry provides the author's or agency's name, the title, the date, a brief description of the contents of the document (indicating whether it contains a bibliography, maps, or index), when and where the research was conducted and who sponsored it, the price, and how to order a copy. Starting with 1996, MOCAT was put on CD-ROM, and an on-line version is on the Web. Figure 9.7 shows you how to interpret a sample entry.

2. *U.S. Government Periodical Index.* Published quarterly since March 1994 by the commercial Congressional Information Service (CIS), this index covers over 180 federal publications "that have major research, reference, or general interest value" and that appeal "to a broad cross-section of researchers." Each quarter over 2,500 more articles are included. "Topics range from the use of satellites by air traffic controllers, to improving management skills, to testing for lead pollutions." Some sample titles and the U.S. government periodicals in which they appear are found below.

- Lifetime Risk of Developing Breast Cancer
  Journal of the National Cancer Institute
- Effects of Prenatal Exposure to Alcohol
  Alcohol Health & Research World
- FDA Reports on Pesticides in Food
  FDA Consumer

FIGURE 9.7 Key to MOCAT entries.

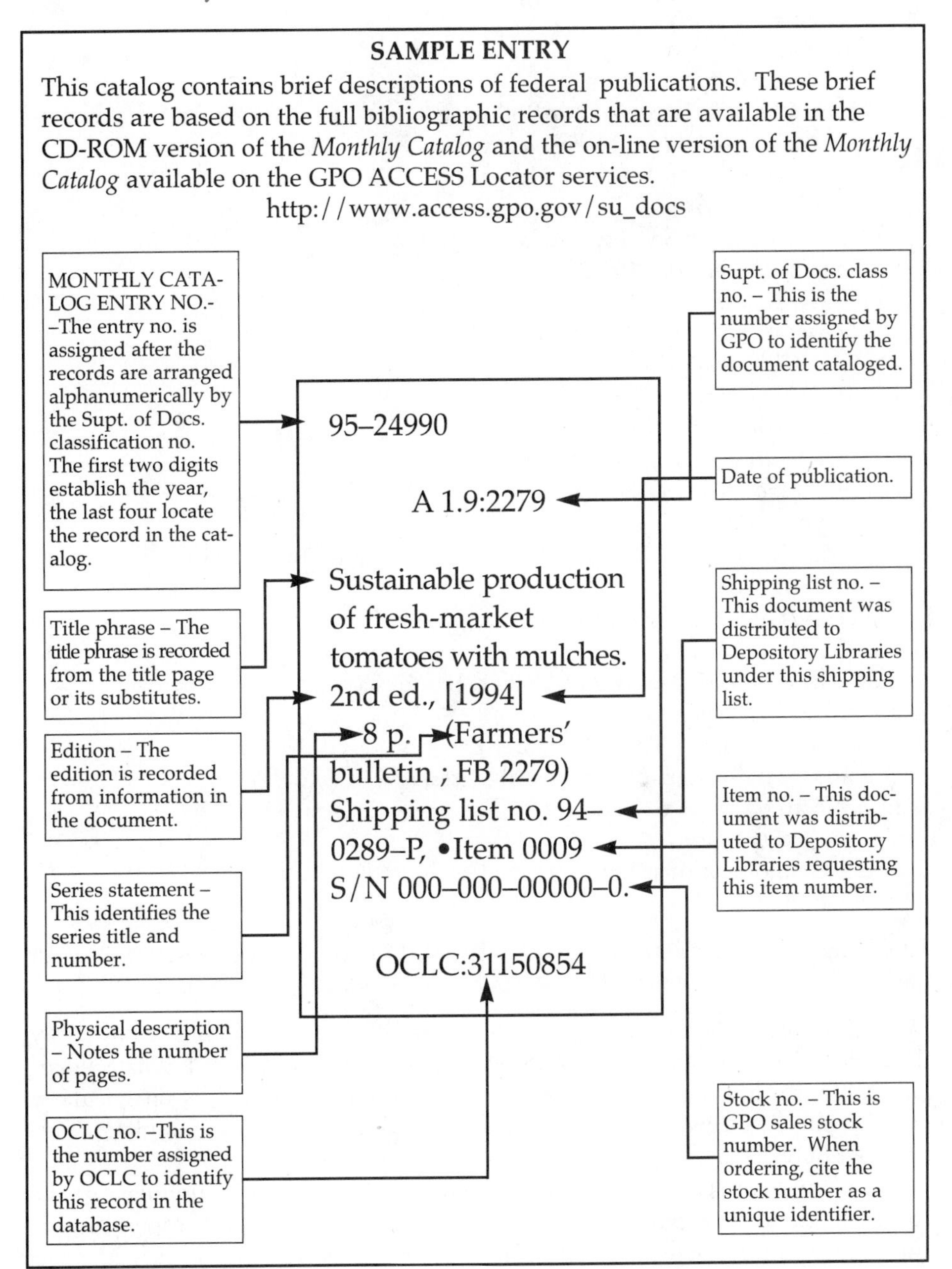

SAMPLE ENTRY
This catalog contains brief descriptions of federal publications. These brief records are based on the full bibliographic records that are available in the CD-ROM version of the *Monthly Catalog* and the on-line version of the *Monthly Catalog* available on the GPO ACCESS Locator services.
http://www.access.gpo.gov/su_docs
MONTHLY CATALOG ENTRY NO.--The entry no. is assigned after the records are arranged alphanumerically by the Supt. of Docs. classification no. The first two digits establish the year, the last four locate the record in the catalog.
Title phrase – The title phrase is recorded from the title page or its substitutes.
Edition – The edition is recorded from information in the document.
Series statement – This identifies the series title and number.
Physical description – Notes the number of pages.
OCLC no. –This is the number assigned by OCLC to identify this record in the database.
95–24990
A 1.9:2279
Sustainable production of fresh-market tomatoes with mulches. 2nd ed., [1994]
8 p. (Farmers' bulletin ; FB 2279)
Shipping list no. 94–0289–P, •Item 0009
S/N 000–000–00000–0.
OCLC:31150854
Supt. of Docs. class no. – This is the number assigned by GPO to identify the document cataloged.
Date of publication.
Shipping list no. – This document was distributed to Depository Libraries under this shipping list.
Item no. – This document was distributed to Depository Libraries requesting this item number.
Stock no. – This is GPO sales stock number. When ordering, cite the stock number as a unique identifier.

- How Can We Measure Leadership Performance?
  Program Manager
- Contingency Forces
  Military Review
- Human Health Effects of Air Pollution
  Environmental Health Perspectives
- Terrestrial Ages of Antarctic Meteorites
  Antarctic Journal
- Neuroimaging in Schizophrenia Research
  Schizophrenia Bulletin
- Pretrial Release and Detention and Pretrial Services
  Federal Probation
- When Cultures Meet
  Folklife Center News
- Should You Build a Future as a Construction Worker?
  Occupational Outlook Quarterly
- GPS: Revolution in Navigation
  Flying Safety
- Marine Recreation and Tourism: Dimensions and Opportunities
  Trends
- Hazards of Geomagnetic Storms
  Earthquakes & Volcanoes
- Malaria Field Studies at NMARU-1, Indonesia
  Navy Medicine
- Loss of Shrimp by Turtle Excluder Devices
  in Coastal Waters of the United States,
  North Carolina to Texas
  Fishery Bulletin
- US-China Relations: The Strategic Calculus
  Parameters

This index is available in print and CD-ROM. For coverage before 1994, consult the *Index to Government Periodicals* (published from 1970 to 1992).

3. *CIS/Index to Publications of the United States Congress.* Published monthly since 1970, this reference work indexes and abstracts House and Senate documents, reports, hearings, investigations, and other publications. It is a particularly valuable guide to legislative investigations and decisions. Annual cumulative volumes are published.

States and counties also engage in research and publish their findings. At the county level, practical publications are available on a wide range of topics in agriculture, education, food science, housing, and water resources. Check with your county agent for copies of relevant publications.

Many government documents are also available on the Internet. Figure 9.8 contains the GPO Access page, which also gives information about on-line services.

FIGURE 9.8 GPO Access page on the Internet.

Electronic Information: Online, On-Demand and Locator Services Through *GPO Access*

The *Federal Register, Congressional Record, Congressional Bills,* and other Federal Government information are available online via ***GPO Access***, a service of the U. S. Government Printing Office (GPO). Public access is available through the Federal Depository Library, or directly from GPO.

- Connection to the Online Databases
- Search GILS Records
- Search the Monthly Catalog of U.S. Government Publications (MOCAT)
- The Federal Bulletin Board (FBB)
- List and Descriptions of the Online Databases
- Access Methods for Online Databases
- Free Public Access Through a Federal Depository Library Gateway
- Free Public Access On Site at a Federal Depository Library
- ***GPO Access*** Federal Locator Services
- ***GPO Access*** On-Demand Delivery Services

Questions or comments about this service? Contact the ***GPO Access* User Support Team** by Internet e-mail at *gpoaccess@gpo.gov*

Page #AACES001 May 15, 1997

The Popular Press

The "popular press" includes reading material written in nontechnical language for the general public. Brochures, consumer documents, and newspapers constitute the popular press. Brochures and manuals are readily available from many sources.

The following section will show you how to locate and find copies of newspapers. Keep in mind that daily issues of many prominent newspapers worldwide (*The New York Times, Boston Globe, Chicago Tribune, Jerusalem Post*) can be found on the Internet.

Indexes to Newspapers

A thorough guide to the newspapers published in the United States and Canada is the *Gale Directory of Publications.* This work, issued annually, lists the date a paper began publication; its current address, rates, and circulation; and its religious or political preference. The directory also includes a capsule history of the community served by the paper.

Unfortunately, the *Gale Directory* will not refer you to specific stories. For that information you need to consult an index. *The New York Times Index,* which classifies stories by subject and author, helpfully reprints the photographs, maps, and other illustrations that accompanied some of the stories. Still another advantage *The New York Times Index* offers is that once you find the date of an event, you can then use that date to see how a local newspaper covered the story. Many libraries have back issues of *The New York Times* on microfilm.

In addition to *The New York Times Index,* look at the indexes for stories published in newspapers representing other areas of the country: the East (*Washington Post*), the Midwest (*Chicago Tribune*), the South (*New Orleans Times-Picayune*), and the West (*Los Angeles Times*). Also consult the indexes to the *Wall Street Journal, Boston Globe,* and *Christian Science Monitor.*

NewsBank

Begun in 1982, NewsBank is an extremely useful newspaper reference service that both collects newspaper stories and indexes topics in them from over 500 newspapers across the United States. It is available on microfiche (transparent four-by-six inch cards on which newspaper pages are reduced) and CD-ROM. The CD-ROM version is increasingly full-text. Also, it offers a General File and a Business File.

NewsBank makes stories from across the country easy to obtain and to read and allows the researcher to see how a story may be reported differently from one section of the country to another. While you will not find stories from the readily available *New York Times, Wall Street Journal,* or *Christian Science Monitor* in NewsBank, you will be able to locate articles that have appeared in major big-city newspapers from Atlanta, Cincinnati, Houston, Kansas City, Las Vegas, and many others.

NewsBank covers a wide range of topics, including business and economic development, consumer affairs, employment, the environment, films and television, health, housing and land development, the performing arts, social relations, and transportation.

The *NewsBank Index,* issued monthly, offers a multilevel index system to allow researchers to scan the material for different aspects of a subject. The index is organized by subject headings and by geographic regions. Keep in mind, though, that the particular citations from NewsBank in the index refer to the NewsBank numbers and not to the page numbers or sections from the original newspapers.

In addition to the index, NewsBank's *Reviews in the Arts* and *Names in the News* (for individuals and groups) are essential guides to the collection.

Audiovisual Materials

Audiovisual materials include records, cassettes, tape recordings, audio compact disks, photographs, films, and microforms. These materials are shelved in a separate part of the library that contains the proper equipment to store and to use them. You will find audiovisual titles listed in the library's on-line catalog. One particular type of audiovisual material, microforms, deserves special attention.

Microforms—microfilm and microfiche—reduce a great deal of information from its original size and store it compactly on film or tape. Microforms save the library space, increase document durability, and offer libraries a wider range of titles for far less money. Microforms are used frequently in business and industry as well.

A **microfilm** is a strip of black-and-white film that stores reduced images of book pages, back issues of periodicals, newspapers, directories, government documents, court proceedings, and so forth.

The advantages of microfilm are many. Without it, no library could possibly save all the back issues of a local newspaper, let alone a large one like *The New York Times.* Using microfilm, a library can gather a newspaper's daily issues from an entire month on one reel. A special machine called a **microfilm reader** enlarges the images for reading. Your librarian can show you how to operate one.

The Internet: The Virtual Library

What Is the Internet?

If the vast assortment of print and on-line research materials covered so far has not exhausted your curiosity and persistence, we turn now to the Internet. The Internet makes the virtual library—the library without walls—possible. Nothing can equal the Internet as a search tool. It is not just one computer system but a vast interconnection of computers and computer networks, talking to each other and gathering and exchanging information. The Internet is synonymous with *cyberspace,* a word coined to describe the power and control of information. Some have called the Internet "a network of networks" interlinked to bring users information. In this largest electronic network, over 200 million individuals are connected to over 3 million main computer networks. The Internet is classified as a Wide Area Network (WAN) since the computers on it cover a wide geographical area—the whole world. Internet users are important citizens in a global electronic village. Each day the net is growing, expanding at about 1,000 new users every hour.

History of the Internet

The idea for the Internet began during the late 1950s when the U.S. military connected various researchers via computer to protect Department of Defense information systems in case of a nuclear war. If one computer was knocked out, information would be duplicated by another computer and communications could travel by an alternative route. The idea of having so many vital, helpful links gave birth to the Internet of today.

The Accessibility of the Internet

The Net is accessible to anyone with a computer, a modem, and an Internet connection, usually available through schools, universities, companies, or access providers such as America Online, CompuServe, Prodigy, or the Microsoft Network (MSN). The charges for these services can be expensive, but local connection directly to the Internet is usually inexpensive, about $20 a month.

You can take advantage of immense amounts of information on the Net without ever traveling to a library; you don't have to leave your personal computer or laptop. With a printer, you can obtain hard copies of articles from journals on the Net as well as other reference materials. You don't have to master a series of complex commands, nor do you have to do a lot of elaborate keyboarding. You can point and click your way along the information superhighway.

The World Wide Web

One of the main services of the Net is the World Wide Web (WWW). In the World Wide Web, or just the Web, computers on the Internet provide information in a form that employs color, graphics, sound, and video. The Web enables a user to move quickly and easily from one set of information to another related set.

The Web is enormous, and the computers on it make available huge amounts of extremely varied and helpful data. You will never again be limited to just one library or the libraries in your town. You can search for information in libraries around the globe. And data are always available—books, reports, and journals are never checked out as in a library.

Here are some kinds of data on the Web that, in the past, you would have to look all over a library, or even several libraries, to find.

- hundreds of databases, collections, and other information services worldwide
- tables of contents for technical and trade journals
- copies of articles from popular magazines to technical journals and newsletters that are published only on-line. The *Internet Press* is a guide to on-line

journals dealing just with the Internet; also look at *Internet World*—http://www.mecklerweb.com. Journal Online News will give you information about accessing journals on the net (*Journal of Air Transportation World Wide,* for example), as well as new journals that have recently come to the Internet.

- popular newspapers from around the world—Sydney to Tokyo to New York to Jerusalem—and all points in between
- the latest edition of a leading encyclopedia, like the Britannica Web site, with entries updated monthly
- transcripts of speeches by major figures given a few miles or continents away
- technical manuals, guides, and product descriptions
- video clips and soundbites from today's most popular movies
- the latest statistical data on stocks and bonds, populations, just about anything that's counted

On the Web, in a matter of minutes, you can find out the exchange rate of the yen to the dollar, make airline reservations, or compile a list of 100 recent and relevant articles on the most specialized topic you are researching. With so much information at your fingertips, you can download computer software to make your researching easier.

TECH NOTE

The Web is only one feature of the Net from which you can gather information. The Net also provides other research services through e-mail and user groups, e.g., listservs. Becoming a part of these special interest discussion groups can greatly enhance the way you obtain information and do research. Some researchers estimate that there are as many as 40,000 news groups asking for and providing information on an endless list of specialized subjects. Once you tap into such a network, your possibilities for research expand phenomenally.

Locating Information on the Web

In order to reap any research benefits from the Web, you need to know something about the way it works.

Hypertext—What Is It?

The Web contains "pages," but they are not like the pages you see in a book or printed magazine. A page on the Web can be a block of text, but it can also include graphics, sound, and animation. These pages are written in **Hyper Text Markup**

Language (HTML). Hypertext is a system whereby particular words in a text, or areas in a picture, are highlighted, underlined, or colored differently from most of the other words. These colored or highlighted words (or graphics) are the connections—links—to other pages on the Web. Because hypertext is crucial to the way the Web functions, Web addresses often start off **http://www.**, which means **h**ypertext **t**ransmission **p**rotocol://**w**orld **w**ide **w**eb. Knowing an Internet address, or a universal resource locator (URL), allows you to go directly to that site. Examples are *http://www.sony.com* for the Sony Corporation home page, *http://www.econet.apc.org/econet/en.issues.html* for current information from the Environmental Issues Resource Center, and *http://www.indy.radiology.uiowa.edu* to make a visit to a virtual hospital.

How Hypertext Works

When you place your cursor on one of these hyperlinks—colored words or graphics—and click the mouse button, the page connected to that word is then displayed. One page on the Web is thus linked to every other; one page is connected (or placed on top of or next) to another. Think of hypertext as a gigantic company in which every department is linked directly to every other department. Hypertext pages do not have to be read in any specified order, front to back. In your pointing and clicking, you have many choices about what information you want presented and in what order. You can move easily in any direction from one page of information to another.

Take a look at Figure 9.9, which shows you a topics home page from the Library of Congress listing the various options, or hypertext layers, you can select. If you click on the last link of that home page—Explore the Internet—you would find four other sites including Learn More about the Internet, which promises guides, tools, and training materials. Choosing this hypertext link takes you even further into this specialized subject by displaying sixteen more sites (Figure 9.10) ranging from Books, Magazines, and Journals about the Internet to Doing Business on the Internet, Internet Statistics, and Internet Guides and Tutorials.

Let's say you want to learn more about the Internet through the Library of Congress's tutorials. You click on Internet Guides and Tutorials and find more than forty different sites to choose from. You can read a brief description of each of these forty sites; for example, for An Introduction to the Internet and WWW, you find that a "Powerpoint viewer for viewing the slides is available for downloading at this site"; and BCK2SKOL: The Electronic Library Classroom 101 presents "A beginner's course in the Net and its various tools targeted toward librarians and other information professionals." Layers of information unfold before you as you click on one topic after another.

Unlike flipping the pages of a book back and forth, hypertext allows researchers to connect and retrieve pages from computer networks all around the world. Information flows to you from many different paths—time, countries, companies, individuals—and in whatever sequence you want. For this reason you might enjoyably spend hours pointing and clicking—**surfing**—your way all over the Web. But as a research tool hypertext serves a valuable function because its

FIGURE 9.9 Topics page for the Library of Congress on the World Wide Web.

Choose a topic below, see what's new, or search our Web pages and Gopher menus.

General Information and Publications
Find out about the Library and its mission, special programs and services, information for visitors, publications (including Library Associates and *Civilization Magazine*), employment opportunities, and other general information.

Government, Congress, and Law
Search THOMAS (legislative information), access services of the Law Library of Congress (including the Global Legal Information Network), or locate government information.

Research and Collections Services
Browse historical collections for the National Digital Library (American Memory), visit Library Reading Rooms, access special services for persons with disabilities, and read about Library of Congress cataloging, acquisitions, and preservation operations, policies, and related standards.

Copyright
Learn about the U. S. Copyright Office and the registration process, access copyright information circulars and form letters, and read about many other copyright-related topics.

Library of Congress Online Services
Search Library of Congress databases and online catalog (including LOCIS) or connect to the Library's Gopher (LC MARVEL).

Events and Exhibits
Read about Library events, conferences, and seminars or view electronic versions of major exhibits.

Explore the Internet
Search the Internet, browse topical collections of Internet resources organized by Library of Congress subject specialists, and learn more about the Internet and the World Wide Web.

Library of Congress
Comments: lcweb@loc.gov (04/23/97)

FIGURE 9.10 Explore the Internet options from Library of Congress home page.

Internet Resources: Guides, Tools and Services

A LIBRARY OF CONGRESS INTERNET RESOURCE PAGE

- Internet Resources Meta Index
  (National Center for Supercomputing Applications)

- Announcements of New Internet Resources
- Books, Magazines, and Journals about the Internet
- Doing Business on the Internet
- History of the Internet
- Internet Access Providers
- Internet Conferences and Events
- Internet Guides and Tutorials
- Internet Networks and Networking
- Internet Policies
- Internet Resources by Subject
- Internet Resources by Tool or Type
- Internet Security
- Internet Software
- Internet Statistics
- Internet Trainers' Resources
- Wide World Web "How-to" Resources and Guides

Link to LC MARVEL Internet Information

cross-referencing allows you to link one part of a topic to another, thereby increasing the layers or clusters of information you can search.

The elaborate organizational strategy behind hypertext to link and connect sources goes far beyond the research possibilities contained in a single library, database, or index. Unlike an on-line catalog, which only points you to the information, hypertext links on the Web actually bring the information to you on your screen. With the Web you'll never have to worry that something is checked out. You can also scan the Web more quickly to see whether something is relevant for your research than you could with an on-line catalog. Finally, some argue that the Web is more user-friendly than on-line catalogs in that it gives easy and fast summaries.

Precautions in Using the Web

While the Web offers unprecedented research advantages, it also presents some problems. One researcher put the main problem of using the Web this way:

> Accessing information on the Web is like going to a bookstore. You have no idea what's very good and what's not. While a library may have a lot of older items and not nearly as many new ones as the Web, the items in a library have at least been edited by a publishing company and have been chosen over less desirable works to be printed. With the Web you are in many cases forced to be your own publisher and editor.

The Web's strengths—its accessibility and its variety—can therefore also be its weaknesses.

Here are some precautions to keep in mind when you do research on the Web.

- You may not always find the best information. Use the Web along with CD-ROM databases, books, and other reference materials.
- Don't expect to get every article from every professional (or popular) journal on the Web. Not all periodicals are on-line. And many periodicals that *are* on the Web start with the most recent issues; back issues are not yet available. Moreover, many periodicals on the Web are either subscription-based or have selected highlights from the actual magazine (for example, only the table of contents or abstracts).
- Because the Web is so vast, you may have to experiment with different search strategies and try a variety of addresses.
- Not everything is free. Accessing some documents on the Web may require passwords that are available only to subscribers.
- Make a hard copy of any material you intend to cite in your paper or report. A Web site can change, be "under construction," or vanish. You need hard evidence.

The Dynamic Organization of the Web

The Web is constantly changing and expanding. Each day new sites are added and previous ones are modified—supplemented, combined, updated, and replaced. In 1999 there will be an estimated 3.5 million Web sites. Because the Web is so vast, there could never be one comprehensive index or guide to its resources. Unlike the traditional library, therefore, which offers patrons limited, precisely defined information contained in bound volumes, CD-ROMs, or electronic databases, the virtual libraries of the Web are potentially limitless. They can be accessed in a variety of ways and through a variety of searches not possible with the reference works described in the earlier sections of this chapter.

Surfing Versus Searching

Surfing the Web describes wandering from one Web site to another without any particular topic or goal in mind. In our example from the Library of Congress in Figures 9.9 and 9.10 above, a user could easily click on general information, copyrights, or events, and retrieve a wide variety of information. When you surf, you "float" through cyberspace just to see what is out there. While enjoyable and entertaining, this kind of activity will not help you to write a restricted report and can be very expensive and time-consuming.

When you *search* the Web, on the other hand, your approach must be more focused, more narrowed, as you labor to find that needle in the haystack. Your search skills have to be careful and precise.

Guidelines for Searching the Web

When you search the Web, keep these four points in mind.

- As you encounter a variety of sources, databases, home pages, and virtual libraries, try not to be overwhelmed; you *can* and *will* find something pertinent to your topic.
- Keep your search strategies flexible and open. Information can be indexed and posted in many ways on the Web. Look under different headings and categories. For example, if you have to search for something about computer software, you may start your search with either a "Software" heading, or the name of a specific software vendor.
- Verify information you find in Web sites by consulting, through a key word search, other independent Web pages and non-Web sources.
- Don't get sidetracked. You will see items that intrigue you and arouse your curiosity. Mark them ("bookmarks") for later exploration and stay focused on your current goals. Searching the Web requires discipline.

TECH NOTE

Just as you slip a sheet of paper into a book or turn down the page of a magazine to mark your place, you can use a Web browser's bookmark feature to keep track of your favorite sites on the Internet and to return quickly to them later. In Netscape Navigator you choose the command "Add Bookmark," and in American Online's browser you click on a heart-shaped icon in the corner of the Web page. When you want to return to that site later, activate the bookmark, and your Web browser will go on-line automatically and take you directly to that page.

One of the most productive ways to accomplish research on the Web is to use a search engine.

Search Engines

A search engine is a guide to the resources on the Web. Basically it is a software program that identifies, describes, and assesses the information on the Net that you need. Another term for a search engine is WAIS (**W**eb **A**rea **I**nformation **S**erver), pronounced *ways*. Think of it as a computerized scout who goes through the diverse networks on the Web and brings back a list of relevant documents. Search engines can look through electronic citation databases such as those on page 333, newswires (Reuters, AP, UPI) periodicals, diverse indexes, abstracts and summaries of articles, articles, advertising information, home pages, and user-group lists. Remarkably, it can search all of these resources simultaneously. A search engine goes through an enormous database including information found across as many as 30 million Web sites. When you request specific information on the Web, a search engine "looks through" the database and returns the information that most closely matches your search request.

Examples of Search Engines

Just as there is no single index to the Web, there is no one best search engine. There are many of them, as you will see below. If you use the Netscape Web browser and click on "Net Search," you will be presented with one of these engines. Like the Web itself, search engines are constantly growing and improving. They vary in what they cover and what they cost. Some are much faster than others, and some incorporate others. To some extent all of them overlap. No one search engine is comprehensive, though some claim to search 30 billion pages. You may be limited by what is available to you through your library or employer.

Expect a tradeoff between how fast a search engine works and how good it is. The fastest search engines tend to look only at titles or key words on a few pages, whereas the slower ones will conduct a more detailed look through many more pages.

- Use a fast search engine first and see how much it gives you. Then, as your research requires, turn to a slower, more exhaustive engine.
- Use several search engines. Even though some of the information you receive may be duplicated, you still increase your chances of obtaining precisely the information that you need about your topic.

Below are some frequently used search engines; except for NLightN, all these are free, but keep in mind that because Internet technology is rapidly changing, some of these search engines may be replaced by newer, more efficient ones.

1. **Alta Vista** has a very large database (30 million Web pages) and may easily give hundreds of thousands of hits for broad key words. For those users who have more experience, it has an advanced search capability that lets you use "and," "or," "not," "near," and parentheses to qualify your search words and so hone in on exactly those information sources that are most applicable to your needs.

2. **Excite** is a general-purpose engine that is a good place to start when you are not precisely sure what you are looking for. It gives percentage values to each of the sites, or hits, that are most likely to match the kinds of information you request through your key word search. The search engine orders each of these sites according to how closely it matches your needs, placing the most likely ones first. Excite also has some ancillary databases such as the City Net, which gives information on cities around the world.
3. **Hotbot** has the ability to search for not only key words but also phrases, people, and Internet addresses. It claims to search the entire Web.
4. **Infoseek Guide,** a standard search engine, offers a special feature called Infoseek Personal, which allows you to describe information (research topics, news, sports) in which you have an ongoing interest. Infoseek Personal will then retrieve related items for you as they become available.
5. **Lycos** has a large database, though smaller than that of Alta Vista. As well as indexing the title, headings, and subheadings in a page, **Lycos** (Greek for "spider") also searches the first twenty lines of text. It displays a relevancy rating for each hit.
6. **NLightN,** the only fee-charging search engine listed here, uses Lycos, multiple newswire services, over 500 bibliographic databases, and a number of other sources, and it searches them all simultaneously.
7. **Open Text** differs from the other search engines in that it pays attention to all words, including "the," "to," and "a," that would be ignored by other engines. Thus a search for "To be or not to be" would be entirely reasonable with Open Text but a waste of time with other search engines.
8. **Yahoo!,** one of the first Web search engines, shines in having a very good directory structure. It is excellent for browsing. If you need to do a directory search **Yahoo!** is the place to start (see Figure 9.11).

Using Key Words with a Search Engine

A successful search starts and proceeds with an appropriate description of precisely what you are looking for. To use a search engine, you must enter key words that best describe your topic. That description is the fuel that drives and guides the search engine.

A search engine functions like an index, but you have to supply the appropriate terms or key words. Then the search engine explores the Web files to see how often and where these key words appear, in what order, and sometimes in what contexts. Finding these key words, the search engine then gives you a ranked list of the documents, with Web addresses, it finds in the files containing the words you specified. You can go back and forth from the list of sources to the files as you modify, restrict, and refine your search. Of course, if you know the URL, you don't need a search engine to direct you to that particular Web site.

Guidelines for Selecting Restricted Key Word Searches

Selecting valid, restricted key words is crucial for a successful search. These guidelines will help you.

1. **To locate information about a product or service, simply enter the name of that company, product, or service.** For example, you might enter *Sunbeam-Oster, Wells Fargo, Brown and Root,* or *Citicorp* to begin your search. But your search for Wells Fargo may also, for example, pull up articles and home pages for Wells Engineering or Fargo, North Dakota. Experts advise putting names in quotation marks to limit your search to these.
2. **Don't use just one key word.** Choose at least two or three significant ones to specify your search. For example, keying in just the word *virus,* you might receive vast amounts of information about human and animal illnesses as well as violations of computer systems. Specify precisely *computer virus* to narrow your search.
3. **To narrow your search further, identify a precise subject.** Don't confuse a broad subject with a more limited heading within it. For example, don't start off with a broad subject such as **computer security.** You might be presented with everything from computer theft at airports to encryptions that protect messages. As in number 2 above, start with **computer virus,** which is a much more focused subject under the larger topic of *computer security.*
4. **Be prepared to refine your key word choice(s)** as entries come up on the screen. For instance, you might replace *computer virus* with *computer disinfect program.*
5. **Link several key words to pinpoint your search.** *Computer virus causes* or *computer virus controls* will net you more precise, pertinent information.
6. **Select a synonym if your initial key word search does not produce results.** In example 4 above, you might substitute the synonym *software* for *program,* which may retrieve additional, relevant information not generated by the key word search using *program.*
7. **Consider using delimiters** such as the words *and* or *or* to further your search, but never use *and/or;* it must be one or the other. Of course, this usage will vary by database. For example, rather than just keying in *language disorders* you might key in *language disorders or speech disorders.* (Actually you may need to key in *language near disorders* or *speech near disorders.*) Delimiters increase your chances of finding more relevant and useful items on the net.
8. **Further refine your search by including *not* as part of your key word search.** Excluding certain words will make your search engine work even more effectively to identify only those sources you need to consult and to reject extraneous ones. Keying in *telecommuting benefits, not environmental* signals that you are not interested in material dealing with how telecommuting saves energy or reduces pollution. Instead you are focusing your search on other kinds of benefits—to the employee and employer.

What if My Search Engine Produces Long Lists and Duplications?

A search engine looks for key words in titles, abstracts, first pages, and in any type of cross-reference. Armed with your key words or phrases (note that engines treat a phrase as a number of distinct words), a search engine sorts through its database and lists what it regards as the most relevant matches for your request.

If you have not refined your search and selected restricted key words, you could initially be overwhelmed with a list of 3,000 or even 30,000 items. Such a huge list signals that you probbaly have not done enough homework about your topic. For example, keying in just *women's golf* or *laser surgery* might return thousands of citations. However, don't despair. Look at a few of the documents this initial search brought you to see whether you can identify there are key subtopics, phrases, patterns, or research areas that you might use to refine your key words search.

If you still are confronted with a long list, focus on the beginning of the list. A search engine assists you by presenting items in order according to how well they match your topic. The more times the engine finds the key words or phrases you searched for, the closer the document will be to the top of the list of references. Typically, the first ten to twenty-five references (called "hits") displayed on your screen are the most promising places to begin your investigation—the best matches. Many lists also display a percentage next to the individual references to indicate probability, or level of confidence you can have, that the reference is appropriate for your search.

Start with the first screenful to see how well they meet your needs. Do not be surprised to see duplicates or to find some of the hits irrelevant. Ignore these. You do not have to spend time studying everything a search engine returns. There is no substitute for your own critical discernment. After looking at the first ten hits, you may decide to continue to the following screen showing the next ten. Or you may decide to refine your search—narrowing it or even broadening it.

Two Major Types of Web Searches

Topic or Directory Searches

A topic or directory search is the easiest way to get started on the Web if you do not already have a carefully focused topic, or when you have only a general topic in mind. Searching through hypertext actually assists your search for a restricted topic. Pointing and clicking through the Web, going back and forth from one Web site to another, you will eventually discover how a subject is restricted and where the most relevant information can be found.

Case Study: Lyme Disease Information on the Internet

Let's say you have to write a report for your class in community health but have no specific topic in mind. Choose one of the search engines provided by your Web browser program. If you use Yahoo!, you would first see a screen, shown in Figure 9.11, that brings up a number of general categories and headings, such as **Arts, Entertainment, Health,** and **Sports.** A directory will file hundreds of thousands of documents under these large headings. The most pertinent general category for your purposes is **Health,** and so you would click on it. You would then see the Health screen subdivided into numerous categories including **Diseases and Conditions.**

Going to that screen, you would find a specific reference to **Lyme Disease,** carried by ticks in wooded areas. You may have heard of this disease or recently

FIGURE 9.11 Net Search screen for the Yahoo! search engine.

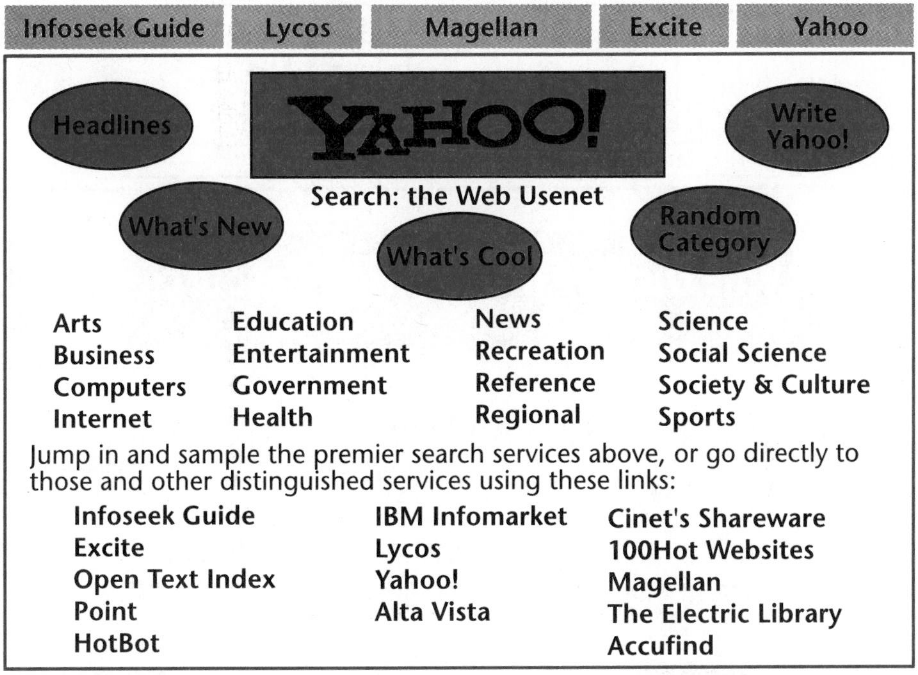

Infoseek Guide | Lycos | Magellan | Excite | Yahoo

Headlines

YAHOO!

Write Yahoo!

Search: the Web Usenet

What's New

What's Cool

Random Category

Arts	Education	News	Science
Business	Entertainment	Recreation	Social Science
Computers	Government	Reference	Society & Culture
Internet	Health	Regional	Sports

Jump in and sample the premier search services above, or go directly to those and other distinguished services using these links:

Infoseek Guide	IBM Infomarket	Cinet's Shareware
Excite	Lycos	100Hot Websites
Open Text Index	Yahoo!	Magellan
Point	Alta Vista	The Electric Library
HotBot		Accufind

read a newspaper article about it. Seeing it listed, you decide to learn more about it. By clicking on Lyme Disease, you are on your way to a number of helpful sources of information via hypertext language. Figure 9.12 represents a much abbreviated version of the various hypertext pages on Lyme Disease you would see in the search described below.

The Diseases and Conditions: Lyme Disease page offers four choices, and not knowing much about the topic, you decide to click on the first—which takes you to the home page of the **American Lyme Disease Foundation.** The foundation might be a place to contact via e-mail for some basic information and further leads. Equally important for your search, this home page suggests useful categories that might lead to a focused topic, including **Management, Precautions,** and **Vaccines.** You decide to pursue the idea of **Management** of the disease and want to see what type of (and how much) information is available and pertinent for your class report.

You recall the heading **Lyme Disease Resource** on the **Health** and **Conditions** page, and so now you go back to that page by clicking on the **back** button at the top of the screen. You now select **Lyme Disease Information Resource** displayed in Figure 9.12; there you get a set of further choices including **General Information, Clinical Information,** and **Search Databases.**

FIGURE 9.12 Lyme Disease Information Resource page on the Web.

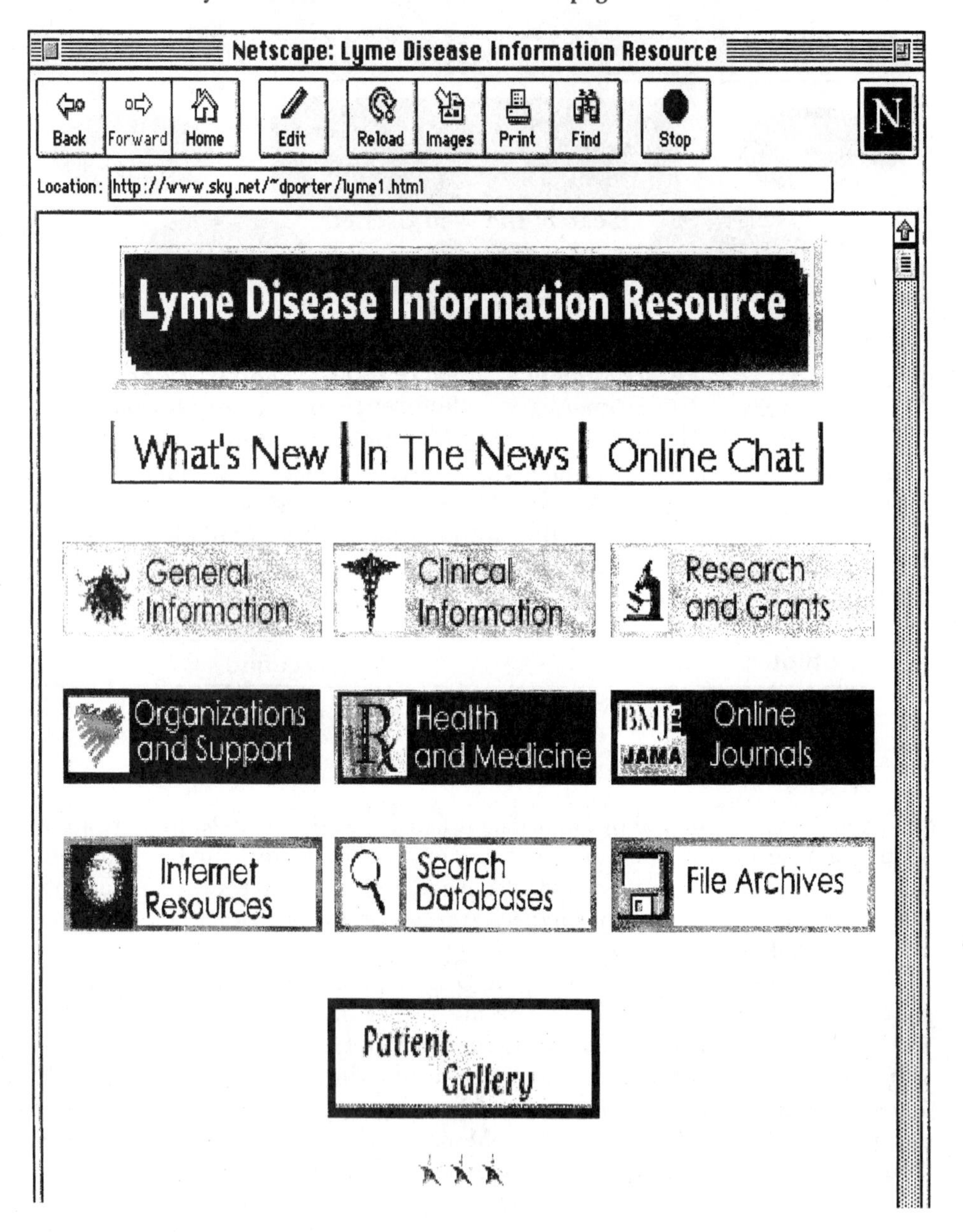

Continued

FIGURE 9.12 (Continued)

Upcoming Events

Emerging Tick-Borne Diseases in the Western United States
Lyme Disease Resource Center Conference
September 13, 1997

Hyperbaric Oxygen Study
Project for fhe Lyme community - LASCM

Send E-mail To: Donna Herrrell <dporter@solar.sky.net>

Comments, Questions or Suggestions

Disclaimer: The Lyme Disease Information Resource shall serve as a clearing house for information about Lyme Disease and/or other illness. Some of the information contained herein, is intended to help patients make informed decisions about their health. However, this resource and its contributors does not offer medical advice and assumes no responsibility for any medical treatment or other activity undertaken by readers of this source. For medical advice please consult with your physician.

060379 times according to CountMan

Copyright © 1995, 1996, 1997 All Rights Reserved.

Reprinted with permission of Lyme Disease Foundation, Inc. and Donna Herrell.

earch Databases you now have several options from which to lect the subset **Medline,** the on-line database of *Index Medicus* (see g to do with genetics. This is the most appropriate subset of the large base, since it deals with a specific subdivision here. You decide to restrict you earch to just treatment of the disease and so for your **Query Terms** you key in *Lyme & Disease & Treatment.*

As a result of your search, you find thirty-seven citations from highly specialized medical journals. Taking a look at a few abstracts, available from *Medline Reports,* you realize at once that this information is for geneticists, far too technical for your purposes. At the same time, you conclude that there must be some other options on the Web you need to pursue.

Since you do not want to abandon the idea of writing on the treatment of Lyme disease completely, you decide to restrict your topic *not* by focusing on treatment of the disease in general but by its impact on a specific population—children. Going back to the Lyme Disease Information Resource page (Figure 9.12) you click on **Clinical Information** and key in *Lyme Disease—Children.* Then you see on your screen an initial list of ten items, sorted by confidence and indicated by percentages, that may be relevant to your topic. Going through these ten hits, you suspect that some are too general but may still be useful for research, especially the first two, including the *Lyme Disease Newsletter.* You also discover that four items are highly relevant on the disease and children—*Managing the Disease, Neurological Manifestations, Lyme Disease and Preschoolers,* and *Adults and Pregnancy*—and you click to get copies of these articles.

Key Word Searches

A key word search—as opposed to a directory search—may be the more expedient search strategy if you have a restricted topic in mind. Your success here depends on finding the key words that will unlock the resources on the Web that you need to write your paper or report. For example, if you need information on Lyme Disease you would search for "lyme disease." Using quotes narrows the topic and limits the search by weeding. If you did not use quotes, search engines would look for every mention of lyme and disease, giving you lots of unrelated material.

Note-Taking

Once you have consulted the appropriate library and on-line sources, you need some systematic way to record relevant information from them to use in preparing your paper or report. Before you can begin drafting your work, you must be able to organize and classify data from your research efficiently. (Even after you start to draft your report, you will probably need to continue your research and take additional notes.)

Note-taking is the crucial link between finding sources and writing a report. Never trust your memory to keep all your research facts straight. Taking notes is time well spent. Do not be too quick in getting it done or too eager to begin writing your paper. Careless note-taking could lead you to omit key words in a quotation or even worse, misrepresent or contradict what the author has said. Carelessness could result in crediting one author with another's work.

Basically, you will be preparing two kinds of notes. One type, on the sources you read, will become your working bibliography. The other type will be reserved for the specific information you take from these sources in the form of direct quotations, paraphrases, or summaries.

Traditionally, researchers have used 3" × 5" and 4" × 6" index cards for their note-taking. Some researchers prefer spiral-bound notebooks. A laptop computer or computerized notebook can greatly help you store and retrieve your notes.

How to Prepare Bibliography Notes

Look at the sample card shown in Figure 9.13. To record accurate and meaningful information on cards, follow these practical guidelines:

1. Use only 3" × 5" index cards. Do not be tempted to use slips of paper or looseleaf notebook paper; cards are less likely to get lost, and they will help you to organize, alphabetize, and label material.
2. Write on only one side of the card. It is easier to copy and check information when you list it all on one side.
3. Put only one title (article or book or film) on each card. These cards will later have to be arranged in alphabetical order for your Works Cited page (see

FIGURE 9.13 Card containing bibliographic information on a source.

Jenkes — Food and Nutrition

Jenkes, Thomas H. "Predicament of
Food and Nutrition."
Food Technology 44
(Oct. 1979): 45–46

pp. 411–413); if you place two titles one one card, you could run the risk of omitting one of the titles.

Whether you use cards or a laptop, follow these guidelines:

1. Include full bibliographic information for each source. For books, list author, title, chapter titles, edition, date, city of publication, publisher, and page numbers; for articles, supply author, title, journal, volume number, date, and page numbers. Keep a record of Internet addresses you used as well.
2. Decode and spell out any periodical index or journal abbreviations. Record periodical entries with their full and accurate bibliographic information. (See Chapter 10 on documentation.) The abbreviated entries in indexes such as *Applied Science & Technology Index* or in on-line databases are not in acceptable format for parenthetical documentation or Works Cited page entries in your paper or report.
3. Record the call numbers of any book (in the upper left-hand corner, if you use cards) so you will know where to find the source again if you need it. If you obtained a copy of an article or other source from an on-line search, note from which database it came.

How to Prepare Information Notes

While your bibliography notes record important facts about authors, titles, and publication data, your information notes contain the specific details from these sources that you need to write your paper. Your notes will contain direct quotations from, or paraphrases of, your sources. Quotations and paraphrases are discussed below. Here are some guidelines to follow when preparing note cards like those shown in Figures 9.14 and 9.15.

FIGURE 9.14 Note card containing a direct quotation.

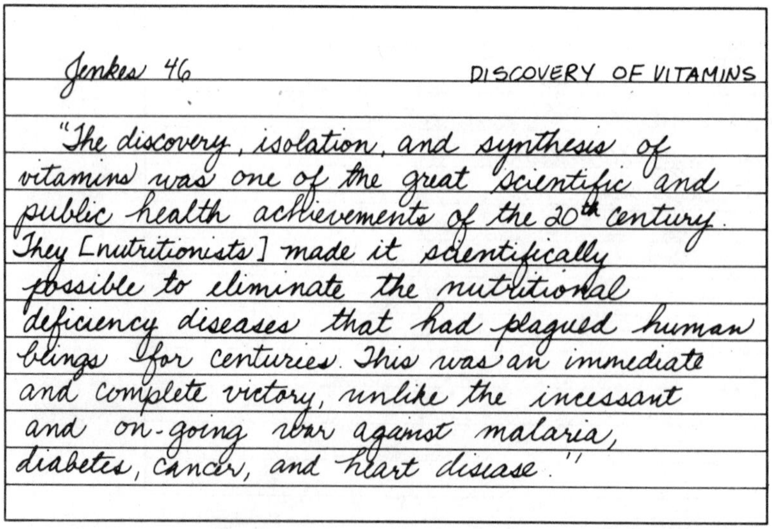

Jenkes 46 DISCOVERY OF VITAMINS

"The discovery, isolation, and synthesis of vitamins was one of the great scientific and public health achievements of the 20th century. They [nutritionists] made it scientifically possible to eliminate the nutritional deficiency diseases that had plagued human beings for centuries. This was an immediate and complete victory, unlike the incessant and on-going war against malaria, diabetes, cancer, and heart disease."

FIGURE 9.15 Note card containing a paraphrase.

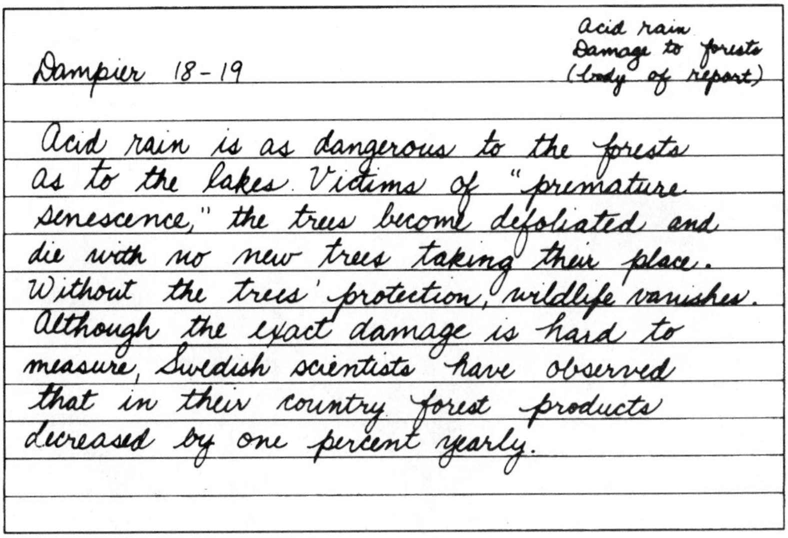

Acid rain
Damage to forests
(body of report)

Dampier 18–19

Acid rain is as dangerous to the forests as to the lakes. Victims of "premature senescence," the trees become defoliated and die with no new trees taking their place. Without the trees' protection, wildlife vanishes. Although the exact damage is hard to measure, Swedish scientists have observed that in their country forest products decreased by one percent yearly.

1. Use 4" × 6" cards to keep them separate from the 3" × 5" bibliography cards.
2. Write on only one side of the card. It will be easier for you to arrange the cards in the order that makes sense for your paper.
3. Don't include notes from two different sources on one card.

Again, whether you use cards or a laptop, follow these guidelines:

1. Copy names, facts, dates, and statistics accurately from the source. Be sure, too, that you record the author's words correctly; you may want to quote these words verbatim in your paper. Always compare what you have written with the original or the copy you may have obtained from an on-line search or from a search on the Internet.
2. Make sure that you distinguish quotations from your own paraphrase by placing quotation marks around any words and sentences you record directly from a source.
3. Write a code word or phrase (in the upper right-hand corner, if you use cards) to identify the topic treated by the source or the information written on the card from that source. Using code words such as "characteristics," "function of," "history of," or "location of" will help you to organize when you write your paper. Figure 9.14 shows how one writer used these code words.
4. You might indicate in the note why the material is significant to your argument or where you might include such information in your report or paper. In Figure 9.15 the writer has identified a possible use for the Dampier paraphrase. You might also write your response to a quotation to help you later organize your paper. But be careful not to include any of your comments within a quotation or paraphrase.

5. Write a short title of the book or journal article (in the upper left-hand corner, if you use cards) or include the author's last name so that the information note is coded to your bibliography note. If your notes extend to a second card, make sure you include the source, with appropriate page number(s) on this second card.

TECH NOTE

Software of all kinds is available to replace index cards and help you quickly locate, update, cite, format, and integrate footnotes and bibliographic entries into your writing projects. High-end word-processing programs like Microsoft Word, WordPerfect, and Nisus Writer keep track of all citations and let you edit (and renumber) notes in place; they also offer capabilities to ease such tasks as outlining, indexing, cross-referencing, and preparing tables of contents. In addition, of course, their search functions will quickly locate all references to particular text within a document and will even revise it as you specify.

If you do a lot of professional writing and need to manage a large number of bibliographies, consider purchasing a dedicated program like TakeNote, EndNote Plus, and Bookends Pro. Some of these work in tandem with the word processors named above and can import and export citations, making it a snap to amass and select from your own up-to-date bibliographies—without a trip to the library. All these programs can format citations in the styles of particular associations (Council of Biology Editors, Modern Language Association, Chicago Manual, American Psychological Association) or up to hundreds of professional publications. Some bibliography managers also link up with the Internet to locate Web site addresses (URLs) and enter them into bibliographies for later research and citation.

To Quote or Not to Quote

Before recording information from sources, ask yourself these three questions:

1. How much should I take down?
2. How often should I copy the author's words verbatim to use direct quotations?
3. When should I paraphrase or summarize?

A safe rule to follow is this: Do not be a human photocopying machine. If you write down too many of the author's own words, you will simply be transferring the author's words from the book or article to your paper. That will show that you have read the work but not whether you have evaluated its findings. Do not use direct quotations simply as a filler.

Direct quotations should be used sparingly and saved for when they count most. When an author has summarized a great deal of significant information concisely into a few well-chosen sentences, you may want to quote this summary verbatim. Or if a writer has clarified a difficult concept exceedingly well, you may want to include this clarification exactly as it is listed. And certainly the author's chief statement or thesis may deserve to be quoted directly. Figure 9.14 contains such an important statement. Just be careful that you do not quote verbatim all of the evidence leading to that conclusion. The conclusion may be pointedly expressed in two or three sentences; the evidence could cover many pages.

If you are worried about exactly how much to quote verbatim, keep in mind that no more than 10 to 15 percent of your paper should be made up of direct quotations. Remember that when you quote someone directly, you are telling your readers that these words are the most important part of the author's work as far as you are concerned. Be a selective filter, not a large funnel.

Sometimes a sentence or passage is particularly useful, but you may not want to quote it fully. You may want to delete some words that are not really necessary for your purpose. These omissions are indicated by using an *ellipsis* (three spaced dots within the sentence to indicate where the words are omitted). Here are some examples.

Full Quotation: "Diet and nutrition, which researchers have studied extensively, significantly affect oral health."

Quotation with Ellipsis: "Diet and nutrition . . . significantly affect oral health."

When the omission occurs at the end of the sentence, you must include the end-of-sentence punctuation after the ellipsis. In the following example, note how the shortened sentence ends with four spaced dots: the three dots for the ellipsis and the closing period.

Full Quotation: "Decisions on how to operate the company should be based on the most accurate and relevant information available from both within the company and from the specific community that the establishment serves."

Quotation with Ellipsis: "Decisions on how to operate the company should be based on the most accurate and relevant information available. . . ."

At times you may have to insert your own information within a quotation. This addition, known as an *interpolation,* is made by enclosing your clarifying identification or remark within brackets inside the quotation; for example, "It [the new transportation network] has been thoroughly tested and approved." In other words, anything within brackets is not part of the original quotation.

Most of your note-taking will be devoted to paraphrasing rather than writing down direct quotations. A *paraphrase* is a restatement in your own words of the author's ideas. Even though you are using your own words to translate or restate, you still must document the paraphrase because you are using the author's facts and interpretations. You do not use quotation marks, though. When you include a paraphrase in your paper, you should be careful to do four things:

1. Be faithful to the author's meaning. Do not alter facts or introduce new ideas.
2. Follow the order in which the author presents the information.
3. Include in your paraphrase only what is relevant for your paper. Delete any details not essential for your work.
4. Use paraphrases in your report selectively. You do not want your work to be merely a restatement of someone else's.

Paraphrased material can be introduced in your paper with an appropriate identifying phrase, such as "According to Dampier's study," "To paraphrase Dampier," or "As Dampier observes." The note card shown in Figure 9.15 paraphrases the following quotation.

> While the effects of acid rain are felt first in lakes, which act as natural collection points, some scientists fear there may be extensive damage to forests as well. In the process described by one researcher as "premature senescence," trees exposed to acid sprays lose their leaves, wilt, and finally die. New trees may not grow to replace them. Deprived of natural cover, wildlife may flee or die. The extent of the damage to forest lands is extremely difficult to determine, but scientists find the trend worrisome. In Sweden, for example, one estimate calculates that the yield in forest products decreased by about one percent each year. . . .[1]

This chapter has introduced you to some very basic yet essential strategies and tools, including the Internet. Clearly, you need to rely on a host of resources—print, on-line databases, various search engines and types of Internet searches—to conduct effective research. More than ever researchers have the means of finding the most correct, most thorough, and most relevant answers to the questions they are asked to investigate and the problems they need to solve on the job. Exploring (and exploiting) all potential resources of the "Information Age" as described in this chapter will prepare you to write the types of documents—instructions, reports, proposals—discussed in later chapters.

✓ Revision Checklist

Process of Research

- ❑ Identified significant, timely, and limited problem.
- ❑ Selected and evaluated most relevant sources.

Library Search

- ❑ Found where appropriate materials are located.
- ❑ Learned how to use research services.

[1]Bill Dampier, "Now Even the Rain Is Dangerous," *International Wildlife* 10 (Mar.–Apr. 1980): 18–19.

Research Tools

- ❑ Searched library's on-line catalog for major term(s) to start working bibliography.
- ❑ Checked specialized and general indexes.
- ❑ Did preliminary search among relevant indexes to check subtopics and scope of material available.
- ❑ Restricted key words for searches in print and on-line.
- ❑ Determined which CD-ROM and on-line searches library has available and selected most useful ones.
- ❑ Ran an on-line search to compile a list of relevant sources.
- ❑ Read abstracts of relevant items.
- ❑ Obtained copies of full text of relevant articles, conference papers, and the like.
- ❑ Checked relevant reference materials, such as almanacs, abstracts, encyclopedias.
- ❑ Used NewsBank to find appropriate newspaper coverage of my topic.
- ❑ Read and scrutinized articles, books, government documents.

The Internet

- ❑ Discovered which Internet searches are most helpful.
- ❑ Did a directory search on the Web.
- ❑ Searched for pertinent information using at least two search engines.
- ❑ Located relevant home page(s) relating to my topic.
- ❑ Continued to refine search when number of hits was too high.
- ❑ Joined newsgroup relevant to my topic.
- ❑ Combined library search with Internet research when time limits or availability of materials warranted.

Taking Notes

- ❑ Investigated software options.
- ❑ Did not quote excessively or out of context, quoted accurately.
- ❑ Paraphrased fairly, accurately representing original material.
- ❑ Distinguished my comments and responses clearly from sources.
- ❑ Used correct punctuation with direct quotations, especially ellipses and brackets.

Exercises

1. Find out if your library provides a map or description of its holdings. If it does, bring a copy to class. If it does not, draw one yourself, indicating the

location of the circulation desk, the on-line catalog, the reference room, government documents, audiovisuals, and other areas you patronize.

2. Find any book in the library and write a brief description of the steps you took to locate that book—from searching in the on-line catalog to checking the book out at the circulation desk. Refer to the map you used in Exercise 1.

3. Using a subject search on the on-line catalog, find a topic that is subdivided as shown in Figure 9.1. Divide these subdivisions even further until you have a restricted topic and a problem about this topic you can investigate for a report. Bring your topic to class.

4. Using a subject, author, and title search, find four or five books on the topic you selected for Exercise 3. Prepare a bibliography card for each book.

5. Prepare a list (providing full bibliographic information) of fifteen articles for the restricted topic you selected for Exercise 3. Use one of the computerized databases discussed in this chapter (pp. 336–338).

6. Write a short memo (two or three paragraphs) to your instructor describing the types of computer searches your library offers and how such a search assisted you in researching the topic you selected in Exercise 3.

7. Identify three computerized databases that specifically would help you to run an on-line search in your research at school or on the job. List five periodicals that you would be able to find indexed on each of these databases. Do not include the same periodical more than once, even though it may appear in more than one database. Be sure to consult a librarian if you need help; he or she may be able to direct you to lists of periodicals indexed on each database.

8. Using the research tools discussed in this chapter, including the Internet, locate the following items related to your major. Select titles that are most closely related to your major, and explain how they would be useful to you. Prepare a separate bibliography card for each title.
   a. an index to periodicals
   b. titles of three important journals that are available in print and on-line
   c. an abstract of an article appearing in one of these journals
   d. a term in a specialized dictionary
   e. a description or illustration in a specialized encyclopedia
   f. a film or tape recording
   g. two government documents
   h. a story in *The New York Times* or one of the newspapers covered by NewsBank that, in the last year, discussed a topic of interest to students in your major

9. Assume that you have to write a brochure about one of the following topics introducing it to an audience of consumers. Using the resources of the Internet and those of one or two of the databases discussed in this chapter, prepare a working bibliography of relevant materials that contains at least ten sources. After gathering and reading these sources, prepare the brochure and submit it

with your bibliography to your instructor. This assignment may be done as a collaborative writing exercise.

- **a.** virtual reality
- **b.** lipoprotein A
- **c.** fiber optics
- **d.** CAD/CAM applications
- **e.** interactive television
- **f.** avoiding sexist language in the workplace
- **g.** artificial intelligence
- **h.** voice-activated computer programs
- **i.** robotics in medicine
- **j.** the greenhouse effect
- **k.** computer dating
- **l.** banking at home
- **m.** cellular phones
- **n.** any topic your instructor approves

10. Using appropriate references discussed in this chapter, answer any five of the following questions. After your answer, list the specific works you used. Supply complete bibliographic information. For books, indicate author or editor, title, edition, place of publication and publisher, date, and volume and page number. For journals and magazines, include volume and page number; for newspapers, precise date and page number. For Internet sites, provide complete http://www.addresses.

- **a.** What is biomass?
- **b.** How many calories are there in an orange?
- **c.** List three interviews that Bill Clinton granted between 1996 and 1998.
- **d.** What is the boiling point of coal tar?
- **e.** What was the headline in *The New York Times* the day you were born?
- **f.** List three publications issued by the U.S. Department of the Interior from 1995 to the present on outdoor recreation.
- **g.** What was the population of Spokane, Washington, in 1990?
- **h.** List three articles published between 1994 and 1996 on the advantages of teleconferencing.
- **i.** Who discovered the neutrino?
- **j.** What is the first recorded (printed) use of the word *ozone*?
- **k.** Who edited the *Encyclopedia of Psychology,* 2nd ed., published in 1994?
- **l.** Give the title, date, and page number and author (if listed) of a story in your local newspaper that focused on child abuse in the last year.
- **m.** List the titles of three articles on the abuse of credit cards that have appeared in professional journals within the past two or three years.
- **n.** What is a high key photograph?
- **o.** Name five plants that have the word *fly* as part of their common name.
- **p.** What are the names and addresses of all the four-year colleges in the state of South Dakota?
- **q.** Who is the current head of state of Nigeria?

r. What is the current membership of the American Dental Association?
s. What are the names of all the justices who currently serve on the U.S. Supreme Court?

11. Write a paraphrase of two of the following paragraphs.
    a. Deep-fat frying is a mainstay of any successful fast-food operation and is one of the most commonly used procedures for the preparation and production of foods in the world. During the deep-frying process, oxidation and hydrolysis take place in the shortening and eventually change its functional, sensory, and nutritional quality. Current fat tests available to food operation managers for determining when used shortening should be discarded typically require identification of a change in some physical attribute of the shortening, such as color, smoke, foam development, etc. However, by the time these changes become evident, a considerable amount of degradation has usually already taken place.[2]
    b. Ponds excavated in areas of flat terrain usually require prepared spillways. If surface runoff must enter an excavated pond through a channel or ditch, rather than through a broad shallow drainageway, the overfall from the ditch bottom to the bottom of the pond can create a serious erosion problem unless the ditch is protected. Scouring can take place in the side slope of the pond and for a considerable distance upstream in the ditch. The resulting sediment tends to reduce the depth and capacity of the pond. Protect by placing one or more lengths of rigid pipe in the ditch and extend them over the side slope of the excavation. The extended portion of the pipe or pipes may be either cantilevered or supported with timbers. The diameter of the pipe or pipes depends on the peak rate of runoff that can be expected from a 10-year frequency storm. If you need more than one pipe inlet, the combined capacity should equal or exceed the estimated peak rate of runoff.[3]
    c. *Using Your ATM Card to Shop*

       Matching the Logos: Just as the various logos that appear on ATM cards tell you where they can be used to get cash or make banking transactions at ATMs, they also indicate where your card can be used to make purchases. Simply match the logos on your card with those you see displayed at the entrance to the store or at the cash register. Or just ask whether the store accepts your ATM card.

       Depending on which logos you find on your card and whether the store has installed PIN pads, your purchases can be handled in one of two ways: either you will punch in your PIN, just as you would at an ATM, or you will sign for the purchase, as you would with a credit card.

[2]Vincent J. Graziano, "Portable Instrument Rapidly Measures Quality of Frying Fat in Food Service Operations," *Food Technology* 33 (Sept. 1979): 50. Copyright © by Institute of Food Technologists. Reprinted by permission.

[3]U.S. Department of Agriculture, Soil Conservation Service. *Ponds for Water Supply and Recreation* (Washington, D.C.: U.S. Department of Agriculture Handbook No. 387, 1971): 48.

Making a Purchase: Let's say you've planned to buy a desk lamp. You need all your cash for other things and don't have your checkbook with you. At the entrance to the store, you notice an ATM network logo that matches the logo on your card. You decide to use your ATM card to pay.

When you present the lamp to the cashier, you will be asked how you would like to pay for the purchase. You offer your ATM card. The cashier will confirm that your card is accepted by the store, and if it is, the following will occur: 1) You will be asked to slide your card through a slot that reads the information contained in the magnetic stripe on the back of your card; 2) the cashier will then enter the amount of the purchase; 3) you will punch in your PIN, or secret code; and 4) the cashier will press a key that initiates an automatic phone call to your bank or credit union. This confirms that the money is available in your account. Once confirmed, your bank or credit union automatically deducts the purchase amount from your account, just like a check. You will receive a receipt of the transaction, if you want one, when the sale is completed. Make sure you record and subtract this amount from your account immediately.

When a Major Credit Card Logo Is on Your ATM Card: If you have an ATM card that also has on it one of two of the major credit card logos mentioned previously, your purchase will be handled as if you were using a credit card, except for three important differences:

- First, the purchase amount will be deducted automatically from your account—like when you write a check—rather than being billed to you at the end of the month.
- Second, typically, you'll pay no interest charges, since you're using your own money on deposit, not borrowing it. (However, there may be other fees associated with using this card, an issue addressed later in this brochure.)
- Third, you will usually sign for the purchase instead of punching in your PIN. However, since this is your ATM card, if a store has installed PIN pads to accept your PIN, and it accepts one of the other logos on your card, the store clerk may ask you to use your PIN instead of signing.

CHAPTER 10

Documenting Sources

Documentation is at the heart of all the research you will do at school or on the job. To document means to furnish readers with information about the materials (books, Web sites and other Internet sources, articles, brochures, films, interviews, questionnaires) you have used for the factual support of your statements. Without proper documentation, you will not be able to persuade a customer to buy your company's product or service and you will not convince your boss that you are doing your best work.

This chapter will give you practical and precise directions on what to document and how to do it efficiently and consistently. Of the various systems (or formats) of documentation, parenthetical documentation is preferred over documentation through footnotes or endnotes. For this reason, this chapter will emphasize the parenthetical documentation methods advocated by the Modern Language Association (the MLA) and the American Psychological Association (the APA). The sample research paper about telecommuting at the end of this chapter uses the MLA style of parenthetical documentation; the long report on e-cash in Chapter 17 (pp. 631–652) follows APA style.

The Whys and Hows of Documentation

Before looking at the specific techniques you need to use when you document, it's necessary to understand why documentation is so important and the major role it plays in your writing strategies.

Why Is Documentation Important?

Documentation is important for at least three reasons.

1. Documentation informs your readers that you have done your homework. It proves that you have consulted experts on the subject and have relied on the most current and authoritative sources to build your case.
2. Documentation gives proper credit to these sources. Citing works by name is not a simple act of courtesy; it is an ethical requirement and, because so much

of this material is protected by copyright, a point of law. By documenting your sources, you will avoid being accused of *plagiarism*--that is, stealing someone else's ideas and listing them as your own. Plagiarism involves not only using someone else's words without quoting them directly but also misquoting a source by omitting or altering words or statistics to suit your own purpose and **patchworking**, using bits and pieces of information and passing them off as your own. Such misrepresentation is unethical and illegal. If you are found guilty of plagiarism, you could be expelled from school or fired from your job.

3. Documentation informs readers about a specific book or article you used. They may want to read it themselves for additional information or to verify the facts you have listed from that source. If your documentation is incorrect or incomplete, your readers will not be able to locate your sources.

What Must Be Documented?

This question often puzzles writers. If you document the following materials, you will be sure to avoid plagiarism and to assist the reader of your research paper or report.

- any direct quotation(s), even a single phrase or key word. Quotations from the Bible, from Shakespeare, or from any literary text should be identified according to the specific work (*Exodus, Merchant of Venice*) and the exact place in that work (for example, act 3, scene 4, line 23, listed as 3.4.23)
- any paraphrase or summary of another individual's written work or from an oral report or presentation
- any opinions—expressed verbally or in writing—that are not your own or any views that you could not have reached without the help of another source
- any statistical data that you have not compiled yourself
- any visuals that you have not prepared yourself—photographs, tables, charts, graphs, drawings (If you construct a visual based on someone else's data, you must acknowledge that source.)
- any software programs that you did not develop yourself

Of course, do not document obvious facts, such as normal body temperature; well-known dates (the first moon landing in 1969); historical information (Ronald Reagan was the fortieth president of the United States); formulas (H_2O; the quadratic formula); or proverbs from folklore ("The hand is quicker than the eye").

Documentation in the Writing Process

Documentation is a vital activity in the process of writing a research paper or report. Documentation begins as soon as you start researching your topic and continues during the course of organizing, drafting, revising, and even editing your work. You need to keep a careful record of the sources you use and the exact material you take from them.

TECH NOTE

It is relatively easy to transfer bibliographic information—Works Cited pages and other bibliographic citations—into your word-processing program from the Net or a CD-ROM. If you are using a Web browser, go to the file menu and click on "save as." You will see a box in which you can give this information a file name and save this new file in your word-processing program. If you want to incorporate bibliographic information into another document—you may be using many of the same sources—open the document into which you want to incorporate the new material. Choose "insert" from the options at the top of your screen and give the new material a file name. Then a copy of the new bibliographic information will be part of the document you are working on. Keep in mind that you will have to press "save" after each step in order not to lose the material.

Throughout various drafts you need to be sure you document precisely what you are using and then add every source you refer to in the text of your paper on the Works Cited page (discussed below). During revision (and editing) be careful to check and double-check the accuracy of your documentation. This means making sure names, Web sites, page numbers, dates, and quotations are correct and that bibliographic information mentioned in the text of your paper matches that on your Works Cited page precisely.

Parenthetical and Footnote Documentation

Numerous formats exist for documenting sources. Two of these formats are *parenthetical documentation* and *footnote documentation.* The following section will introduce you to parenthetical documentation by contrasting it with the footnote method.

Parenthetical Documentation

One widely used system of parenthetical documentation is found in the *MLA Handbook for Writers of Research Papers,* 5th ed., edited by Joseph Gibaldi (New York: Modern Language Association, 1997). The MLA system is used primarily by individuals in the humanities and other related disciplines; APA is used in psychology, nursing and allied health disciplines, the social sciences, and some technological fields. Basically, both MLA and APA

- do *not* recommend footnotes or endnotes to document sources
- do *not* contain a bibliography of works the writer may have consulted but has not actually cited directly in the paper

Instead, the MLA and the APA use parenthetical, or in-text, documentation. That is, the writer tells readers directly in the text of the paper, at the moment the acknowledgment is necessary, what reference is being cited. MLA style, for example, includes the author's last name in parentheses together with the appropriate page number(s) from which the information is borrowed.

> Creating an effective Web site was among the top three priorities businesses had over the last two years (Morgan 205).

The citation (Morgan 205) lets the reader know that the writer has borrowed information from a work by Morgan, specifically from page 205 of that work. Such a source (author's last name and page number) obviously does not provide sufficient documentation. Instead, the parenthetical reference refers readers to an alphabetical list of works that appears at the end of the paper. This list—called "Works Cited" in MLA or "References" in APA—contains full bibliographic data—titles, dates, Web sites, page numbers, and so on—on all the sources cited in the paper.

Footnote/Endnote Documentation

As you may recall from other writing courses, when you document using footnotes or endnotes, you insert a slightly raised numeral, called a superscript, immediately after the information you wish to document, like this: [1]. Then you provide source information either in a footnote at the bottom (or "foot") of the page, preceded by the same raised numeral [1], or on an endnotes page at the end of the entire paper. The order in which the endnotes are listed must correspond exactly to the order in which the information is cited within the paper. When readers see a [7], for instance, they expect to find information about the particular source for that information at [7] on the endnotes page. Footnotes and endnotes provide the same details: author's name, title of the work, place of publication, publisher, date of publication, and page numbers.

Figure 10.1 An example of an endnotes page showing the documentations of sources.

Endnotes

[1] Julie Teunissen, "Opportunities for Technical Writers," Computer Outlook 17 (1998): 43.

[2] George Tullos, "Technical Writers and the Importance of Online Documentation," Journal of Computer Operations 8 (1997): 15.

[3] Mary Bronstein, The New Generation of Technical Writers (San Francisco: FTP Systems, 1999): 107.

Figure 10.2 Documentation of sources on a "Works Cited" page.

Works Cited

Bronstein, Mary. The New Generation of Technical Writers San Francisco: FTP Systems, 1999.

Teunissen, Julie. "Opportunities for Technical Writers." Computer Outlook 17 (1998): 42-45.

Tullos, George. "Technical Writers and the Importance of Online Documentation." Journal of Computer Operations 18 (1997): 15.

Comparing the Two Methods

Figure 10.1 shows a section of the endnotes page containing information about the sources that have been cited within a paper. Figure 10.2 shows how the same sources are documented for a paper using the parenthetical documentation format. Figure 10.2 shows the relevant section of the Works Cited page.

To provide accurate parenthetical documentation for your readers, you must first prepare a careful Works Cited page and then include the documentation in the right form and at the right place in your text. Preparing the Works Cited page and documenting within the text of a paper are discussed in the next two sections.

Preparing the Works Cited Page

Before you can document your sources parenthetically you must first establish what those sources are. Even though the list of references cited comes at the end of your paper, it is important that you prepare this list *before* you start to document. As we saw, just to give an author's name and a page number in parentheses does not provide readers with adequate publication information. But preparing the list first, you will know what sources you must cite and what page numbers you must list. You will also avoid accidentally omitting a source. And you can use your Works Cited page to verify information listed in the text of your paper.

When you prepare your list of references for print sources, you must include the information in the following order in accordance with the MLA style:

Books	Articles
author(s) or editor(s)	author(s)
title (underscored or in italics)	title of article (put in quotation marks)
edition (if second or subsequent)	name of journal (underscored or in italics)
place of publication	volume number (in arabic numerals)
publisher's name	date of publication
date of publication	page number(s)

In Works Cited lists, sources appear in alphabetical order according to authors' last names. Use periods between the elements in the citation. The examples below show you how to list different types of books and articles.

- *Book by one author*

    Braun, Eric. The Internet Directory. New York: Fawcett, 1993.

    Note that no page numbers are listed in this citation because the appropriate page numbers to Braun's book would be included parenthetically in the paper.

- *Two or more books by the same author*

    Hordeski, Michael F. Microprocessors in Industry. New York: Van Nostrand, 1984.

    ---. Personal Computer Interfaces: Macs to Pentiums. New York: McGraw, 1995.

    When you cite two or more works by the same author, do not repeat the author's name in subsequent reference(s). Type three hyphens in place of the name and then a period. (List the works in alphabetical order.)

- *Book by two or three authors*

    Muggins, Carolyn, and Keith Applebauer. Electronic Advertising: Principles and Practices. Chicago: General Books, 1999.

    Both authors' names are listed in the order they appear on the title page, not in alphabetical order. The first author's name is listed in reverse order, last name first, and the second author's name in normal order.

- *Book by more than three authors*

    Dossey, Barbara Montgomery, et al. Critical Care Nursing: Body, Mind, Spirit. 3rd ed. Philadelphia: Lippincott, 1992.

    When there are more than three authors, list only the first author's name in reverse order and add "et al." ("and others") after the comma following the first author's name. Note that when a book has a subtitle, you must include it. Separate the title and subtitle by a colon, as in the Dossey entry. When a book goes into a second or subsequent edition, list that fact after the title, as in the Dossey book, but do not underline the edition.

- *Corporate author*

Educational Foundation of the National Restaurant Association. Applied Foodservice Sanitation. 3rd ed. New York: Wiley, 1987.

A corporate author refers to an organization, society, association, institution, or government agency that publishes a work under its own name—for example, the Federal Aviation Administration. In the example above, the Foundation (often cited as EFNRA) wrote the book. The name of the state is not needed after well-known cities such as New York, Boston, Chicago, or San Francisco. The state is given after a smaller city to tell readers that the book was published, for instance, in Lexington, Massachusetts, as opposed to Lexington, Kentucky, or Lexington, Virginia.

- *Edited collection of essays*

Tyson-Jones, Sandra, ed. Secure Mutual Funds Investments. New York: Merrimack, 1998.

The abbreviation "ed." for *editor* follows the editor's name listed in reverse order.

- *Essay included within a collection*

Holcomb, Barry T. "No Load Mutuals: A Continuing Investment Opportunity." Secure Mutual Funds Investments. Ed. Sandra Tyson-Jones. New York: Merrimack, 1998. 321-29.

The name of the author of the article in this collection comes first—in reverse order—and then the title of the article in quotation marks. Next comes the title of the collection underscored. The editor's name is listed after the title, with "Ed." before her name to indicate that she is the editor. Do not list the editor's name in reverse order. Note that page numbers for the essay within the collection conclude the entry.

- *Professional journal article*

Hansen, Melissa. "Merging Databases After a Merger." Bank Marketing 28.5 (1996): 31-35.

Note how a reference to a journal article differs from one citing a book in MLA. The title of the article is in quotation marks, not underscored; no place of publication is listed. The volume number immediately follows the title of the journal with no intervening punctuation. And the page number(s) on which the article is found follow the colon placed after the publication date within parentheses.

▪ *Signed magazine article*

Weiner, Leonard. "How I Bought My New Car--Online." <u>US News & World Report</u> 6 May 1996: 75.

Unlike the more scholarly journal articles, popular and frequently issued magazines (such as *Business Week, TIME, U.S. News & World Report*) are listed by date and not volume number. Note again, the page number(s) following the date; no "p." or "pp." is used with page numbers in MLA style.

▪ *Unsigned magazine article*

"Can Phone Lines Catch Up to Cable?" <u>Business Week</u> 1 Apr. 1996: 87.

Many magazine articles do not carry an author's name (or by-line) because they are written by staff members of the magazine. If this is the case, begin with the title of the article. Unsigned works are always listed according to the first word of their title (excluding *a, an,* or *the*).

▪ *Newspaper article*

Wittington, Delores. "The Dollar Buys More Vacation Overseas This Year." <u>Springfield Herald</u> 30 Mar. 1999, late ed., sec. 2:10.

Give the title of the newspaper as it appears at the top of the first page of the newspaper, including the name of the city if it is part of that title. If the name of the city does not appear in the title, place it in square brackets immediately following the title, for example, *Reporter* [Allendale]. List the article by day, month, and year, *not* according to the cumbersome volume and issue numbers. Identify section, page, and edition information for readers. In the example above, readers know that the story appeared in the late edition on page 10 in section 2. Sometimes the story you cite will not require these details. The example below cites an article found on page B1 and B5 of a paper that issues only one edition per day.

Agins, Terri. "The Status of Denim: Designer Jeans Make a Comeback." <u>Wall Street Journal</u> 2 July 1996: B1,5.

▪ *Encyclopedia article*

Truxal, John G. "Telemetering." <u>McGraw-Hill Encyclopedia of Science and Technology</u>. 7th ed. 1992.

Because it is a multivolume, alphabetical work, only the particular edition and year of an encyclopedia have to be listed on the Works Cited page. If you cite the name of the author of an article in an encyclopedia, begin your reference with his or her last name. Some encyclopedia articles are not signed.

- ***Pamphlet or brochure***

National Institute on Aging. Bound for Good Health: A Collection of Age Papers. Bethesda: National Institute on Aging, 1991.

Document a pamphlet or brochure the same way you would a book. Note here that the corporate author is also the publisher of the book.

- ***Film***

Understanding AIDS. Video. Philadelphia: Health Care Media, 1997. 37 min.

Underscore the title of a film and include the medium, distributor, and date. Other information, such as individual contributors to the film, may be added after the title, if significant, to your in-text use of the source. If you indicate the length of the film, include this information last.

- ***Radio or television program***

60 Minutes. CBS News. 16 Oct. 1995.

"The Dilemma." Rich Man, Poor Man. PBS. WTQA, Lincoln. 13 Sept. 1997.

Underscore the title of a program but put an individual episode in a series within quotation marks, as in the title from *Rich Man, Poor Man* above. (In your Works Cited page, *60 Minutes* would be listed under S for "Sixty.")

- ***Cartoon or advertisement***

Lees, Charlotta. Cartoon. Miami Magazine. Aug. 1998: 36.

American Resort Council. "Leisure Life Pays Off." Advertisement. Vacation Life Feb. 1998: 71.

- ***Published interview***

Zeluto, Thomas. "Interview with Former Budget Director." Findlay Magazine Oct. 1997: 2-4.

Begin with the name of the individual being interviewed. Then indicate the title of the interview.

- ***Unpublished interview***

Cilwik, Martin. CEO, Emerson Plastics. Personal interview. 7 Aug. 1996.

Jensen, Barbara. Professor of Physics, Northwest College. Telephone interview. 15 May 1999.

Begin with the name of the individual—in reverse order—and then indicate how and when the interview was conducted.

- *Questionnaire*

```
Questionnaire for Med Techs. Distributed between 16-20
     Nov. 1998. Southwest Labs.
```

- *Lecture*

```
Melka, Mary. "The Gopher Turtle--An Endangered Species
     in Southeast Pilsen County." Lecture at Franzen
     State University, 21 Mar. 1998.
```

How to Alphabetize the Works in Your Reference List

Your list of references must be in alphabetical order to enable readers to find an entry quickly. Here are some guidelines to follow when you alphabetize your list.

1. Make sure that each author's name is in correct alphabetical sequence with the author's (or the first of multiple authors') name in reverse order. Thus, you would have Jones, Sally T., not Sally T. Jones.

2. Hyphenated last names should be alphabetized according to the first of the hyphenated names.

   Grundy, Alex H.
   Mendez-Greene, A. Y.
   Mundt, Jill

3. List corporate authors as you would names of individuals, but do not invert the corporate name.

   Marine Fisheries Association
   Nally, Mark
   National Bureau of Standards
   Nuttal, Marion

4. List names beginning with the same letters according to the number of letters in each name—the shorter names precede the longer ones.

   Lund, Michael
   Lundford, Sarah
   Lundforth, Jeffrey

5. Disregard the article (*a, an, the*) when you list an unsigned article or a film.

   The Cable Television Guidebook (an unsigned pamphlet, listed under C on the Works Cited page)
   The Godfather, Part III (film, listed under G)
   "An Improved Means of Detecting Computer Crime" (unsigned article, listed under I)

Documenting Within the Text

The Works Cited list does not, of course, tell readers what you actually borrowed from your sources or where that information is located within a source. To give readers that information, you must include documentation within the text of your paper or report.

What to Include and Why

As you write and revise your draft(s), make sure that you insert the author's name and appropriate page number(s) for each one of the sources that you use. Be sure also that the information you include parenthetically within the text—names, page numbers, and sometimes short titles—precisely matches the information you supplied under Works Cited at the end of your paper or report. Short titles within parenthetical citations should begin with the first key word used in the title. Double-check the spelling of names, page numbers, and publication data against the titles on your Works Cited page. Remember, if you fail to document within your text, you are guilty of plagiarism; and if your documentation is incorrect or incomplete, readers will have trouble finding the source and may doubt the reliability of your work.

How to Document: Some Guidelines

Parenthetical, or in-text, documentation is relatively simple. Keep your documentation brief and to the point so you do not interrupt the reader's train of thought. In most cases, all you will need to include is the author's last name and appropriate page number(s) in parentheses, usually at the end of the sentence. For unsigned articles or radio and television programs, you would use a shortened title in place of an author's name. Note in the following example that no mark of punctuation appears before the citation and that a period follows it. Also, no "page" or "p." or comma comes between the author's name and the page number.

```
About 5 percent of the world's population has diabetes
mellitus, and 25 percent of the world's population acts
as carriers of the disease (Walton 56).
```

Seeing this parenthetical documentation, the reader will expect to find the title and publication information about Walton's study correctly listed under Walton on your Works Cited page.

```
Walton, J. H. Common Diseases of the World. New York:
     Medical Books, 1998.
```

The number after Walton's name in the parenthetical documentation refers to the page number where the information you cite can be found.

You may refer to the same work more than once in your paper or report. For second or subsequent references you will use the same method of documentation. As in the Walton example above, you will place Walton's name and the appropriate page number—even if it is the same as in the previous reference(s)—within parentheses following the borrowed information. Of course, if you include Walton's name within the sentence, there is no need to repeat it in the parentheses; all you need do is give the page number. The exact placement of the author's name within the sentence is discussed below.

If you are using a work that has two authors, list both their last names parenthetically.

```
Tourism has increased by 21 percent this last quarter,
thanks to individuals passing through our state on their
way to the World's Fair (Muscovi and Klein 2-3).
```

If one of the works you use has three or more authors, list just the first author's last name in parentheses followed by et al. and the page number(s).

```
The principles of ergonomics have revolutionized the
design of office furniture (Brodsky et al. 345-47).
```

If the work you are borrowing from has a corporate author, use a shortened version of the name within the parentheses, as in the following example in which "Commission" replaces "Commission on Wage and Price Control."

```
Salaries for local electricians were at or above the
national average (Commission 145).
```

In all of the preceding examples, the names of the authors have appeared in parentheses. However, you may mention the author's name within your text. If so, there is no need to repeat it parenthetically. Include only the appropriate page number(s) within parentheses. The following examples show three different, acceptable ways of citing an author's name in the text and indicate how writers can document page references.

```
Clausen sees the renovation of downtown areas as one of
the most challenging issues facing city governments
today (29).
```

```
As Clausen notes, the renovation of downtown areas is
one of the most challenging issues facing city govern-
ments today (29).
```

```
The renovation of downtown areas, according to Clausen,
is one of the most challenging issues facing city
governments today (29).
```

Similarly, if you list the title of a reference or anonymous work within the text of your paper, you need not repeat it for your parenthetical documentation.

```
According to the Encyclopaedia Britannica, Cecil B.
deMille's King of Kings was seen by nearly 800,000,000
individuals (3:458).
```

The first number in parentheses refers to the volume number of the *Encyclopaedia Britannica;* the second points to the page number of that volume. In this case, the writer wisely gives both volume and page numbers to indicate that the information is listed under deMille and not the title of the film.

If you are citing information from two or more works by the same author, you will have to inform readers clearly from which work a particular fact or opinion comes. Let's say that you used information from the following works by the same author:

```
Howe, Grace. Networking in the Information Age. New
     York: Business Publications, 1998.

---. "Systems Control for Small Businesses." The
     Electronic Workplace 15 (1997): 67-81.
```

You have a number of ways to tell readers from which specific work by Grace Howe you are borrowing material.

1. Cite the author's name, short title, and page number parenthetically.

   ```
   Communication checkpoints are necessary in any business
   to provide a maximum flow of information (Howe,
   Networking 132).
   ```

   Use a comma after the author's name.

2. Mention the author's name in your sentence and use a short title and page number within parentheses.

   ```
   Howe thinks communication checkpoints in any business
   are necessary to provide a maximum flow of information
   (Networking 132).
   ```

3. Give the author's name and a shortened title in the text with only the page number included parenthetically.

   ```
   According to Howe's article, "Systems Control,"
   productivity increases by at least 20 percent after
   each training session involving communication networking
   techniques (71).
   ```

Occasionally you will have to cite two sources at the same time to document a point. Include the names of the authors of both sources just as if you were listing them individually but insert a semicolon between sources.

```
The use of salt domes to store radioactive wastes has
come under severe attack (Jelinek 56-57; McPherson and
Chin 23-26).
```

Be careful, though, that you do not overload readers by including a long string of references in your parenthetical documentation.

```
Wind energy has been successfully used in both rural
and urban settings (Bailey 34; Henderson 9; Kreuse 78;
```

Mankowitz 98-99; Olsen 456-58; Vencenti 23; Walker and Smith 43).

Rather than interrupting the reader and crowding references together, consider revising your sentence to make the subject more precise and the references more restricted.

Wind energy has long benefited the farm community (Kreuse 78; Walker and Smith 43). But recent experiments in New York City have shown the effectiveness of this form of energy for apartment dwellers, too (Bailey 34; Henderson 9). Similar experiments in San Francisco also show how wind power helps urban residents (Mankowitz 98-99; Olsen 456-58; Vincenti 23).

If you include a quotation, place the parenthetical documentation at the end of the sentence containing the quotation.

Pilmer has observed that coffee "is only mildly addictive in the sense that withdrawal will not harm you or produce violent symptoms" (16).

Note that the period follows the parentheses, not the quotation marks. Even if the quotation is short and appears in the middle of the sentence, place the documentation at the end of the sentence.

Alvin Toffler uses the phrase "third wave" to characterize the scientific and computer revolution (34).

If the material you quote runs to more than four typed lines, set the quotation apart from the text by indenting it *ten spaces* on the left-hand side and by eliminating the quotation marks. Double-space the quotation. Place the parenthetical documentation after the quotation and outside the period as in the following example.

L. J. Ronsivalli offers this interesting analogy of how radiation can penetrate solid objects:

> One might wonder how an X-ray, a gamma ray, or a cosmic ray can penetrate something as solid as a brick or a piece of wood. We can't see that within the atomic structures of the brick and the wood there are spaces for the radiation to enter. If we look at a cloud, we can see its shape, but because distance has made

> them too small, we can't see the droplets of moisture of which the cloud is made. Much too small for the eyes to see, even with the help of a microscope, the atomic structure of solid materials is made of very small particles with a lot of space between them. In fact, solids are mostly empty spaces. (20-21)

If you omit anything from a quotation, follow the rules governing ellipses on page 367.

If you take a quotation from any place but the original source (if, for example, the quotation you want to use is included in the book you are citing but originally came from another book or article), you should document that fact by including "qtd. in" in your parenthetical documentation.

> The monthly business meeting serves a number of valuable functions. In fact, perhaps the most important one is that "chain-of-command meetings provide the opportunity to pass information up as well as down the administrative ladder" (qtd. in Munroe 87).

This documentation lets readers know that you found the quotation in Munroe, not in the original work from which these words come.

Documentation in Scientific and Technical Writing

The MLA style is, of course, not the only method for documenting sources. Numerous other formats exist. Many professions publish their own style guide or book to provide such instruction; some of the better known are the *Publication Manual of the American Psychological Association* (4th ed.), the *Council of Biology Editors' Style Manual* (5th ed.), and the American Institute of Physics's *Style Manual for Guidance in the Preparation of Papers*. Other professions recommend that writers follow the format used in a specific technical or scholarly journal. Every professional group, however, would advise writers *against* using the formats and abbreviations found in various databases. The way information is listed in these sources is not offered as a model of documentation. Before you write a paper or report, ask your instructor or employer about the format he or she prefers.

Many professions use the *author-date method of documentation* or the *ordered references method*. Both methods rely on parenthetical documentation; that is, rather than using footnotes or endnotes with superscripts, the particular information about the source is placed directly in the text within parentheses. The MLA method just discussed is one such format. Another is the APA method, which differs slightly from MLA style.

How APA Differs from MLA Parenthetical Documentation

APA, like MLA, uses a simplified form of parenthetical documentation. But in APA the publication date is much more prominent than in MLA. Such an emphasis is understandable given the rapidly changing discoveries in and requirements of the world of science and technology. Every parenthetically documented source in APA includes (1) the author's name, (2) the date of the work, and (3) the page number or numbers preceded by a "p." or "pp."

> The theory that new housing becomes increasingly expensive as buyers move farther north has been recently advanced (Jones, 1999, p. 13).

The complete bibliographic information is assembled in a References page, similar to the MLA's Works Cited page. But, unlike MLA, APA requires writers to do the following when they list their references:

> Jones, T. (1999). The cost of housing on Lincoln's north side. Urban Studies, 7, 10-24.

1. Use just an author's surname and initials rather than spelling out his or her first name—Jones, T., rather than Jones, Tracy.
2. Place the date of publication of the book or article in parentheses immediately after the author's inverted name: Jones, T. (1999).
3. Capitalize only the first word in a title of an article or book (except for proper names in the title).
4. Underline or italicize the title of a book, journal, or magazine; write out in full the names of the months; and underline or italicize the volume numbers.
5. Do not enclose the title of a journal, magazine, or encyclopedia article in quotation marks.
6. Use a comma to separate the volume number of a journal from the page numbers for that article, but do not use "p." or "pp." for journal page numbers.
7. Indent the first line of each citation five spaces.

- *Book with a single author*

MLA: Zednick, Dorothea. Optical Properties of Insulators. New York: Scientific Publishing, 1996.

APA: Zednick, D. (1996). Optical properties of insulators. New York: Scientific Publishing.

- *Book with multiple authors*

MLA: Smith, Frank S., et al. Global Environmental Change. Chicago: The Fermi Institute, 1997.

APA: Smith, F.S., Pisanto, G., Nicholson, R., & Mellor, A. (1997). Global environmental change. Chicago: The Fermi Institute.

Note that in APA style up to six authors' names are listed; APA does not use the abbreviation *et al.* but uses the ampersand (&) in place of the conjunction *and.*

- ***Book with a corporate author***

MLA: American Institute of Banking. Electronic Tracking Systems. Pittsburgh: Economics, 1995.

APA: American Institute of Banking. (1995). Electronic tracking systems. Pittsburgh, PA: Economics Press.

- ***Edited book with individual essays***

MLA: Katz, Roberta. "Computer-Based Writing Aids." Eds. Yvonne Dietrich and Mark Hunt. Social Implications of Computing. Los Angeles: Southeastern UP, 1999. 72-79.

APA: Katz, R. (1999). Computer-based writing aids. In Y. Dietrich & M. Hunt (Eds.), Social implications of computing (pp. 72-79). Los Angeles: Southeastern University Press.

- ***Journal article***

MLA: Perez, Pedro, and Theresa Vali. "The Role of the Soundtrack in Three Recent Films." Studies in Film and Culture 14.11 (1998): 54-73.

APA: Perez, P., & Vali, T. (1998). The role of the soundtrack in three recent films. Studies in Film and Culture *14*(11), 54-73.

Note that APA italicizes the volume number of a journal.

- ***Newspaper article***

MLA: Lai, Neelou. "New Law Affects Local Businesses." The Springfield Herald 13 July 1996: D2.

APA: Lai, N. (1996, July 13). New law affects local businesses. The Springfield Herald p. D2.

- ***Unsigned article***

MLA: "Hottest Internet Sites of the Month." The Net 31 Mar. 1998: 12.

APA: Hottest Internet sites of the month. (1998, March 31). The Net 13, p. 12.

- *Encyclopedia article*

MLA: "Topology." The Encyclopedia of Mathematics and Statistics. 1999 ed. Detroit: Professional Books.

APA: Topology. (1999). In The encyclopedia of mathematics and statistics (Vol. 8, p. 283). Detroit, MI: Professional Books.

In-Text Documentation Following the APA Method

Like MLA, APA uses in-text parenthetical documentation. But as you will see from the following examples, the APA system differs from the MLA in a number of ways. A writer using APA style puts the author, the date, and page number, with a "p." or "pp.", in parentheses when citing quotations.

The theory that new housing becomes increasingly expensive as buyers move father north has been recently advanced (Jones, 1999, p. 13).

The reader sees that Jones developed this theory on page 13 of a work written in 1999. If Jones's name were mentioned in the text, "Jones advances the theory that new housing . . . ," only (1999, p. 13) would be listed. To find Jones's work, readers turn to an alphabetical References page at the end of the paper, report, or article where, under Jones, they find a bibliographic entry for the work. As in the MLA system, only works actually cited in a paper are listed in the reference list. If two works by Jones were cited, the references for both are given. If they were done in the same year, they are differentiated in the text *and* in the list of references by a lowercase letter *a* and *b*.

Housing is increasingly more expensive on the north side than on the south (Jones, 1999a, p. 22).

A recent study has established a demographic pattern for small cities in the Midwest (Jones, 1999b, p. 73).

In the accompanying list of references, the two works by Jones might be listed as follows:

Jones, T. (1999a). The cost of housing on Lincoln's northside. Urban Studies 72, 10-24.

Jones, T. (1999b). Demographic density in three Midwestern small cities. Cincinnati, OH: Western Press.

The sample long report on e-cash in Chapter 17 uses the APA method of documentation.

Documenting Electronic Sources

Up to this point we have been discussing documenting materials that you research in print. But as part of your research you will also be consulting electronic sources. As we saw in Chapter 9, there are a variety of media through which you can access electronic sources—CD-ROM, on-line databases, the Internet. The MLA distinguishes between "portable databases"—those like CD-ROMs that are self-contained and updated periodically—and "on-line databases"—those like CARL UnCOVER that are continuously available and updated.

Some General Principles

When you document electronic materials, keep the following general principles in mind.

1. With electronic sources, as with printed sources, you are obligated to acknowledge (a) what you used, (b) where you found it, and (c) how others can obtain a copy.
2. Note that an electronic source has different characteristics from a print version that your readers need to be aware of—for example, Internet addresses, dates when the material was first posted and when you actually accessed it.
3. Some of your sources may be available in both printed and electronic versions. If so, you have an additional responsibility to inform your readers of this fact.

The sections below will help you document accurately any material you use from CD-ROM and on-line databases and any information you access through the Internet.

Documenting CD-ROM and On-Line Database Material

The first thing you have to establish is whether the CD-ROM information is also available in print. If a printed version exists, inform readers of that fact first and then cite the CD-ROM source you used. Where the information has no print counterpart, you do not have to worry about this.

Basically, for material taken from a CD-ROM, list information in a bibliographic citation in this order:

- author's name (last name first)
- title of the material
- publication data for printed version
- title of the CD-ROM database (underscored)
- publication medium (CD-ROM)
- name of the vendor (if available)
- electronic publication date

- *Work on CD-ROM with printed source or analogue*

Stephenson, Jason. "Sick Kids Find Help in a Cyberspace
World." JAMA [Journal of the American Medical

Association] 274 (27 Dec. 1995): 1899-901. MEDLINE Express. CD-ROM. 11 Aug. 1996.

Wood, S. H., and V. J. Ransom. "The 1990s: A Decade for Change in Women's Health Care Policy." JOGNN [Journal of Obstetric, Gynecological, and Neonatal Nursing] 23 (Feb. 1995): 139-43. CINAHL. CD-ROM. 10 July 1996.

- *Work on CD-ROM without a printed source or analogue*

"Access Health, Inc.: Annual Cash Flow Statement, 1995-96." Compact Disclosure. CD-ROM. 1 June 1996.

Cazzin, Julie. "Kids, Cash, and Capitalism." Maclean's 6 May 1996. CD-ROM. Periodical Abstracts. UMI Company. 5 Nov. 1996.

- *Article on on-line database with a printed source*

Carey, Catherine, and Dawn Langkamp Bolton. "Brand Versus Generic Advertising and the Decision to Advertise Collectively." Review of Industrial Organization 11 (Feb. 1996): 93-105. EconLit. On-line. CARL UnCOVER. 12 Apr. 1996.

Hayes, Robert D., and K. W. Hollman. "Managing Diversity: Accounting Firms and Female Employees." CPA Journal 66 (5): 36. 1 May 1996. On-line. CARL UnCOVER. 1 June 1996.

Smith, Douglas K., and Barbara J. Minnick. "Electronic Teacher-Student Communication." Business Communication Quarterly 59 (Mar. 1996): 74-81. ABI/Inform. On-line. ProQuest. 16 July 1996.

- *Article on on-line database without a printed source*

Chinnock, Chris. "Virtual Reality Goes to Work." Mar. 1996. Applied Science. On-line. Wilsonline. 12 Apr. 1997.

Citing Locations on the Internet—Beyond MLA

The 1997 edition of the *MLA Handbook* covers materials found on the Internet and the problems researchers face in citing them. As we saw in Chapter 9, the Internet is a major research tool that you will no doubt frequently use and document.

A very helpful guide to documenting materials on the Internet is Andrew Harnack and Gene Kleppinger's "Beyond the MLA Handbook: Documenting Electronic Sources on the Internet." The following discussion on citing Internet locations is based on Harnack and Kleppinger's work. You may access Harnack and Kleppinger's Web site directly at http://falcon.eku.edu/honors/beyond-mla/#citing_sites.

How an Internet Source Differs from a Print Source

Citing locations on the Internet poses problems for researchers that print sources do not, for the following reasons.

1. An Internet source is documented with a different set of symbols and punctuation marks than are used with a print source.
2. An Internet location has no page numbers.
3. An Internet source has no fixed publication date; sites can change over weeks or months or in the matter of an hour. (This is why so many Internet locations are frequently "under construction.") They can also be deleted.
4. A researcher can access an Internet site through many pathways—different access providers, different search engines, and so on.
5. An Internet source can have many different addresses.
6. An Internet source can be changed or revised at any time. This is what Harnack and Kleppinger call *invisible revocability,* "the process by which an electronic document can be altered, moved, or deleted by the author or by the computer system managers, without any publicly-accessible trail of evidence."
7. An Internet source can be accessed at any time of the day, week, month, or year. Before or after it is accessed, a site may open, change, or close.

Because of these features, an Internet location requires a different documentation format than a print or CD-ROM source does.

Documenting Different Types of Internet Sites

There are five types of Internet sites that you need to know how to document.

1. File Transfer Protocol (FTP)
2. World Wide Web (WWW)
3. Listserv/Newsgroup
4. E-mail

The following discussion explains how to cite materials from each one.

1. File Transfer Protocol (FTP) sites. File Transfer Protocol, or FTP, allows you to transfer material from the Internet without having to change disks and download information. To cite FTP sites, include the following information:

- the author's name
- the full title of the document, in quotation marks
- the date document was posted (if available)
- the FTP address and path, enclosed in angle brackets (⟨ ⟩)
- the date you accessed the site, in parentheses

```
Exploratorium Science-at-Home Team. "Do You Want to Be
     an Exploratorium Scientist at Home?" ⟨ftp://ftp.
```

exploratorium.edu/events/science-fun-at-home⟩ (4 Oct. 1996).

2. World Wide Web (WWW) sites. The World Wide Web is an Internet service that makes browsing, or "surfing the Net," both possible and exciting. The Web uses hypertext markup language (HTML) to link documents to each other and to add graphics to plain text. The result is that you can click your way through colorful links to arrive at the page that interests you most. (And if you click on the bookmark feature when you get there, you can return just as easily.) To provide a bibliographic citation to a WWW site, list the following information in this order:

- the author's full name (if available) or the name of the corporate author
- the title of the WWW document, in quotation marks
- the title of the work, in italics or underlined
- the date of publication, or the date the site was last updated or revised
- the uniform resource locator (URL) address—that is, the http://address—in angle brackets
- the date you accessed the site, in parentheses

Here are a few sample WWW citations to locations you are likely to use in your research.

- *Newspapers on the WWW*

Behr, Peter. "U.S. Slows D.C. Area Purchases." Washington Post. 1 July 1996. ⟨http://www.washingtonpost.com/wp-srv/business/daily⟩(10 July 1996).

Cleary, Paul. "Welfare Boom for Rich." Sydney Morning Herald. 11 July 1996. ⟨http://www.smh.com.au/daily/pageone/960711-pageone.html⟩(17 Sept. 1996).

Lane, Polly. "New Area Codes Coming: Thank Your Fax, Modem." Seattle Times. 9 July 1996. ⟨http://www.seattletimes.com/topstories/browse/html/code⟩(3 Oct. 1996).

- *Electronic journal or newsletter*

Hughes, John H., et al. "A Comparison of Rolling vs. Non-Rolling Cultures for the Early Detection of Viruses in Clinical Specimens." World Wide Web Journal of Biology. June 1996. ⟨http://epress.com/w3jbio/wh6.html⟩(8 Dec. 1996).

- *Dictionary or encyclopedia*

"Magnetic Disks." NASA Thesaurus. ⟨http://www.sti.nasa.gov/thesaurus/M/word8907.html⟩.

"Computer." Hypertext Webster Interface. Aug. 1995. ⟨http://gs213.sp.cs.cmu.edu/prog/webster?⟩(3 June 1996).

- *Home page for a company product or service*

"About Our Hospital." Aultman Hospital, Canton, OH. April 1996. ⟨http://www.aultman.com/aultman/aultman.html⟩(13 Sept. 1996).

"News and Facts." Bankers Trust, New York, N.Y. 9 July 1996. ⟨http://www.bankerstrust.com⟩(4 Aug. 1996).

"Welcome to Erich Tech." Erich Tech Industrial Co., Ltd., Taipei, Taiwan. June 1996. ⟨http://www.tradewatch.com/eti/indes.html⟩(3 Oct. 1996).

"Welcome to Kitchen Sink Software Online." Kitchen Sink Software, Inc., Westerville, Ohio. July 1996. ⟨http://www.kitchen-sink.com⟩(10 Nov. 1996).

3. **Listserv/newsgroup sites.** To cite a newsgroup source, include

- the author's name
- the author's e-mail address, in angle brackets
- the posting's subject line, in quotation marks
- the date of the posting
- the newsgroup name in angle brackets
- the date you accessed the posting, in parentheses

Slade, Robert. ⟨res@maths.bath.ac.uk⟩ "UNIX Made Easy." 26 Mar. 1996. ⟨alt.books.review⟩(31 Mar. 1996).

Kennedy, Sean. ⟨skennedy@utkvx.utk.edu⟩ "FBI and Military Secrets." 30 July 1996. ⟨alt.alien.visitors⟩(5 Oct. 1996).

A listserv provides information on a variety of topics—some highly technical, some consumer-oriented—to individuals who subscribe to that particular service. To cite a listserv reference, you should include

- the author's name
- the author's e-mail address, in angle brackets
- the subject line, in quotation marks
- the date of the posting
- the listserv address, in angle brackets
- the date you accessed the message, in parentheses

Mielziner, Jennifer. ⟨mielziner@canton.org⟩ "Lipoprotein A--the New Enemy." 17 Oct. 1996. ⟨biochem@harrison.prods.harrison.edu⟩(6 Dec. 1996).

4. E-mail. Even though e-mail messages are regarded as unpublished correspondence like memos or letters, you still need to document them by giving the following information.

- the author's name (last name first)
- the author's e-mail address, in angle brackets
- the subject line (regarded as the title) found in the posting, in quotation marks
- the date the e-mail was sent
- the type of e-mail (personal; inter- or intra-office communication)
- the date you accessed the e-mail, in parentheses

Jerach-Gordon, Tina. ⟨jerach-gor@aol.com⟩ "Hosco Software Update Problems." 30 Apr. 1997. Personal e-mail. (5 May 1997).

LaPine, Susan. ⟨lapine@tuco.com⟩ "Spring Newsletter Design" 15 Feb. 1998. Inter-office e-mail. (1 Mar. 1998).

Sample Research Paper Using MLA In-Text Documentation

The rest of this chapter consists of a research paper on the advantages of telecommuting. Study the paper to see how the student author has successfully used the MLA system of documentation to cite print, CD-ROM, and Internet sources. Compare the references mentioned in the text with the Works Cited page to see how the writer has handled documentation appropriately. (The sample long report in Chapter 17 (pages 631–652) follows the APA system of documentation. You might want to compare these two papers to become even more familiar with these two methods of parenthetical documentation.)

1" down

Henrietta Holland
Professor Meyer
Business Writing 301
12 April 1997

The Advantages of Telecommuting in the Information Age

Today telecommuting is transforming the world of work. Perhaps the best definition of telecommuting comes from the New York Telecommuting Advisory Council:

> Telecommuting is using telecommunications technology to replace traditional forms of commuting. Employees work all or part of the time outside the traditional office, at remote work locations, which may include the home. The work goes to the worker rather than the worker to the work. People work where they are most effective. ("What Is Telecommuting")

Telecommuters travel to work on the information superhighway. Any individual whose job depends on information technology can be a telecommuter, from skilled professionals such as computer programmers and analysts, architects, and documentation specialists to telemarketing representatives and travel agents who process information over the Internet or telephone (Steve 37).

Telecommuting became a significant part of the American workplace in 1992 when AT&T's telecommuting program began with a handful of employees and grew to 35,000 in the first three years (Hannon 87). This rapid change reflects a national trend. Telecommuting has grown at a rate of about 20 percent each year, and by the year 2001 Myra Perez-Hoya believes that approximately 40 million workers will use this alternative work arrangement ("Workers" 102). Even more optimistically, Link Resources, a New York research group, predicts that by early 1999, "the number of telecommuters will jump to 60 million" (qtd. in Bredin 3). While the United States is a world leader in telecommuting, McQuarrie emphasizes that Canada, Australia, Japan, and Scandinavia also have enthusiastically endorsed the concept (80).

But telecommuting is much more than the substitution of technology for travel to and from work. The practice involves important changes in management policies as well as in the way people work and view their careers. This paper will survey the growth of telecommuting by discussing (1) the ways it differs from the traditional office, (2) the advantages telecommuting has for employees, (3) the advantages it offers employers, and (4) the impact telecommuting will have on the workplace in the 21st century.

Telecommuting Transforms the Traditional Office

According to Greg Bahue, the publisher of High Technology Careers Magazine, "The telecommuting trend can be likened to the change from telegraph to telephone, horse and carriage to automobile, black-and-white to color television" ("The Age of Telecommuting"). Telecommuting is a technology whose time has come. As society switches from an industrial/manufacturing economy to one based on information, telecommuting will become the "new world of work in the electronic infrastructure" (Perez-Hoya, "Telecommuting" 8). Advances in communication technology in this "new world"--fiber optics, modems, notebooks, semiconductors--make telecommuting possible. Thanks to these technological advances, telecommuting is radically altering the traditional idea of work schedules and workplaces.

The traditional office functions on a fixed schedule. From the days of Scrooge demanding that Bob Cratchet not be late for work to the Dolly Parton film Nine to Five, the workday has been run on a "ruthless timeclock" ("Learning New Work" 12). Employees often had to stay late or come in early to do (or keep) their jobs. As one former 9-to-5er put it, "When a big order had to go out, we all worked frantically to beat the clock" (Vint). Communicating electronically, a telecommuter can conduct business on-line at any time, day or night, weekends, and holidays. No set hours dictate when something must be done. Gone, too, are the days when an employee had to take work home. As Bahue aptly notes, "home-based work has become an extension of the workday" ("Age of Telecommuting").

Telecommuting also transforms the idea of the workplace--a centralized office where a job is done usually in a large downtown building where employees are restricted to their desks. Telecommuters can work at home, at remote sites, or while traveling. Obviously, telecommuting's flexibility of place is inappropriate for manufacturing, health care, or other positions that require specialized machines or conditions (Barnes 10). But for millions of telecommuters work is accomplished through technologies that are blind to their surroundings. E-mail sent from the central office to a customer is identical to e-mail sent from a sales representative's home or from her notebook at an airport.

Telecommuting does not eliminate the need for a central office, however. All those modems, phone lines, computer networks, and the staff to operate and maintain them have to reside somewhere, and the organization needs a mailing address. Customarily, telecommuters work at home three days a week but must be in the office the other two days for meetings and when emergencies arise (Bredin 27; Lewis and Zhang 71). Patel stresses that in his company telecommuters are expected to come to the central workplace to attend seminars, meet visitors, and discuss significant personnel issues with management.

Advantages of Telecommuting for Employers

Telecommuting offers many advantages for companies whose overriding corporate concern is "productivity" (Bredin 28-29; Lewis 50). Managers have repeatedly acknowledged that the quality and quantity of an employee's work improve with telecommuting. Many managers report as well that, in Steve's words, "telecommuters have more energy working at home and are more accurate" (37). According to numerous studies (Fougere and Behling 26; McNerney 4), employers further verify that telecommuters have a more positive attitude, which also increases productivity and company loyalty.

Many large corporations (General Dynamics, International Business Technologies) also happily found that telecommuters took less leave time. Allowed to work at home, telecommuters can still care for sick children, spouses, or parents. Telecommuting also helps companies retain valued, skilled employees who, as Cooper notes, might otherwise leave due to a spouse's transfer or other change (11). Faced with less staff turnover, a company will find its production schedules, marketing plans, and customer orders less frequently interrupted or delayed. As one manager of Southwest CyberSystems observed, "Before telecommuting, when we lost a key person, everything went offcenter. Too many times highly qualified people left not because of salary or benefits but because of schedules. Switching to telecommuting was one of the smartest moves the corporation ever made" (qtd. in Lewis and Zhang 53).

Telecommuting saves a business money, too. By gaining access to a much larger labor pool, even to a global labor market, as we'll see below, firms escape paying relocation expenses (Korzeniowski 47). Companies like AT&T with large numbers of telecommuters recoup a great deal on facilities costs. Fewer employees permanently assigned to a central workplace means smaller facilities can suffice. Hannon estimates that companies can reduce their office space costs by at least $4,000 to $6,000 per telecommuter (87). The accounting firm of Ernst & Young cut its annual real estate budget by $25 million when it began telecommuting (McNerney 4). And when reduced costs for power, light, security, and support staff are factored in, telecommuting is, according to Alicia Lewis, "a bargain worth the time and energy that go into implementing it" (51).

Even more significantly, telecommuting helps a company protect the environment by complying with the 1990 Clean Air Act. This law, passed to improve air quality in smog-plagued cities, requires a company with more than 100 employees in any one of thirteen specific areas to "curtail work-related vehicle miles traveled by their employees" by at least 25 percent or risk hefty fines (Bredin 24; Perez-Hoya, "Telecommuting" 8). By encouraging carpooling and permitting telecommuting, a company can adhere to the letter as well as the spirit of the law while enhancing its public image.

As beneficial as telecommuting can be for the employer, the practice presents some challenge. As Lewis points out, employers have feared losing management or supervisory control (42). But such worries are more potential disadvantages than actual liabilities. By planning telecommuting programs carefully, employers can avert problems before they occur and by exploring different management styles companies can better motivate employees. The many positive experiences firms see from telecommuting have all but removed the old management fear that telecommuters will "watch Oprah rather than work" (Korzeniowski 46).

Advantages of Telecommuting for Employees

Telecommuting employees also benefit by not working fixed hours in a traditional office. Gaining more personal time for self and family was the chief advantage telecommuters cited most often in a recent survey (Korzeniowski 47). Telecommuting saves workers enormous commuting time. "The president of the California Chamber of Commerce estimates that Californians alone spend 300,000 hours each day in traffic delays" (Bahue). A veteran telecommuter, Alice Bredin enthusiastically observes in her influential Virtual Office Survival Handbook:

> Working at least part-time away from a traditional work site means you can complete your work efficiently and still have some time left over. It means you can drop your kids off a little later at day care and pick them up a little earlier. You can finally find time to stay in shape because you aren't giving a couple of hours a day to a commute. It means you can do the quality of work you

know you are capable of because you have interruption-free time. (3-4)

Reinforcing Bredin's strong sentiments, Cooper reported that telecommuters experienced "improved family functioning," "increased civic involvement," and did much more "volunteerism" (11). Overall, telecommuters achieve a better quality of life.

Telecommuting, moreover, saves employees money. Not having to travel back and forth to work each day, five days a week, significantly reduces transportation costs. On the average, an employee travels at least 20 miles a day to work (Lewis and Zhang 42). Taking into account the expense of gasoline, parking, and upkeep of a vehicle (or public transportation fees), a telecommuter can save $4,000 to $5,000 a year. Other savings result from reduced costs for food (the commuter will eat out less often), clothes, and dry cleaning. Any expenses incurred by telecommuting are small in comparison to the savings a worker receives. As Myra Perez-Hoya maintains, telecommuters should have minimal overhead in establishing a home office, most often set up in a spare bedroom, since employers often pay for additional equipment ("Telecommuting" 106).

As a worker, too, the telecommuter reaps rewards. Having flexible schedules, telecommuters can produce at the times most desirable for them. By starting the work day earlier or later than at a traditional office, telecommuters work when their energy levels are the highest. Being at home also lets the telecommuter take longer, and perhaps more frequent, breaks without worrying that "he or she will have to stay at the office until midnight" to finish the job (Perez-Hoya, "Workers" 103). Barnes observes that many telecommuters feel far less frustrated because they are less frequently "under the penalty of heavy deadlines" (9).

Yet another benefit of telecommuting is opening the job market to individuals who might otherwise not be able to compete. "Telecommuting offers possibilities for some individuals to earn a living in a rural community independent of the local economy" (New York Telecommuting Council). For example, when a paper mill closed in rural Vermont, unemployed workers became telecommuters for a sports clothing company that sold most of its products over the telephone.

In addition, telecommuting offers employment opportunities to handicapped or chronically ill persons (McQuarrie 80). Sight or speech impaired individuals also can, with proper equipment, function superbly as telecommuters. Other types of workers benefit, too. The Best Western motel chain has hired women incarcerated in a correctional facility as reservation takers. Telecommuting thus offers job training and the possibility of rehabilitation for these individuals (Hannon 87).

Although telecommuting is highly productive, it has some drawbacks. Cooper fears it can encourage people to overwork when they are no longer locked into the 9-to-5 routine (12). More commonplace, though, is the sense of isolation that some telecommuters express (Fougere and Behling 28). Yet many of these potential disadvantages can be reduced, if not eliminated, if the telecommuter practices time management. Feeling "out of the loop" by working alone at home is a normal reaction telecommuters can overcome by seeing themselves as part of a global village, as Myra Perez-Hoya maintains ("Workers" 102). Finally, since most firms require telecommuters to make regular visits to the office, these "separation problems" can be minimized (Baig 104; Patel).

What the Future Holds for Telecommuting

The future looks promising for telecommuting. As Bahue enthusiastically predicts, "Telecommuting . . . will become as ubiquitous as your VCR, microwave, and bank Versateller card" ("Age of Telecommuting"). At the Spring 1997 Festival of Telecommuting, Agnes Reynolds similarly foresaw many good things: "Imagine our big cities with no traffic jams, far less pollution, happier and more productive workers freed from undue stress, and a better-informed and more cohesive society--all these are possible through more telecommuting." Unquestionably, telecommuting has far-reaching implications for workers, businesses, and society in general.

Future technological advances will make the telecommuter's job even easier. The technology telecommuters now use--computers, modems, cellular phones--is getting better and faster. As Lewis and

Zhang stress, "the hardware for effective telecommuting is in place; all that has to be developed is more efficient software" (74). Manufacturers are now experimenting with faster modems, which will considerably reduce down or lost time. Faster modems will mean that e-mail and other electronic communications may be instantaneous. Telecommuters will soon be able to get their e-mail through cable television (Lewis 51).

Videoconferencing will enhance telecommuting, too. Sprint has the technology in place to link telecommuters to each other and to their employers even better. "For collaborative efforts, look for software with dataconferencing features that allow workers to have meetings with each other through their home computers" ("Technology Tips"). Faster technology for activities such as dataconferencing will make every business communication easier and cost efficient. Full-motion video seems destined to enhance even the telecommuter's limited trips to an employer's centralized office.

Expansion of the Internet will surely broaden telecommuting. Telecommuters will become better informed citizens of the world economy by easily gathering crucial business data from around the globe ("Learning New Work Technologies" 10). They will not be limited to reading just one or two local newspapers, but can scan stories from the world press. Even beyond that, telecommuters through global collaboration will have colleagues--"intelligent agents" as Myra Perez-Hoya perceptively labels them ("Workers" 105)--who report and summarize, over the Internet, relevant business news along with interpretations.

Telecommuting will also transform the relationship between employees and managers. For Lewis and Zhang, many distinctions between worker and manager will dissolve in the future because of telecommuting (80-81). Telecommuters will become much more independent, being able to change jobs without changing residences or expertise. "Rather than being part of a company's workforce fixed on the payroll, telecommuters will be more like private contractors in the future responding to the most lucrative proposals" (Perez-Hoya, "Telecommuting" 10). The role of managers, too, will undergo major changes, according to several experts (Lewis 46; Patel). A manager will become more like a collaborator than a supervisor, working in a group of individuals each responsible for a different complementary, not subordinate, segment of a job. "The roles of telecommuter and manager will merge by the year 2001" (Vint).

Conclusion

Telecommuting is rapidly transforming the world of work and the way employers and employees think of their jobs. Presenting a challenging though profitable alternative to the traditional office, telecommuting is a highly effective way to conduct business in the Information Age. As one telecommuting proponent put it, "Ever since the introduction of the PC in the early 1980s, the business world has been moving toward the greater flexibility and productivity that telecommuting offers" (Reynolds).

As we have seen, both employers and employees profit professionally and personally from telecommuting. Employers gain a more productive and loyal workforce while saving money on office space and protecting the environment. Similarly, telecommuters can work at their most productive times during the day (or night) and still meet family and personal obligations. Telecommuting also opens the world job market to individuals who in the past may have been excluded, one of the most worthwhile effects of this growing technology. As technological advances make telecommuting even more attractive and feasible, current limitations or problems will surely fade away. The world of business is on the threshold of a new frontier because of telecommuting. Considering that telecommuting has been practiced for less than a decade, telecommuters and their employers have traveled a long distance on the Information Superhighway.

Works Cited

Bahue, Greg. "The Age of Telecommuting." On-line. Internet. 23 Jun. 1996. ⟨http://www.careerexpo.com/pub/docs/telecom.html⟩(25 Jan. 1997).

Baig, Edward C. "Welcome to the Wireless Office." Business Week 26 June 1995: 104-106.

Barnes, Kathleen. "The New Workforce: Tips for Managing Telecommuters." HR Focus 15 (Nov. 1994): 9-10. UMI-ProQuest. CD-ROM. Jan. 1997.

Bredin, Alice. The Virtual Office Survival Handbook: What Telecommuters and Entrepreneurs Need to Succeed in Today's Nontraditional Workplace. New York: Wiley, 1996.

Cooper, Richard C. "Telecommuting: The Good, The Bad, and The Particulars." Supervision Feb. 1996: 10-12, 19.

Fougere, Kenneth T., and Robert P. Behling. "Telecommuting Is Changing the Definition of the Workplace." Journal of Computer Information Systems 36.2 (1995-96): 26-29.

Hannon, Kerry. "A Long Way from the Rat Race." U.S. News & World Report 30 Oct. 1995: 86-87.

Korzeniowski, Paul. "Telecommuting--A Driving Concern." Business Communications Review 13 (Feb. 1997): 45-48.

Holland 15

"Learning New Work Technologies." Business Trends Feb. 1997: 8-15.

Lewis, Alicia. "The Business of Telecommuting." Journal of Contemporary Business 11 (1997): 39-51. On-line. Internet. 20 Feb. 1997. ⟨http://www.contempbus.jrn.com⟩(4 Mar. 1997).

--- and Tim P. Zhang. Telecommuting Practices. New York: Technology P, 1997.

McNerney, Donald J. "Telecommuting: An Idea Whose Time Has Come." HR Focus Nov. 1995: 1, 4-5 UMI-ProQuest. CD-ROM.

McQuarrie, Fiona A. E. "Telecommuting: Who Really Benefits?" Business Horizons (Nov.-Dec. 1994): 79-83.

New York Telecommuting Advisory Council. "What Is Telecommuting?" ⟨http://www.sccsi.com/telecommute/ny/html⟩(Apr. 1996).

Patel, Ivor. Manager, Techworks, Inc. Personal interview. 26 Feb. 1997.

Perez-Hoya, Myra. "Telecommuting and Tomorrow's Workers." Home Office Journal 3 (1997): 7-10.

---. "Workers Become Telecommuters." The Electronic Workplace 21.10 (1997): 101-106.

Reynolds, Agnes. "The Festival of Telecommuting Ushers in New Ways of Doing Business." On-line. Internet. 10 Feb. 1997. ⟨http://www.festival.bus/telecom.com⟩(1 Mar. 1997).

Steve, Bob. "Telecommuting: Concepts and Resources." Business Credit (Jan. 1996): 36-40.

"Technology Tips." Home Computers, Erie, PA. Mar. 1997. <http://www.homecomp./com/homecomp/html> (19 Mar. 1997).

Vint, Barbara. "Re: Telecommuting at Pacific Software Systems." 21 Mar. 1997. Personal e-mail. (23 Mar. 1997).

✓ Revision Checklist

- ❑ Gave full and proper credit to sources consulted and those used for the preparation of my work.
- ❑ Recorded all direct quotations accurately.
- ❑ Paraphrased information correctly and acknowledged rightful sources.
- ❑ Double-checked spelling of authors' and publishers' names and accuracy of all pertinent publication information.
- ❑ Followed MLA or APA documentation method consistently in preparing Works Cited or References pages.
- ❑ Included all necessary in-text (parenthetical) references; cited each parenthetical reference in Works Cited list or References pages.
- ❑ Indicated clearly which work I used when sources included more than one work by same author.
- ❑ Included only the works referred to in my paper in Works Cited or References pages.
- ❑ Cited all CD-ROM and on-line database material completely and consistently.
- ❑ Alphabetized Works Cited and References entries correctly.
- ❑ Documented all Internet sources properly, giving both posting and access dates.

Exercises

1. Ask a professor in your major what he or she regards as the most widely respected periodical in your field. Find a copy of this periodical and explain its method of documentation (providing examples). How does it differ from the MLA method?

2. Put the following pieces of bibliographic information in proper form according to the MLA method of documentation for Works Cited.
   - **a.** New York, Hawthorn Publishing Company, John Anderson, 1997, pages 95–97, second edition, *A New Way to Process Film.*
   - **b.** Margaret Mannix, Amy Bernstein, Mary Kathleen Flynn. *U.S. News & World Report,* pages 58–60; "The Internet: Best Web Sites for Women." 1 July 1996; located on InfoTrac or ProQuest on 3 Sept. 1996.
   - **c.** *Hartford Messenger;* 6 October 1997; page 3; Business NewsBank. CD-ROM; November 1997; Petula D'Orsay-Baldwin: "Stock Levels Rise After Global Meeting." Accessed December 1998.

d. *American Journal of Nursing,* Karen E. Forbes and Shirlee A. Stokes, July 1984, pages 884–88, "Saving the Diabetic Foot." Vol. 84.
e. *Air Quality Management,* the Environmental Protection Agency, Office of Air and Radiation, the Government Printing Office, available after November 1991, free of charge, page 8.
f. On-line. Posted 29 Mar. 1998. Home page. Pacific Technologies, Inc. "Buying the Smart Modem This Year." Internet. Accessed 18 Apr. 1998. http://pacifictech.com
g. *HealthWatch: The Online Journal for Health Conscious Internet Users;* 16 par. On-line. 5 June 1997; available AOL; "Carpal Tunnel Syndrome: Early Warning Signs." Hattie Grandville and Paul Hall. 2.4. 1997. Prodigy. Internet.
h. ProQuest. ABI/Inform. Stephanie Stahl. *Informationweek.* Mar. 11, 1996. p. 89. "Groupware for Exchange Server."
i. horvath@spot.colorado.edu. "Re: Preparation of Computer Graphics." E-mail to Gayle Lowery, Regional Supervisor. 3 May 1998. Joan Horvath, sender.
j. *Academic American Encyclopedia.* "Computers." Compuserve. 9 Oct. 1995. Posted 8 Aug. 1995.
k. *Global Encyclopedia.* Posted 8 November 1996. Internet. "Olympics." http://www.halcyon.com/jensen/encyclopedia/
l. An interview with your local police chief that took place in the college auditorium after she delivered a talk on crime prevention on Wednesday, 23 January 1998. Her name is Tina B. Holmes.
m. Today's editorial in your local newspaper.
n. Pyramid Films, Inc., *Pulse of Life,* 1996, Santa Monica, California, order number 342br.
o. *Essays on Food Sanitation,* John Smith (editor), Framingham, Massachusetts, 3rd edition, pages 345–356, Mary Grossart (author), Albion Publishing Company, "Selection of Effective Chemical Agents," 1997.
p. Today's editorial in *Figaro,* a leading newspaper in Paris.
q. January to February of 1998, pages 11–13, "Light Sensors and Smoke Detectors," Mathis, Patricia, *Fire Science Quarterly.*
r. 7th ed. of *McGraw-Hill Encyclopedia of Science & Technology,* "Lasers."
s. "Fax Boxes Keep Your Laser Printer Busy," page 127, Stan Miastkowski, vol. 16, *Byte,* Feb. 1991.
t. *Pro-Cite* (Personal Bibliographic Software), version 3.4, by Victor Rosenberg et al. Copyright 1996. For IBM-AT with DOS 2.0 or higher. Disk format.

3. Put the bibliographic references you listed in MLA format in Exercise 2 into APA format.

4. Convert the bibliographic information given for one article in the periodical you chose for Exercise 1 to the MLA parenthetical style.

5. The following passage contains mistakes in the MLA method of documentation. Find these mistakes and explain how to correct them.

More and more companies are allowing employees to "telecommute" (see Smith; Dawson; Brown; Gura and Keith; and Allen). One expert defines telecommuting as "home-based work" (13). Having terminals in their homes "allows employees to work at a variety of jobs" ("New Employment Opportunities"). It has been estimated that currently 900,000 employees work out of their homes (Pennington, p. 56). That number is sure to increase as computer-based businesses multiply in the late 1990s (Brown). In one of her recent articles on telecommuting, Holcomb (167) found that "in the last year alone 43 companies in the metropolitan Phoenix area made this option available to their employees."

Employees who telecommute cite a variety of benefits for such an arrangement (see in particular articles by Gura, Smith, and Kaplan). One employee of a mail order company whose opinion was quoted observed that "I can save about 15-17 hours a week in driving time" (from Allen). Working at home allows the telecommuting employee to work at his or her optimum times ("The Day Does Not Have to Start at 9:00 A.M."). Also, in articles by Kaplan and Keith the benefits of not having to leave home are emphasized: "A telecommuting parent does not have to worry about child care" (39). Telecommuting may "be here to stay" (quoted in a number of different Web site sources).

6. Submit your preliminary list of references (your tentative Works Cited page) for a research paper or long report to your instructor.

Summarizing Material

A summary is a brief restatement of the main points in a book, report, article, laboratory test, meeting, or convention. A summary saves readers hours of time because they do not have to study the original or attend a conference. A summary can reduce a report or article by 85 to 95 percent (or even more) or capture the essential points of a three-day convention in a one-page memo. Moreover, a summary can tell readers whether they should even be concerned about the original; it may be irrelevant for their purposes. Finally, since only the most important points of a work are included in a summary, readers will know they are given the crucial information they need.

The Importance of Summaries

Summaries can be found all around you. Television and radio stations regularly air two-minute broadcasts—sometimes called "newsbreaks"—to summarize in a few sentences the major stories covered in more detail on the evening news. Popular news magazines such as *TIME, Newsweek,* and *U.S. News & World Report* have a large readership because of their ability to condense seven days of news into short, readable articles highlighting key personalities, events, and issues.

Newspapers also employ summaries for their readers' convenience. Daily newspapers such as the *Wall Street Journal* or weekly papers such as *Barron's* or the *National Law Journal* print on the first page brief summaries of news stories that are discussed in detail elsewhere in the paper. Some newspapers simply print a column entitled "News Summary" on the first or second page of an issue to condense major news stories. *Facts On File's News Digest* comprehensively summarizes world news every week. Figure 11.1 illustrates a summary that appeared in *Facts On File* about copyright status for computer-colored films. Note how a few paragraphs capture and emphasize the most significant data that may have taken weeks and numerous reports or stories to record. The *Reader's Digest,* one of the most widely read publications in America, is devoted largely to condensing articles, stories, and even books while still preserving the essential message, flavor, and wording of the original.

FIGURE 11.1 A summary of the copyright status of colorized films.

Copyrights Approved for "Colorization." Computer-colored films would be granted copyright status if those tinted versions revealed a certain degree of human creativity and were produced by existing computer-coloring technology, the U.S. Copyright Office of the Library of Congress ruled June 19.

The ruling came after months of debate surrounding the "colorization" process, a debate that had reached all the way to the U.S. Senate.

The copyright office ruled that computer-colored versions of black-and-white films were entitled to copyright protection as "derivative works." Such protection gave a studio the right, generally for 75 years, to distribute a colored version of a movie through broadcast and cable-TV outlets and to sell videotape copies. The studio could also collect damages from companies that copied the colored films.

Films would not be eligible for copyright protection if the tinting "consists of the addition of only a relatively few number of colors to an existing black-and-white motion picture."

TECH NOTE

Summaries are widely used at Internet Web sites to encourage users to go further into the subject. Each home page is in essence a summary of the various links to which it is connected. Summaries are also a vital part of the search strategies on the Net. When search engines retrieve positive "hits," a short summary accompanies each citation to help users determine whether the material is relevant to their needs.

As a student, you probably already know how critical it is to be able to write summaries when you gather and record research. Summaries are crucial in highlighting key ideas from material that students have to include, and frequently evaluate, in research papers.

On the job, writing summaries for employers or co-workers is a regular and important responsibility. Each profession has its own special needs for summaries.

Chapter 16 discusses a variety of reports—progress, sales, periodic, trip, test, and incident—whose effectiveness depends on a faithful summary of events. You may be asked to summarize a business trip lasting one week into one or two pages for your company or agency. A busy manager may ask you to read and condense a ninety-page report so that she will have a knowledgeable overview of its contents. Acute-care nurses must write a one- or two-page discharge summary for patients who are being referred to another agency (home health, nursing home, rehabilitation center). These nurses must read the patient's record carefully and summarize what has happened to the patient since admission to the hospital—surgeries, treatments, diagnoses, and prognoses, and indicate necessary follow-up treatments (medications, office visits, outpatient care, X-rays).

You may have to write summaries for individuals outside your company or agency to inform them about the work your employer is doing. Public relations news releases or newsletters, the subject of Chapter 8, are written for the news media to give them the most up-to-date information about important events at your company or agency. Figure 11.2 contains a summary of ongoing research being done at the National Institute of Standards and Technology, an agency of the U.S. Commerce Department's Technology Administration. This summary, written for editors and others outside the agency, presents basic information on the objectives, methods, findings, and funding of important research on refrigerants that can help protect our environment.

FIGURE 11.2 A news release summarizing a research investigation written for individuals outside the company/agency.

NEWS RELEASE

Media Contact: Jan Kosko, 301 / 975-2767

Two Mixtures Could Replace Banned Refrigerant

NIST researchers say two refrigerant mixtures appear promising as environmentally safe replacements for R22, a refrigerant widely used in residential heat pumps. The Clean Air Act of 1990 calls for hydrochlorofluorocarbons (HCFCs), such as R22, to be phased out starting in 2015. HCFCs belong to a family of chemicals believed to be damaging to the Earth's atmosphere. The two mixtures, named R32/R134a and R32/R152a, ***do not contain chlorine or bromine,*** the two main catalysts some believe are destroying the Earth's ozone layer. The researchers used a NIST-developed computer simulation program, called CYCLE11, and a laboratory version of a heat pump to examine how the mixtures would perform in the machine. The NIST study showed that the two mixtures could perform up to 15 percent better than R22. NIST is currently conducting flammability testing to evaluate the one possible drawback, the fact that both mixtures contain at least one flammable component. This research is being funded primarily by the Electric Power Research Institute with some additional support from the U.S. Environmental Protection Agency.

NIST Update May 11, 1992: 1–2.

Contents of a Summary

The chief problem in writing a summary is deciding what to include and what to omit. As we have just seen, a summary, after all, is a much abbreviated version of the original; it is a streamlined review of *only* the most significant points. You will not help your readers save time by simply rephrasing large sections of the original and calling the new version a summary. That will simply supply readers with another report, not a summary.

Make your summary lean and useful by briefly telling readers about the main points: the purpose, scope, conclusions, and recommendations. A summary should concisely answer the readers' two most important questions.

1. What findings does the report or meeting offer?
2. How do these findings apply to my business, research, or job?

Note how the summary in Figure 11.2 successfully answers these questions by indicating why R32/R134a and R32/R152a are safer, how they can be used, and who has funded the research.

How long should a summary be? While it is hard to set down precise limits about length, effective summaries are generally 5 to 15 percent of the length of the original. The complexity of the material being summarized and your audience's exact needs can help you to determine an appropriate length. To help you know what is most important for your summary, the following suggestions will guide you on what to include and what to omit.

What to Include in a Summary

1. **Purpose.** A summary should indicate why the article or report was written or why a convention or meeting was held. (Often a report is written or a meeting is called to solve a problem or to explore new areas of interest.) Your summary should give the reader a brief introduction (even one sentence will do) indicating the main purpose of the report or conference.

2. **Essential specifics.** Include only the names, costs, codes, places, or dates essential to understanding the original. To summarize a public law, for example, you need to include the law number, the date it was signed into law, and the name(s) of litigants.

3. **Conclusions or results.** Emphasize what was the final vote, the result of the tests, and the proposed solution to the problem.

4. **Recommendations or implications.** Readers will be concerned especially with important recommendations—what they are, when they can be carried out, and why they are necessary.

What to Omit from a Summary

1. **Opinion.** Avoid injecting opinions—your own, the author's, or a speaker's. You distract readers from grasping main points by saying that the report was too long or that it missed the main point, or that a salesperson from Detroit monopo-

lized the meetings, or that the author in a digression took the Land Commission to task for failing to act properly. A later section of this chapter will deal with evaluative summaries.

2. New data. Stick to the original article, report, book, or meeting. Avoid introducing comparisons with other works or conferences, because readers will expect a digest of only the material being summarized.

3. Irrelevant specifics. Do not include any biographical details about the author of an article. Although many journals contain a section entitled "notes on contributors," this information plays no role in the reader's understanding of your summary.

4. Examples. Illustrations, explanations, and descriptions are unnecessary in a summary. Readers must know outcomes, results, and recommendations, not the illustrative details supporting or elaborating on those results.

5. Background. Material in introductions to articles, reports, and conferences can usually be excluded from a summary. These "lead-ins" prepare the reader for a discussion of the subject by presenting background information, anecdotes, and details that will be of little interest to readers who want a summary to give them the big picture.

6. Reference data. Exclude information found in footnotes, bibliographies, appendixes, tables, or graphs. All such information supports rather than expresses conclusions and recommendations.

7. Jargon. Technical definitions or jargon in the original may confuse rather than clarify the essential information for the general reader.

The Process of Preparing a Summary

To write an effective summary, you need to proceed through a series of steps. Basically, you will have to read the material very carefully, making sure that you understand it thoroughly. Then you will have to identify the major points and put aside everything else. Finally, you will have to put the essence of the material into your own words. The process of writing a summary demands an organized plan. Follow these steps to prepare your summary effectively.

TECH NOTE

Your word-processing program can help you in a number of ways as you prepare your summary. You might download the material (article, report, technical paper) you need to summarize and then as you read through it the first time on your screen you can cut out any extraneous, nonessential material. It is easier to cut out what you don't want this first time through than it is to select exactly what you do want. To pare down material, simply select what you don't want by running your cursor across the material or, to eliminate an entire paragraph or paragraphs, down the side of the material; then press "delete."

Another way to use your computer when you have to summarize is to highlight the key points as you read through the material you have downloaded. Highlighting during the first pass through the material will guide you on your second reading as you attempt to include only relevant material for your summary.

You can also take a notebook with you to a conference or talk and key in the most important points as you listen to the speakers.

1. **Read the material once in its entirety to get an overall impression of what it is about.** Become familiar with large issues, such as the purpose and organization of the work, and the audience for whom it was written. Look at visual cues—headings, subheadings, words in italic or boldface type, notes in the margin—that will help you to classify main ideas and summarize the work. Also see whether the author has included any mini-summaries within the article or report or whether there is a concluding summary at the end.

2. **Reread the material.** Read it a second time or more often if necessary. To locate all and only the main points, underline them in the work; or, if the work is a book or an article in a journal that belongs to your library, you may find it easier to work with a photocopy so that you can underline. To spot the main points, pay attention to the key transitional words, which often fall into predictable categories.

- *Words that enumerate:* first, second, third, initially, subsequently, finally, next, another
- *Words that express causation:* accordingly, as a result, because, consequently, therefore, thus
- *Words that express contrasts and comparisons:* although, by the same token, despite, different from, furthermore, however, in contrast, in comparison, in addition, less than, likewise, more than, more readily, not only . . . but also, on the other hand, the same is true for, similar, unlike
- *Words that signal essentials:* basically, best, central, crucial, foremost, fundamental, indispensable, in general, important, leading, major, obviously, principal, significant

Pay special attention to the first and last sentences of each paragraph. Often the first sentence of a paragraph contains the topic sentence, and the last sentence summarizes the paragraph or provides some type of transition to the next paragraph.

You also have to be alert for words signaling information you do *not* want to include in your summary, such as the following:

- *Words announcing opinion or inconclusive findings:* from my personal experience, I feel, I admit, in my opinion, might possibly show, perhaps, personally, may sometimes result in, has little idea about, questionable, presumably, subject to change, open to interpretation
- *Words pointing out examples or explanations:* as noted in, as shown by, circumstances include, explained by, for example, for instance, illustrated by, in terms of, learned through, represented by, such as, specifically in, stated in

3. Collect your underlined material or notes and organize the information into a draft summary. At this stage do not be concerned about how your sentences read. Use the language of the original, together with any necessary connective words or phrases of your own. Keyboard the draft into your computer. You will probably have more material here than will appear in the final version. Do not worry; you are engaged in a process of selection and elimination. Your purpose at this stage is to extract the principal ideas from the examples, explanations, and opinions surrounding them.

4. Read through and revise your draft(s) and delete whatever information you can. As you revise, see how many of your underlined points can be condensed, combined, or eliminated. You may find that you have repeated a point. Check your draft against the original for accuracy and importance. Make sure that you are faithful to the original by preserving its emphases and sequence. Using a word processor will make it easy for you to make changes and deletions without rekeying the entire summary.

5. Now put the revised version into your own words. Again, make sure that your reworded summary has eliminated nonessential words. Connect your sentences with words that show relationships between ideas in the original (*also, although, because, consequently, however, nevertheless, since*). Compare this version of your summary with the original material to double-check your facts.

6. Do not include remarks that repeatedly call attention to the fact that you are writing a summary. You may want to indicate initially that you are providing a summary, but avoid such remarks as "The author of this article states that water pollution is a major problem in Baytown"; "On page 13 of the article three examples, not discussed here, are found."

7. Edit your summary to make sure it reads clearly and concisely. Check to be certain it is coherent, too. Tell the reader how one point flows into another. Also proofread your summary carefully.

8. Identify the source you have just summarized. Do this by including pertinent bibliographic information in the title of your summary or in a footnote or endnote. This gives proper credit to the original source and informs your readers where they can find the complete text if they want more details.

Figure 11.3, a 2,500-word article entitled "Virtual Reality: The Future of Law Enforcement Training," appeared in the *FBI Law Enforcement Bulletin* and hence would be of primary interest to individuals in law enforcement administration.

FIGURE 11.3 The original article with important points underscored for use in a summary.

Virtual Reality

The Future of Law Enforcement Training

By
JEFFREY S. HORMANN

Delete scenario–example of background; an opener

A late night police pursuit of a suspected drunk driver winds through abandoned city streets. The short vehicle chase ends in a warehouse district where the suspect abandons his vehicle and continues his flight on foot. Before backup arrives, the rookie patrol officer exits his vehicle and gives chase. A quick run along a loading dock ends at the open door to an apparently unoccupied building. The suspect stops, brandishes a revolver, and fires in the direction of the pursuing officer before disappearing into the building. The officer, shaken but uninjured, radios in his location and follows the suspect into the building.

Include important observation

Did the officer make a good decision? Probably not by most departments' standards. Whether the officer's decision proves right or wrong, the training gained from this experience is immeasurable, that is, provided the officer lives through it. Fortunately for this officer, the scenario occurred in a realistic, high-tech world called virtual reality, where training can have a real-life impact without the accompanying risk.

TRADITIONAL TRAINING LIMITATIONS

Important distinction

Experience may be the best teacher, but in real life, police officers may not get a chance to learn from their mistakes. To survive, they must receive training that prepares them for most situations they might encounter on the street. However, because many training programs emphasize repetition to produce desired behaviors, they may not achieve the intended results, especially after students leave the training environment. Thus, the more realistic the training, the greater the lessons learned.

Delete explanation and example

Include significant qualification

Additionally, even some in law enforcement may fall prey to the effects of what has come to be termed "The MTV Generation."[1] As products of this generation, today's young officers purportedly have short attention spans requiring new, nontraditional training methods. The key to teaching this new breed is to provide fast-paced, attention-getting instruction that is clear, concise, and relevant.[2]

Continued

FIGURE 11.3 (Continued)

TRAINING WITH VIRTUAL REALITY

Emphasize author's main point

Virtual reality can provide the type of training that today's law enforcement officers need. By completely immersing the senses in a computer-generated environment, the artificial world becomes reality to users and greatly enhances their training experiences.

Important reason for its neglect by law enforcement

Although considerable research and development have been conducted in this field, only a limited amount has applied directly to law enforcement. The apparent reason simply is that, for the most part, law enforcement has not asked for it.

Restatement of main point above; note parallel items with key words signaling important applications

Because virtual reality technology is relatively new, most law enforcement administrators know little about it. They know even less about what it can do for their agencies. By understanding what virtual reality is, how it works, and how it can benefit them, law enforcement administrators can become significantly involved in the development of this important new technology.

WHAT IS VIRTUAL REALITY?

Include definition

Simply stated, virtual reality is high-tech illusion. It is a computer-generated, three-dimensional environment that engulfs the senses of sight, sound, and touch. Once entered, it becomes reality to the user.

Important explanation

Within this virtual world, users travel among, and interact with, objects that are wholly the products of a computer or representations of other participants in the same environment. Thus the limits of this virtual environment depend on the sophistication and capabilities of the computer and the software that drives the system.

HOW DOES VIRTUAL REALITY WORK?

Significant phrase

Delete specific pieces of equipment

Based on data entered by programmers, computers create virtual environments by generating three-dimensional images. Users usually view these images through a head-mounted device, which, for instance, can be a helmet, goggles, or other apparatus that restricts their vision to two small video monitors, one in front of each eye. Each monitor displays a slightly different view of the environment, which gives users a sense of depth.

Delete example

Another device, called a position tracker, monitors users' physical positions and provides input to the computer. This information instructs the computer to change the environment based upon users' actions. For example, when users look over their shoulders, they see what lies behind them.

Continued

FIGURE 11.3 (Continued)

Because virtual reality users remain stationary, they use a joy stick or trackball to move through the virtual environment. Users also may wear a special glove or use other devices to manipulate objects within the virtual environment. Similarly, they can employ virtual weapons to confront virtual aggressors.

To enhance the sense of reality, some researchers are also experimenting with tactile feedback devices (TFDs). TFDs transmit pressure, force, or vibration, providing users with a simulated sense of touch.[3] For example, a user might want to open a door or move an object, which in reality, would require the sense of touch. A TFD would simulate this sensation. At present, however, it is important to remember that these devices are crude and somewhat cumbersome to use.

Delete further examples

Major conclusion

USES FOR VIRTUAL REALITY

In today's competitive business environment, organizations continuously strive to accomplish tasks faster, better, and inexpensively. This especially holds true in training.

Major value to audience of administrators

Delete example

Virtual reality is emerging rapidly as a potentially unlimited method for providing realistic, safe, and cost-effective training. For example, a firefighter can battle the flames of a virtual burning building. A police officer can struggle with virtual shoot/don't shoot dilemmas.[4]

Emphasize significant advantages in training situations

Within a virtual environment, students can make decisions and act upon them without risk to themselves or others. By the same token, instructors can critique students' actions, enabling students to review and learn from their mistakes. This ability gives virtual reality a great advantage over most conventional training methods.

Use only main points relevant to target audience of law enforcement administrators

The Department of Defense (DOD) leads public and private industry in developing virtual reality training. Since the early 1980s, DOD has actively researched, developed, and implemented virtual reality to train members of the armed forces to fight effectively in combat.

DOD's current approach to virtual reality training emphasizes team tactics. Groups of military personnel from around the world engage in combat safely on a virtual battlefield. Combatants never come together physically; rather, simulators located at various sites throughout the world transmit data to a central location, where the virtual battle is controlled. Basically, it costs less to move information than people. Consequently this form of training has proven quite cost-effective.

Continued

FIGURE 11.3 (Continued)

An additional benefit to this type of training is that battles can be fought under varying conditions.

Virtual battlefields re-create real-world locations with interchangeable characteristics. To explore "what if" scenarios, participants can modify enemy capabilities, terrain, weather, and weapon systems.

Note main military advantage

Delete examples

Virtual reality also can re-create actual battles. Based on information from participants, the Institute for Defense Analyses re-created the 2nd Armored Cavalry Regiment Offensive conducted in Iraq during Operation Desert Storm. The success of the virtual re-creation became apparent when, upon viewing the simulations, soldiers who had fought in the actual battle reported the extreme accuracy of the event's depiction and the feeling of reliving the battle.[5]

Major conclusion signaled by key word clearly

Clearly, virtual reality holds great potential for accurate review and analysis of real-world situations, which would be difficult to accomplish by any other method.

Omit example

Preliminary studies, for instance, show that military units perform better following virtual reality training.[6] Even though virtual environments are only simulations, the complete immersion of the senses literally overwhelms users, totally engrossing them in the action.

Note key word "major"; idea relevant to law enforcement

This realism presumably plays a major role in the program's success and likely will prove positive in future endeavors.

Omit military application

In fact, due to its success in training multiple participants in group combat situations, DOD plans to train infantry personnel individually with virtual reality fighting skill simulators.[7]

LAW ENFORCEMENT TRAINING

Key point

While virtual reality has proven its value as a training and planning tool for the military, applications for this technology reach far beyond DOD.

Major parallel points

In varying but key ways, many military uses can transfer to law enforcement, including training in firearms, stealth tactics, and assault skills.

Delete statistics

Unfortunately, few organizations have dedicated resources to developing virtual reality for law enforcement. According to a recently published resource guide, more than 100 companies currently are developing and/or selling virtual reality hardware or software. However, none of these firms mentioned law enforcement uses.[8]

Subordinate idea

Further, a review of relevant literature revealed numerous articles on virtual reality technology, but only a few addressed law enforcement applications. Yet, virtual real-

Continued

FIGURE 11.3 (Continued)

Restatement of major point

ity clearly offers law enforcement benefits in a number of areas, including pursuit driving, firearms training, high-risk incident management, incident re-creation, and crime scene processing.

PURSUIT DRIVING

Include application but omit example

Pursuit driving represents one area in which virtual reality application has become reality for law enforcement. Law enforcement personnel identified a need and provided input to a well-known private corporation that developed a driving simulator equipped with realistic controls.

Delete specific mechanism and explanation of operation of screen mechanism

The simulator provides users with realistic steering wheel feedback, road feel, and other vehicle motions. The screen possesses a 225-degree field of view standard, with 360-degree coverage optional. As noted in demonstrations, simulations can involve one or more drivers, and environments can alternate between city streets, rural back roads, and oval tracks. The vehicle itself can change from a police car to a truck, ambulance, or a number of others.

Note cost efficiency again

Virtual reality driving simulators provide police departments invaluable training at a fraction of the long-term cost of using actual vehicles. In fact, the simulator is being used by a number of police departments around the country.

Include major advantage but exclude specific example

Note major distinction for training purpose

During the past year, for example, the Los Angeles County Sheriff's Office Emergency Vehicle Operations Center (EVOC) has used a four-station version of the driving simulator to train its officers. The simulators help students develop judgment and decision-making skills, while providing an environment free from risk of injury to students or damage to vehicles. Still, as the EVOC supervisor cautions, virtual reality training should complement, not replace, actual behind-the-wheel instruction.[9]

FIREARMS TRAINING

New subtopic; include advantage but delete examples

In another way, virtual reality could greatly enhance shoot/don't shoot training simulators currently in use, such as the Firearms Training System, a primarily two-dimensional approach that possesses limited interactive capabilities. A virtual reality system would allow officers to enter any three-dimensional environment alone or as a member of a team and confront computer-generated aggressors or other virtual reality users.

Include significant points on advantages

Evaluators could specifically observe the training from any perspective, including that of the officers or the

Continued

FIGURE 11.3 (Continued)

next three reasons to use virtual reality signaled by key words in addition, also, and likewise

Training scenarios could involve actual building floor plans or local city streets, and criteria such as weather, number of participants, or types of weapons could be altered easily.

HIGH RISK INCIDENT MANAGEMENT

Delete examples

In addition to weapons training, virtual reality could prove invaluable for SWAT team members before high-risk tactical assaults. Floor plans and other known facts about a structure or area could be entered into a computer to create a virtual environment for commanders and team members to analyze prior to action.

INCIDENT RE-CREATION

Law enforcement agencies could also collect data from victims, witnesses, suspects, and crime scenes to recreate traffic accidents, shootings, or other crimes. The virtual environment created from the data could be used to refresh the memories of victims and witnesses, to solve crimes, and ultimately, to prosecute offenders.

CRIME SCENE PROCESSING

Virtual reality crime scenes could likewise be used to train both detectives and patrol officers. First, students could search the site and retrieve and analyze evidence without ever leaving the station. Then, actual crime scenes could be re-created to add realism to training or to evaluate prior police actions.

IS VIRTUAL REALITY VIRTUALLY PERFECT?

Crucial qualification and justification for using virtual reality in law enforcement training

Though virtual reality may appear to be the ideal law enforcement tool, as with any new technology, some drawbacks exist. Currently, areas of concern range from cumbersome equipment to negative physical and psychological effects experienced by some users. Fortunately, however, the field is evolving and improving constantly, and as virtual reality gains widespread use, most major concerns should be dispelled.

PHYSICAL LIMITATIONS AND EFFECTS

Delete explanations of limitations/effects

Because computers currently are not fast enough to process large amounts of graphic information in real time, some observers describe virtual environments as "slow-moving."[10] The human eye can process images at a much faster rate than a computer can generate them. In a virtual environment, frames are displayed at a rate of about 7 per second, an extremely slow speed when compared to television, which generates 60 frames per second.[11] Users find the resulting choppy or slow graphics less than appealing.

Adapted from: *FBI Law Enforcement Bulletin* 64, no. 7 (July 1995): 7–12.

FIGURE 11.4 A working draft summary of the "Virtual Reality" article in Figure 11.3.

Law enforcement officers put their lives on the line every day, yet their training does not fully allow them to anticipate what they will find on the streets. Virtual reality will give them realistic, high-tech benefits of encountering criminals without any risks. Traditional training methods, which work through repetition, cannot equal the advantages of virtual reality when it comes to teaching officers the lessons they must learn to survive in the field. This new breed of officers is demanding the attention-getting, highly realistic training that virtual reality affords them. Virtual reality translates the artificial world of the computer into the real world. Yet even though much research has been done on virtual reality, it is new to law enforcement officials. Moreover, manufacturers have not marketed their technology to them. It is essential that these administrators know how virtual reality works and what it can do for them. Virtual reality has been defined as high-tech illusion through the computer user's interaction with the real world. Working through sophisticated software, virtual reality gives users a three-dimensional (hearing, feeling, and seeing) view of the things and people around them. Virtual reality requires specific equipment including goggles/headsets, a tracker, a trackball, and special gloves. But these devices do have problems; at present, they are crude and can be cumbersome. Even so, virtual reality provides cost-effective and life-saving benefits for law enforcement administrators. Thanks to this technology, students will be able to make quicker and better decisions in the field. Virtual reality has already been tried by the Department of Defense; the armed forces have used it to re-create battlefield conditions, helping the troops better understand the enemy and

Continued

FIGURE 11.4 (Continued)

its position. Yet virtual reality holds great appeal for other real-world applications, especially law enforcement. Unfortunately, the 100 companies that manufacture virtual reality equipment have neglected these law enforcement applications. Yet virtual reality easily accommodates law enforcement instruction. Driving simulators help officers prepare for high-speed chases. In Los Angeles County, such simulators complement more traditional training. Virtual reality can help officers in a variety of training missions—firearms, high-risk incidents, re-creating crimes, understanding the crime scene. Using virtual reality, officers never have to leave the station. Admittedly, virtual reality has drawbacks, but as this new technology improves, users should face fewer problems.

Assume that you are asked to write a summary of this article for your boss, a police chief in a medium-size city who might be interested in incorporating virtual reality into the police academy training program. By following the steps outlined previously, you would first read the article carefully two or three times, underscoring or highlighting the most important points, signaled by key words. Note what has been underscored in the article. Also study the comments in the margins; these explain why certain information is to be included or excluded from the summary.

After you have identified the main points, extract them from the article and, still using the language of the article, join them into a coherent working draft summary as in Figure 11.4. This working draft then has to be shortened and rewritten in your own words to produce the compact final version of your summary, as in Figure 11.5. Only 162 words long, this summary is 12 percent of the length of the original article and records only major conclusions relevant to the audience for the article.

To further understand the effectiveness of the summary in Figure 11.5, review the wordy and misleading summary of the same article in Figure 11.6. The latter not only is too long but also dwells on minor details at the expense of major points. It includes unnecessary examples, statistics, and names; it even adds new information while ignoring crucial points about the applications of virtual reality to law enforcement officials. But even more serious, this summary distorts the meaning and

FIGURE 11.5 A final, effective summary of Figure 11.3.

Virtual reality offers benefits for law enforcement training that traditional methods cannot provide. This computer-generated technology simulates and recreates real-life crime scenes without placing officers at risk. Thanks to virtual reality's three-dimensional world of sight, sound, and touch, officers enter the criminals' world to gain invaluable experience interacting with them. Because virtual reality is new and has not been marketed for law enforcement use, administrators may not know about it. Yet it provides a cost-effective, realistic way to enhance training programs. The applications of virtual reality far exceed its military use of simulating battlefield conditions. Virtual reality allows administrators to give trainees hands-on experience in pursuit driving, firearms training, SWAT team assaults, incident re-creation, and crime location processing. Officers can investigate a crime without ever leaving the station. Although virtual reality is an emerging technology with limitations, it is quickly improving and rapidly expanding. Administrators need to incorporate it into their curriculum to give officers field-translatable experiences.

the intention of the original article. The reader concludes that the article says virtual reality is not very valuable for law enforcement administrators and officers—just the opposite of the point the article makes. You can avoid such mistakes by deemphasizing minor points, by making sure that all parts of your summary agree with the original, and by not letting your own opinions distort the message of the original article.

Evaluative Summaries

To write an evaluative summary, also called a **critique,** follow all the guidelines previously discussed in this chapter except one. The one major exception is that an evaluative summary includes your *opinion,* or view, of the material.

Your instructors and employers may often ask you to summarize and assess what you have read. In school you may have to write a book report or compile

FIGURE 11.6 A misleading summary of Figure 11.3.

Nonessential introductory material	A rookie police officer makes many mistakes in pursuing subjects. Training can cover many realistic situations, but young officers in the MTV Generation have short attention spans.
Distorts article which advocates the benefits of virtual reality for law enforcement	Given the research so far on virtual reality, it holds little promise for law enforcement use.
Dwells on specific virtual reality equipment at the expense of the main advantages	Virtual reality has too many limitations, but it works interestingly through gloves, helmets, and goggles, and with a position tracker users can see over their shoulders. It even has a joy stick (like those in an amusement park) and a crude device—a TFD—that simulates touch (nice to have in a horror movie).
Overlooks the significance of virtual reality for trainees	Instructors can gain much from virtual reality because they can better criticize their trainees.
Reverses chronology of events; misrepresents the role of virtual reality	In the early 1980s, the DOD used virtual reality to duplicate battlefield conditions.
How relevant to law enforcement? Delete unit's name	The 2nd Armored Cavalry Regiment Offensive won the Iraqi War because of virtual reality.
Again, distorts intention of article	But companies manufacturing virtual reality technology are not interested in law enforcement applications, another indication of its limitations.
Delete specifications	The Los Angeles Sheriff's EVOC used a driving simulator--offering a 225-degree field of view but it can be ordered with a 360-degree field--but expressed their caution about it.
One-sided; omits success of simulation	
Focuses on limits rather than usefulness	There have been limited interactions in the use of virtual reality for firearms training,
Fails to attribute these benefits to virtual reality	though floor plans might have helped SWAT teams. Witnesses may need to refresh their memories with virtual reality.
Does not subordinate flaws	Again drawbacks exist. Computers are not as fast as the human eye in processing information.

a critical, annotated bibliography commenting on the usefulness of the material you found in those sources. On the job your employer may ask you to condense a report and judge the merits of that report, paying special attention to whether the report's recommendations should be followed, modified, or ignored. Your company or agency may also ask you to write short evaluative summaries of job applications or sales proposals it has received or of conferences you have attended.

Characteristics of a Successful Evaluative Summary

In order to write an effective evaluative summary, make sure that you

- keep the summary short—not more than 5 to 10 percent of the length of the original
- blend your evaluation with your summary; do not save your evaluations until the end of the summary
- place your evaluation next to the summary of the points to which it applies so readers will see your remarks in context
- include a pertinent quotation from the original to emphasize your recommendation
- comment on both the content and style of the original

Here are some questions on content and style that you should answer for readers of your evaluative summary.

Evaluating the Content

1. **How carefully is the subject researched?** Is the material accurate and up-to-date? Are important details missing? Exactly what has the author left out? Where could the reader find the missing information? If the material is inaccurate, will the whole work be affected or just a part of it?

2. **Is the writer or speaker objective?** Are conclusions supported by evidence? Is the writer or speaker following a particular theory, program, or school of thought? Is this fact made clear in the source? Has the author or speaker emphasized one point at the expense of others? What are the writer's qualifications and background?

3. **Does the work achieve its goal?** Is the topic too large to be adequately discussed in a single talk, article, or report? Is the work sketchy? Are there digressions, tangents, or irrelevant materials? Do the recommendations make sense?

4. Is the material relevant to your audience? How would that audience use it? Is the entire work relevant or just part of it? Why? Would this work be useful for all members of your agency (or community) or only for those working in certain areas? Why? What answers offered by the work would help to solve a specific problem you or others have encountered on the job?

Evaluating the Style

1. Is the material readable? Is it well written and easy to follow? Does it contain helpful headings, careful summaries, and appropriate examples?

2. What kind of vocabulary does the writer or speaker use? Are there many technical terms or jargon? Is it written for the layperson? Is the language precise or vague? Would your audience have to skip certain sections that are too complicated?

3. What visuals are included? Charts? Graphs? Photographs? How are they used? Are they used effectively? Are there too many or too few of these visuals?

Figures 11.7, 11.8, and 11.9 contain evaluative summaries. Note how the writer's assessments are woven into the condensed version of the original. Figure 11.7 contains a student's opinions of an article summarized for a class in office

FIGURE 11.7 An evaluative summary of an article.

Pastor, Joan and Risa Gechtman, "Delegate to Get Ahead," The Secretary (Jan. 1992): 15-16.

According to this helpful article, office personnel can keep ahead of their workload by delegating responsibilities to subordinates, to peers, and to supervisors. Traditional delegation—to a subordinate—is most effective if the superior selects a staffer capable of performing the task and monitors progress with reasonable deadlines in mind. The authors list several other useful guidelines for successful "downward" delegation. Delegating to peers, a more complicated process, succeeds only if coworkers can see the benefits, to themselves and to the company, of sharing the workload. Workers can also encourage "sideways" delegation by showing appreciation to coworkers and offering them assistance in return. When supervisors overload assignments, subordinates may need to delegate "upwards." A worker can shift responsibility back to supervisors by diplomatically asking them to set priorities for the projects they have assigned. This readable and innovative article can help prevent workers from falling behind in their projects and, consequently, move their company forward.

FIGURE 11.8 A collaboratively written evaluative summary of a seminar.

Sabine County Hospital
Sabine, TX 77231
512-555-6734
FAX 512-555-6789

TO:	Marge Geberheart, M.S.N. Director of Nurses	SUBJECT:	Evaluation of Physical Assessment Seminar
FROM:	Judith Kim, R.N. JK Lee Schoppe, R.N. LS	DATE:	September 14, 1998

On September 7, Doris Gandy, R.N., and Rick Vargass, R.N., both on the staff of Houston Presbyterian Hospital, conducted a practical and beneficial seminar on physical assessment. The one-day seminar was divided into three units: (1) **Techniques of Health Assessment**, (2) **Assessment of Heart and Lungs**, and (3) **Assessment of the Abdomen**.

Techniques of Health Assessment

Four procedures used in physical assessment—inspection, percussion, palpation, auscultation—were defined and demonstrated. Return demonstrations, used throughout the seminar, meant we did not have to wait until we went back to work to practice our skills. The instructors also stressed the proper use of the stethoscope and the ways of taking a patient's medical history. We were also asked to take the medical history of the person next to us.

Assessment of the Heart and Lungs

After we inspected the chest externally, we discussed the proper placement of hands for percussion and palpation and the significance of various breath sounds. Using stethoscopes, our instructors helped us identify areas of the lung. We listened to and identified heart sounds. The film on examining the heart and lungs was ineffective since it included too much detailed information for seminar participants.

Continued

FIGURE 11.8 (Continued)

2.

Assessment of the Abdomen

The instructors warned that the order of examination of the abdomen differs from that of the chest cavity. Auscultation, not percussion, follows inspection so that bowel sounds are not activated. The instructors then clearly identified how to detect bowel sounds and how to locate the abdominal and palpate organs.

Recommendations

We strongly recommend a seminar like this for all nurses whose expanding role in the health care system requires more physical assessments. Although the seminar covered a wealth of information, the instructors admitted that they discussed only basics. In the future, however, it would be better to offer follow-up seminars on specific body systems (chest cavity, abdomen, central nervous system) instead of combining these topics because of the amount of information involved and the time required for demonstrations.

management procedures. Figure 11.8 is an evaluative summary in memo format collaboratively written by two employees who have just returned from a seminar. They have divided their labor, one writing the opening paragraph and the summary of "Techniques of Health Assessment" and the other writing the summaries of "Assessment of the Heart and Lungs" and "Assessment of the Abdomen." Together they drafted and revised the "Recommendations" and prepared the final copy of the memo.

Another kind of evaluative summary—a book review—is shown in Figure 11.9. Many journals and Web sites print book reviews to inform their professional audiences about the most recent studies in their field. Reviews condense and assess books, reports, government studies, tape cassettes, films, and other materials. The short review in Figure 11.9 comments briefly on usefulness to the intended audience and on style, provides clarifying information, and explains how the book is developed. For further examples, you may want to look at *Book Review Digest,* which prints excerpts from reviews in many areas.

A book review, or evaluative summary, includes the most important and useful—to a key audience—information about a book or report. Reviews are also important for the kinds of information that they *do not* include—details and irrelevant

FIGURE 11.9 A book review.

Book Reviews

By Richard Power

Protection and Security on the Information Superhighway

By Frederick B. Cohen
John Wiley & Sons (New York, NY), softcover, 301 pages, $24.95

The general populace's love affair with the Internet, coupled with increasingly daring escapades on the part of the electronic underground, has propelled the subject of information security into the mass media. The weekly news magazines and major newspapers have begun to cover (albeit rather ineptly) the saga of the cyberspace frontier. Meanwhile, losses from computer crime and telecommunications fraud have been rising for many years. And the most serious dimension of the threat—information warfare waged by rival corporations and governments competing for hegemony in the new global economy—has hardly been addressed at all.

In **Protection and Security on the Information Superhighway**, Cohen has delivered a sober, thorough and detailed study of the grim realities behind the hype of computer crime and information warfare.

Unlike some other books available on information warfare, this tome doesn't indulge in flights of imagination and cries of "Fire!" Instead, Cohen carefully documents the dangers to U.S. corporations, government agencies, financial institutions and other entities with real-world case studies and factual references. He also offers savvy evaluations of information security technologies and practical suggestions for developing comprehensive information security strategies.

To get up to speed on the current and future danger to the entire information infrastructure—from your home PC to the Pentagon War Room—read this book.

Computer Security Journal • Volume XI, Number 1, 1995

Computer Security Journal 11, no. 1 (1995).

(for the audience) information that would only clog a summary. For example, the review in Figure 11.9 indicates that the book documents the dangers of computer crime to corporations and governments but does not go into detail describing the security strategies recommended by the author.

Abstracts

The Differences Between a Summary and an Abstract

The terms *summary* and *abstract* are often used interchangeably, resulting in some confusion. This problem arises because there are two distinct types of abstracts:

descriptive abstracts and *informative abstracts.* The informative abstract is another name for a summary; the descriptive abstract is not. Why? An informative abstract (or summary) gives readers conclusions and indicates the results or causes. Look at the summary in Figure 11.5. It explains why virtual reality should be included in law enforcement training: because virtual reality gives officers field-translatable training. Informative abstracts are found at the beginning of long reports. Descriptive abstracts do not give conclusions.

All abstracts share two characteristics: the writer never uses "I" and avoids footnotes.

Writing the Informative Abstract

As a part of your course work or on your job you will probably have to write informative abstracts for long reports (see Chapter 17). One way to approach writing the abstract of a report is to think of it as a table of contents in sentence form. The table of contents is, in effect, the final outline; it is easily fleshed out into an abstract, as the following example shows. On the left is the table of contents, and on the right is the abstract written from that outline.

Table of Contents	*Abstract*
Need for Genetic Counseling Definition of Genetic Counseling Statistics on Genetic Counseling	Genetic counseling is a service for people with a history of hereditary disease. One in 17 births contains some defect; one-fourth of the patients in hospitals are victims of genetic diseases (including diabetes, mental retardation, and anemia). One of every 200 children born has chromosome abnormalities.
Purpose of Genetic Counseling	Genetic counseling offers parents an alternative to giving birth to children with genetic diseases and assistance for those with children already afflicted.
The Counseling Process Evaluating the Needs of the Counselees Taking a Family History Estimating the Risks Counseling the Family	The first step in counseling is to evaluate the needs of the parents. A family history is prepared and risks of future children being afflicted are evaluated. The life expectancy and possible methods of treatment of any afflicted child also can be determined. Alternatives are presented.
Determination of a Genetic Disorder Amniocentesis Karyotyping Fluorescent Banding Staining	Four prenatal tests are used to determine if a genetic disorder is present: amniocentesis, karyotpying, fluorescent banding, and staining.
Advantages of Genetic Screening Lower Cost Increased Availability	The development of these four relatively simple methods has lowered the cost of genetic counseling and increased its availability.

By permission of Professor Mary Scotto.

This system works only if your table of contents is neither too detailed nor too skimpy. Starting off with a good outline of an article or a report provides the best beginning for your abstract. Make sure your sentences are complete and grammatical. Do not omit verbs, conjunctions, and articles. Proper subordination is essential. You should expect to condense whole paragraphs of the original to a sentence, and individual sentences to a phrase, phrases to a single word.

Writing the Descriptive Abstract

A descriptive abstract is short, usually only a few sentences. Because it does not go into any detail or give conclusions, it is not actually a summary. A descriptive abstract provides information on what topics a work discusses, but not how or why they are discussed. Busy readers rely on a descriptive abstract to decide whether they want or need to consult the work itself. Here is a descriptive abstract of the article summarized in Figure 11.5 (p. 432).

> Virtual reality can be used to teach law enforcement officers firearms training, SWAT team assaults, incident re-creation, and crime location processing. This new training technology will be of interest to law enforcement administrators.

Figure 11.10 contains a group of descriptive abstracts about books of interest to professionals in public welfare. In a few words each abstract tells what kinds of information the books contain but does not reveal the solutions, plans, views, or recommendations that the authors of these books advance. A descriptive abstract is never a substitute for the report itself.

Where Abstracts Are Found

1. At the beginning of an article, company report, or conference proceedings; on a separate page; or on the title page of the report.
2. On the table-of-contents page of a magazine, briefly highlighting the features of the individual articles in that issue, or *at the beginning* of the chapter in a book or in advertisements for publications.
3. In reference works devoted exclusively to publishing collections of abstracts of recent and relevant works in a particular field (e.g., *Abstracts of Hospital Management Studies, Science Abstracts, Women's Studies Abstracts*). Figure 11.11 reproduces some descriptive abstracts from *Library & Information Science Abstracts.* Note that letters after each abstract refer to the initials of the individuals who prepared the abstract. See Chapter 9 for a discussion of the usefulness and scope of these types of abstracts.

✓ Revision Checklist

- ❑ Read and reread the original thoroughly to gain clear understanding of purpose of work.

FIGURE 11.10 Descriptive abstracts of books.

Adolescents in Foster Families
Edited by Jane Aldgate, Anthony Maluccio, and Christine Reeves. Chicago: Lyceum Books, 1989. 192 pp. $25.95; $14.95 paper.
Explores British and American perspectives on foster care, including foster parent assessment and training, strategies for successful placement, and preparation for independent living.

Children of Color: Psychological Interventions with Minority Youth
By Jewelle Taylor Gibbs, Larke Nahme Huang and Associates. San Francisco: Jossey-Bass, 1989. 423 pp. $27.95.
Presents comprehensive guidelines for the treatment of minority children and adolescents.

Housing Issues of the 1990s
Edited by Sara Rosenberry and Chester Hartman. New York: Praeger Publishers, 1989. 395 pp. $55.
Discusses national housing goals, populations with special housing needs, and public policies that would provide housing for those unable to acquire it on their own.

Lifestyles of the Elderly: Diversity in Relationships, Health, and Caregiving
Edited by Linda Ade-Ridder and Charles B. Hemon. New York: Plenum Publishing, 1989. 262 pp. $34.50.
Examines older people's diverse approaches to long-term marriage, family and friends, health maintenance, and caregiving.

Understanding Race, Ethnicity, and Power
By Elaine Pinderhughes. New York: Free Press, 1989. 269 pp. $24.95.
Examines the influence of racial and ethnic identity on the psychological and social dynamics of interactions between individuals with diverse backgrounds.

Public Welfare Vol. 4, no. 4 (Fall 1989).

- ❑ Used word-processing program to develop summary.
- ❑ Underlined key transitional words, main points, significant findings, applications, solutions, conclusions, recommendations.
- ❑ Separated main points clearly from (a) minor ones, (b) background information, (c) illustrations, and (d) inconclusive findings.
- ❑ Excluded examples, explanations, and statistics from summary or abstract.
- ❑ Deleted information not useful to audience—information too technical or irrelevant.
- ❑ Changed language of original to my own words so that I am not guilty of plagiarism.
- ❑ Made sure that emphasis of summary matches emphasis in original.

FIGURE 11.11 Descriptive abstracts of journal articles.

Library & Information Science Abstracts

5 May 1996

Abstracts 5464–5474

5464
Community networks: new frontiers, old values. K. G. Schneider. *American Libraries*, 27 (1) Jan 96, p. 96.
Increasingly, librarians in the USA have been using new technologies to develop or collaborate on community networks. These community networks are often created in collaboration with other agencies and advocacy groups, weaving libraries more tightly into the community organism. Gives examples which illustrate some of the roles and relationships libraries are carving out with respect to community networks. EB

5465
Talking to the people: making the most of Internet discussion groups. K. L. Robinson. *Online*, 20 (1) Jan/Feb 96, p. 26-32. refs.
Offers advice on making the best of the knowledge sharing aspects of electronic conferences or discussion groups on the Internet and reducing the stress that sometimes comes with e-conference membership. Explains the different types of e-conferences available, how they are organized, how to locate and select them, how to cope with mailing list overload, and suitable netiquette. SE

5466
Teaching electronic information literacy. C. Towney and D. A. Barclay. New York, Neal Schuman, 1996, 150 p. ISBN 1-55570-186-8.
Focuses on ways of teaching and training library users in the environment of electronic information, such as the Internet and electronic classrooms. LT

5473
The Internet and Eastern Europe. T. Konn. *Information World Review*, (109) Dec 95, p. 24-5. il.
Presents an overview of information on the countries of Central and Eastern Europe (CEE) accessible through the Internet, both from CEE servers and Western servers. Discusses news services, business information, legal and related information, and academic material. A list of key Web sites is included. JP

5474
Engineers on the Internet. S. Thomas. *Information World Review*, (109) Dec 95, p. 25-6.
Presents an overview of engineering resources on the Internet. Includes manufacturing; professional institute activities; commercial companies, industry areas and patents; indexes and search tools; and newsgroups and mailing lists. Provides a list of key engineering information sites. JP

Library & Information Science Abstracts, Volume 5, May 1996 Abstracts 4772–5953.

- ❑ Determined that sequence of information in summary follows sequence of original.
- ❑ Added necessary connective words that accurately convey relationships between main points in original.
- ❑ Edited to eliminate wordiness and repetition from summary.
- ❑ Cited source of original correctly and completely.
- ❑ Avoided phrases that draw attention to the fact that I am writing a summary or abstract.

- ❑ Summarized material objectively without adding commentary (for informative summary).
- ❑ Commented on both content and style (in evaluative summary).
- ❑ Interspersed evaluative commentary throughout summary so that assessments appear near relevant points.
- ❑ Included a direct quotation to illustrate or reinforce my recommendation (for evaluative summary).
- ❑ Ensured that descriptive abstract is short, to the point, and does not offer a judgment.

Exercises

1. Summarize a chapter of a textbook you are now using for a course in your major field. Provide an accurate bibliographic reference for this chapter (author of the textbook, title of the chapter, title of the book, place of publication, publisher's name, date of publication, and page numbers of the chapter).

2. Summarize a lecture you heard recently. Limit your summary to one page. Identify in a bibliographic citation the speaker's name, date, and place of delivery.

3. Listen to a television network evening newscast and also to a later news update on the same station. Select one major story covered on the evening news and indicate which details from it were omitted in the news update.

4. Write a summary of the research paper on telecommuting (pp. 398–413) or on e-cash (pp. 631–652).

5. Bring to class an article from the *Reader's Digest* and the original material it condensed, usually an article in a journal or magazine published six months to a year earlier. In a paragraph or two indicate what the *Digest* article omits from the original. Also point out how the condensed version is written so that the omitted material is not missed and the condensation does not misrepresent the main points of the article.

6. Assume that you are applying for a job and that the personnel manager asks you to summarize your qualifications for the job in two or three paragraphs. Write those paragraphs and indicate how your background and interests make you suited for the specific job. Mention the job by title at the outset of your first paragraph.

7. Write a summary of one of the following articles.
   a. "Microwaves," in Chapter 1.
   b. "Videoconferencing: Ready for Prime Time?" on pages 444–445.

8. Write a descriptive abstract of the article you selected in Exercise 7.

Videoconferencing: Ready for Prime Time?

The goal for videoconferencing hasn't changed in more than a decade: Increase the productivity of scattered individuals and groups by enhancing simultaneous, real-time information sharing through voice, data, and video communication. At last it seems that videoconferencing technology is finally poised to fulfill its promise to users.

Advances in hardware, software, and transmission technologies have combined with market factors to bring a broad spectrum of applications much closer to reality than they've ever been. These applications fall within three broad categories: point-to-point, room-to-room, and interactive corporate broadcasting.

Point-to-Point

When most businesspeople think of videoconferencing they're thinking about point-to-point (desktop-to-desktop) communication. Increasingly common in very large corporations, the hardware normally consists of proprietary cards that plug into a desktop computer, a modem, and tiny video camera. The software that drives this setup is also proprietary. The video, data, and voice transmission usually uses a company's wide area network (WAN) or digital phone lines (ISDN).

Although older systems only permitted two-way communication links, most new systems allow several people at different locations simultaneous access through a central "bridge." This means that a live picture of each person appears in a window on the monitor screen. Data can be displayed, downloaded, or uploaded upon request while conversations take place.

Video quality is a good news/bad news situation. The good news is that the problems of "Max Headroom" jerkiness and seeming delay between the sound and picture have been solved. The bad news is that as a practical matter it may not be available to you.

"The problem isn't hardware, software, or even bandwidth," notes Tony Paradiso, director of marketing for Picture Tel (Danvers, MA), the company that has captured more than half of the point-to-point hardware/software market. "The real problem," Paradiso continues, "is the lack of generally accepted common standards within the videoconferencing industry. Everyone is still producing proprietary support systems."

Paradiso believes that there are two key reasons why this doesn't help users. First, the nature of developing proprietary systems means that some users wind up buying systems that are (or soon will be) obsolete, hurting both the customers and the industry. Second, users are accustomed to total transparency with phones and faxes, and expect as much from videoconferencing.

Room-to-Room

In this form of videoconferencing a small group of people (three to six) gather in a specially equipped room—either inside or outside the company—to communicate with another group of people in a similarly equipped room. This is a way for teams of people to get together without the cost of airfares or hotel rooms for either group.

The real saving, however, is in productivity. Neither group has to spend two or more days away from their primary work location, and they can still share all the needed information. It's important to factor these elements into your calculations, because the cost of setting up a room-to-room facility is at least $13,000 for hardware and software alone. It can be more than worth the cost if you otherwise have to send six or seven people by air away from home for joint team meetings.

Transmission is over ISDN lines, and is often projected onto a screen. The quality of the video, however, is controlled by 384KB bandwidth, which doesn't account for more than 20 percent of the total number of room-to-room and point-to-point systems currently in use.

"There's a natural migration toward higher bandwidths and broadcast-quality video," states Bob Boughton, Eastern regional manager for videoconferencing sales for TIE/communications (Overland Park, KS). The company has focused on providing hardware and software for room-to-room communicating, and to some extent, on large-audience interactive broadcasting. "We still need to resolve the matter of transmission and reception standards with proprietary capabilities," adds Boughton.

Large-Audience Broadcasting

Large-audience corporate interactive broadcasting is the high end of videoconferencing, and in many ways it resembles a commercial television broadcast. All the same capabilities exist for interactivity and data transmission, but the nature of the event and its support systems are such that video transmissions are usually of commercial broadcast quality at 30 frames per second.

"Corporate interactive broadcast is almost exclusively used by Fortune 500 companies," states Steve Holtzman, CEO of Flying Squirrel Production, Inc. (Cherry Hill, NJ), a conferencing company that specializes in technical production of multisite broadcasts. "Audiences at each site are usually large, more than 50 and often more than 200. The meeting itself is more structured than either point-to-point or room-to-room videoconferences. There are frequently large stage sets, props, special lighting and sound requirements, broadcast-quality video cameras, as well as producers, directors, and technicians. In addition to digital land lines, it frequently involves satellite, microwave, and fiber optic transmission."

Boughton comments, "Central capabilities for large-audience videoconferencing put an incredible communications tool in the hands of management. There's no doubt, however, that you'd better know what and how you plan to communicate."

Paradiso sees interactive broadcasting occupying a narrow market niche. "Large- audience videoconferencing doesn't replace the annual sales meeting or stockholders' meeting for big companies, but it may form elements of them. It will replace the 'all hands' meeting when management has to get information out to everyone fast. We still haven't figured out how to exchange information in this setting."

Holtzman is less dubious. "We've seen hot issues being discussed during broadcasts. The questions aren't planted, the responses aren't canned, and the impact isn't faked. That's one of the reasons multiple-city press conferences are often done by interactive broadcasts. Large-audience videoconferencing gives all members of the press access to top executives within a single time frame. This helps build the company's credibility."

PART IV

Preparing Documents and Visuals

CHAPTER 12

Designing Successful Documents

The success of your document—letter, brochure, proposal, report—depends as much on how it looks as what it says. A report filled with nothing but thick paragraphs of type, with no visual cues to break them up or to make the information in them stand out, is sure to intimidate and turn readers away. They will conclude that your work is too complex and not worth the effort. They may also be offended that you did not keep in mind their needs to find information quickly.

Your documents need to look reader-friendly. They should signal to readers that the message is (a) easy to read, (b) easy to follow, and (c) easy to recall. Don't bog readers down with long unbroken paragraphs; break information up into smaller units that are visually appealing. Help readers find key points at a glance through **chunking** (using smaller paragraphs), lists, boldfacing, and bullets. To assure readers that you are aware of their busy schedules, make sure you highlight and distinguish major points for them. That way readers can find them easily the first time through or on a second reading if they have to double back to verify a point.

Organizing Information Visually

Take a quick look at Figures 12.1 (p. 456) and 12.2 (pp. 457–459). The same information is contained in each figure. Which appeals most to you? Which do you think would be easier to read? As these two figures show, the design or layout of your document plays a crucial role in your audience's overall acceptance of your work. Information is organized (and perceived) graphically as well as verbally. A document that looks logical, easy to read, and is clearly organized with the reader in mind will increase your credibility. If you offer readers a densely packed document, with no visual road signs to help them navigate through it, they will conclude that you are poorly organized and that your thinking is not very clear or direct. And if it is one thing that the world of work dislikes, it is someone who has trouble getting to the main point—the bottom line. A poorly designed document may tell your readers that you can't make up your mind or can't reason logically.

Your company, too, will win or lose points because of document design. A visually appealing document will enhance a company's reputation and improve its sales. A poorly designed one will not. Consumers will think your firm is unbending, hard to do business with, and does not care about specific problems customers may have over a policy or a set of instructions. Your company could easily get a reputation for not respecting the individuals it does business with or for being old-fashioned, hard-nosed, or just plain unprofessional.

Chapter 12 will explain the tools you need and show you how to design professional-looking, reader-friendly documents.

Characteristics of Effective Design

As you read this chapter and apply its principles to your own written work, make sure that your documents offers readers the following qualities:

- visual appeal
- logical organization
- clarity
- accessibility
- variety
- audience relevance

By incorporating these characteristics into your documents, you guarantee that your work will be well received.

Tools for Designing Your Documents

In the world of work you will have access to computers, word-processing software, and high-memory laser-quality printers. This affordable and accessible technology makes it easy for you to **design, illustrate, edit, format, store, retrieve, transfer,** and **print** documents. Thanks to this technology, you or your team can design almost any document discussed in *Successful Writing at Work.* In efforts similar to the collaborative writing discussed in Chapter 3, people in business frequently use the flexibility in automation systems to collaborate on document design.

Three Basic Tools

The three categories of automation tools available for document design include

1. **Computer hardware.** Critical hardware components include the central processing unit (CPU) or microprocessor, monitor, and printer. The CPU houses the "brains" of the computer, while the monitor and printer help you to visualize what is inside the CPU. To design a professional-looking document you need hardware that consists of the following basic features and components.

- Pentium or power PC
- 8–16 MB RAM
- 1GB hard drive
- 3.5" disk drive

- CD-ROM drive
- high-resolution monitor
- fax modem
- sound card

2. **Software.** Software tools are operating and program applications. The two primary operating systems are Mac OS and Windows, with Windows by far the more widely used of the two. The basic types of software include word processing, spreadsheet, database, graphics, and communications. Helpful software packages available for page designing include

- word-processing programs such as Microsoft Word and WordPerfect
- desktop publishing programs such as Quark XPress, Microsoft Publisher, and Pagemaker
- graphics programs such as Harvard Graphics, Printmaster, and Adobe Illustrator

Word-processing programs offer a variety of options that make document design fast and easy to develop and just as easy to change.

TECH NOTE

It is important to distinguish between **word-processing programs** (such as WordPerfect or Microsoft Word) and **desktop publishing programs** (such as Quark XPress or Aldus Pagemaker) since they perform many of the same functions.

Word processing is the fastest way to handle text. It is excellent for business correspondence, proposals, reports, and other documents in general. Word-processing programs are suitable for materials requiring a simple design and layout.

Desktop publishing programs can also be used for text, but often the text is imported from a previously created word-processing file. The desktop publishing programs are used to produce documents requiring much more sophisticated design and typography, such as newsletters, brochures, and booklets. Desktop publishing programs allow greater control, with more choices, over typography, page layout, and graphic elements. Most documents that are going to be commercially printed, especially those that are being printed in color, should be done with desktop publishing software.

3. **Printers and scanners.** Dot matrix printers are rapidly being replaced by two types of high-resolution printers—laser and ink-jet or bubble-jet—that are graphic-capable, letter-quality, and user-friendly. These high-resolution printers offer

quality text and graphic image printing in a relatively short time. Both types of printers are affordable and easy to operate; both come in monochrome or color versions.

Once documents (text or graphic pictures) are placed in a scanner, the scanner then makes a digital image, and the information is placed in a graphics file. Scanners are discussed more in the next chapter.

Desktop Publishing

Desktop publishing programs, sometimes referred to as **page layout software,** provide an inexpensive alternative to a professional print shop. Because desktop publishing software permits users to design page layout, include visuals, and produce high-quality final copies, you can professionally create printed documents right in your own home or office.

With desktop publishing you can

- delete, insert, and move entire blocks of text
- take advantage of numerous typefaces
- integrate various changes in typeface—like bold, italics, and underlining
- vary type sizes
- justify margins
- change line spacing
- center words, titles, or lines of text
- break and number pages
- arrange text in multiple columns
- add headers and footers (printed words that appear at the top and bottom, respectively, of each page of text)
- blow up quotations
- import graphics, such as drawings, photographs, and logos

Desktop Publishing Departments

Desktop publishing is such an important tool in the visual design of your work that many companies rely on this automation to promote their image and publish their message cost-effectively. In fact, many businesses have their own in-house Desktop Publishing Departments, usually referred to as DTPs, which produce handsome newsletters, brochures, reports, and manuals. DTPs can include subdivisions such as production design, art direction, binding, and marketing.

Elements of Desktop Publishing

Type

There are many different styles of type (see pp. 460–464). All computers come equipped with a small number of typefaces (usually a dozen). Companies with DTP departments often have a large library of typefaces that can be downloaded as needed.

Templates
Desktop publishing software is equipped with predesigned templates. Templates are patterns and predesigned text columns that offer numerous page layout formats for reports, newsletters, brochures, and other marketing/communications documents. A template for a report, for example, would contain all the headings and divisions you need. You can also create and save your own template for an original format you use frequently.

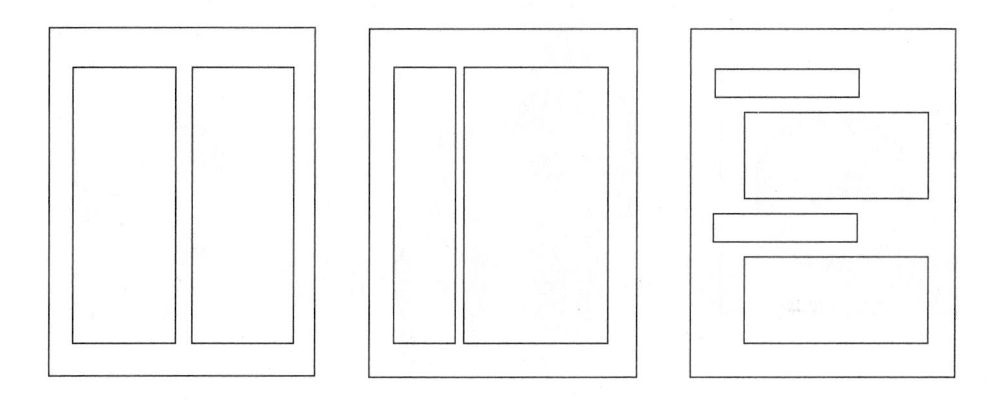

Graphics
A graphics program supplies shapes and lines that can be manipulated (skewed, enlarged) and offers options for sophisticated use of color and shading. There are drawing programs to create diagrams, charts, and illustrations that can be saved as electronic files and imported into word-processing or desktop publishing documents.

You can move and place graphics anywhere within the document. Graphic design programs provide customized graphs, charts, tables, and expanded font (design) sizes and shapes. These programs are ideal for designing charts in black and white or color for group presentations.

Here are some types of graphics available to you.

1. Drawing tools. Desktop publishing programs contain tools that allow you to create a variety of shapes, rules, borders, and arrows. Having drawn a box, circle, triangle, or whatever, you can use other tools (such as paintbrush, eraser, or text tool) to fill in or alter the appearance of the shape or to add words to it; then you can rearrange all the elements into a graphic that communicates quickly and effectively.

2. Icons. Icons are symbols or visual representations of concepts or actions. The skull and crossbones on a container of poison is an icon that warns us of danger. Many highway signs are icons that tell us quickly what to expect ahead: an S-curve, merging traffic, a railroad crossing. *Graphic icons* like these are simply pictures that communicate directly. No matter what language we speak—and even if we cannot read—icons tell us at a glance which restroom to use or how to fasten the seat belt in an airplane. A nation's flag is an icon, and so are many religious symbols and most company logos.

Computer software often relies on icons to help us perform common actions without remembering keyboard commands. Arrows in the scroll bars move us easily around a "window"; a file folder helps us group related documents; and a trash can holds files to be deleted later. Icons in the menu bar make it easy to print a file, open an address book, search for specific text, cut some copy, or dial the phone.

3. Clip art. Clip art comprises a library of simple drawings, often classified by themes, that can be imported (copied) from floppy disk or CD to your document. Some clip art software packages offer as many as 60,000 images, arranged into such diverse categories as animals, holidays, famous people, food, and various technologies. Clip art is widely used to make business documents attractive and appealing. If the graphic is symbolic or represents something unique about the business, it can also be the basis for an icon or a company logo.

4. Stock photos/Stock art. Assembled photos and art on CD can be imported, sometimes with a permission fee, sometimes free with the purchase of the CD.

The ABCs of Document Design

The basic features of document design are

(a) page layout
(b) typography
(c) graphics

Each of your pages should integrate these three elements. The proper arrangement and balance of type, white space, and graphics involve the same level of preparation that you would spend on your research, drafting, revising, and editing. You need to create a document that cooperates with—not detracts from—your message. Just as you research your information, you have to research and experiment in order to adopt the most effective design for your document.

Page Layout

Each of your pages needs to coordinate space and text pleasingly. Too much or too little of one or the other can jeopardize your reader's acceptance of your message. To make sure you design an effective and attractive page layout, pay attention to the following elements.

1. White space. The proper combination of white space and text is essential to maintain your reader's interest and investment in what you have to say. White space refers to the open areas on the page—free of text and visuals. Readers are initially drawn to text on a page because of the pleasant and professional use of white space. White space can entice, comfort, and appeal to the reader's "psychology of space" by

- attracting the reader's attention
- assuring the reader that information is presented logically
- announcing that information is easy to follow
- assisting the reader to organize information visually
- allowing the reader to forecast and highlight important information

Compare Figures 12.1 and 12.2. Which document was designed by someone who understands the importance of white space?

2. Margins. Use wide margins, usually 1 to $1\frac{1}{2}$ inches, to "frame" your document with white space surrounding text and visuals. Most documents easily accommodate equal margins on all four sides. This technique creates balance and prevents the document from looking cluttered or overcrowded. If your document requires binding, you may have to leave an even wider left-hand margin (2 inches).

3. Line length. Most readers find a text line of 10 to 14 words, or 50 to 70 characters (depending on the type size you choose), comfortable and enjoyable reading. Line length, or "line measuring," depends, of course, on font type and size, number of columns, width of margins, and space between words. It goes without saying that you should never exceed your margin settings. In the example below, note how the extra-long lines unsettle your reading and tax your eye movement; they signal rough going.

In order to succeed in the world of business, workers must learn to brush up on their networking skills. The network process has many benefits that you need to be aware of. These benefits range from finding a better job to accomplishing your job more easily and efficiently. Through networking you are able to expand the number of contacts who can help you. Networking means sharing news and opportunities. The Internet is the key to successful networking.

Conversely, do not print a document with too short or extremely uneven lines.

> In order to succeed in the world of
> business, workers must learn
> to brush up on their networking skills. The
> network process has many
> benefits you need to be
> aware of.

Readers will suspect your ideas are incomplete and that your approach to the topic is superficial or even simpleminded.

FIGURE 12.1 A poorly designed document.

The results for the recent cholesterol screening at our company's Health Fair were distributed to each employee last week. Many employees wanted to know more information about cholesterol in general, the different types of cholesterol, what the results mean, and the foods that are high or low in cholesterol.

We hope the information provided below will help each employee better answer their questions concerning cholesterol and our cholesterol screening program.

High cholesterol, along with high blood pressure and obesity, is one the primary risk factors that may contribute to the development of coronary heart disease and may eventually lead to a heart attack or stroke. Cholesterol is a fatty, sticky substance found in the bloodstream. Excessive amounts of the bad type of cholesterol can deposit on the walls of the heart arteries. This deposit is called plaque and over a long period of time plaque can narrow or even block the blood flow through the arteries.

Total cholesterol is divided into three parts—LDL (low-density lipoprotein) or bad cholesterol, HDL (high-density lipoprotein) or good cholesterol, and VLDL (very low-density lipoprotein), a much smaller component of cholesterol you don't have to worry about. Bad (LDL) cholesterol forms on the walls of your arteries and can cause a lot of damage. Good cholesterol, on the other hand, functions like a sponge, mopping up cholesterol and carrying it out of the bloodstream.

You should have received three cholesterol numbers. One is for your HDL, or good cholesterol, and the other is for your LDL, or bad cholesterol, reading. These two numbers are added to give you the third, or composite, level of your total cholesterol. You are doing fine.

As you can see, a total cholesterol reading of 200 or below is considered safe. Continue what you have been doing. If your reading falls in the moderate risk range of 200–239, you need to modify your diet, get more exercise, and have your cholesterol checked again in six months. If your reading is above 240, see your doctor. You may need to take cholesterol-lowering medication, if your doctor prescribes it. Reducing your total cholesterol by even as little as 25% can decrease your risk of a heart attack by 50%.

The Surgeon General recommends that your LDL, or bad, cholesterol should be below 130. And your HDL, or good, cholesterol needs to be at least above 36. Ideally, the ratio between the two numbers should not be greater than 5 to 1. That is, your HDL should be at least 20% of your LDL. The higher your HDL is, the better, of course. So even if you have a high LDL reading, if your HDL is correspondingly high you will be at less risk.

One of the easiest ways to decrease your cholesterol is to modify your diet. Cholesterol is found in foods that are high in saturated fat. Saturated fat comes from animal sources and also from certain vegetable sources. Foods high in bad cholesterol that you should restrict, or avoid, include whole milk, red meat, eggs, cheese, butter, shrimp, oils such as palm and coconut, and avocados. Generally, Food groups low in cholesterol include fruits, vegetables, and whole grains (wheat breads, oat meal, and certain cereals), lean meats (fish, chicken), and beans.

The goal of our cholesterol screening is to help each employee lower his or her cholesterol level and eventually reduce the risk of heart disease. Besides the advice given above, you can do the following: get regular aerobic exercise—bicycling, brisk walking, swimming, rowing—for at least 30 minutes 3–4 times a week. But get your doctor's approval first. Eat foods low in cholesterol but high in dietary fiber (beans, oatmeal, brown rice). Maintain a healthy weight for your frame to lower your body fat. Minimize stress, which can increase cholesterol. Learn relaxation techniques.

FIGURE 12.2 An effectively designed document with the same text as Figure 12.1.

Cholesterol Screening

The results for the recent cholesterol screening at our company's Health Fair were distributed to each employee last week. Many employees wanted to know more information about cholesterol in general, the different types of cholesterol, what the results mean, and the foods that are high or low in cholesterol. We hope the information provided below will help each employee better answer their questions concerning cholesterol and our cholesterol screening program.

Determining Risk Factors

High cholesterol, along with high blood pressure and obesity, is one the primary risk factors that may contribute to the development of coronary heart disease and may eventually lead to a heart attack or stroke. Cholesterol is a fatty, sticky substance found in the bloodstream. Excessive amounts of the bad type of cholesterol can build up a deposit on the walls of the heart arteries. This deposit is called plaque and over a long period of time plaque can narrow or even block the blood flow through the arteries.

Separating Types of Cholesterol

Total cholesterol is divided into three parts: (1) LDL (low-density lipoprotein), or bad cholesterol, (2) HDL (high-density lipoprotein), or good cholesterol, and (3) VLDL (very low-density lipoprotein), a much smaller component of cholesterol you don't have to worry about. Bad (LDL) cholesterol forms on the walls of your arteries and can cause a lot of damage. Good cholesterol, on the other hand, functions like a sponge, mopping up cholesterol and carrying it out of the bloodstream.

Continued

FIGURE 12.2 (Continued)

Understanding Your Cholesterol Results

You should have received three cholesterol numbers. One is for your HDL (or good cholesterol) and the other is for your LDL (or bad cholesterol) reading. These two numbers are added to give you the third, or composite, level of your total cholesterol.

Cholesterol levels can be classified as follows:

Minimal Risk	Moderate Risk	High Risk
below 200	200–239	above 240

As you can see, a total cholesterol reading of 200 or below is considered safe. You are doing fine. Continue what you have been doing. If your reading falls in the moderate risk range of 200–239, you need to modify your diet, get more exercise, and have your cholesterol checked again in six months. If your reading is above 240, see your doctor. You may need to take cholesterol-lowering medication, if your doctor prescribes it. Reducing your total cholesterol by even as little as 25 percent can decrease your risk of a heart attack by 50 percent.

Relationship Between Bad/Good Cholesterol

The Surgeon General recommends that your LDL (or bad) cholesterol should be below 130. And your HDL (or good) cholesterol needs to be at least above 36. Ideally, the ratio between the two numbers should not be greater than 5 to 1. That is, your HDL should be at least 20 percent of your LDL. The higher your HDL is, the better, of course. So even if you have a high LDL reading, if your HDL is correspondingly high you will be at less risk.

Recognizing Food Sources of Cholesterol

One of the easiest ways to decrease your cholesterol is to modify your diet. Cholesterol is found in foods that are high in saturated fat. Saturated fat

Continued

FIGURE 12.2 (Continued)

comes from animal sources and also from certain vegetable sources. Foods high in bad cholesterol that you should restrict include:

1. whole milk
2. red meat
3. eggs
4. cheese
5. butter
6. shrimp
7. oils such as palm and coconut
8. avocados

Generally, food groups low in cholesterol include fruits, vegetables, and whole grains (wheat breads, oatmeal, and certain cereals), lean meats (fish, chicken), and beans.

Realizing It Is Up to You

The goals of our cholesterol screening are to help each employee lower his or her cholesterol level and eventually reduce the risk of heart disease. Besides the advice given above, you can do the following:

- Get regular aerobic exercise—bicycling, brisk walking, swimming, rowing—for at least 30 minutes 3–4 times a week. But get your doctor's approval first.
- Eat foods low in cholesterol but high in dietary fiber (such as beans, oatmeal, and brown rice).
- Maintain a healthy weight for your frame to lower your body fat.
- Minimize stress, which can increase cholesterol. Learn some relaxation techniques.

The author is indebted to Sgt. Mannie E. Hall of the U.S. Army for creating this document.

4. Columns. Document text usually is organized in either single-column or multicolumn formats. Memos, letters, and books are usually formatted without columns, whereas documents that intersperse text and visuals (such as newsletters and magazines) work better in multicolumn formats. (See examples of multicolumn brochures and newsletters in Figures 8.5 and 8.6.)

Typography, or Type Design

1. Selecting the Right Typeface

Readability of your text is crucial. Select a typeface, therefore, that ensures your text is

- legible
- attractive
- functional
- appropriate for your message
- complementary with accompanying graphics

The most familiar typefaces are Times Roman, Arial, and Helvetica, although other very useful typeface styles are available with WordPerfect, MS Word, and other software packages. Other typefaces include Script, Modern, Old Style, Decorative, and Traditional.

Times Roman	Frutiger
Helvetica	Palatino
Alexa	**STENCIL**

Type is also classified as having serif or sans serif fonts. **Serif fonts** appear to be the most readable in the body of the text. These fonts are distinguished with tail features or crossbars at the ends of the letters. Serifs add flair, arouse the reader's interest, and increase readability. Letter strokes are of varying widths and sometimes tapered.

Stone Serif	Sabon Roman
Courier	Janson Text
Times	Palatino

Sans serif fonts are recommended for headlines or subheadlines. Having no tails or crossbars, sans serif fonts provide a clear, crisp, legible image ideal for short messages.

Futura	Stone Sans
Univers Oblique	Geneva
Gill Sans Extra Bold	TradeGothic

2. Type Size
Type size options are almost unlimited, depending again on your software package and printer capabilities. Type size is measured in units called **points,** 72 points to the inch. The bigger the point size, the larger the type. Most business and educational documents use from 10-point to 12-point type, although the range today varies from 6 to 72 points (and beyond). Newspaper classified ads are in small 6-point to 8-point type, while headlines are set in much larger 30-point to 36-point type. This textbook is set in 10-point type.

Times 7 point

Univers 10 point

Palatino 14 point

Frutiger 22 point

Stone Serif 36 point

TECH NOTE

Be consistent in differentiating between **typeface** and **type style.**

A **typeface** is a specific family of type, such as Times Roman or Helvetica. The word *typeface* is sometimes used synonymously with *font*, a slightly more specific term that refers to the family as well as the size of the type.

Type style, especially as used by software manufacturers, refers to attributes such as bold, italics, shadowed, and other variations.

3. Type Styles

Also known as **attributes,** these include boldface, italics, underlining, and shading.

Boldface
Italics
Outline
Shadow
Underlining
Small Caps
Shading

4. Line Spacing

The amount of white space between lines of your text also affects how your document will be perceived. Most word-processing programs refer to line spacing as **leading.** Leading is also measured in points, the same as type size. Standard line spacing should be 2 points more than type size, so 10-point type would use 12-point leading, 12-point type would use 14-point leading, and so on.

5. Justification

Sometimes referred to as alignment, justification consists of left, right, full, and center options. Left-justified (also called **unjustified** or **right ragged**) is the preferred method because it allows for space between words in a line of text to remain constant, while fully justified alignment (both left *and* right) creates unequal space gaps between words in a line of text. Left justification gives a document a less formal look than full justified. Narrow columns of text should be set left-justified to avoid awkward gaps between words and excessive hyphenation.

Our new website offers consumers a mall on the Internet. It gives shoppers access to our products and services and makes buying easy and fun. Our new website offers consumers a mall on the Internet. It gives shoppers access to our	Our new website offers consumers a mall on the Internet. It gives shoppers access to our products and services and makes buying easy and fun. Our new website offers consumers a mall on the Internet. It gives shoppers access to our
Left justified	Right justified
Our new website offers consumers a mall on the Internet. It gives shoppers access to our products and services and makes buying easy and fun. Our new website offers consumers a mall on the Internet. It gives shoppers access to our products and	Our new website offers consumers a mall on the Internet. It gives shoppers access to our products and services and makes buying easy and fun. Our new website offers consumers a mall on the Internet. It gives shoppers access to our
Full justified	Centered

6. Headings and Subheadings

Use brief descriptive words or phrases to introduce or summarize a document, or new section or subsection within a document. Your headings, called **heads** and **subheads,** should be grammatically parallel, and not wordy. Note how the headings below from a poorly organized proposal from the Acme Company are non-parallel.

- What Is the Problem?
- Describing What Acme Can Do to Solve the Problem
- It's a Matter of Time . . .
- Fees Acme Will Charge
- When You Need to Pay
- Finding Out Who's Who

Revised, the heads are parallel and easier for a reader to understand and follow.

- A Brief History of the Problem
- A Description of What Acme Can Do to Solve the Problem
- A Timetable Acme Will Follow
- A Breakdown of Acme's Fees
- A Payment Plan
- A Listing of Acme's Staff

Heads and subheads immediately attract attention and quickly inform readers about the function, scope, purpose, or contents of the document or section. Additionally, heads and subheads introduce helpful white space to separate text and organize your document. The space around a heading is like an oasis for the reader, signaling both a rest and a new beginning.

Subheads allow the reader to skim or review a document and its content in an "outline type" format. Headings and subheadings follow an established order or hierarchy.

In designing a document with heads and subheads, follow these guidelines.

- Use larger type size for heads than for text; major heads should be larger than subheads. If your text is in 10-point type, your headings may be in 16-point type and your subheadings may be in 12- or 14-point type.

> **Sixteen-Point Head**
>
> **Subheading in 12 Point**
>
> Use larger type for headings than you do for text; major headings should be larger than subheadings. If your text is in 10-point type, your heading may be in 16-point type. A subheading may be in 12 or 14.

- Modify type to differentiate sections. For example, for headings and subheadings use uppercase versus lowercase, bold type, underlining, and changes in font style or type.
- Establish a horizontal position for a head, such as centered or aligned left, and keep it consistent throughout the document
- Allow additional space, or leading, above a head to set it off from the preceding section. You may also want to allow additional space below a head.
- If you are using a color printer, consider using a second color for major headings.

7. Lists

Placing items in a list helps readers by dividing, organizing, and ranking information. Lists emphasize important points and contribute to an easy-to-read page design. Lists can be (a) numbered, (b) lettered, or (c) bulleted. Take a look at the reports, proposals, and memos in other chapters that effectively use lists.

8. Captions

Used to accompany, explain, highlight, or reference pictures or other graphics (like charts and graphs), these titles help a reader to identify a visual and quickly explain the nature of the picture or other graphic. (Chapter 13 covers using captions with visuals in a document.)

Graphics

Like other visuals, graphics work in conjunction with your words. A document without visuals or graphics may look boring, confusing, or unattractive. You will have to judge how and when a graphic can improve your message and convince your reader. The list below identifies some common graphics included in DTP software packages.

1. **Clip art.** See page 454 above.
2. **Boxes.** These lines isolate or highlight text or visuals. Tables 13.1, 13.2, and 13.3 use boxes.
3. **Rules.** These lines are classified as either vertical or horizontal. Vertical rules are used to separate columns of text, while horizontal rules separate sections introduced with subheads.
4. **Letterheads and logos.** A company's corporate image is represented and symbolized by its letterhead. A letterhead will usually consist of a graphic icon integrated with a type treatment of the company name. Often the company's address, e-mail address, and Web site are included in a letterhead. Company letterhead conveys the firm's message and expresses its character. Typically, logos and sometimes the entire letterhead are imported as graphic files. A letterhead/logo should creatively set one company apart from another.
5. **Color.** When properly used, color may add visual appeal and be a welcome relief for readers accustomed to black text on white paper. Color draws attention to your message and makes your document more interesting. Again, computer software packages, color laser printers, and even color copying machines make the widespread use of color feasible. But do not overuse color just to decorate; make sure it is functional. For example, don't use a different color for each page of a report just to prove to readers you have the technical capability. Your report will look like a rainbow popsicle.

Poor Document Design: What *Not* to Do

Up to this point we have introduced the various elements of effective document design. Now, in contrast, we'll look at what you need to avoid. By knowing what looks bad from a reader's point of view, you will be much better able to design a document that works well visually for your reader.

Figure 12.1, a poorly designed document, illustrates many of the following common mistakes in document design. But note how Figure 12.2, which contains the same information as Figure 12.1, incorporates many of the effective techniques just described. Avoid the following errors when you design your document.

1. Insufficient white space. Narrow margins and limited space between headings and text are classic mistakes. Skimping on white space can frustrate your reader, who is eager to locate key parts of a document to identify and process them easily. A lack of white space between paragraphs or heads or around the borders of a document sends a negative message to your audience. Your work tells a reader, "Roll up your sleeves: this will be tough, unenjoyable reading."

2. **Inappropriate line length.** Excessively long lines are hard to read and signal to your audience that your work is highly complex and unrewarding.

3. **Overuse of visuals.** Too many visuals (boxes, rules, and clip art images) can create barriers and confusion and will crowd your pages. Establish a balance between text and visuals. Also, use visuals for a specific purpose, not just for decorations or to fill space. (For additional discussion of the use of visuals, see Chapter 13.)

4. **Mixing typefaces.** Select one typeface, usually serif, for text body throughout the document and stick with it. You can then choose a contrasting typeface—say, a sans serif one—for heads. But try not to mix different serif typefaces or different san serif faces. It's tricky to make them work together and the result often looks like a printing mistake.

5. **Too few or no heads and subheads.** Without these useful guideposts your document will seem unorganized and illogical. As we saw in Figure 12.2, heads and subheads are typographical markers that signal starting points and major divisions; they thus provide helpful landmarks for readers charting their course through your document.

6. **Excessive spacing.** Too much space also distorts the document and its message. Leaving three or four spaces between consecutive lines of a text signals to readers that your ideas may be lightweight, not very significant. Also, including too much space around a visual or around the borders of your text can call into question the overall professional status of your work.

To avoid these errors, follow these tips:

- Use only one space after a period—not the traditional two spaces.
- Don't indent the first line of text after a heading or subheading.
- Avoid full justification on narrow columns.
- Eliminate excessive spacing in lists between bullets, numbers, or symbols and the actual text entries.

7. **Misusing capitals, boldface, or italics.** Printing an entire document with capitals (large or small) will make a document hard to read. But make sure to print heads or subheads in capitals to differentiate them from your text. Similarly, avoid overusing boldface or italics. Not only will too many special effects make your work harder to read, but you will lose the dramatic impact these attributes have to distinguish and emphasize key points that *do* deserve boldface or italic type.

✓ Revision Checklist

- ❑ Kept reader's busy schedule and reading time limits in mind when designing a document.

- ❑ Learned options of desktop publishing program.
- ❑ Arranged information in the most logical, easy to grasp order.
- ❑ Included only relevant visuals—clip art, icons, stock photos.
- ❑ Left adequate, eye-pleasing white space.
- ❑ Provided adequate margins to frame document.
- ❑ Justified margins, right, left, or center, depending on document and reader's needs.
- ❑ Maintained pleasing, easy to read line length.
- ❑ Kept line spacing consistent and easy on the reader's eyes.
- ❑ Chose appropriate typeface for message and document.
- ❑ Selected serif or sans serif fonts depending on message and reader's needs.
- ❑ Did not mix typefaces.
- ❑ Used effective type size, neither too small (under 10-point) nor too large (over 12-point), for body of text.
- ❑ Incorporated appropriate visual cues (attributes) for readers.
- ❑ Inserted heads and subheads to organize information for reader.
- ❑ Made all heads and subheads parallel and grammatically consistent.
- ❑ Supplied lists, bullets, numbers to divide information.

Exercises

1. Find an example of an effectively designed document, according to the criteria discussed in this chapter. This could be a memo, a brochure, a newsletter, a report, a set of instructions, a section of a textbook, even a Web site. E-mail or write a short (one-page) memo to your instructor describing the document's design and explaining why it works. Attach a copy of the document to your e-mail or memo.

2. Working with a team of three or four students, bring poorly designed documents to class. As a collaborative venture determine which of the documents the group submits is the hardest to follow, the most unappealing, and the least logically arranged. After selecting that document, collaboratively write a memo to your instructor on what is wrong with the design and what you would do to improve its appearance and organization.

3. Redesign (reformat; add headings, spacing, and visual clues; include clip art; and so on) the document your group selected for Exercise 2 and submit it to your instructor.

4. Find an ineffectively designed document—a form, a set of instructions, a brochure, a section of a manual, a story in a newsletter—and assume that you are a document design consultant. Write a sales letter to the company or agency that prepared and distributed the document, offering to redesign it and

any other documents they have. Stress your qualifications and include a sample of your work. You will have to be convincing and diplomatic—precisely yet professionally persuading your readers that they need your services to improve their corporate image, customer relations, and sales or services.

5. Redesign the following document to make it conform to the guidelines specified in this chapter.

7

TTI's in the loop on effective detector placement

Ever sit in bumper-to-bumper traffic and wish they'd widen the roads so people could get through more quickly? Well, that costs a lot of money. Which is why transportation engineers who deal with traffic congestion and the problems it causes look for more cost-effective alternatives to get you where you're going—and faster.

TTI researchers recently completed a TxDOT/FHWA-sponsored study entitled *Effective Detector Placement for Computerized Traffic Management.* The research sought to expand and improve the use of inductance loop detectors (ILDs) to complement traffic signals, signal systems and other advanced traffic management systems. This is a cheaper congestion solution than building or widening a road.

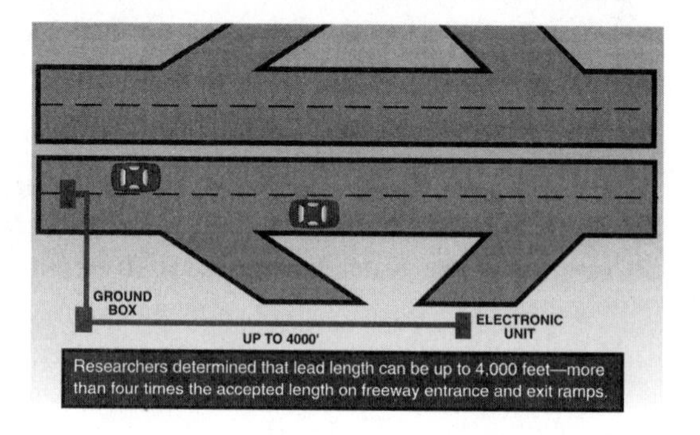

Researchers determined that lead length can be up to 4,000 feet—more than four times the accepted length on freeway entrance and exit ramps.

An ILD is an electrical circuit containing a loop of copper wire embedded in the pavement. As a vehicle passes over the wire loop, it takes energy from the loop. If that change is large enough, a detection is recorded. Thus, we are able to collect data on the movement or presence of vehicles on the roadway. Advanced traffic management systems operate best with accurate information on how many vehicles are present and how fast they are traveling.

The primary goal of the recent project was to use loop detectors as an integral part of the congestion-reduction system. Traditional problems with ILDs were addressed—like crosstalk, or interference between two adjacent loops—and innovative new applications for ILDs in advanced traffic systems—like detecting wrong-way HOV-lane movements.

Other applications include using ILDs to move traffic more efficiently at diamond interchanges, at high-volume, high-speed approaches and on the freeway entrance ramps. The long-range contribution of the study is a set of guidelines for using ILDs in the situations listed above. As freeway management systems continue to evolve, the guidelines developed through the nine study reports will provide designers with practical information on the most effective placement of ILDs.

A major finding of the research deals with lead length, or the length of wire necessary to connect the loop to the detector electronic unit. The study showed that the loop can be placed more than 4,000 feet from the point of control—four times the currently accepted distance. This information will give traffic designers much more flexibility when integrating ILDs into their traffic system designs.

The researchers also made some important discoveries about using ILDs to measure speed. They found that the best speed trap is nine meters (two loops interconnected with a timing device and spaced nine meters apart). They also determined that to get reasonably accurate and consistent speeds with an ILD, some things must be the same between a pair of loops: make, type or model of the detector units, sensitivity settings and loop configuration.

The findings from this research facilitate the use of loop detectors in managing traffic. And better management of driver frustration—just as important, even if less measurable than the congestion that causes it—is bound to follow.

Ultimately the three watchwords for this project were optimization, innovation, and implementation. Taking the tried-and-true and finding a better way to use it is, after all, the underlying building block for all engineering endeavors.

To order TTI Research Report 1392-9F, see the back page order form of this issue. For more information on loop detectors, contact Don Woods, 409/845-5792, FAX 409/845-6481 (E-mail: d-woods@tamu.edu).

Source: Texas Transportation Institute's *Researcher;* article author, Chris Pourteau. Reprinted with permission of the Texas Transportation Institute.

CHAPTER 13

Designing Visuals

Experts estimate that as much as 80 percent of our learning comes through our sense of sight. The written word, of course, forms a large part of our visual information. In conjunction with words, though, **visuals** convey a large share of the facts we receive. Visuals are especially useful on the job because they help readers see what you are discussing.

Chapter 13 surveys the kinds of visuals you will encounter most frequently and shows you how to read, construct, and write about them. It also describes the types of visuals and visual configurations you can create and copy with graphics software packages. One software manufacturer advertises that users can select more than 500 different visuals configurations and can make numerous changes in them. The impact and importance of visuals in document design is nowhere better illustrated than in the "virtual library" of the Internet, where every successful Web page exploits graphics to capture attention.

TECH NOTE

Most computer systems sold today arrive "bundled" with at least basic graphics software (such as MS Paint, Print Shop, Print Artist, and Arts and Letters) that lets you create simple pictures easily. You also can use graphics software to import clip art and other existing images from disks or the Internet. And today's ink-jet and laser printers produce remarkable results—in black and white or full color—easily and economically. Computer-savvy writers who want visual impact but can't afford the services of a designer, an artist, and a print shop are now limited only by their own imaginations and effort.

A discussion of visuals, however, is not confined to just this chapter. They are important in preparing successful instructions, proposals, and written and oral reports.

The Purpose of Visuals

How can visuals improve your work? Here are several reasons why you should use them. Each of the points is graphically reinforced in Figure 13.1.

1. *Visuals arouse readers' immediate interest.* Because many readers are visually oriented, visuals unlock doors of meaning. Readers who place great emphasis on visual thinking will pay special attention to the visuals you use. Visuals catch readers' eye quickly by setting important information apart and by giving them relief from looking at sentences and paragraphs. Visuals will also help you maintain your readers' interest. Because of their size, shape, color, and arrangement, visuals are dramatic. Note the eye-catching quality of the visual in Figure 13.1.

FIGURE 13.1 A well-designed "visual" that accomplishes all the purposes of a visual.

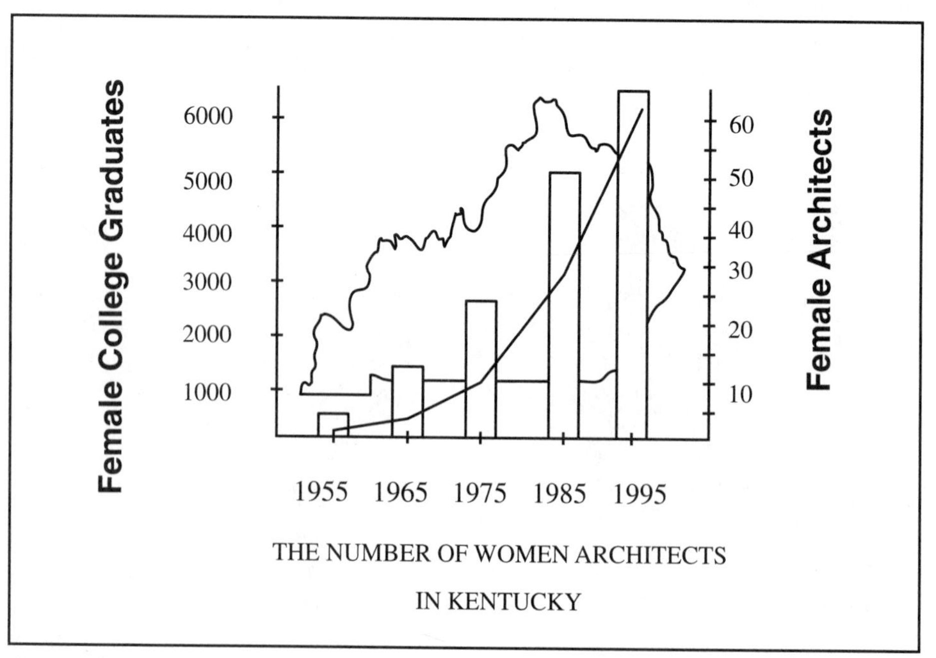

Courtesy of Lucy Cowan

2. *Visuals increase readers' understanding by simplifying concepts.* A visual *shows* ideas whereas a verbal description only *tells* them. Words without images may be less forceful or more complex. Visuals are especially important and helpful if you have to explain a technical process to a nonspecialist audience. Moreover, visuals can simplify densely packed statistical data, making a complex set of numbers easier to comprehend. Visuals help readers see percentages, trends, comparisons, and contrasts. Figure 13.1, for example, shows at a glance the change in the number of female architects in Kentucky.

3. *Visuals are especially important for ESL and multicultural audiences.* Visuals speak a universal language and so can readily be understood. Because visuals pose fewer problems in interpretation, they can help reduce ambiguities and misunderstanding. Given the international audience for many business documents, you will need to communicate with a variety of readers whose first language is not English. Visuals can make your communication with diverse audiences easier and clearer.

4. *Visuals emphasize key relationships.* Through their arrangement and form, visuals quickly show contrasts, similarities, growth rates, downward and upward movements as well as fluctuations in time, money, and space. Pie and bar charts (discussed on pp. 487–491), for example, show relationships of parts to the whole; and an organizational chart (p. 494) can graphically display the hierarchy and departments of a company or agency.

5. *Visuals condense and summarize a large quantity of information into a relatively small space.* Enormous amounts of statistical or financial data, over many weeks, months, and even years, can be incorporated concisely into one compact visual. A visual also allows you to streamline your message by saving words. It can record data in far less space than it would take to describe these facts in words alone. With a visual a reader can grasp the significance and relationship of multiple items in a single glance. Note how in Figure 13.1 multiple visuals express the increase in the number of female architects over the years.

6. *Visuals are highly persuasive.* Placed in appropriate sections of your work, visuals can capture the essence of your thinking to convince a reader to buy your product or service or to accept your point of view. A visual can graphically display, explain, and reinforce the benefits and opportunities of the plan you are advocating. A visual can also help you introduce a subject or assist you in developing a winning recommendation section of a report. Readers are far more likely to recall the visual than they might be a verbal description or summary of it.

Choosing Effective Visuals

Select your visuals very carefully. Special computer software programs allow you to select, create, and introduce visuals. The following suggestions will help you to choose effective visuals.

TECH NOTE

The capabilities of a computer to make visuals are almost endless. All you have to do is put the raw numerical data into your computer's database program and then, with a few simple keyboard commands, select the most appropriate visual shape (bar or pie chart, pictogram, graph) to display this statistical data. Following your commands, the computer will analyze and plot the data you keyed in into a proportionately accurate chart or graph. The particular visual will then appear on the computer screen, and with the assistance of a printer you can produce it on hard copy (paper) in black and white or in color. You do not have to be an experienced computer scientist or mathematician to achieve high-quality results.

1. Use visuals only when they are relevant for your purpose and audience. A visual should contribute to your text—not be redundant. Do not use a visual if your text is absolutely clear, and never include a visual simply as a decoration. A short report on fire drills, for example, does not need a picture of a fire station; instructions on how to prepare a company report do not require a picture of an individual at a computer terminal.

2. Consider how a specific visual will help your readers. Ask yourself these three questions before you include a visual.

- What do my readers need to know visually?
- What type of visual will best meet my readers' needs?
- How can I create this visual—scan it, import it, or make it myself?

Elaborate graphs are unnecessarily complex for a community group interested only in a clear-cut representation of the rise in food prices over a three-month period. Include a detailed, complex visual only for a technical reader. In general, though, keep your visuals simple and direct.

3. Use visuals in conjunction with, not as a substitute for, written work. Visuals do not always take the place of words. In fact, you may need to explain information contained in a visual. A set of illustrations or a group of tables alone may not satisfy readers looking for summaries, evaluations, or conclusions. Note how the visual in conjunction with the description of a magnetic resonance imager (MRI) in Figure 13.2 makes the procedure easier to understand than if the writer had used only words or only a visual. This visual and verbal description are appropriately included in a brochure teaching patients about an MRI procedure.

4. Use a visual when it would be more difficult to rely on words alone. A verbal description is sometimes more difficult to follow than a visual presentation. In such cases, a visual gives readers significant information that the text could not

FIGURE 13.2 A visual used in conjunction with written work.

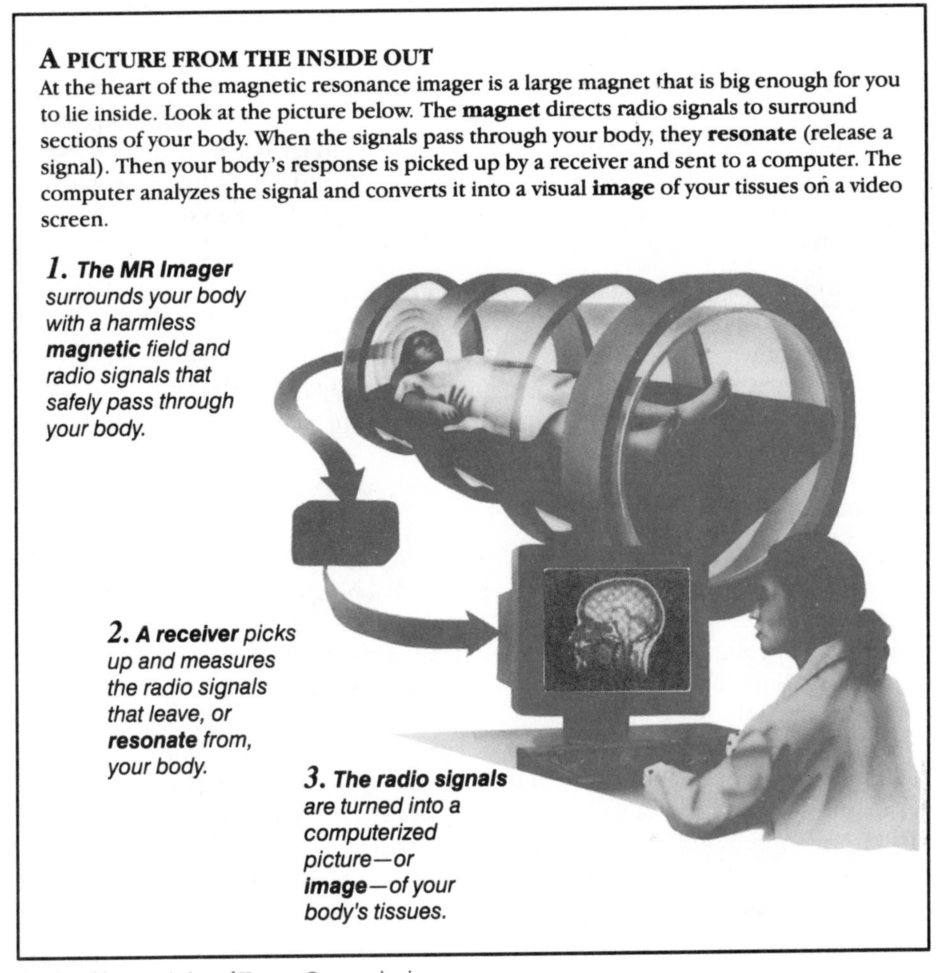

A PICTURE FROM THE INSIDE OUT
At the heart of the magnetic resonance imager is a large magnet that is big enough for you to lie inside. Look at the picture below. The **magnet** directs radio signals to surround sections of your body. When the signals pass through your body, they **resonate** (release a signal). Then your body's response is picked up by a receiver and sent to a computer. The computer analyzes the signal and converts it into a visual **image** of your tissues on a video screen.
1. The MR Imager surrounds your body with a harmless magnetic field and radio signals that safely pass through your body.
2. A receiver picks up and measures the radio signals that leave, or resonate from, your body.
3. The radio signals are turned into a computerized picture—or image—of your body's tissues.

Reprinted by permission of Krames Communications.

easily convey. Use visuals when you have to describe a piece of equipment with many (or concealed) parts, to explain an involved concept, or to account for a process.

5. **Experiment with several visuals.** Evaluate a variety of options before you select a particular visual. For instance, a graphics software package will allow you to represent statistical data in a number of ways. Preview a few different versions of a visual or even different types of visuals to determine which one would be best.

6. **Be prepared to revise and edit your visuals.** Just as you draft, revise, and edit your written work to meet your audience's needs, create several versions of your visual to get it right. Expect to change shapes or proportions, experiment with different colors and shadings and labels, and try various sizes before selecting the

most appropriate visual. Also check each visual against the data you want it to display to make sure it is accurate and ethical, not unclear or distorted. (Review Chapter 1 on the ethics of using visuals.)

7. **Always use high-quality visuals.** Your visuals should be clear, easy to read, and relevant. If readers have trouble understanding its function and arrangement, your visual is probably not appropriate and you need to change or revise it. Do not assume that readers will have magnifying glasses on their desks or that they will tolerate a crowded drawing. If you photocopy or scan a visual, make sure that the copy is clear and readable and does not cut off any part of the original.

8. **Consider how your visuals will look on the page.** Visuals should add to the overall appearance of your work, not detract from it. Don't cram visuals onto a page or allow them to spill over your text or margins. Observe generous margins. Many computer programs will let you move your visuals directly into your word processor document so that you can insert them properly. See pages 455–460 on effective page layouts.

Downloading and Scanning Visuals

Thanks to computer technology you can incorporate sophisticated visuals that once had to be created by an artist or a professional printer or were the province of special effects departments or film or marketing companies. Sitting in front of your personal computer (provided you have the right software), you can download professional-looking graphics in a striking array of colors, sizes, shapes, and dimensions in only a few minutes.

The Internet has had a big impact in changing the quality of visuals you can include in your work, whether for college courses or the corporate world. Because of the variety of companies and individuals setting up Web pages, the variety of visual information, and hence the means to represent it, are almost limitless. You can get creative new ideas for graphics by searching the Web. Maybe the best place to start is with some of the "Top 100 Sites" that are usually accessible through your Web browser.

If you see a graphics image in a Web site that you would like to use in a report or paper, it is fairly easy to save that image on your disk. But always remember

TECH NOTE

Once you have found an image on the Web, put your mouse cursor on the image and press the mouse button. (*Note:* the image may not have a hyperlink so your cursor may not change shape; that does not matter in this case.) A small pop-up menu box will appear:

Back
Forward

View this image
Save this image as . . .
Copy this image location

Use your cursor to select the "Save this image as . . ." option. The normal "Save as" menu box will appear and you can select a name for the file where the image will be saved as well as the disk drive and the directory. This process is exactly the same as choosing "Save as" from the "File" menu at the top of the screen.

Now that the image is on your disk, you can run your word processor and get into your document. Place your cursor at the point in the document where you want the image to appear, and then select the "Insert" option at the top of the screen and choose "picture." The image will then be copied to your word-processing file.

Images on computers are saved in a number of different formats; that is, the information the computer will use to produce the picture can be stored in different ways. Sometimes the format of the image you saved from the Web is not a format that your word processor can accommodate. In this case, you need a simple graphics program such as Graphic Workshop (downloadable from the Web) to change the format before you can insert it into your document. These graphics programs will also let you edit an image, such as changing its size so that it better fits the space where you want to place it.

that because the images you save from the Web were put there by someone else, you should always reference your source. And if your work is to be published, or if you intend to put it on the Web yourself, be sure to get written permission for any copyright images you want to include.

Scanning

In addition to downloading graphics, you can copy and save them with a scanner. Basically, a scanner is a device used to digitize an image electronically that you can then transfer, edit, and store in your computer. A scanner works the same way a fax or photocopier does.

A scanner can be very useful, particularly if you want a graphic image in your work that you are not capable of drawing yourself but that you do have a picture of.

Writing About Visuals: Some Guidelines

Using a visual requires more of you as a writer than simply inserting it in your written work. You need to use visuals *in conjunction with* what you write. The following guidelines will help you to (1) identify, (2) insert, (3) introduce, and (4) interpret visuals for your readers. By observing these guidelines, you can use visuals more effectively and efficiently.

Reference Visuals

Always mention in the text of your paper or report that you are including a visual. If you don't alert readers to a specific visual, they may skip it or wonder why it is there.

Identify Visuals

Each visual should have a number and a caption (title) that indicates the subject or that explains what the visual illustrates. An unidentified visual is meaningless. A caption helps your audience to interpret your visual—to see it with your purpose in mind. Inform your readers about what you want them to look for.

- Use a different typeface and size in your title than what you use in the visual itself.
- Include key words about the function and subject of your visual in a caption.
- Make sure any terms you cite in a caption are consistent with the units of measurement and scope of your visual.

Tables and figures should be numbered separately throughout the text—Table 1 or Figure 3.5, for example. (In the latter case, Figure 3.5 is the fifth figure to appear in Chapter 3.)

Table 2. Paul Jordan's Work Schedule, January 15–23
FIGURE 4.6 The proper way to apply for a small business loan.
Figure 12. Income Estimation Figures for North Point Technologies.

Cite the Source for Visuals

If you use a visual that is not your own work, give credit to your source (a newspaper, magazine, textbook, company, federal agency, individual, or Web site). If your paper or report is intended for publication, you must obtain permission to reproduce copyrighted visuals from the copyright holder.

Insert Visuals Appropriately

Here are some rules to keep in mind.

- Place visuals as close as possible to the first mention of them in text. By inserting an appropriate visual near the beginning of your discussion, you help readers understand the discussion better than if you placed the visual near the end.
- Never introduce a visual *before* a discussion of it; readers will wonder why it is there. Be sure to tell readers where the visual is found—"below," "on the following page," "to the right," "at the bottom of page 3."
- Visuals are most effective at either the top or bottom of a page.
- If the visual is small enough, insert it directly in the text rather than on a separate page. If your visual occupies an entire page, place that page containing your visual immediately after the page on which the first reference to it appears.
- Don't put a visual one or more pages after the discussion to which it pertains.

- Never collect all your visuals and put them in an appendix. Readers need to see them at that point in your discussion where they are most pertinent.

Introduce Your Visuals

Refer to each visual by its number and, if necessary, mention the title as well. In introducing the visual, though, do not just insert a reference to it, such as "See Figure 3.4" or "Look at Table 1." Relate each visual to the context of your discussion of it. Here are three ways of writing a lead-in sentence for a visual.

> **Poor:** Our store saw a dramatic rise in the shipment of electric ranges over the five-year period as opposed to the less impressive increase in washing machines. (See Figure 3.)

This sentence does not tie the visual (Figure 3) into the sentence where it belongs. The visual just trails insignificantly behind.

> **Better:** As Figure 3 shows, our store saw a dramatic rise in the shipment of electric ranges over the five-year period as opposed to the less impressive increase in washing machines.

Mentioning the visual in Figure 3 at the beginning is distracting. Readers will want to stop and look at the visual immediately before they know what it is or how you are using it.

> **Best:** Over the last five-year period, our store realized a dramatic rise in the shipment of electric ranges as opposed to the less impressive increase in the shipment of washing machines, as shown in Figure 3 below.

This last sentence is the best of the three because the figure reference and the explanation are in the same sentence, but the reference is not a distraction.

Interpret Your Visuals

Sometimes you may need to interpret a visual in context. In a study on the benefits of vanpooling, one writer supplied the following visual:

TABLE 1. Travel Time (in minutes): Automobile Versus Vanpool

Private automobile	*Vanpool*
25	32.5
30	39.0
35	45.5
40	52.0
45	58.5
50	65.0
55	71.5
60	78.0

Source: U.S. Department of Transportation. *Increased Transportation Efficiency Through Ride-sharing: The Brokerage Approach* (Washington, D.C., DOT-OS—40096): 45.

Explaining the table, the writer called attention to it in the context of the report on transportation efficiency.

> Although, as Table 1 above suggests, the travel time in a vanpool may be as much as 30 percent longer than in a private automobile (to allow for pickups), the total trip time for the vanpool user can be about the same as with a private automobile because vanpools eliminate the need to search for parking spaces and to walk to the employment site entrance.[1]

Alert Readers to Content

Tell readers why a visual is there and what specifically to look for. Of course, you should not spend time repeating information that is obvious from reading the text or looking at the visual. But occasionally you will have to tell your readers what is most significant in or about your visual. A director of an alumni association, eager to sell alumni life insurance, used the table below and then supplied a "sales" conclusion for it.

> Consider this table based on the U.S. Department of Labor Consumer Price Index for the past fifteen years. The value of insurance-benefit dollars decreases right along with dollars used in everyday expenses.

Average Annual Inflation Rates

Year	*Inflation Rate*	*Relative Dollar Value*
1982	6.2	$1.00
1983	3.2	0.97
1984	4.3	0.92
1985	3.6	0.88
1986	1.9	0.86
1987	3.6	0.82
1988	4.1	0.77
1989	5.1	0.7
1990	4.7	0.67
1991	4.7	0.64
1992	2.7	0.62
1993	3.0	0.63
1994	2.6	0.61
1995	2.9	0.59
1996	2.4	0.55

> If you haven't looked at your life insurance coverage recently, you may be surprised. Benefit levels thought sufficient just a few years ago may be inadequate for current and future needs.[2]

---

[1]James A. Devine, "Vanpooling: A New Economic Tool," *AIDC Journal* 15 (Oct. 1980): 13.
[2]Adaption courtesy of Northwestern University Alumni Affairs Office.

FIGURE 13.3 An ineffective visual: too much information is crowded into one graphic.

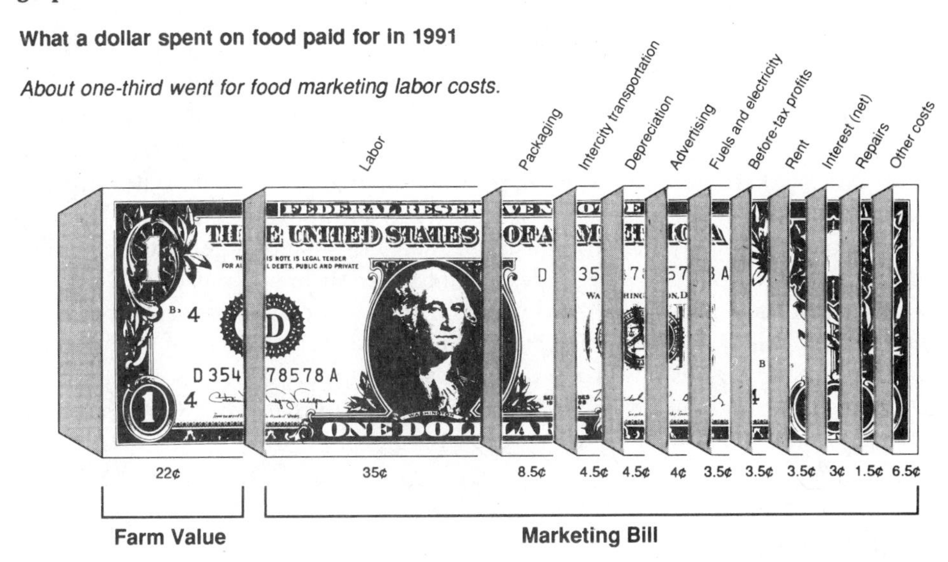

What *Not* to Do with a Visual

- Avoid visuals that include more details than you discuss or that your audience needs.
- Never use a visual that presents information that contradicts your work.
- Never distort a visual for emphasis or decoration.
- Be careful that you don't omit anything when you reproduce an existing visual.
- Never use visuals that discriminate or stereotype (for example, avoid pictures of a workforce that excludes female employees).
- Avoid visuals that would be misunderstood or be regarded as offensive in another culture.
- Don't offend ESL readers by using a culturally biased color (for example, reconsider using red for warning signals or danger; red in China signals happiness and good fortune and is used at weddings).

See how Figure 13.3 violates many of these rules. It divides the dollar bill into too many slices. Confronted with so many different wedges the reader would have trouble identifying, separating, comparing, and understanding the costs. It would be better to use a different visual to avoid so much clutter.

Two Categories of Visuals

Visuals can be divided into two categories—**tables** and **figures.** A table arranges information—numbers and/or words—in parallel columns or rows for easy comparison of data. Anything that is not a table is considered a figure. Figures include graphs, circle charts, bar charts, organizational charts, flow charts, pictograms, maps, photographs, and drawings.

Tables

Tables are parallel columns or rows of information that present data. The information is organized and arranged into categories to show changes in time, distance, cost, employment, or some other distinguishable or quantifiable variable. Tables allow readers to compare a great deal of information in a compact space. Electronic spreadsheet programs heavily rely on tables to convey information, as in Table 13.1. Tables also summarize material for easy recall—causes of wars, symptoms of diseases, provisions of a law, troubleshooting commands, as in Table 13.2. Observe how the table in Table 13.3 easily condenses much information and arranges it in quickly identifiable categories.

TECH NOTE

Most word-processing programs include a function to help you create and position a table within your text. Using the "Table" command, specify the number of rows and columns the table will contain, and the word-processing software will create an empty table for you to fill in with data or will convert your running text into a table. You can also edit such tables and insert or remove lines separating rows and columns. Other features include commands to widen columns and to include various spreadsheet templates (such as schedules, investments, and profit margins).

Parts of a Table

To use a table properly, you need to know the parts that constitute it. Refer to Table 13.3 as you read the following:

- The main **column** is "Years Attending," and the **subcolumns** are the years (1985, 1990, 1995, 2000) for which the table gives data.
- The **stub** refers to the first vertical column on the left-hand side. The stub column heading is "Period of Service." The stub lists the items (the wars and conflicts in which the Lincoln-area veterans participated) for which information is broken down in the subcolumns.
- A **rule** (or line) across the top of the table separates the headings from the body of the table.

TABLE 13.1 A Spreadsheet Table

	A	B	C	D	G	H	
5	Acct#	Account Description	YTD	June	March	February	Januar
6							
7	Employee Costs						
8	110	Payroll	$171,465	$29,750	$28,547	$27,540	
9	120	IRS/FICA/Wk comp/State/SDI	$46,295	$8,032	$7,707	$7,435	
10	130	Commissions	$56,436	$9,520	$8,875	$9,825	
11	140	Retirement Plan	$20,575	$3,570	$3,425	$3,304	
12	150	Insurance	$9,000	$1,500	$1,500	$1,500	
13							
14	Subcontractors & Services						
15	201	Telecommunication Services	$2,616	$436	$436	$436	
16	202	Design Consultants	$875	$500	$375	$0	
17	203	Photo/Video Services	$535	$45	$325	$45	
18	250	Graphic Services	$2,957	$765	$95	$375	
19	251	Photo/Stats	$1,612	$568	$755	$0	
20	252	Typesetting	$1,453	$388	$325	$195	
21	253	Printing Services	$6,186	$951	$849	$325	
22	254	Legal & Accounting					
23							
24	Supplies and Materials						
25	301	Office Supplies	$875	$500	$732	$433	
26	302	Office Postage	$535	$45	$255	$325	
27	303	Office Equipment & Furniture	$2,957	$765	$78	$21	
28	304	Miscellaneous Supplies	$1,612	$568	$49	$36	
29							
30	Facilities Overhead						
31	405	Plant	$1,612	$568	$1,700	$1,700	
[illegible]	410	Office Postal	[illegible]	[illegible]	[illegible]	[illegible]	

Labeling a Table

Label the categories appropriately and consistently. If some form of measurement is consistently involved, include the unit of measurement as part of the column heading—Weight (in pounds), Distance (in miles), Time (in hours), Quantity (in dozens). The unit of measurement should not be repeated for each entry in a column. Also, units should be consistent; do not jump from miles to meters, pounds to ounces.

Wrong:	**Weight**	**Height**
	120 lbs	165 cm
	132 lbs	5'9"
	122 lbs	5'7"
	58.5 kg	172 cm

Correct:	**Weight (in kilograms)**	**Height (in centimeters)**
	54.0	165
	55.2	176
	54.9	166
	58.5	172

If something in the table needs to be explained or qualified, put a footnote (often signaled by a small raised letter: *a* or *b*) below the table where the signaled information is further identified or qualified. In Table 13.3 the *a* after the title

TABLE 13.2 Informational Table Showing Causes and Remedies of Display-Related Problems

Symptom	*Possible Cause*	*Corrective Action*
No display.	No power to computer.	Turn the computer and monitor on. Make sure the system unit and monitor are plugged into a power outlet.
No display; computer beeps three times when power is applied.	Video board not recognized.	Make sure the video board is configured. Check jumper and switch settings. Check interrupt, memory, DMA channel, etc. See Appendix A and the instructions that came with your video board for information.
No display; computer beeps once; floppy drive light flashes.	No video signal to monitor.	Make sure cabling between monitor and video board is correct; check connectors. Check brightness and contrast settings on monitor.
Screen goes blank.	A TSR, such as a screen saver, is interfering with display.	Disable the terminate-and-stay-resident program (TSR) by removing the statement that loads it from your system files, or for screen savers, turn them off from the Desktop application in Control Panel.
Video distorted; characters unreadable; system hangs.	Device drivers incompatible with video board or not installed.	Reinstall video device drivers from Windows Setup or select a different display driver from the options provided.

Packard Bell, Microsoft Multimedia Pack, *User's Guide: Version 1.0 For Windows,* 1993. Reprinted with permission of Microsoft Corporation.

points to the qualification that three communities were not considered in the Lincoln area when data for the table were gathered. The [b] after the Vietnam entry in the stub clarifies the official dates for that conflict.

Guidelines for Using Tables

When you include a table in your work, make sure you follow these guidelines:

- Number each table according to the order in which it is discussed (Table 1, Table 2, Table 3).
- Give each table a concise and descriptive title.
- Label all parts of your table.
- Provide stubs for names; but put numbers under column headings.
- Never show readers your table before you discuss it.
- Put your table on one page; it is hard for readers to follow a table spread across different pages.
- If possible, place the table vertically, not horizontally, on the page.

Table number

TABLE 13.3 Veterans Attending Lincoln-Area VFW Posts[a]

Period of service	*Years Attending*			
	1985	*1990*	*1995*	*2000*
World War II	59	55	22	6
Korean Conflict	330	309	100	96
Vietnam[b]	240	230	205	186
Gulf War	0	0	217	203

← *Column heading*
← *Subcolumn heading*
Stub

Source: Lincoln VFW Association ← *Origin of data*
[a]Does not include Bayside, Morton, or Westover
[b]August 1964–May 1975
Footnotes

- Arrange the data you want to compare vertically; it is easier for an audience to read down than across a series of rows.
- Arrange the information in a table in descending order; that is, list higher values first.
- Place tables at the top (preferable) or bottom of the page and center them in the type area.
- Leave at least one inch of white space between text and table.
- Don't use more than four or five columns; tables wider than that are more difficult for readers to use.
- Round off any numbers in your columns to the nearest whole number to assist readers in following and retaining information.
- Always give credit to the source (the supplier of the statistical information) on which your table is based.

Figures

As we saw, any visual that is not a table is classified as a **figure.** The nine types of figures we will examine here are

- line graphs
- pie charts
- bar charts
- organizational charts
- flow charts
- pictographs
- maps
- photographs
- drawings

Presentation graphics software (discussed on pp. 504–505) will allow you to produce the visuals in this section of Chapter 13 very easily.

Line Graphs

Graphs transform numbers into pictures. They take statistical data presented in tables and put them into rising and falling lines, steep or gentle curves.

Functions of Line Graphs

Graphs vividly portray data that changes, such as

- cycles
- fluctuations
- trends
- distributions
- increases and decreases in profits
- employment
- energy levels
- temperatures

Graphs are often used in business communications. The ups and downs in the graph line depict changes in sales, profits, production costs, manufacturing output, expenses, staffing, and much more. Such graphs not only describe past and current situations but also forecast trends.

Graphs and Tables

Because graphs actually show change, they are more dramatic than tables. You will make the reader's job easier by using a graph rather than a table. Many financial Web sites and publications—the *Wall Street Journal,* for example—open with a graph for the benefit of busy readers who want a great deal of financial information summarized quickly.

A Simple Graph

Basically, a simple graph consists of two sides—a **vertical axis** and a **horizontal axis**—that intersect to form a right angle as in Figure 13.4. The space between these two axes contains the picture made by the graph—in Figure 13.4 the amount of snowfall in Springfield between November 1998 and April 1999. The vertical line represents the dependent variable—the snowfall in inches, the horizontal line, the independent variable—time in months. The dependent variable is influenced most directly by the independent variable, which is almost always expressed in terms of time or distance. Hence, in Figure 13.4, the given month affects the amount of snow Springfield received. The vertical axis is read from bottom to top; the horizontal axis is read from left to right.

When the dependent variable occurs at a particular time on the independent variable (horizontal line), the place where the two points intersect is marked, or plotted, on the graph. It is called the **data point.** After all the points are plotted, a

FIGURE 13.4 A simple line graph of the amount of snowfall in Springfield from November 1998 to April 1999.

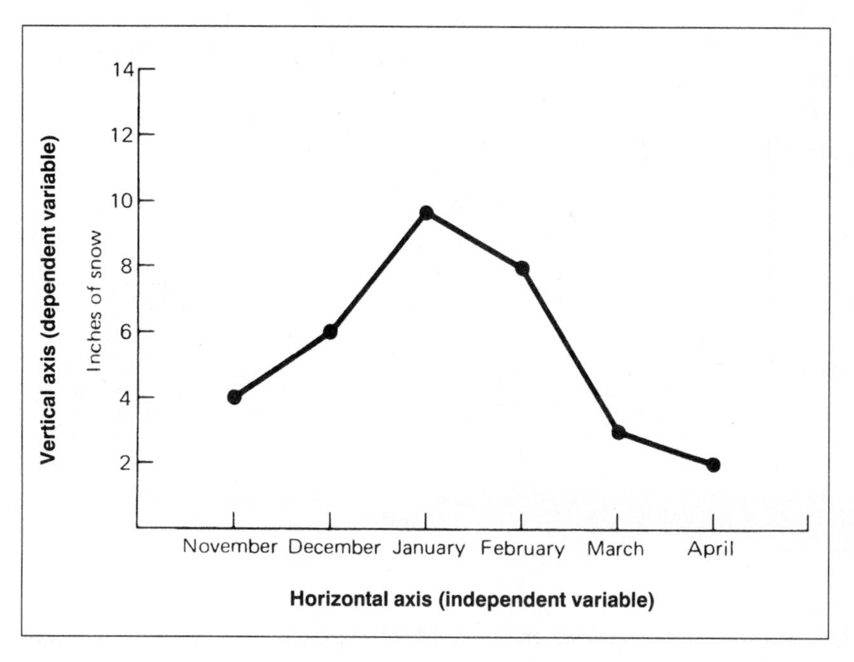

line is drawn to connect them; the resulting curve gives a picture of the overall pattern of snowfall in Springfield during the winter of 1998–1999.

The way in which scales are set up is crucial to the success of a graph. Many graphs may not have ranges indicated by equally spaced lines **(tick marks).** But on most of the graphs you construct, you should use tick marks as a scale to show values, distances, or time. The topic will dictate the intervals to use. The vertical axis tick marks in Figure 13.4 indicate 2 inches of snow. The time scale (independent variable on horizontal axis) can be calculated in minutes, hours, days, years, or, as in Figure 13.4, months.

Multiple-line Graphs

The graph in Figure 13.5 contains only one line per category. But a graph can have multiple lines to show how a number of **dependent variables** (conditions, products) compare with each other.

The six-month sales figures for three salespeople can be seen in the graph in Figure 13.5. The graph contains a separate line for each of these three salespersons. At a glance readers can see how the three compare and also how many dollars each salesperson generated per month. Note how the line representing each person is clearly differentiated from the others by means of dots, dashes, or an unbroken line. You can also use color to distinguish lines. Additionally, each line is clearly labeled and that the labels do not cover any lines or data points.

FIGURE 13.5 A multiple-line graph showing sales figures for the first six months of 1998 for three salespeople.

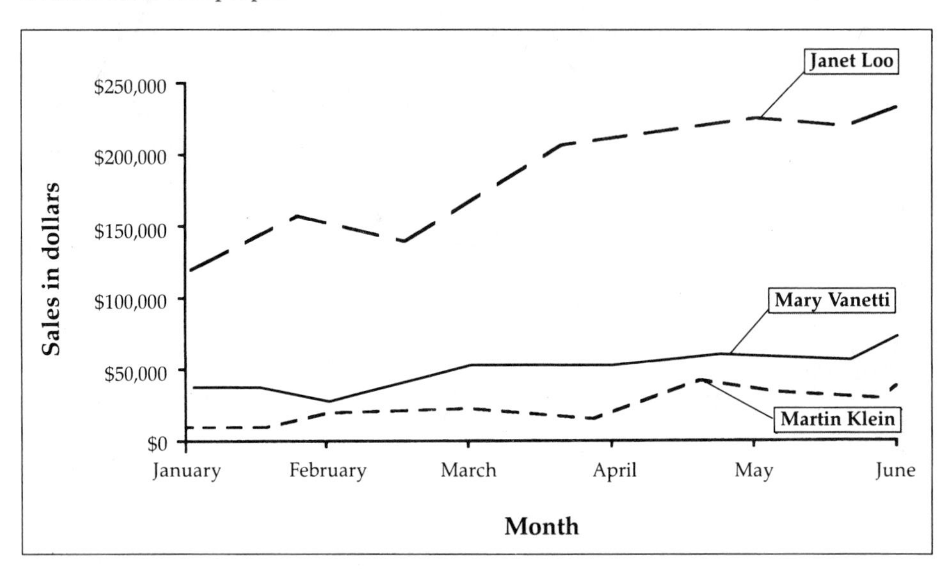

If the lines do run close together, a **legend,** or explanatory key, underneath or above the graph is a better way to identify individual lines. Although some graphs may contain as many as five or six lines, it is better to limit multiple-line graphs to two or three lines (or dependent variables) so that readers can interpret the graph easily.

Guidelines on Creating a Graph

1. Make a rough pencil sketch of your graph first (as a draft) to identify the axes, scale, and how many data points you need to plot.
2. Use no more than three lines in a multiple-line graph.
3. Label each line to identify what it represents for readers.
4. Keep each line distinct in a multiple-line graph by using different colors or dots or dashes.
5. Make your graph data points large enough to show a reasonable and ethical number of plotted points (using only three or four data points may distort the evidence for readers).
6. Keep the scale consistent and realistic. If you start with hours, do not switch to days or vice versa. If you are recording annual rates or accounts, do not skip a year or two in the hope that you will save time or be more concise. Do not jump from 1988–1989, 1991, 1993, 1996, 1997, to 1999–2000. Include all the years you are surveying, or equal multiples of them (such as 1988, 1990, 1992, 1994, 1996, 1998).
7. For some graphs there is no need to begin with a zero. This is called a suppressed zero graph which automatically begins with a larger number when it

would be impossible to start with zero. For others, you may not have to include numbers beyond seven or eight. Your subject and the ranges you are showing will determine what value you give your tick marks.

Charts

Although charts and graphs may seem similar, there is a big difference between them. Because graphs are more complex, charts are preferable when you are communicating with a consumer audience. A graph is plotted according to specific mathematical coordinates. Charts, on the other hand, do not display exact and complex mathematical data. Instead, charts basically give readers the benefit of a significant visual impact, presenting an overall picture of how individual pieces of data (from a graph or table) fall into place to express relationships. A chart is often constructed from data contained in spreadsheets and other statistical instruments.

Among the most frequently used charts are (1) circle, or pie, charts, (2) bar charts, (3) organizational charts, and (4) flow charts, all discussed below.

Circle Charts

Circle charts are also known as **pie charts,** a name that descriptively points to their construction and interpretation. The circle chart is one of the most easily understood

FIGURE 13.6 A 3-D circle chart showing the breakdown by department of proposed Midtown city budget for 1997.

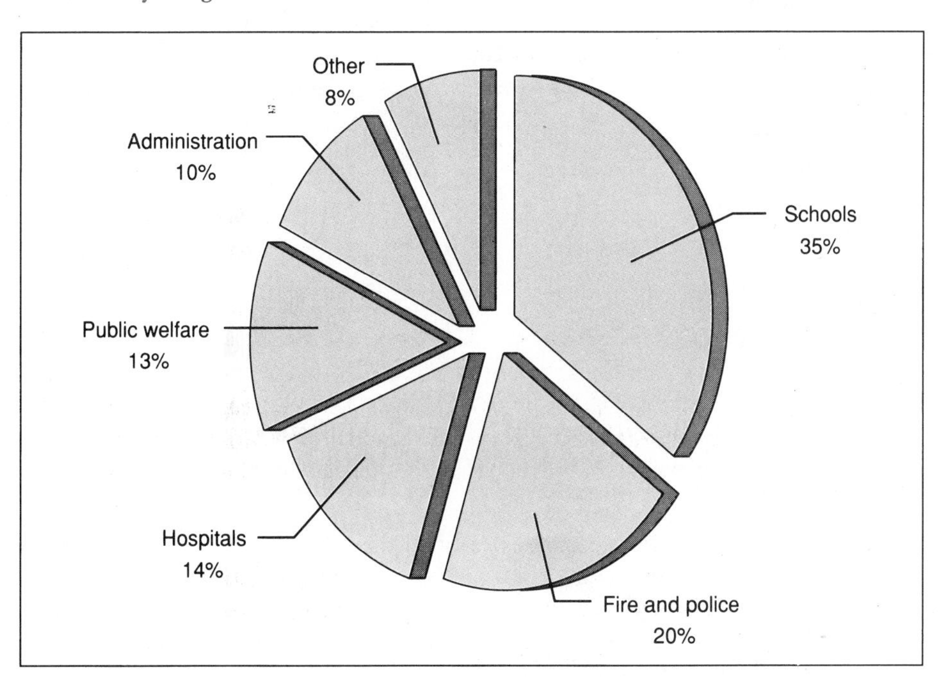

illustrations; Figure 13.6 shows an example of one. Because of its simplicity, the pie chart is popular in government documents (especially the Bureau of the Census), financial reports, and advertising messages.

The full circle, or pie, represents the whole amount (100 percent) of something: family income, population group, area of land, resources of an organization of institution. Each slice or wedge represents a percentage of portion of the whole. The full circle can stand for the entire budget of a company or a family, or it can represent just a single dollar of that budget and show how it is broken down for various expenses. Each slice of the pie, then, stands for a part of the whole.

A circle chart effectively allows readers to see two things at once: the relationship of the parts to one another and the relationship of the parts to the whole.

Preparing a Circle Chart

Follow these six rules to create and present your circle chart.

1. **Keep your circle chart simple.** Don't try to illustrate technical statistical data in a pie chart. Pie charts are primarily used for general audiences.

2. **Do not divide a circle, or pie, into too few or too many slices.** If you have only three wedges, consider using another type of visual to display them (a bar chart, for example, which will be discussed later). If you have more than seven or eight wedges, you will divide the pie too narrowly, and overcrowding will destroy the dramatic effect. Instead, combine several slices of small percentages (2 percent, 3 percent, 4 percent) into one slice labeled "Other," "Miscellaneous," or "Related Items."

3. **Make sure the individual slices total 100 percent.** For example, if you are constructing a circle chart to represent a family's budget, the breakdown percentages might be as follows:

Category	Percentage	Angle of slice
Food	22%	90.0°
Housing	25%	90.0°
Energy	20%	79.2°
Clothes	13%	46.8°
Health Care	12%	43.2°
Miscellaneous	8%	28.8°
Total	100%	360°

4. **Put the largest slice first, at the 12 o'clock position, and then move clockwise with proportionately smaller slices.** By beginning with the biggest slice, you call attention to its importance.

5. **Label each slice of the pie horizontally.** If the slice is large enough, key in the identifying term or quantity inside, but make sure the label is big enough to read. Do not put in a label upside down or slide it in vertically. If the individual slice of the pie is small, do not try to squeeze in a label. Draw a connecting line from the slice to a label positioned outside the pie.

6. Shade, color, or cross-hatch slices of the pie to further separate and distinguish the parts. But be careful not to obscure labels and percentages; also make certain that adjacent slices can be distinguished readily from each other.

Bar Charts

A bar chart consists of a series of vertical or horizontal bars that indicate comparisons of statistical data. For instance, in Figure 13.7 vertical bars show increases in Internet users; Figure 13.8 uses horizontal bars to depict increases in numbers of working mothers. The length of these bars is determined according to a scale that your computer software can easily compute.

Bar charts make a dramatic visual impression on the reader. They are valuable tools in sales meetings to demonstrate how well (or poorly) a product, a service, or a particular section of the company has done. You often see bar charts at an office recording the financial goals for charitable drives and on home computer screens showing the breakdown of budgets or the relationship of a consumer's income to liabilities.

Advantages of Bar Charts

Bar charts are less exact than tables, as Figures 13.7 and 13.8 show. Note that the number of individuals using the Internet or the percent of mothers working outside the home are not expressed in exact figures, as they would be in a table, but are approximated by the length of the bar. Do not assume, however, that the bar chart is inaccurate; many bar charts are based on tables. What you lose in precision with a bar chart, you gain in visual flair. Finally, a bar chart is much more fluid and dynamic than a circle. The circle is static; the bar chart (like the graph) presents a moving view.

FIGURE 13.7 A vertical bar chart.

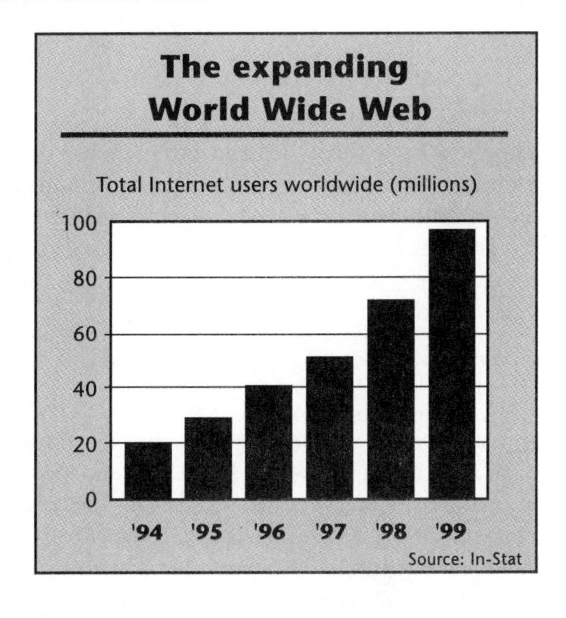

FIGURE 13.8 A horizontal bar chart with labeled bars.

Bar Charts, Graphs, and Tables—Which Should You Use?

When should you use a bar chart rather than a table or line graph? Your audience will help you decide. If you are asked to present statistics on costs for the company accountant, use a table. That reader demands a precise listing. However, if you are presenting the same information to a group of stockholders or to a diverse group of employees, a bar chart may be more relevant because these readers are more interested in seeing the statistics in action—the effects of change—than with the underlying causes and precise statistical details. Since it is limited to a few columns, however, a bar chart cannot convey as much information as a table or graph.

Types of Bar Charts

- *A simple, vertical bar chart*

  Figure 13.7 is the most basic form of bar chart. Each bar represents the number of Internet users, and the height of the bar corresponds to the number of individual users (in millions) for a given year. To read the chart effectively, note where the top of the bar is in relation to the vertical scale on the left. The chart compares one type of data (the number of Internet users) over a period of time (each year from 1994 to 1999).

- *A multiple-bar chart*

  Figure 13.9 shows a variation of the simple, vertical bar chart. Four differently shaded bars are used for each year to represent the amount of money spent on different types of advertising in 1985, 1990, 1995, and 2000. A legend at the top of the chart explains what each bar stands for. The multiple-bar chart is useful for showing a variety of comparisons. A word of caution is in order about multiple-bar charts: never use more than four bars in a group for any one year. Otherwise, your chart will become crowded and difficult to read.

FIGURE 13.9 A multiple-bar chart showing advertising expenditures by selected media: 1985–2000.

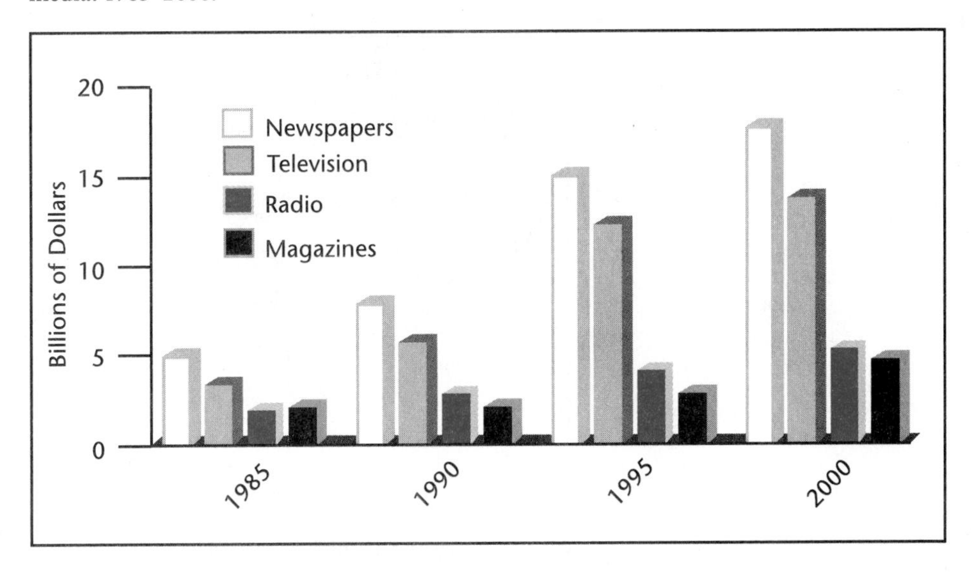

- *A segmented, or cross-hatched (divided), bar chart*

  To show the different components that constitute a measured whole, use a segmented bar chart like the one in Figure 13.10. A single segmented bar lists the travel expenses of Weemco, a small firm, in October 1998. The entire bar equals the travel total—$137,000—which was spent in four areas: airfare, ground transportation, lodging, and meals. Each of these expenses is represented by a different type of shading on the single column. Combined, the multiple sources account for all travel expenses Weemco had in one month. A group of segmented bars can be used to show multiple comparisons among many categories, as in Figure 13.11, which depicts energy consumption levels and types in five states.

Organizational Charts

Unlike the charts discussed so far, an **organizational chart** does not display statistical data, nor does it record movements in space or time. Rather, it pictures the chain of command in a company or agency, with the lines of authority stretching down from the chief executive, manager, or administrator to assistant manager, department heads, or supervisors to the workforce of employees. Figure 13.12 shows a hospital's organizational chart for its nursing services.

Functions of Organizational Charts

Organizational charts

- inform employees and customers about the makeup of a company
- show the various offices, departments, and units through which orders and information flow in the company or agency

FIGURE 13.10 A segmented bar chart representing total travel expenditures for Weemco Industries for October 1998.

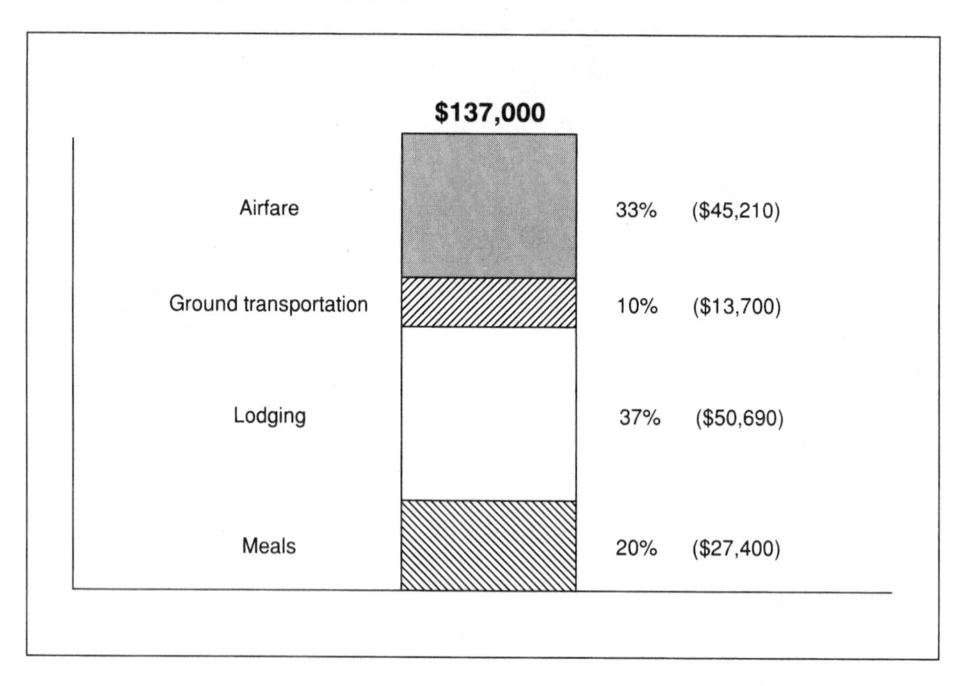

- coordinate employee efforts in routing information to appropriate departments

Structuring an Organizational Chart

An organizational chart shows relationships by using lines to connect rectangles or circles to each other, starting at the top with the chief executive and moving down to lower-level employees. (Sometimes the name of the individual holding the office is listed in addition to the title of the office.) In the organizational chart in Figure 13.12, positions of equal authority are on parallel lines, and all jobs under the supervision of one individual are joined by bracketing lines. Individuals who serve in advisory capacities or who are indirectly responsible to a higher administrator are listed with broken or dotted lines, as depicted in the unit secretary positions in Figure 13.12.

Flow Charts

Like an organizational chart, a **flow chart** does not present statistical information, but as its name implies, a flow chart does show movement. It displays the stages in which something is manufactured, accomplished, develops, or operates. Flow charts can also be used to plan the day's or week's activities.

A flow chart tells a story with arrows, boxes, and sometimes pictures. Boxes are connected by arrows to visualize the stages of a process. The presence and di-

FIGURE 13.11 A multiple-bar, segmented bar chart showing energy consumption by sector in five states that consumed the most energy in a given year.

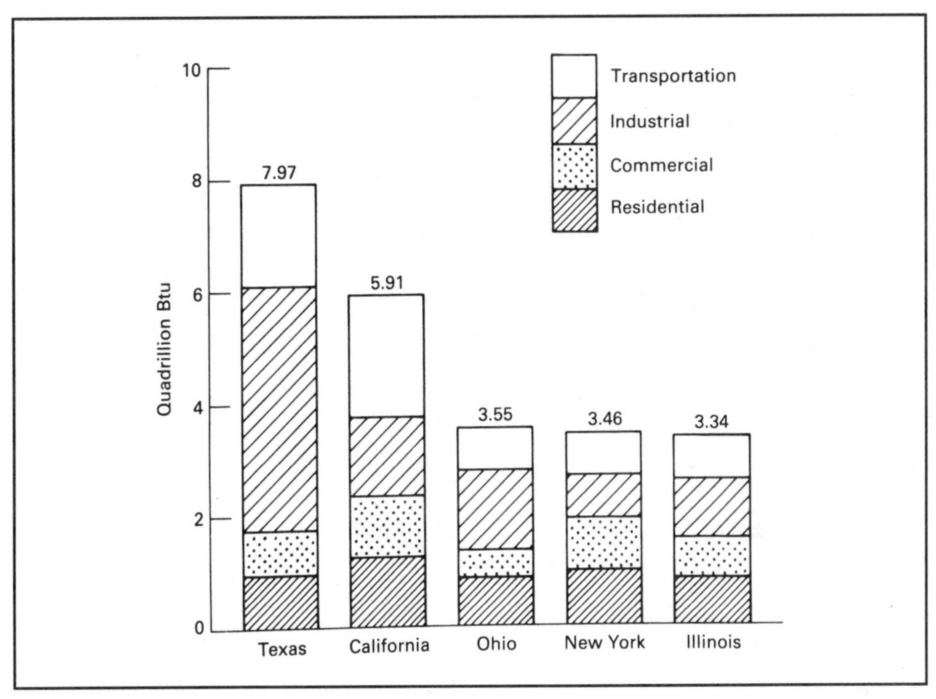

Source: U.S. Energy Information Administration.

rection of the arrows tell the reader the order and movement of events involved in the process. Flow charts often proceed from left to right and back again, as in the flow chart below, showing the steps to be taken before graduation.

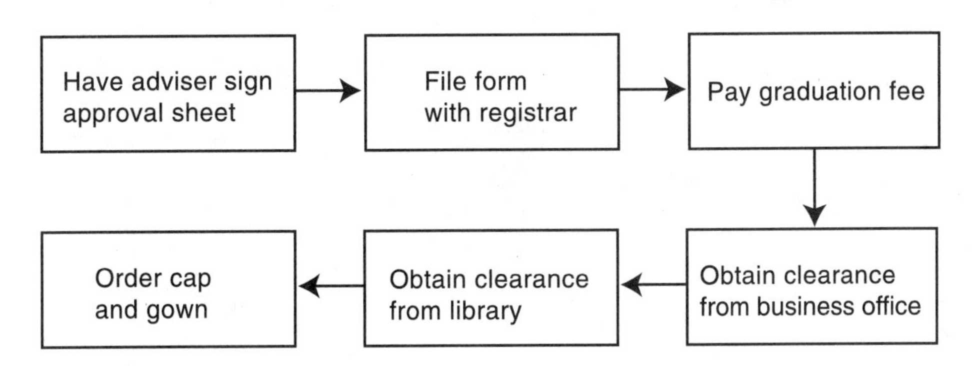

Flow charts can also be constructed to read from top to bottom. Computer programming instructions frequently are written this way. See, for example, Figure 13.13, which uses a computer programming flow chart to show the steps a student must follow in writing a research paper.

FIGURE 13.12 An organizational chart representing critical care nursing services at Union General Hospital.

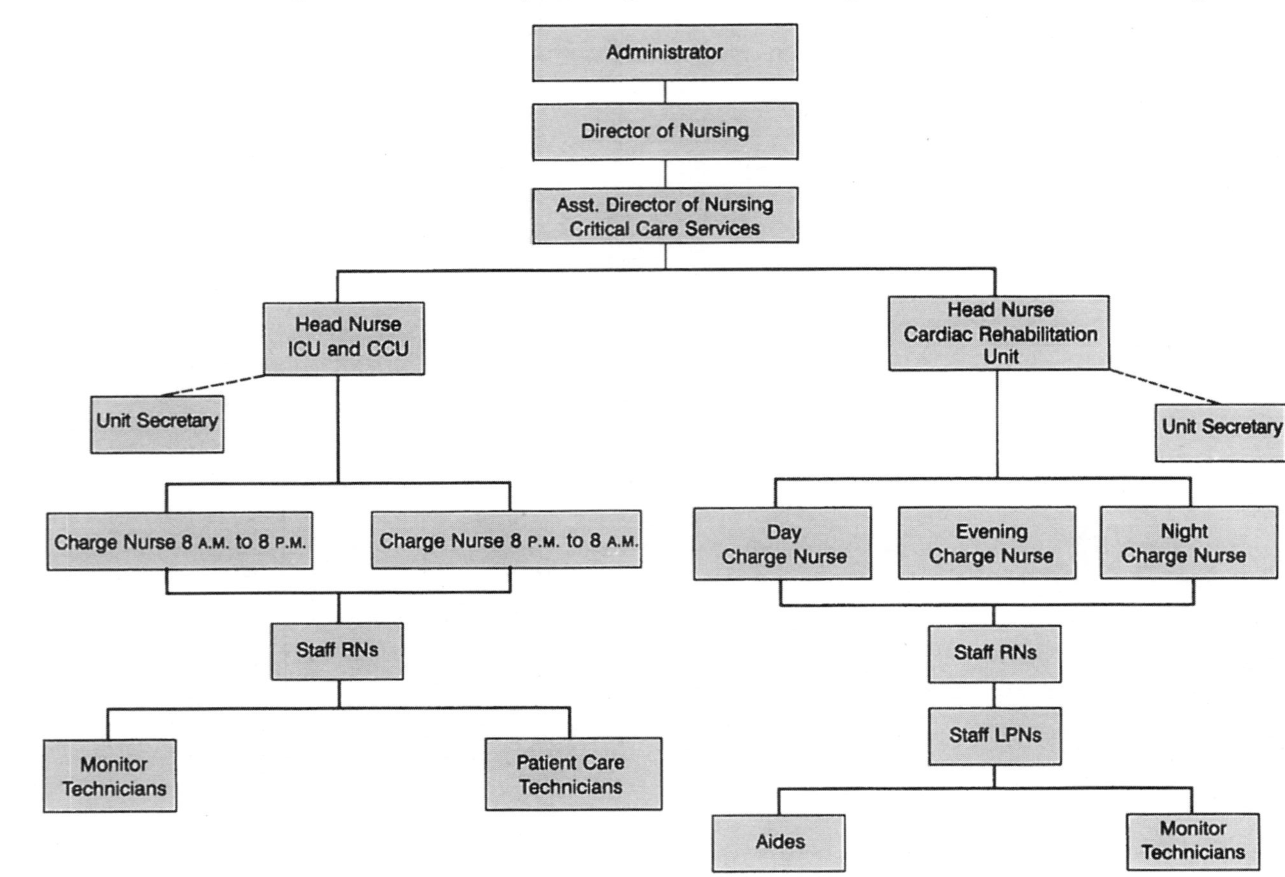

FIGURE 13.13 A computer programming flow chart showing steps in writing a research paper.

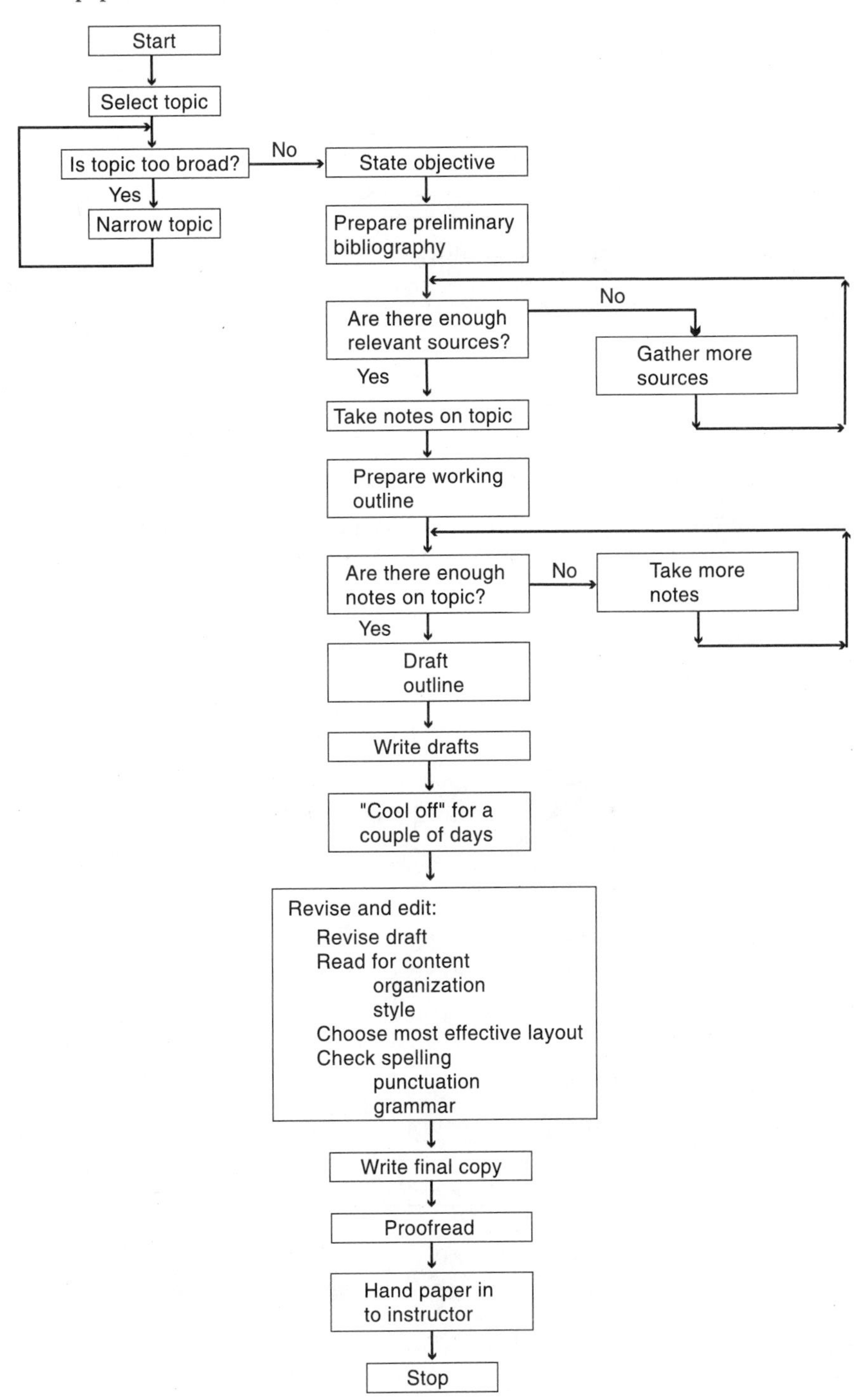

A flow chart should clarify, not complicate, a process. Do not omit any important stages, but at the same time do not introduce unnecessary or unduly detailed information. Show at least three or four stages, however, and make sure the various stages appear in the correct sequence.

Pictographs

Similar to a bar chart, a pictograph uses picture symbols (called pictograms) to represent differences in statistical data, as in Figure 13.14. A pictograph repeats the same symbol or icon to depict the quantity of items being measured. Each symbol stands for a specific number, quantity, or value. Pictographs are visually appealing and dramatic and are far more appropriate for a nontechnical than a technical audience.

Guidelines on Creating Pictographs

When you create a pictograph, follow these four guidelines:

1. Choose an appropriate symbol for the topic—such as computer screens to represent the increase in the number of fax modems or barns for the number of farms.

FIGURE 13.14 An example of a pictograph.

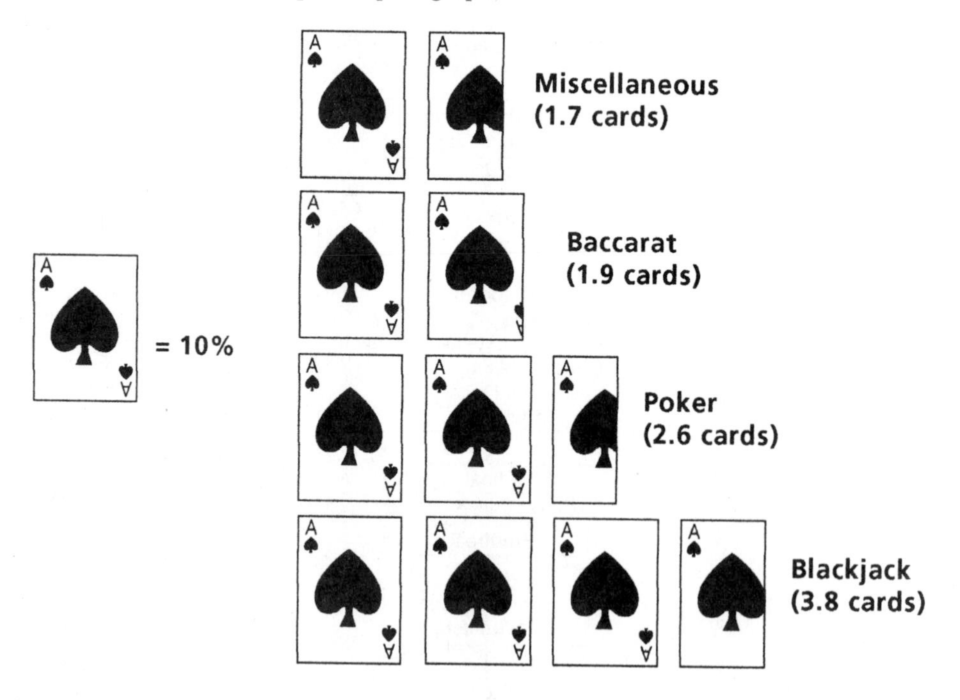

2. Always indicate the precise quantities involved by placing numbers after the pictures so that the reader knows exactly how much the total number of pictograms represents.
3. Increase the number of symbols rather than their sizes because differences in sizes are often difficult to construct accurately (even with a graphics software program) and hard for the reader to interpret.
4. Avoid crowding too much information into a pictograph such as the one in Figure 13.3, in which a dollar bill is divided into too many wedges showing the budget breakdown.

Maps

The maps you use on the job may range from highly sophisticated and detailed geographic tools to simple sketches. You may need a **large-scale map** that displays a good deal of social, economic, or physical data (such as population density, location of retail businesses, hills, expressways, or rivers) for a small area. That kind of detail is given in the map used by the Smithville Water Department in Figure 13.15.

FIGURE 13.15 A large-scale map showing location of Smithville Water Department's filter plants and pumping stations.

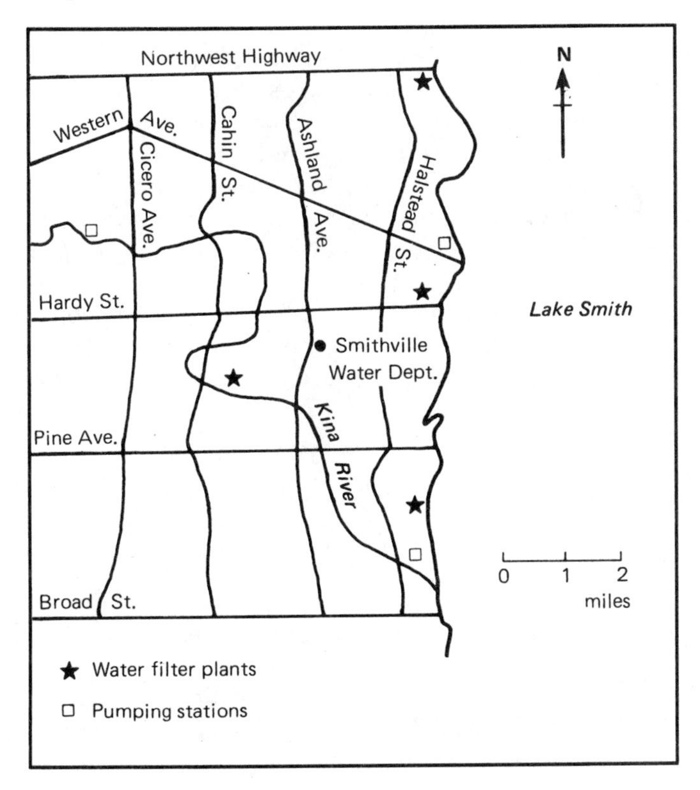

TECH NOTE

You can easily create very detailed road maps with Mapquest software. First you select a country, say, the United States, then below the map display area, you point and click on "Zoom in." Next point to the area on the map that you are interested in and click again; each time you point and click, a new map will be drawn that is more detailed than the previous one and that is centered on the place you last pointed to. You just continue to point and click until you have the desired level of detail.

In addition to roads, Mapquest has a number of "Places of Interest" categories that you can select to be displayed. As you are zooming in you can go at any time to the bottom of the screen and select whatever categories of interesting places you want to add to your map. Categories include *Personal* (ones you add yourself), *Dining, Lodging, Entertainment, Education, Recreation, Shopping,* and *Health Care*. Each of these categories has a selection of subcategories from which you can further choose so that you will see displayed exactly what you need.

Your job requirements will dictate how detailed your map should be. Architects and builders need extremely detailed maps showing the location of pipes, telephone cables, and easement lines. An urban planner involved in developing a new community or an employee submitting site plans for a company's new location will use less detailed maps. Between these two extremes, the individual working for a government agency investigating fire or flood damage to a neighborhood may require a map that indicates individual houses without presenting detailed features of those dwellings.

You may have to construct your own map or scan one in a published source (a government document, an atlas, or an auto club publication) or on the Internet. If you scan a published map, be sure to obtained permission to use it from the copyright holder.

Guidelines for Using a Map

Follow these steps when you use a map:

1. Always acknowledge your source if you did not construct the map yourself.
2. Use dots, lines, colors, symbols, or shading to indicate features. Markings should be clear and distinct.
3. If necessary, include a legend, or map key, providing an explanation of dotted lines, colors, shading, or symbols. (Note the legend for water filter plants and pumping stations in Figure 13.15.)

4. Exclude features (rivers, elevations, county seats) that do not directly depict your topic. For example, a map showing the crops grown in two adjacent counties need not show all the roads and highways in those counties. A map showing the presence of strip mining needs to indicate elevation, but a map depicting population or religious affiliation need not include topographical (physical) detail.
5. Indicate direction. Conventionally, maps show north, often by including an arrow and the letter *N*: *N*↑.

Photographs

Correctly taken or scanned, photographs are an extremely helpful addition to job-related writing. The photograph's chief virtues are realism and clarity. To a reader unfamiliar with an object or a landscape, a photograph may provide a much more convincing view than a simple drawing. Photographs of "before and after" scenes are especially effective.

You will have to use special care when you take a picture. The most important point to remember is that what you see and what the camera records might be two different sights. Before you take a picture, decide how much foreground and background image you need. Include only the details that are *necessary* and *relevant* for your purpose. Inexperienced photographers need to remember the following four points.

1. **Focus your camera.** Make sure that you have proper lighting.
2. **Select the correct angle.** Choose a vantage point that will enable you to record essential information as graphically as possible.
3. **Give the right amount of detail.** Pictures that include clutter compete for the reader's attention and detract from the subject. A real estate agent wanting to show that a house has an attractive front does not need to include sidewalks or streets. At the other extreme, do not cut out a necessary part of an object or landscape. Avoid putting people in a photograph when their presence is not required to show the relative size or operation of an object.
4. **Take the picture from the right distance.** The farther back you stand, the wider your angle will be, and the more the camera will capture with less detail. If you need a shot of a three-story office building, your picture may show only one or two stories if you are standing too close to the building when you photograph it. Standing too far away from an object, however, means that the photograph will reduce the object's importance and record unnecessary details.
5. **Choose the right film.** Depending on your printing capabilities and intentions, select color or black-and-white film, of the proper speed for your subject and setting.

To get a graphic sense of the effects of taking a picture the right and wrong way, study the photographs found in Figures 13.16 through 13.19. A clear and useful picture of a hydraulic truck (often called a "cherry picker") used to cut high branches can be seen in Figure 13.16. The photographer rightly placed the

FIGURE 13.16 An effective photograph—truck in foreground, enough background information, and a worker to show size and function of truck.

Photograph by David Longmire.

truck in the foreground, but included enough background information to indicate the truck's function. The worker in the bucket helps to show the truck in operation.

In Figure 13.17 everything merges because the shot was taken from the wrong angle. The reader has no sense of the parts of the truck, their size, or their function. Another kind of error can be seen in Figure 13.18. Here the photographer was more interested in the person than in the equipment. But looking at this picture, the reader has no idea where the worker came from or what he is doing up there. Finally, the reader looking at Figure 13.19 has a view of work going on, but no idea that the worker is performing the job from a truck.

Drawings

Drawings can show where an object is located, how a tool or machine is put together, or what signals are given or steps taken in a particular situation. By studying your drawing and following your discussion, readers will be better able to operate, adjust, repair, or order parts for equipment. Drawings are especially helpful when you are giving instructions (see Chapter 14).

FIGURE 13.17 A poor photograph—taken from the wrong angle so that everything merges and becomes confusing.

Photograph by David Longmire.

FIGURE 13.18 A poor photograph—focus is on worker, but there is nothing else to identify person or work going on.

Photograph by David Longmire.

FIGURE 13.19 A poor photograph—focus is on work going on, but there is nothing to indicate that worker is operating from a hydraulic truck.

Photograph by David Longmire.

Advantages of Drawings over Photographs

Drawings generally have two advantages over ordinary photographs:

1. You can include as much or as little detail as necessary in a drawing. The eye of the camera is not usually so selective; it tends to record everything in its path, including details that may be irrelevant for your purpose.
2. A drawing can show interior as well as exterior views, a feature that is particularly useful when the reader must understand what is going on under the case, housing, or hood.

Types of Drawings

A drawing can be simple, such as the one in Figure 13.20, which shows readers exactly where to place smoke detectors in a house.

A more detailed drawing can reveal the interior of an object. Such sketches are called **cutaway drawings** because they show internal parts normally concealed from view. The underground pipes and service lines in a sanitary sewer are shown in Figure 13.21. The earth banking left in the foreground of the sketch shows where these pipes are buried.

Another kind of sketch is known as an **exploded drawing.** This drawing blows the entire object up and apart to show how the individual parts, each kept in order, are arranged. An exploded drawing of a chair is seen in Figure 13.22. Figure 13.22 also uses **callouts,** or labels, identifying the components of what is being depicted. These labels are often attached to the drawing with arrows or lines. As the name suggests, the labels "call out" the parts so readers can identify them quickly.

FIGURE 13.20 A drawing showing where to place smoke detectors in a house.

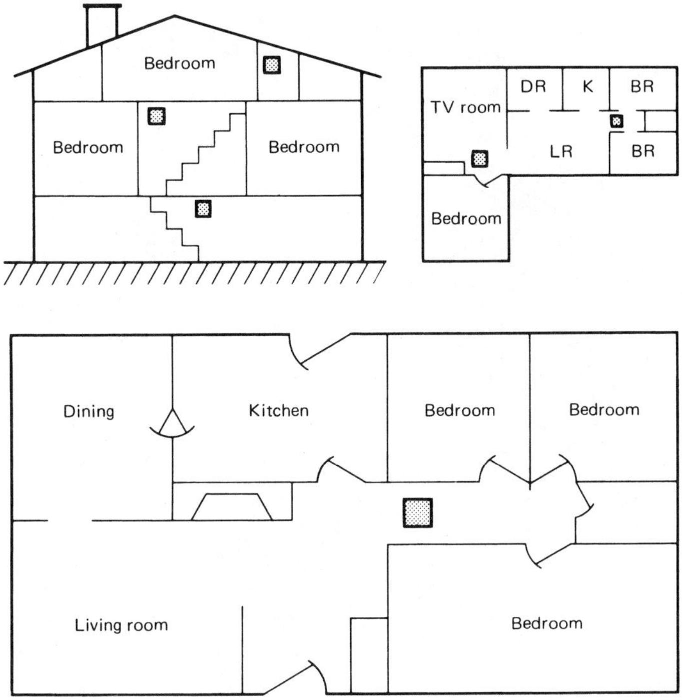

Where to place smoke detectors . . .

In a single level home (bottom), a smoke detector (square) should be located outside the sleeping area. A smoke detector should be provided to protect each sleeping area in multi-level homes (top) or homes where the sleeping areas are separated. In addition, a detector should be placed at the head of the basement stairs.

Southern Building (Dec. 1978–Jan. 1979): 10. Reprinted by permission.

Guidelines for Using a Line Drawing

1. Keep your drawing simple. Include only as much detail as your reader needs to understand to assemble or operate the mechanism. Do not include any extra details. Even a line or two might distort the reader's view. And don't add decorations to make your drawing look fancy.
2. Clearly label all parts so your reader can identify and separate them.
3. Decide on the most appropriate view of the object you want to illustrate—aerial, frontal, lateral, reverse, exterior, interior—and indicate which one it is in your title.
4. Keep the parts of the drawing proportionate unless you are purposely enlarging one section. If you change any part(s) of a mechanism in your drawing, indicate where and why.

FIGURE 13.21 A cutaway drawing showing construction of sanitary sewers with a steel trenching box.

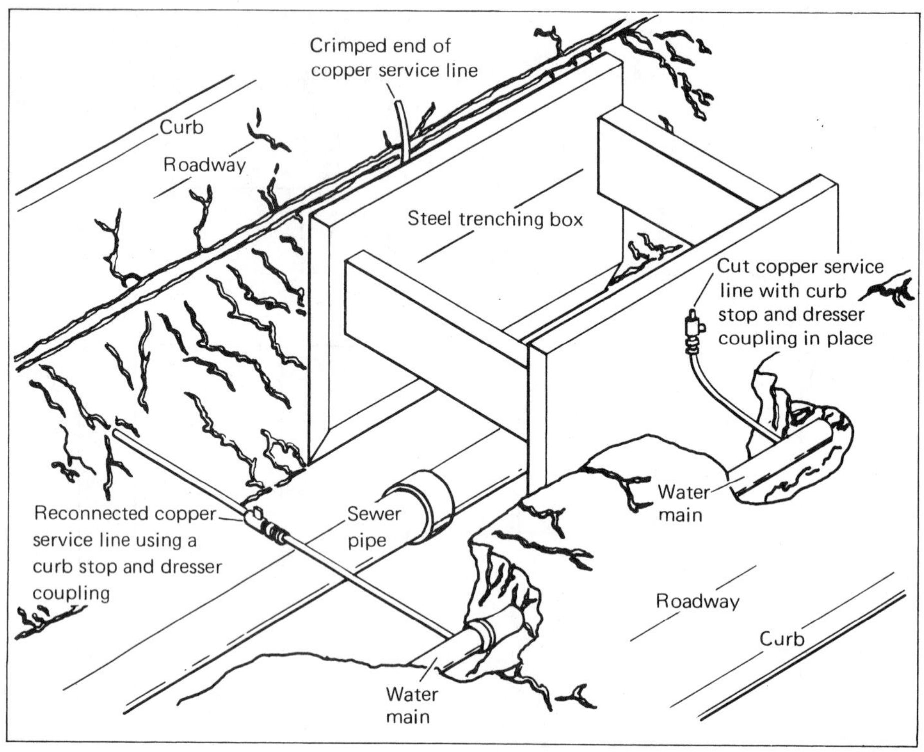

August A. Guerrera, "Grounding of Electric Circuits to Water Services: One Utility's Experience." Reprinted from *Journal of the American Water Works Association* 72 (Feb. 1980): 86. Reprinted by permission. Copyright 1980, the American Water Works Association.

Computer Graphics

With computer graphics software, you can locate, create, edit, and position in your documents virtually all of the kinds of visuals discussed in this chapter. Charts and graphs that illustrate, or *present,* data values are sometimes called **presentation graphics.** Other visuals, such as photos, drawings, and preproduced images or icons, are *representational,* as opposed to presentational, in that they are pictures that look like what they are depicting.

Presentation Graphics

A presentation graphics package is designed specifically to produce the kinds of charts and graphs discussed in this chapter. It guides you through the process of inputting your data and selecting the way you want them displayed. You simply plug your raw data into the computer software, or use data already stored in a database or from a spreadsheet program. (The presentation graphics software will even store this information so you do not have to reenter it when you want to

FIGURE 13.22 An exploded drawing of a chair showing the relationship of the parts.

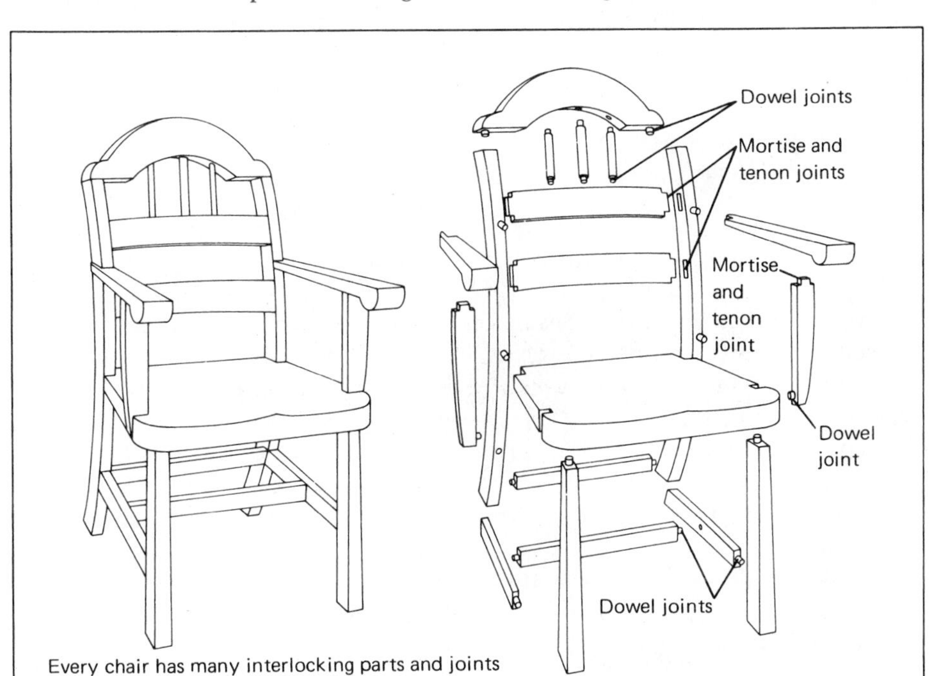

From Reader's Digest *Fix-It Yourself Manual:* 67. Courtesy of Reader's Digest Association, Inc.

change or update a visual.) Your computer then will plot the data you entered into a proportionately accurate chart, graph, or table. In fact, the computer can suggest the most relevant visual representation (graph, pie chart, bar chart, or table) to display your data. This software also lets you customize every other aspect of your visual—what title to give it, what scale to follow, what annotations to include, and even what colors to use. The computer gives you more choices than you would ever imagine possible.

Presentation graphics start with a set of data that needs to be represented. This data must first be typed into a spreadsheet program such as Lotus 1-2-3. Once you enter the data and data labels through a few clicks of the mouse button, the program then allows you to look at a variety of charts and graphs, which you can accept or adapt.

Once you have decided on the graph you want, you can print it or put it into your word-processing document. If you put it in your word-processing program you have two options: (1) put a copy of the graph as it is directly into your document, or (2) link the graph to the document.

Graphics Software Options

Other graphics software comes with a wide range of capabilities, from the simple to the highly technical. At the simple end of the range, you can use a graphics package for straightforward drawing or, more frequently, to adjust to your requirements an image that was created by someone else.

TECH NOTE

Depending on their features and complexity, computer graphics programs offer options to help you create, enhance, and manipulate drawings in documents. After you draw or import an image, you can add detail or emphasize the relationships among its elements in countless ways: select and enlarge a portion of it, relabel or reposition its components, present it in mirror image or in three dimensions, rotate or stretch its elements, combine it with text or photographs, and so forth. You can also repeat any image in multiple renderings—for example, contrasting a black-and-white version with one in color, or changing its texture or proportions to create different effects, or presenting a line drawing as a painting.

When the image is customized to your satisfaction, you can either "drag" it directly to where you want it in that document, or you can "drop" it (import it, really) into any number of documents. If necessary, you can revise and update the customized image as time passes, using it again and again.

There are many images available for computer use. The only major caution is that you must be careful of copyright protection. Do not use an image that belongs to someone else without prior written approval. There are still millions of images not under copyright that are available to you, or you can create your own images with pictures and a scanner as the student did in Figure 13.2 .

Many graphics packages come with comprehensive clip art libraries, such as those seen in Figure 13.23. These are small drawings of almost everything you can think of. To help you select the clip art you need, most packages include a manual with full-color pictures of all their clip art images grouped under useful headings.

As the graphics abilities of computers and graphics software increase, it will become easier to produce professional-looking, three-dimensional images of both real and projected objects. Another graphics feature that is increasing in use and popularity is animation. As more communications are presented on-line, readers will expect to see movement in the images. Thus, for example, a bar graph that shows different values of items for a particular year could have a time component added so that the heights of the bars change for different years, giving an extra dimension to the information being presented.

FIGURE 13.23 Clip art examples.

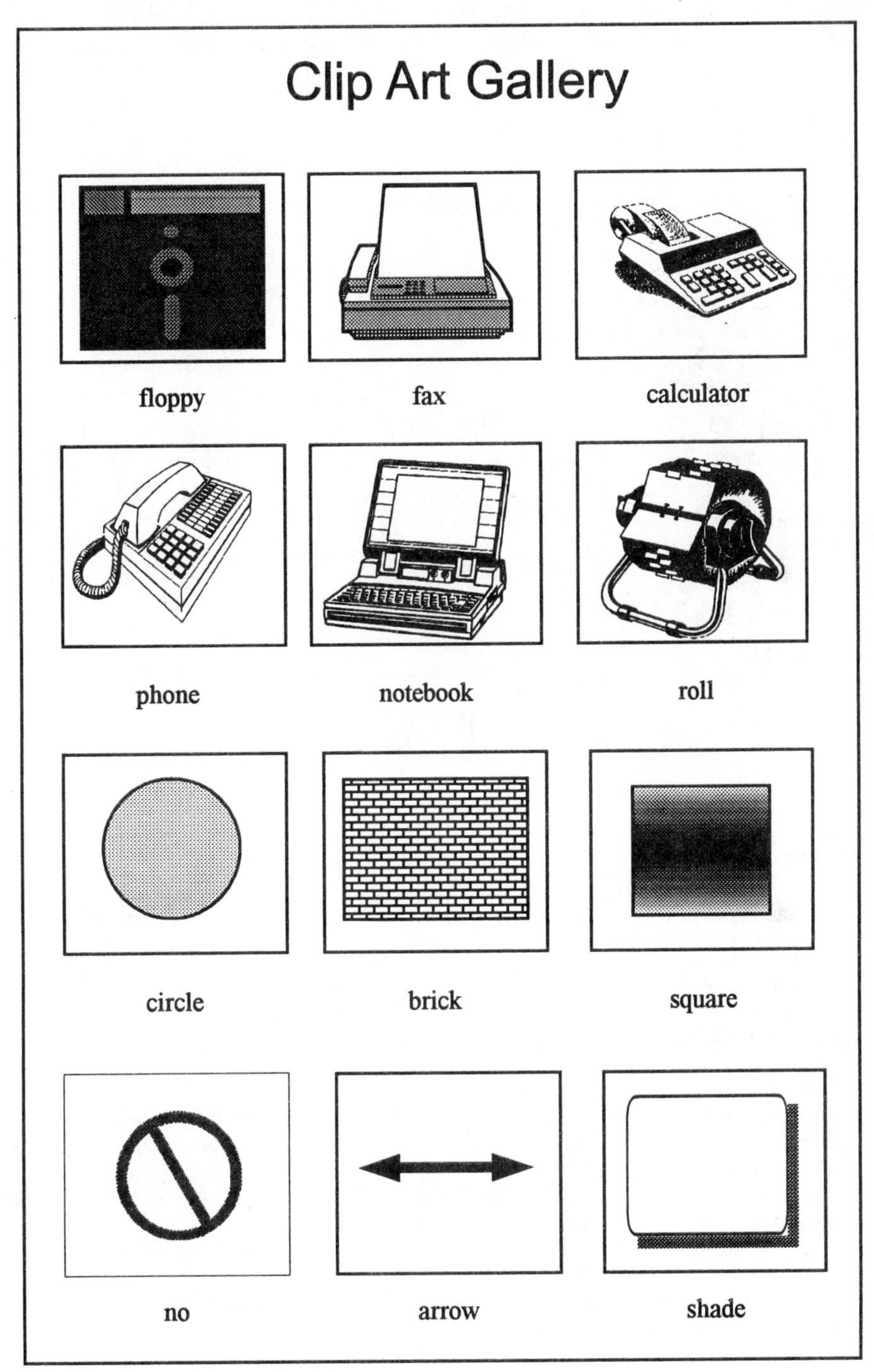

Courtesy of Lucy Cowen.

✓ Revision Checklist

- ❑ Located all places where a visual would help readers better understand my message.
- ❑ Used Internet as virtual library of visuals.
- ❑ Chose most effective visual (table, chart, graph, drawing, photograph) to represent information the audience needs.
- ❑ Experimented with different software programs.
- ❑ Drafted and edited visual until it meets readers' needs.
- ❑ Verified statistical data that visual portrays.
- ❑ Made sure that visual does not simply repeat information in my written work.
- ❑ Selected right amount of technical detail to include in visual.
- ❑ Ensured every visual is attractive, clear, complete, and relevant.
- ❑ Gave each visual a title, number, and where necessary, a legend and call-out.
- ❑ Inserted visual near the written description or commentary to which it pertains.
- ❑ Introduced and interpreted each visual in appropriate place in report or paper.
- ❑ Acknowledged sources for any copyrighted visuals and gave credit to individuals whose statistical data are basis of a visual.

Glossary

By way of review, the following is a glossary of terms used in this chapter:

bar chart a visual using vertical or horizontal bars to measure different data in space and time; bars can also be segmented to show multiple percentages within one bar. Bar charts are used to show a variety of facts for easy comparison.

callouts labels that identify the parts of an object in a visual.

captions titles or headings for visuals.

circle chart a visual shaped in a circle, or pie, whose slices represent the parts of the whole. Circle charts portray budgets, expenditures, shares, and time allotments.

computer clip art ready-made electronic images, symbols, and pictures available in a computer library.

computer graphics a variety of visuals generated by a computer; software (programs) and hardware (computer screens, scanners) create these visuals.

cropping the process of eliminating unnecessary, unwanted details of a photograph by reproducing only the desired portion.

cross-hatching the process of marking parts of a visual with parallel lines that cross each other obliquely; used to differentiate one bar or slice of a circle chart from another.

cutaway drawing a sketch in which the exterior covering of an object has been removed to show an interior view.

data point the intersection of the vertical and horizontal axes on a graph to plot the occurrence of statistical data.
dependent variable the element (cost, employment, energy) plotted along the vertical axis of a line graph and most directly influenced by the independent variable.
drawing software packages computer graphics software that allows users to create visual representations of data either plotted for accuracy or drawn freehand (charts, graphs, general illustrations).
exploded drawing a sketch of an entire object that has been blown up and apart to show the relationship of parts to one another.
figures any visuals that are not tables—charts, drawings, graphs, pictograms, photographs, maps.
flow chart a sketch with boxes and arrows revealing the stages in an activity or process.
graph a picture that represents the relationship of an independent variable to one or more dependent variables; produces a line or curve to show their movement in time or space. Graphs are used to depict figures that change often—temperatures, rainfall, prices, employment, productions, and so forth.
independent variable the element, plotted along the horizontal axis of a graph, which most directly and importantly affects the dependent variable; most often, the independent variable is time or distance.
large-scale map a map that shows a great deal of detail, whether physical (elevations, rivers), economic (income levels), or social (population, religious affiliation).
legend the explanation, or key, indicating what different colors, shadings, or symbols represent in a visual.
organizational chart a visual showing the structure of an organization from the chief executive to the workforce of employees. An organizational chart reveals the chain of command with areas of authority and responsibility.
pictograph a visual showing differences in statistical data by means of pictures varying in size, number, or color.
pie chart see **circle chart.**
presentation graphics software computer graphics software that allows its users to assemble visual images in digital form for presentation purposes (e.g., audiovisual presentation at a sales conference).
spreadsheet an electronic table generated by a computer graphics package.
stub the first column on the left-hand side of a table; contains line captions listing those units to be discussed in the columns.
table a visual in which statistical data or verbal descriptions are arranged in rows or columns.
tick marks equally spaced marks drawn on the vertical or horizontal scale of a graph to show units of measurement.

Exercises

1. Bring to class three or four home pages from the Internet that use especially effective visuals. In a short memo (three or four paragraphs) to your instructor, indicate why and how each visual is appropriate for and convincing to a particular audience. What would each home page look like without its visual?

2. Record the highest temperature reached in your town for the next five days. Then collect data on the highest temperature reached in three of the following cities—

Boston, Chicago, Dallas, Denver, Los Angeles, Miami, New Orleans, New York, Philadelphia, Phoenix, Salt Lake City, San Francisco, Seattle—over the same five days. (You can get this information from a printed newspaper or one on the Internet.) Prepare a table showing the differences for the five-day period.

3. Go to a supermarket and get the prices of four different brands of the same product (hair spray, aspirin, a soft drink, a box of cereal). Present your findings in the form of a table.

4. One government agency supplied the following statistics on the world production of oranges (including tangerines) in thousands of metric tons for the following countries during the years 1993–1996: Brazil, 2,005, 2,132, 2,760, and 2,872; Israel, 909, 1,076, 1,148, 1,221; Italy, 1,669, 1,599, 1,766, 1,604; Japan, 2,424, 2,994, 2,885, 4,070; Mexico, 937, 1,405, 1,114, 1,270; Spain, 2,135, 2,005, 2,179, 2,642; and the United States, 7,658, 7,875, 7,889, 9,245. Prepare a table with this information and then write a paragraph in which you introduce and refer to the table and draw conclusions from it.

5. Keep a record for one week of the number of miles you walk, ride, or drive each day. Then prepare a line graph depicting this information.

6. Prepare a table to show the following statistical data: According to the 1990 census, the town of Ardmore had a population of 34,567. By the 2000 census the town had decreased its population by 4,500. In the 1990 census the town of Morrison had a population of 23,809, but by the 2000 census the population had increased by 3,689. The 2000 census figure for the town of Berkesville was 25,675, which was an increase of 2,768 from the 1990 census.

7. Prepare a line graph for the information in Exercise 6.

8. Prepare a bar graph for the information in Exercise 6.

9. Write a paragraph introducing and interpreting the table printed below.

Year	*Soft Drink Companies*	*Bottling Plants*	*Per Capita Consumption (Gallons)*
1940	750	750	10.3
1945	578	611	12.5
1950	457	466	18.6
1955	380	407	17.2
1960	231	292	15.9
1970	171	229	15.4
1975	118	197	16.0
1980	92	154	18.7
1985	54	102	21.1
1990	43	88	23.1
1995	45	82	25.3
2000	37	78	27.6

10. Prepare a circle chart showing the breakdown of your budget for one week or one month.

11. According to a municipal study in 1998 the distribution of all companies classified in each enterprise industry in that city was as follows: minerals, 0.4%; selected services, 33.3%; retail trade, 36.7%; wholesale trade, 6.5%; manufacturing, 5.3%; and construction, 17.8%. Make a circle chart to represent this distribution and write a one- or two-paragraph interpretation to accompany (and explain the significance of) your visual.

12. Construct a segmented-bar chart to represent the kinds and numbers of courses you took in a two-semester period or during your last year in high school.

13. Prepare a bar chart for the different brands of one of the products in Exercise 3. Write a paragraph introducing this illustration.

14. Find a pictograph in a math or business textbook, in a magazine (try *Newsweek* or *U.S. News & World Report* in print or on-line), or on the Web, and make a bar graph from the information contained in it. Then write a paragraph introducing the bar graph and drawing conclusions from it.

15. Make an organizational chart for a business or agency you worked for recently. Include part-time and full-time employees, but indicate employees' title or function with different kinds of shapes or lines. Then write a brief letter to your employer explaining why this kind of organizational chart should be distributed to all employees. Focus on the types of problems that could be avoided if employees had access to such a chart.

16. Prepare a flow chart for one of the following activities.
    a. jumping a "dead" car battery
    b. giving an injection
    c. crocheting an afghan
    d. painting a set of louvered doors
    e. making an arrest
    f. putting out an electrical fire
    g. joining a chatgroup on the Internet
    h. preparing a visual using a graphics software package
    i. any job you do

17. Draw an interior view of a piece of equipment you use in your major; then identify the relevant parts using callouts.

18. Prepare a drawing of one of the following simple tools and include appropriate callouts with your visual.
    a. golf club
    b. hammer
    c. pliers
    d. stethoscope
    e. swivel chair
    f. computer terminal, keyboard, mouse, and printer
    g. ballpoint pen
    h. soldering iron
    i. table lamp
    j. pair of eyeglasses

19. Prepare appropriate visuals to illustrate the data listed below. In a paragraph immediately after the visual explain why the type of visual you selected is appropriate for this information.

  a. Life expectancy is increasing in America. This growth can be dramatically measured by comparing the number of teenagers with the number of older adults (over age 65) in America during the last few years and then by projecting these figures. In 1970 there were approximately 28 million teenagers and 20 million older adults. By 1980 the number of teenagers climbed to 30 million and the number of older adults increased to 25 million. In 1990 there were 27 million teenagers and 31 million older adults. By the year 2000 the number of teenagers should level off to 23 million, but the number of older adults will soar to more than 36 million.

  b. Researchers estimate that for every adult in America 3,985 cigarettes were purchased in 1970; 4,100 in 1975; 3,875 in 1980; 3,490 in 1985; 3,210 in 1990; and 3,420 in 1995.

20. Find a photograph that contains some irrelevant clutter. Write a letter to the photography department of a company for which you presumably work that wants to use the photograph. Tell the department what to delete and why.

21. Make a simple line drawing of only the relevant portions of the photograph in Exercise 20. Explain in two paragraphs why the drawing works better than the photograph.

22. You work for a large manufacturer of industrial heat pumps and have been asked to help write a section of a report on the increased business your firm has been doing overseas. Based on the sales figures below for the years 1995, 1996, and 1997 (listed in that order) for each of the following countries, prepare two different yet complementary visuals. Also, supply a one-page description and interpretation of the statistics represented in your visuals. You may work collaboratively with one or more students in your class to prepare the visuals as well as to write the section of the report on international sales.

> Argentina, 45, 53, 34; Australia, 78, 90, 115; Bolivia, 23, 43, 52; Brazil, 29, 34, 35; Canada, 116, 234, 256; Denmark, 65, 54, 87; England, 256, 345, 476; France, 198, 167, 345; Germany, 234, 398, 429; Holland, 65, 80, 89; Italy, 49, 52, 97; Japan, 67, 43, 29; New Zealand, 12, 69, 114; Norway, 33, 92, 104; Switzerland, 164, 266, 306; Sweden, 145, 217, 266; People's Republic of China, 7, 100, 296; Korea, 55, 43, 28.

In your written report, take into account trends, shifts in sales, and possible consequences for further marketing, and conclude with a specific recommendation to your employer.

CHAPTER 14

Writing Instructions

Clear and accurate instructions are essential to the world of work. Instructions tell, and frequently show, how to do something. They indicate how to perform a procedure (draw blood; change the oil in your car); operate a machine (run a fork-lift); assemble, maintain, or repair a piece of equipment (a modem, a Web site). Everyone from the consumer to the specialist uses and relies on carefully written and designed instructions.

Instructions are found everywhere from short product inserts to long, complex manuals. Magazines such as *Internet, Popular Mechanics,* and *Popular Photography,* as well as how-to books offer consumers money-saving instructions on topics ranging from repairing their homes to designing a home page. You might want to read some of these publications and manuals or visit a Web site to see how they identify and meet the needs of the reader trying to perform a certain procedure.

On-line instructions are especially adapted to access from a computer terminal. To assist consumers (and potential buyers) many companies include instructions for their products or services directly on the Internet.

TECH NOTE

On-line instructions offer added benefits. Most computers have the ability to give you pictures and sounds, but these two features can be limited without additional hardware. With a sound card your computer can provide a much wider range of possibilities, including having someone actually talk you through the various steps of a procedure and alert you to special problem areas. If your computer also has a graphics card you can speed up the display of images and take advantage of animation that shows the process being done. Such features—visuals, sound—further help instruction writers meet the precise needs of their readers who may need to perform a trial run before attempting a procedure or who may want to double-check a step before going on.

As part of your job, you may be asked to write instructions, alone or with a group, for your co-workers as well as for the customers who use your company's services or products. When writing long, complex instructions you certainly will be part of a team of engineers, programmers, document design experts, marketing specialists, and even attorneys. But whether the instructions are brief or lengthy, your employer stands to gain or lose much from the quality and accuracy of instructions you prepare.

Chapter 14 will show you how to develop, write, illustrate, and design a variety of instructions.

Why Instructions Are Important

Perhaps no other type of occupational writing demands more from the writer than do instructions because so much is at stake. You cannot afford to be unclear, inaccurate, or incomplete. Instructions are significant for the following reasons.

Safety

Carefully written instructions get a job done without damage or injury. Poorly written instructions can be directly responsible for an injury to the person trying to follow them and may result in costly damage claims or even lawsuits. Notice how the product labels in your medicine chest inform users when, how, and why to take a medication safely. Without these instructions, consumers would be endangered by taking too much or too little medicine or by not administering it properly.

Efficiency

No work would be done if employees did not have clear instructions to follow. Imagine how inefficient it would be for a business if employees had to stop their work each time they did not have, or could not understand, a set of instructions. Or, equally alarming, what if employees made a number of serious mistakes because of confusing directions, costing a business lost sales and increased expenses. Well-written instructions help a business run smoothly and efficiently.

Convenience

Effective instructions also help to ensure a customer's satisfaction by making a job easier to do. Customers would not purchase a product or service if they had trouble understanding how to use it properly or safely. How many times have you heard complaints about a product because instructions on how to assemble or use it were frustratingly unclear or incomplete? In the customer's view, instructions indicate a product's quality and a service or procedure's convenience. Clear and easy-to-follow instructions are also a strong sales point, a vital part of "service after the sale." Owners' manuals, for example, instruct buyers on how to avoid the inconvenience of a product breakdown (and the headaches and expense of starting over) by keeping it in good working order. If your customers, especially ESL readers, cannot readily understand your instructions, they will not purchase your company's products or services again.

FIGURE 14.1 Instructions on how to prepare and administer a sitz bath.

First, adjust water temperature dial to 105–110°F. Then turn on the faucet and fill sitz tub with enough water to cover the patient's hips. Before assisting the patient into the tub, place bath towel in the bottom of the tub. Allow the patient to sit in the tub for 15–20 minutes. At the end of this time, help the patient out of the tub. Then dry the patient thoroughly.

The Variety of Instructions: A Brief Overview

Instructions vary in length, complexity, and format. Some instructions are one word long: *stop, lift, rotate, print, erase.* Others are a few sentences long: "Insert blank diskette in external disk drive"; "Close tightly after using"; "Store in an upright position." Short instructions are appropriate for the numerous relatively nontechnical chores performed every day.

For more elaborate procedures, however, detailed instructions may be as long as a page or a book. When your firm purchases a new mainframe computer or a piece of earth-moving equipment, it will receive an instruction pamphlet or book containing many steps, cautionary statements, and diagrams. Many businesses prepare their own training manuals containing instructions for 200 or 300 different procedures.

Instructions can be given in paragraphs or in lists, and you will have to determine which format is most appropriate for the kinds of instructions you write. Figures 14.1 and 14.2 show two sets of brief instructions written in paragraph

FIGURE 14.2 Instructions on how to repair a halyard.

Easy Temporary Join for Synthetic Ropes

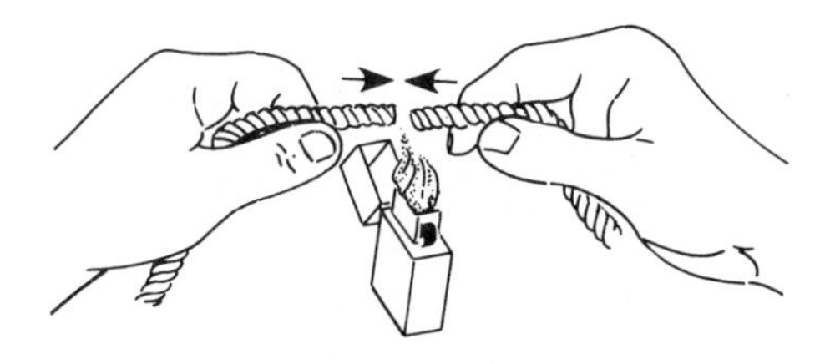

If you are faced with the problem of reeving a new halyard on a flagpole or mast, or through a block or pulley in an inaccessible location, the solution can be easy if both old and new lines are made of nylon or polyester (Dacron, Terylene, etc.). Simply join the ends of the old and new lines temporarily by melting end fibers together in a small flame (a little heat goes a long way). Rotate the two lines slowly as the fibers melt. Withdraw them from the flame before a ball of molten material forms, and if the stuff ignites, blow out the flame at once. Hold the joint together until it is cool and firm.

R. I. Standish, *Parks* 51 (Apr.–June 1980): 21.

FIGURE 14.3 Instructions for starting File Manager.

The File Manager Window

Follow the steps below to start File Manager.

1. Load windows and open the Main window.
2. Double-click on the File Manager icon (it looks like a 2-drawer file cabinet) to display the File Manager window.

NOTE: Your desktop may appear similar to the one shown in Figure 4.1. In Figure 4.1, Windows was loaded from the C drive and the File Manager default settings were used to control the appearance of the windows. Don't be concerned if your windows look a little different from the ones in the text. You will soon learn to control how the windows are displayed.

The **File Manager window,** like the Program Manager window (and other application windows), includes a Control-menu box, title bar, sizing buttons, menu bar, work area for document windows and icons, and adjustable borders. The File Manager window may be moved throughout the desktop. The **status bar,** located along the bottom of the window, provides information about the selected drive, including the number of bytes free, total number of bytes used, and number of files.

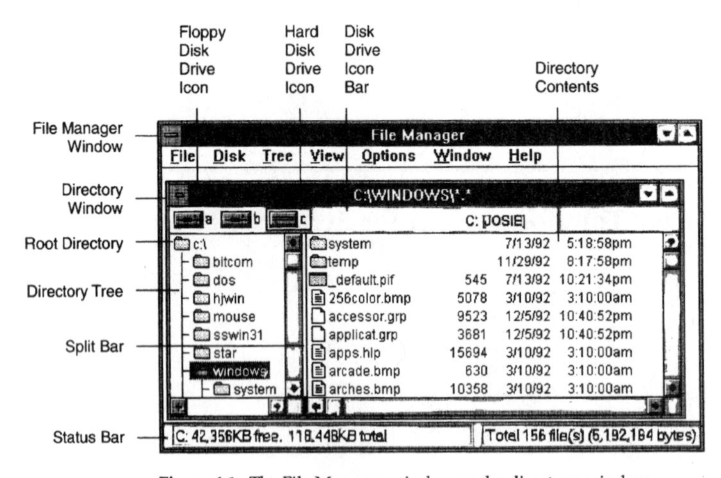

Figure 4.1 The File Manager window and a directory window.

Source: Instruction Manual for Microsoft Windows. Reprinted with permission from Microsoft Corporation.

FIGURE 14.4 Instructions on how to assemble an outdoor grill.

ASSEMBLY INSTRUCTIONS

The instructions shown below are for the basic grill with tubular legs. If you have a pedestal grill, or a grill with accessories, check the separate instruction sheet for details not shown here.

NOTE: Make sure you locate all the parts before discarding any of the packaging material.

TOOLS REQUIRED . . . A standard straight blade screwdriver is the only tool you need to assemble your new Meco grill. If you have a pedestal grill, you will need a 7-16 wrench or a small adjustable wrench.

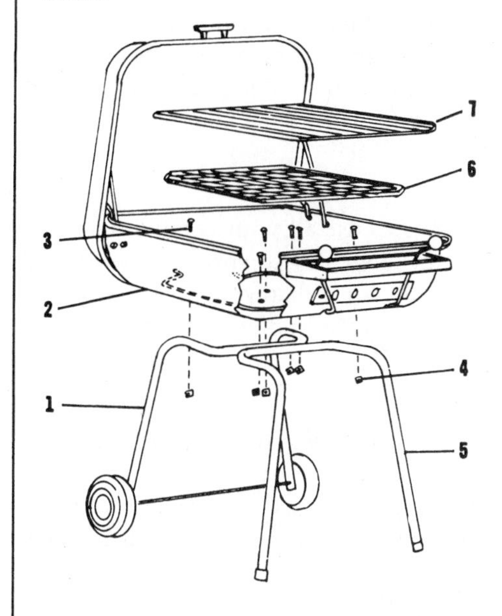

1. Before you start, take time to read through this manual. Inside you will find many helpful hints that will help you get the full potential of enjoyment and service from your new Meco grill.

2. Lay out all the parts.

3. Assemble roller leg (1) to bottom rear of bowl (2) with 1¼″ long bolts (3) and nuts (4).

4. Assemble fixed leg (5) to bottom front of bowl (2) with 1¼″ long bolts (3) and nuts (4).

5. Place fire grate—ash dump (6) in bottom of bowl (2) between adjusting levers.

6. Place cooking grid (7) on top of adjusting levers. Make sure top grid wires run from front to back of grill.

Meco Assembly Instructions and Owners Manual, Metals Engineering Corp., P.O. Box 3005, Greenville TN 37743. Reprinted by permission.

format. In Figure 14.1 hospital employees are told how to prepare and administer a sitz bath; in Figure 14.2 park rangers are given instructions on how to repair a halyard, or tackle, used to raise a flag or move a pulley.

The instructions shown in Figures 14.3 and 14.4 are printed in list form. In Figure 14.3 readers will find directions on how to start File Manager; the directions in Figure 14.4 explain how to assemble an outdoor grill.

Assessing and Meeting Your Audience's Needs

Regardless of their format (paragraphs or lists), instructions must be clear, complete, and easy to follow. Put yourself in the readers' position. In most instances you will not be available for readers to ask you questions when they do not understand something. Consequently, they will have to rely only on your written instructions. Your purpose in writing these instructions is to get them to perform the same steps you followed and, most important, to obtain the same results you did.

Do not assume that members of your audience have performed the procedure before or have operated the equipment as many times as you have. (If they had, there would be no need for your instructions.) No writer of instructions ever disappointed readers by making directions too clear or too easy to follow. Keep in mind that your audience will often include non-native speakers of English, a worldwide audience of potential consumers of your company's products and services.

Writing instructions is like teaching. You must not only understand the material yourself, but you must also know the best way of presenting it. Set goals for your readers the way a teacher sets goals for students in a course: establish what you want them to do, how, and why.

Key Questions to Ask About Your Audience

To prepare effective instructions you will have to use all your resources for analyzing your audience. Learn all you can about them. Try to anticipate your readers' questions and problems, making absolutely sure that your instructions address these concerns. The best way to establish your audience's needs is to ask yourself the following questions:

1. How and why will my readers use my instructions? (Engineers have different expectations than do office personnel, and customers have different expectations still.)
2. How much do my readers know about the procedure they have to follow?
3. How much essential background or introductory information will I have to supply?
4. Where will my audience most likely be following my instructions—in the workplace, outdoors, in a workshop equipped with tools, or alone in their homes?
5. How many resources—such as special equipment or power sources—will my readers need to perform my instructions successfully?

The Process of Writing Instructions

As we saw in Chapter 2, clear and concise writing evolves by following a process. In order to make sure your instructions are accurate and easy for your audience to perform, follow these steps.

Plan

Before writing, you must do some research and completely understand the job or procedure that you are asking someone else to perform. Make sure you know

- the reason for doing something
- the parts or tools required
- the steps to follow to get the job done
- the results of the job

If you are not completely sure about the procedure, ask an expert for a demonstration. Do some background reading and e-mail colleagues who may have written or performed a similar instruction.

If you work collaboratively, everyone in your group needs to understand the procedure, from start to finish. An uneven level of expertise among group members spells trouble; some steps of your instructions will be complete and accurate while others may not be.

Do a Trial Run

Actually perform the procedure (assembling, repairing, maintaining, dissecting) yourself or with all of your writing team present. Go through a number of trial runs. Take notes as you go along and be sure to divide the procedure into simple, distinct steps for readers to follow. Don't give readers too much to do in any one step. Each step should be complete, sequential, reliable, straightforward, and easy for your audience to identify and perform.

Write and Test Your Draft

Transform your notes into a draft (or drafts) of the instructions you want readers to follow. To test your draft(s), ask someone from the intended audience (consumers, technicians) who may never have performed the procedure to follow your instructions as you have written them. Observe where the individual runs into difficulty—cannot complete or seems to miss a step, gets a different result than yours. Do not coach him or her with verbal cues. Your intended reader won't have you there to coach them. Note any places in your instructions that are vague, hard to follow, inconsistent (steps bunched up or out of order), or incomplete. Ask your experimenter-user what stumped him or her.

Revise and Edit

Based on your observations and user feedback, revise your draft(s) and edit the final copy of the instructions that you will give to readers. Also consider whether your instructions would be easier to accomplish if you included visuals and ask where they would be most helpful. If visuals are recommended, incorporate some graphic support (see pp. 521–524).

As you revise and edit, pay special attention to the technical language and the amount of detail you include. Analyzing the needs and background of your audience will help you to determine which words and details are appropriate. A set of instructions accompanying a chemistry set would use different terminology, abbreviations, and detail than would a set of instructions a professor gives a class in organic chemistry.

General audience: Place 8 drops of vinegar in a test tube with a piece of limestone about the size of a pea.

Specialized audience: Place 8 gtts of CH_3COOH in a test tube and add 1 mg of CO_3.

The instructions for the general audience avoid the technical abbreviations and symbols the specialized audience requires. If your reader is puzzled by your directions, you defeat your reasons for writing them.

Using the Right Style

To write instructions that readers can understand and turn into effective action, observe the following guidelines.

1. Use verbs in the present tense and imperative mood. Imperatives are commands that have deleted the pronoun *you.* Almost all of the instructions in Figures 14.1 through 14.4 contain imperatives—"adjust water temperature" instead of "you adjust the water temperature." Deleting the *you* is not discourteous, as it would be in a business letter or report. The command tells readers, "These steps work, so do it exactly as stated." Don't ever water down your instructions with "you might" or "you could" ("You might want to ignite the fire next" does not have the authority of "ignite the fire next.") Instead, choose action verbs such as those listed in Table 14.1.

2. Write clear, short sentences in the active voice. Since readability is especially important in instructions, keep sentences short and uncomplicated. This advice is especially important when you are writing for ESL readers. Sentences under twenty words (and preferably under fifteen) are easy to read. Note that the

TABLE 14.1 Some Helpful Imperative Verbs Used in Instructions

adjust	delete	insert	open	roll	tear
apply	determine	inspect	point	rotate	tie
blow	dig	lift	pour	rub	tighten
boot up	drain	load in	press	save	trim
call up	drill	loosen	prevent	saw	turn
change	drop	lower	print	shift	twist
check	ease	lubricate	pull	shut off	type
choose	eliminate	measure	push	slide	unplug
clean	enter	mount	raise	slip	verify
close	freeze	move	release	spread	wash
connect	hold	notify	remove	start	wind
cut	increase	oil	review	switch	wipe

sentences in Figures 14.1 through 14.4 are, for the most part, under fourteen words.

Avoid the passive voice. In place of "The air blower is to be used last," write "Use the air blower last." Do not write a sentence that sends the reader the opposite message from what you intend. A direction such as "Before using the soldering iron on metal, clean it with Freon" may mislead the inexperienced welder to put Freon on the iron rather than on the metal that is to be cleaned.

3. Use precise terms for measurements, distances, and times. Indefinite, vague directions leave users wondering whether they are doing the right thing. The following vague directions are better expressed through precise revisions.

Vague: Turn the distributor cap a little. (How much is a little?)
Precise: Turn the distributor cap three quarters of a rotation.
Vague: Let the contents cool for a while. (How long?)
Precise: Let the contents cool for 10 minutes.

Precise measurement and timing are essential to the success of many kinds of instructions ranging from baking a cake to completing a blood test in a laboratory.

4. Use connective words as signposts. Connective words specify the exact order in which something is to be done (especially when your instructions are written in paragraphs). Words such as *first, then, before* in Figure 14.1, for example, help readers stay on course, reinforcing the sequence of the procedures.

5. Number each step when you present your instructions in a list. You may also use bullets. Plenty of white space between steps also distinctly separates them for the reader.

Using Visuals Effectively

Readers welcome visuals in almost any set of instructions. Visuals are graphic and direct, helping readers to understand what they must do. A visual can

- simplify a process
- reinforce or even save words
- illustrate a "right" and "wrong" way
- increase readers' confidence
- help get a job done more quickly

For example, the illustration in Figure 14.2 reinforces the process of joining the two parts of the halyard by fire. Another frequently used visual in instructions is the exploded drawing, such as the one shown in Figure 14.4, which helps consumers see how the various components of the grill fit together.

The number and types of visuals you include will, of course, depend on the procedure or equipment you are explaining. Some instructions, such as those in Figure 14.5, use a visual to illustrate each step. Others may require only one or two. Instructions for computer programs sometimes show users what they can

FIGURE 14.5 Instructions on inserting and removing videocassettes where a visual illustrates each step.

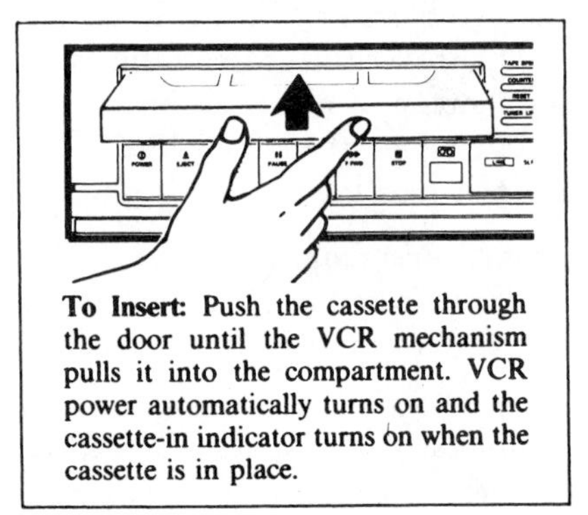

To Insert: Push the cassette through the door until the VCR mechanism pulls it into the compartment. VCR power automatically turns on and the cassette-in indicator turns on when the cassette is in place.

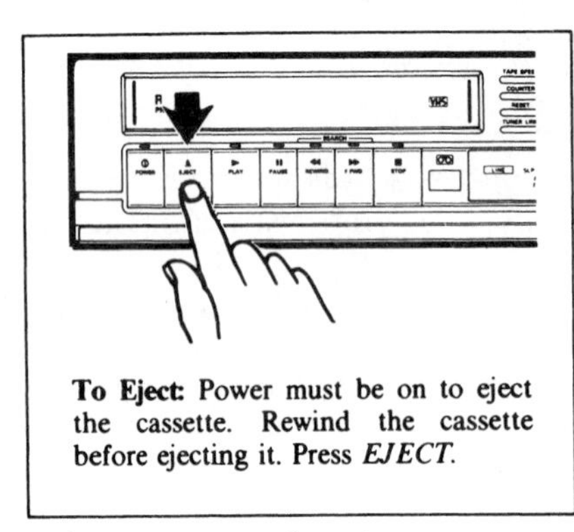

To Eject: Power must be on to eject the cassette. Rewind the cassette before ejecting it. Press *EJECT*.

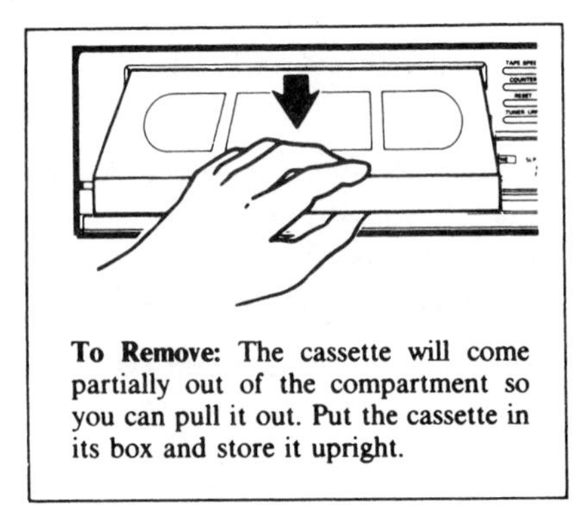

To Remove: The cassette will come partially out of the compartment so you can pull it out. Put the cassette in its box and store it upright.

RCA/Thomson Consumer Electronics.

expect to see on a screen—either to clarify an explanation, as in Figure 14.6, or to identify problems. Gear the type of visual you include to the task you want the reader to perform and to the reader's knowledge of that task.

Regardless of the number and type of visuals you include, follow these guidelines to use them effectively.

1. Whenever possible, place the visual next to the step it illustrates, not on another page or buried at the bottom of the page. To gain the most from visuals, readers must be able to see the illustrations or diagrams immediately before and after reading the accompanying directions.
2. If you are using many visuals, assign each one a number (Figure 1, Figure 2) and refer to the visuals by number in the written instructions to eliminate potential confusion.
3. Make sure that the visual looks like the object the user is trying to assemble, maintain, run, or repair. Don't use a photo of an earlier or different model.
4. Always inform readers if a part of an object is missing or reduced in size in your visual.

FIGURE 14.6 Instructions on using a computer operating system showing what the user will see on the screen.

Dialog Boxes

Dialog boxes typically resemble small windows that appear in the foreground of the desktop to provide information and to request a user response. Dialog boxes are displayed when:

1. the user chooses a command followed by an ellipsis (...). (For example, choosing the FILE-Exit Windows... command displays the Exit Windows dialog box.)
2. additional information is needed to complete a requested command (such as entering the name of a file to be saved).
3. Windows provides a warning message (i.e., quitting an application before saving a new or revised file contained within the application).
4. Windows provides a message explaining why a requested command cannot be completed (i.e., not having the required hardware (such as a modem) needed to run an application).

The majority of the dialog boxes require input on a limited number of decisions (like the Exit Windows dialog box, which asks you to choose either the OK or Cancel button). Other dialog boxes, such as the Desktop dialog box shown in Figure 2.7, may be a little more complex. In the following pages you

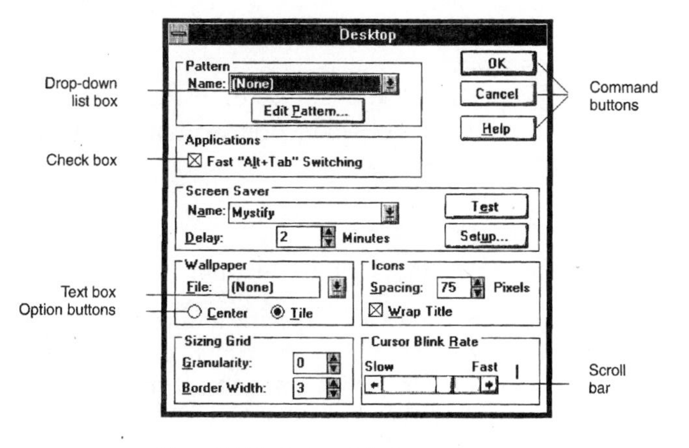

Figure 2.7 The Desktop dialog box.

Source: Instruction Manual for Microsoft Windows 95. Reprinted with permission from Microsoft Corporation

5. Wherever necessary, label parts of the visual. The drawing in Figure 14.4 does this very carefully for readers.
6. Set each visual off with white space so it is easy to find and examine.

The Four Parts of Instructions

Except for very short instructions, such as those illustrated in Figures 14.1 through 14.4, a set of instructions generally contains four main parts: (1) an introduction, (2) a list of equipment and materials, (3) the actual steps to perform the process, and (4) a conclusion (when necessary).

The Introduction

The function of an introduction is to provide readers with enough *necessary* background information to understand why and how your instructions work. An introduction must make readers feel comfortable and well prepared before they turn to the actual steps.

Do I Need an Introduction?

Whether you need an introduction depends on the particular process or machine you are describing. Short instructions require no introduction or only a brief one- or two-sentence introduction. More complex instructions require lengthier introductions. For instance, a ninety-page instruction manual may contain a two- or three-page introduction.

Keep Your Introduction Proportional

An introduction should be proportional to the kinds of instructions that will be given. For example, instructions on how to sand a floor will not need a one-page introduction on how friction works.

What You Include in an Introduction

You can do one or all of the following in your introduction:

1. State why the instructions are useful for a specific audience. Many instructions begin with introductions that stress safety, educational, or occupational benefits. Here is an introduction from a safety procedure describing protective lockout of equipment:

> The purpose of this procedure is to provide plant electrical technicians with a uniform method of locking out machinery or equipment. This will prevent the possibility of setting moving parts in motion, energizing electrical lines, or opening valves while repair, set-up, or cleaning work is in progress.

Figure 14.7 contains a set of instructions for installing a demonstration disk—"The Anaerobe Educator Demo"—intended for microbiology classes as well as professional use in clinical laboratories. Note how the introduction emphasizes the variety of applications of this software for this specific audience.

FIGURE 14.7 Instructions for installing a demonstration disk.

The Anaerobe Educator Demo

The Anaerobe Educator is the latest word in microbiology multimedia software. Although *The Anaerobe Educator Demo* contains only a small portion of the actual program, it should give you a good idea of what the full version offers.

The Anaerobe Educator can be used in many different ways:

- Training clinical lab staff
- Teaching presumptive identification to students in schools
- Quick access online reference for trained professionals
- Easy periodic reviews for trained professionals

The full version features over 250 color photographs of 19 organisms. It also contains the complete version of **The Reference Center** and teaches in conjunction with two of the foremost textbooks in the field:

- *Principles and Practice of Clinical Anaerobic Bacteriology*, by Engelkirk, Duben-Engelkirk, and Dowell.
- *Wadsworth Anaerobic Bacteriology Manual*, by Summanen, Baron, Citron, Strong, Wexler, and Finegold.

The Anaerobe Educator Tour, which you received with the demo disks, will help you explore the software. You can duplicate the tour document and *The Anaerobe Educator Demo* disks, and distribute them freely. Use is restricted to the original intent of demonstrating *The Anaerobe Educator* software. Furthermore, *The Anaerobe Educator Demo* disks must be distributed as is, without the addition or deletion of any files. All other uses of *The Anaerobe Educator Demo* and its documentation are strictly prohibited by copyright law.

Installation (IBM)

To install *The Anaerobe Educator Demo* follow these steps:

1. Insert **Disk 1** of *The Anaerobe Educator Demo* into the 3.5" disk drive.
2. Select **Run** from the **File** menu of **Program Manager**.
3. Type "B: Anedinst.exe" (if your floppy 3.5" drive is designated by a letter other than B, use that letter instead).
4. Press **Return.**
5. Click **Continue** to install the AnEd directory directly onto the root level of drive C, or type in a different path if you wish to install it elsewhere.
6. Click **OK** in the next two windows to install *The Anaerobe Educator*'s data and pictures into their default directories. Do not change the path to these directories, or the program will not function properly.
7. Remove **Disk 1** from the floppy drive, and replace it with **Disk 2** when the program requests the second disk.
8. Click **Yes** if you wish to read or print the README.WRI file. This file contains The Anaerobe Educator Tour, which you also received with the demo disks.
9. Click **OK** when a dialog informs you that installation is complete. Store the disks in a safe place.
10. Select **Exit** from the **File** menu of Write when you are done reading or printing this file.
11. Launch *The Anaerobe Educator Demo* by double-clicking its icon in **Program Manager**.

Anaerobe Systems, San Jose, California.

Figure 14.8 contains the introduction to a guide for nurses who have to be familiar with the features "Set-up, Operation, and Response to Alarms" for an infusion pump. This opening section, "Overview Orientation," highlights the safety and convenience of this piece of equipment for the nursing staff as they meet patients' needs.

2. Indicate how a particular machine, procedure, or process works. An introduction may briefly discuss the "theory of operation," to help readers understand why something works the way your instructions say it should. Such a discussion sometimes describes a scientific law or principle. This introduction to instructions on how to run an autoclave begins by explaining the function of the machine: "These instructions will teach you how to operate an autoclave. The autoclave is used to sterilize surgical instruments through the live additive-free stream."

Figure 14.9 contains a description of a piece of exercise equipment from the owner's instructional manual. Note how precise yet clear the description of the StairMaster is for users. This description focuses on what the equipment does and how it accomplishes it for users.

3. Specify how long it should take the user to complete all steps of the instructions. If users know how long a procedure should take, they will be better able to judge whether they are doing it correctly—if they are waiting too long or not long enough between steps or if they are going too slowly or too quickly: "It should take about three and one-half hours from the time you start laying the floor tiles until the time they dry well enough to walk on." You may also have to advise readers when a certain procedure must be monitored or performed. Instructions for the EPS premium cartridge for laser printers, for example, tell purchasers that they must clean their cartridge corona at the time of installation and after every thousand copies.

4. Stress any advantages or benefits the reader will gain by performing the instructions. Make the reader feel good about buying or using your product by explaining how it will make a job easier to perform, save the reader time and money, or allow the reader to accomplish a job with fewer mistakes or false starts. Note how the following introduction to a set of instructions on using an auto dial/auto answer modem encourages the reader to want to learn how to operate this computer-based telephone system.

> Welcome to high-speed telecommunications and congratulations on choosing the Signalman EXPRESSi. You've made an excellent choice. The EXPRESSi is the ideal link between your computer and the ever-expanding world of information utilities, databases, electronic mail, bulletin boards, computer time sharing, and more.
>
> The EXPRESSi can be used in the IBM Personal Computer. And because the EXPRESSi fits inside your PC, it saves valuable desk space and eliminates expensive, bulky cables.[1]

---

[1]Courtesy of Anchor Automation, Inc., Chatsworth, Calif.

FIGURE 14.8 Introduction to a guide for using an infusion pump.

1 Overview Orientation

Gives function of equipment

The LifeCare PROVIDER 5500 System is a portable infusion pump, specially designed to deliver analgesic drugs, antibiotics, and chemotherapeutics.

The pump can be programmed in either milligrams or cubic centimeters, and in four different delivery configurations for greater nursing convenience and to tailor precisely the most effective regimen for each patient.

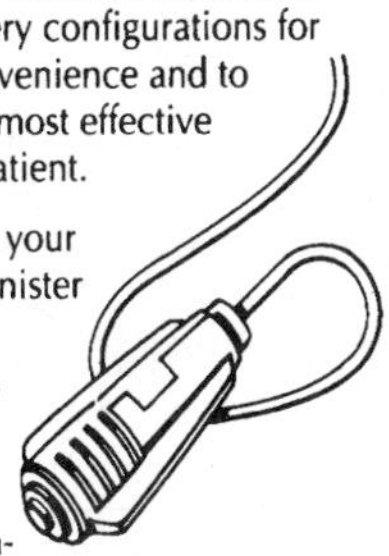

Bolus Mode allows your patient to self-administer analgesia within programmed limits.

This is the traditional PCA delivery, "analgesia-on-demand," based on the patient's need.

Continuous Mode delivers a continuous "background" infusion with no additional PCA doses permitted.

Continuous-plus-Bolus Mode allows the patient to self-administer a Bolus dose in addition to receiving a simultaneous Continuous dose infusion.

Intermittent Mode delivers a specific dose (in cc or mg) at intermittent intervals over 24 hours.

Emphasizes safety features

You can also establish the "lockout" interval, the frequency with which a patient may receive a Bolus dose of analgesic drug.

Saves time

The PROVIDER 5500 System records all settings in memory and can be quickly re-programmed to save nursing time when repeating established protocols or changing fluid reservoirs.

The portable system operates on battery power.

Offers security

To minimize tampering and discourage theft, there is an optional locking security lockbox that also secures the system to an IV pole.

The audible alarm signals in the event of a malfunction, and the digital readout describes the malfunction.

Explains convenience features

- Compact and lightweight.
- Delivery rates between 0.1 cc and 250 cc per hour, in 0.1-cc increments.

Display Panel

- Individual display indicators appear only during programming and operation.
- Only on a *selective* basis.
- Tone sounds when activated.
- Runs on BATTERY POWER ONLY.
- Disposable Primary IV set with integral infusion cartridge.

Reprinted by permission of Abbott Laboratories Hospital Products Division.

FIGURE 14.9 Introduction to an owner's manual for a piece of exercise equipment.

WHAT IS THE STAIRMASTER 4000 PT EXERCISE SYSTEM?

The StairMaster 4000 PT Exercise System is a vertical climbing machine that provides an aerobic workout equivalent to climbing stairs, without the inertial loads and skeletal trauma common to most aerobic activities.

The 4000 PT Exercise System is computer controlled to offer automated, timed workouts from 5 to 45 minutes, as selected by the user. There is a choice of eight preprogrammed workouts, each with ten levels of intensity. In addition, there is also a nonprogrammed (manual) workout that allows you to pace yourself or experiment with the various speeds. Also, users have the option of designing up to ten "customized" workouts that can be created and stored in the computer.

All of the workout programs on the StairMaster 4000 PT Exercise System feature computer-controlled speeds from 26 steps / minute to 138 steps / minute, based on 8-inch steps. The faster the speed, the greater the intensity level at which you are working. At the conclusion of your completed workout, the computer console displays the number of calories you burned while exercising, the equivalent number of floors you climbed, and the equivalent number of miles that you ran. If your machine has Revision 2.1 or 2.2 software, you can request this information and also request elapsed time, KGMs, and watts at any time during your workout and then return to the standard workout display.

From *Owner's Instructional Manual for the StairMaster 4000 PT Exercise System.*

5. **Inform the user about any special circumstances to which the instructions apply.** Some instructions precede others or are used only on special occasions. Readers must be informed about those changes or emergency situations. A supervisor of a large chemical plant sent employees the memo contained in Figure 14.10 to describe operating procedures to be followed during energy shortages. The brief introduction (the opening one-sentence paragraph) emphasizes the special circumstances.

Two Short Case Studies

The instructions contained in the memos in Figures 14.10 and 14.11 are addressed to two different audiences, each with separate needs. The Hercules memo in Figure 14.10 was sent to a specific, technical audience—firefighters and supervisors—who needed instructions on a special process. Cliff Burgess's memo in Figure 14.11, on the other hand, went out to all Burton employees, a more diverse, rather than technical, group of readers. His helpful instructions do not require a list of equipment or materials or a description of steps in a process. Instead, his memo consists of an introduction, three bulleted instructions, and a conclusion. Study the annotations to Figure 14.11 to see how Cliff Burgess met the needs of his general, diverse audience.

FIGURE 14.10 Instructions alerting a technical audience to special circumstances.

TO: All Shift Supervisors
All Firefighters
FROM: Robert Ferguson R.F.
SUBJECT: Operating Procedures During Energy Shortages
DATE: October 10, 1998

The following policy has been formulated to aid in maintaining required pressures during periods of low wood flow and severe natural gas curtailment.

All boilers are equipped with lances to burn residue or no. 6 fuel oil as auxiliary fuels. When wood is short, no. 6 oil should be burned in no. 2 and no. 3 boilers at highest possible rate consistent with smoke standards. To do this, take these steps:

1. Put no. 6 oil on no. 3 boiler lances.
2. Shut down overfire air.
3. Shut down forced draft.
4. Turn off vibrators.
5. Keep grates covered with ashes or wood until ash cover exists.

These steps will result in an output of 25,000 to 35,000 lbs./hr. steam from no. 3 boiler and will force wood on down to no. 4 boiler.

If required, the same procedure can be repeated on no. 2 boiler.

FIGURE 14.11 An instructional memo listing safety precautions for a general audience.

BURTON MANUFACTURING SYSTEMS
St. Louis, Mo 63174

TO: All Burton Employees
FROM: Cliff Burgess, Environmental Safety C.B.
RE: Video Display Terminal (VDT) Safety Precautions
DATE: September 2, 1999

Introduction emphasizes reasons for instructions, gives non-technical explanation, and offers an analogy

You may experience some possible health risks in using your computer video display terminal (VDT). These risks include sleep disorders, behavioral changes, danger to reproductive system, and cancer.

The source of any risks comes from the electromagnetic fields (EMFs) that surround anything that carries an electric current—for example, photocopiers, circuit breakers, and especially VDTs. Your computer monitor is a major source of EMFs. Magnetic fields go through walls as easily as light goes through glass.

Although EMFs may affect your health, you can considerably reduce your exposure to these fields by following these three simple steps:

Explains how to use equipment. Steps stand out through bullets, boldface, and spacing

- **Stay at least three feet (an arm's length) away from the front of your VDT.** (The magnetic field is reduced greatly with this amount of distance.)

- **Stay at least four feet away from the sides and back of someone else's VDT.** (The fields are weaker in the front of the **VDT** but much stronger everywhere else.)

- **Switch your VDT / computer off when you are not using it.** (If the computer has to remain on , be sure to switch off the monitor; screen savers do not affect the exposure to **EMFs**.)

Conclusion reassures readers they are acting safely by following instructions

Our environmental safety team will continue to monitor and investigate any problems. Observing the guidelines above, however, will help you to take all the necessary precautions in order to minimize your exposure to EMFs.

If you have any questions about these procedures or your exposure to EMFs, please e-mail me at burg@burton.com.

Not every introduction to a set of instructions will contain facts in all five categories of information listed above. Some instructions will require less detail. You will have to judge how much background information to give readers for the specific instructions you write.

List of Equipment and Materials

Immediately after the introduction, inform readers of all equipment or materials they will need. This list should be complete and clear. Do not wait until the readers are actually performing one of the steps in the instructions to tell them that a certain type of drill or a specific kind of chemical is required. They may have to stop what they are doing to find this equipment or material; moreover, the procedure may fail or present hazards if users do not have the right equipment at the right time.

Specify Features Such as Size, Shape, and Quantity

Do not expect your readers to know exactly what size, model, or quantity you have in mind. Tell them precisely. For example, if a Phillips screwdriver is necessary to complete one step, specify this type of screwdriver under the heading "Equipment and Materials"; do not just list "screwdriver." Here are some additional examples of unclear references to equipment and materials, with more helpful alternatives listed in parentheses after them: solution (0.7% NaCl solution); pencils (two engineering pencils); electrodes (four short platinum wire electrodes); file (cheese-grater file); sand (10-lb. bag of sand); needle (butterfly needle); water (2 gallons untreated sea water).

Anticipate Any Problems Readers May Have

If you are concerned that readers will not understand why certain equipment or materials are used, give the explanation in parentheses after the item. For example, listing alcohol and cotton as materials needed to take a blood pressure might confuse readers not familiar with the uses of these materials. A clarifying comment after the materials such as "for cleaning stethoscope headphones" would help. The following example, "Instructions for Absentee Voters," contains such helpful comments.

> This absentee ballot package has been sent to you at your request. It contains the following items:
>
> 1. sample paper ballot (for information only)
> 2. official ballot (this is a punch card)
> 3. punching tool
> 4. envelope with attached declaration
> 5. addressed return envelope

The parenthetical information in items 1 and 2 tells voters that the punch card is their official ballot and that it does not look like the sample ballot. The point is reinforced later in the directions by this statement:

IMPORTANT

- Punch only with tool provided, never with a pencil or pen.
- Your vote is recorded by punching this ballot card, not by marking the paper ballot.
- Do not return the sample paper ballot.

The Procedural Steps

The heart of your instructions will consist of clearly distinguished steps that readers must follow to achieve the desired results. To make sure that you help your readers understand your steps, observe the following rules.

1. Put the steps in their correct order. If a step is out of order or is missing, the entire set of instructions can be wrong or, worse yet, dangerous, as the cartoon in Figure 14.12 humorously illustrates. Double-check your steps before you write them down. Each step should be numbered to indicate its correct place in the sequence of events you are describing. Never put an asterisk (*) before or after a step to make the reader look somewhere else for information. If the information is important, put it in your instructions; if it is not, delete it.

2. Group closely related activities into one step. Sometimes closely related actions are grouped into one step to help the reader coordinate activities and to emphasize their being done at the same time, in the same place, or with the same equipment. Study the following instructions listing the receptionist's duties performed by the unit secretary in a hospital.

1. Greet patients warmly and make them feel welcome. Never fall into the trap of groaning and saying "not another one." Remember the patient probably did not want to come to the hospital in the first place.
2. Check the identity bracelet against the summary sheet and addressograph plate. If the bracelet is not on the patient's wrist, place it there immediately. Set about correcting any errors you may find in any of this information at once.
3. If asked, escort the patient to his or her room. Explain the signal light, answer any questions, and introduce the patient and roommate, if any.
4. Notify the head nurse and/or the nurse assigned to that room of the patient's arrival.
5. Notify the admitting physician by phone of the patient's arrival and room. If the patient lacks admission orders, mention this to the physician's secretary at this time.[2]

Note how step 3 contains all the duties the unit secretary performs once patients are taken to their rooms. It is easier to consolidate all the activities that take place in the room—explaining the signal light, answering questions, making introduc-

---

[2]Myra S. Willson, *A Textbook for Ward Clerks and Unit Secretaries.* St. Louis: C. V. Mosby, 1979: 70. Reprinted by permission.

FIGURE 14.12 A cartoon emphasizing the importance of putting instructions in proper order.

Jocular Science
by Bill Lae
© 1988 Bill Lae
AE 300 MANUAL
Excerpts from poorly written user instruction manuals.... *"Finally, push the red button. But first, pull down on the big lever."*

tions—than to list each as a separate step. But be careful that you do not overwork a single step. To combine steps 4 and 5 would be wrong (and impossible) because those two distinct steps require the unit secretary to speak to someone in person and to make a phone call. These two actions are two separate stages in a process.

Similarly, instructions on how to use a fax machine are clearer when distinct steps are stated separately. The first set of instructions below incorrectly tells users how to transmit a fax by combining steps that must be performed separately. Step 2 asks users to pick up the phone and then dial the number—two separate actions. Step 3 asks users to press the button and return the handset, again two actions that cannot be performed simultaneously.

Incorrect: 1. Load the paper into the outgoing document slot, adjusting the paper guides to the appropriate width.
2. Pick up the telephone handset and listen for a dial tone. When you hear the dial tone, dial the number of the receiving fax machine.
3. When the receiving fax machine answers the ring, press the start button. After the transmission is completed, return the handset to its cradle.

Correct: 1. Load the paper into the outgoing document slot, adjusting the paper guides to the appropriate width.
2. Pick up the telephone handset and listen for a dial tone.
3. When you hear a dial tone, dial the number of the receiving fax machine.
4. When the receiving fax machine answers the ring, press the start button.
5. After the transmission is completed, return the handset to its cradle.

Don't divide an action into two steps if it has to be done in one. For example, instructions showing how to light a furnace would not list as two steps actions that must be performed simultaneously to avoid a possible explosion.

Incorrect: 1. Depress the lighting valve.
2. Hold a match to the pilot light.

Correct: 1. Depress the lighting valve while holding a match to the pilot light.

Similarly, do not separate two steps of a computer command that must be performed simultaneously.

Incorrect: 1. Press the CONTROL key.
2. Press the ALT key.

Correct: 1. While holding down the CONTROL key, press the ALT key.

3. Give the reader hints on how best to accomplish the procedure. Obviously, you cannot do this for every step, but if there is a chance that the reader might run into difficulties or may not anticipate a certain reaction, by all means provide assistance. For example, telling readers that a certain aroma or color is to be expected in an experiment will reassure them that they are on the right track. Particular techniques on how to operate or service equipment are also helpful: "If there is blood on the transducer diaphragm, dip the transducer in a blood solvent, such as hydrogen peroxide, Hemosol, etc." If readers have a choice of materials or procedures in a given step, you might want to list those that would give the best performance: "Several thin coats will give a better finish than one heavy one."

4. State whether one step directly influences (or jeopardizes) the outcome of another. Because all steps in a set of instructions are interrelated, you could not (and should not have to) tell readers how every step affects another. But stating specific relationships is particularly helpful when dangerous or highly intricate operations are involved. You will save the reader time, and you will stress the need for care. Forewarned is forearmed. Here is an example.

Step 2: Tighten fan belt. Failure to tighten the fan belt will cause it to loosen and come off when the lever is turned in step 5 below.

Do not wait until step 5 to tell readers that you hope they did a good job in tightening the fan belt in step 2. That information comes after the fact.

Insert Warning, Caution, and Note Statements Where Necessary

A warning statement tells readers that a step, if not prepared for or performed properly, can endanger their safety.

WARNING: UNPLUG THE MONITOR BEFORE CLEANING ANY PART WITH DAMP CLOTH.

DO NOT APPLY PRESSURE UNTIL SAFETY VALVE IS COMPLETELY SEALED.

A caution statement tells readers to take certain precautions: wear protective clothing, check an instrument panel carefully, use special care in running a machine, measure weights or dosages exactly, save a document on a computer.

CAUTION

MAKE SURE BRAKE SHOES WON'T RUB TIRE AND THAT SHOES MATE WELL WITH RIM WHEN BRAKES ARE APPLIED.

BE SURE TO ENTER CORRECT CODE FOR **WORD SMART.** KEYBOARDING THE WRONG CODE WILL ERASE DOCUMENT.

Warning and caution statements are very important for legal and safety reasons. Including them is more than a courtesy for readers and is *not* optional. They are necessary to protect lives and property. In fact, you and your company can be sued if you fail to notify users of your product or service of potentially dangerous conditions. Follow these three guidelines in using warning and caution statements in your instructions.

1. **Put warning and caution statements in the right place.** Place them immediately before the step to which they pertain. When they are out of place, you risk placing your readers and/or their equipment in danger. If you insert a warning or caution statement too early, the reader may fail to remember it when the time he or she comes to the step to which it applies. And if you put the notification too late, you almost certainly expose the reader to danger and the equipment to breakdown.
2. **Put warning and caution statements in the right format.** Because of the extremely important information they impart, warning and caution statements should be graphically set apart from the rest of the instructions. There should be no chance that readers will overlook them. Put these statements in

FIGURE 14.13 Instructions on how to paint a garage floor.

Cleaning the Floor

1. Remove all objects from the floor.
2. Scrape areas where there is old chipped paint with a metal scraper. For hard-to-reach places such as corners or underneath pipes, use a 3-inch wire brush. New paint will not stick to the floor if old paint is not first removed.
3. Sweep the floor first with a broom; then to make sure that all particles of dust and dirt are removed, use a vacuum sweeper.
4. Open all windows for proper ventilation. Fumes from cleaning solution used in Step 5 should not be inhaled.

CAUTION: USE PROTECTIVE EYEWEAR AND RUBBER GLOVES FOR NEXT STEP TO PROTECT SKIN FROM IRRITATION.

WARNING: DO NOT USE DETERGENTS CONTAINING AMMONIA. AMMONIA ADDED TO THESE INGREDIENTS WILL CAUSE AN EXPLOSION.

5. Mop the entire floor with a solution composed of the following ingredients:
   1/3 box of Floorex
   1 quart of bleach
   10 quarts of water
   1 cup powdered detergent
   Note: Do not worry if the mixture does not appear soapy; it does not need suds to work.
6. For any grease spots that remain, sprinkle enough dry Floorex powder to cover them entirely. Scrub these spots with a 5-inch wire brush.
7. Rinse the entire floor thoroughly with water to remove cleanser. Allow floor to dry (**30** min.) before painting.

Painting the Floor

8. Mix the paint with a stirrer ten or fifteen times until the color is even.
   Note: If the paint has not already been shaken by machine before being opened, shake the can for about three minutes, open, and stir the contents for about five to ten minutes or until the paint is mixed.
9. Pour the paint into the paint tray.
10. With the 3-inch paint brush, paint the floor around the baseboard. Come out at least two to three inches so that roller used in the next step will not touch the baseboard.
11. Paint the rest of the floor with a roller attached to an extension handle. A roller handles much more easily than a brush and distributes the paint more smoothly. Move the roller in the same direction each time. Overlap each row painted by one-half inch to avoid spaces between rows of paint.

Continued

FIGURE 14.13 (Continued)

12. Allow two hours to dry. The floor will be ready to walk on.

CAUTION: DO NOT DRIVE VEHICLES ONTO FLOOR FOR 24 HOURS. TIRES WILL PICK UP NEW PAINT.

capital letters, in boldface type, in boxes, in different colors (red is especially effective for warnings, yellow for caution when you are writing for non-ESL readers). Use one or all of these devices. Distinctive symbols, such as a skull and crossbones or an exclamation point inside a triangle, often precede a warning notation.

TECH NOTE

You can use icons to draw readers' attention to warnings, cautions, notes, and even tips. An icon is a small, meaningful picture symbol, such as ♻ to remind users to recycle the container. You can find a wide range of icons with any graphics software package you use. Once you select an icon, it is easy to paste it into any word-processing document. But choose your icons carefully; use them only when they are functional and unambiguous. An icon of a trash can might ambiguously signal that material is to be thrown away as well as to be saved.

3. **Include enough explanation to help readers know what to watch out for and what precautions to take.** Do not just insert the word WARNING or CAUTION. Explain what the dangerous condition is and how to avoid it. Look at the examples of warnings and cautions in Figures 14.13 and 14.14.

Warning and caution statements should not be used just because you want to emphasize a point. Putting too many of these statements in your instructions will decrease the dramatic impact they should have on readers. Use them sparingly—only when absolutely necessary—so readers will not be tempted to ignore them.

Unlike warnings and cautions, a *note* statement simply adds a comment to clarify instructions.

Note: All models in 3500 series have a hex nut in the upper right, not left, corner.

Drive B disk indicator will glow and the drive will make a few clicking sounds as the disk is formatted.

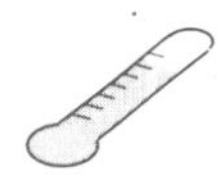

At 20 degrees F, a battery uses about 68 percent of its power.

The instructions in Figure 14.13 on how to paint a garage floor contain steps that offer hints on how best to do the job, comment on how one step affects another, and issue warning, caution, and note statements. Not every instruction requires this amount of detail. Use such comments and signals only when the procedure you are describing calls for them and when they will help your readers. Note how the design of these instructions effectively contributes to their reception and use by readers. The writer has employed plenty of white space, helpful headings, boldface type, and numbered steps.

The Conclusion

Not every set of instructions requires a conclusion. For short instructions containing a few simple steps, such as those in Figures 14.1 through 14.4, no conclusion is necessary. These instructions usefully end with the last step the reader must perform. For longer, more involved jobs, a conclusion can help readers finish the job with confidence and accuracy. For example, a short conclusion like the following might help users who have just followed the step-by-step instructions on painting a garage floor given in Figure 14.13.

> With proper care this procedure should have to be repeated only once every three years. To keep the floor in top shape, sweep it at least twice a month to prevent gritty materials such as sand from scaling the painted surface. If possible, wipe up grease and oil immediately to prevent them from soaking into the paint. Make sure you dispose of rags properly to avoid chance of fire.

When they are necessary, conclusions can help the reader in a variety of ways. They might either provide a succinct wrap-up of what the reader has done or end with a single sentence of congratulations, or they can reassure readers as the conclusion in Cliff Burgess's memo (Figure 14.11) does. A conclusion might also tell readers what to expect once a job is finished, describe the results of a test, or explain how a piece of equipment is supposed to operate. Furthermore, conclusions can give readers practical advice on how to maintain a piece of equipment or how to follow a certain procedure, as does the conclusion for painting a garage floor.

Instructions: Some Final Advice

Perhaps the most important piece of advice to leave you with is this: Do not take *anything* for granted when you have to write a set of instructions. It is wrong and

FIGURE 14.14 Instructions on pipeline safety for a nontechnical audience.

ENTERPRISE PRODUCTS COMPANY
Houston, Texas

Dear Resident:

We operate pipeline through your area, and we work hard to keep them safe and secure. Our pipelines are designed, installed, tested, operated, patrolled, and maintained to promote your safety and to prevent hazards.

Unforeseen damage can cause any pipeline to leak. That is why we want you, our neighbors, to know how to recognize a pipeline leak, what to do if you ever notice a leak, and how to report it quickly so it can be repaired. Our aim is to prevent any hazard to you and your property.

You can help by **first immediately contacting your local police and fire departments**. Then, contact us at the **phone number listed below (and on the handy emergency sticker enclosed).**

You have probably seen signs like the one pictured on this page. They mark our pipelines that operate around the clock, carrying vital products to markets for treatment and distribution. Their uninterrupted operation is important to all of us. If our pipelines are damaged, or if leaks go unreported, a very real hazard may exist. Your attention and assistance will help us protect your safety and our country's energy supplies.

Report any pipeline leak to:

Enterprise Products Company
P.O. Box 4324
Houston, Texas 77210
In Texas phone: 1-800-392-2880
Out of Texas phone: 1-800-231-2809
Chemtrec phone: 1-800-424-9300

Pipeline Markers (Signs)

Take a close look at the sign pictured above. These signs mark our pipeline route and stand either on top of or very near the underground pipeline. They are there to warn the public and to help our company patrol and monitor the pipelines.

Reprinted by permission of Enterprise Products Company.

on occasion dangerous to assume that your readers have performed the procedure before, that they will automatically supply missing or "obvious" information, or that they will easily anticipate your next step. No one ever complained that a set of instructions was too clear or too easy to follow.

To make sure that your instructions are clear and complete, draft, revise, edit, proofread, and test them with your reader in mind. Give the reader all the appropriate information needed for the task—and in the right order. Make sure your instructions cover everything the reader must do to complete the job safely and accurately. Use visuals to clarify a tricky step or to simplify a procedure, and always insert warning, caution, or note statements where needed.

Figure 14.14 contains the final example of instructions in this chapter. Sent to "concerned residents," or a nontechnical audience, these instructions explain the possible safety hazards of a pipeline leak and clearly outline the precautionary measures residents need to take if they recognize such a danger. These instructions—in both content and design—illustrate the guidelines this chapter has emphasized. Study this figure to review these guidelines and to prepare for writing similarly clear and helpful instructions for your audiences.

✓ Revision Checklist

- ❑ Analyzed my intended audience's background, especially why and how they will use my instructions.
- ❑ Took special care with meeting needs of any ESL reader unfamiliar with my culture.
- ❑ Tested my instructions to make sure they include all necessary steps in their proper sequence.
- ❑ Made sure all measurements, distances, times, and relationships are precise and correct.
- ❑ Selected language appropriate for my audience. Avoided technical terms if my audience is not a group of specialists in my field.
- ❑ Used the imperative mood throughout my instructions.
- ❑ Wrote clear, short sentences.
- ❑ Eliminated ambiguity from my instructions.
- ❑ Chose or created effective visuals where necessary and labeled and placed them next to the step(s) to which they apply.
- ❑ Made my introduction proportionate to the length and complexity of my instructions and suited to my readers' needs.
- ❑ Included necessary background, safety, and operational information in introduction.
- ❑ Included a complete list of tools and materials my audience needs to carry out the instructions.

- ❑ Put instructions into easy-to-follow steps and in the right order.
- ❑ Used numbers or bullets to label each step and used connective words to indicate order.
- ❑ Used warning, caution, and note statements where necessary and in a form that makes them easily seen and read.
- ❑ Supplied a conclusion that summarizes what readers should have done or reassures them that they have completed the job satisfactorily.

Exercises

1. Bring to class two examples of short directions that require no introduction, list of materials and equipment, or conclusion. Look for these two examples on labels, carton panels, or backs of envelopes. Evaluate the effectiveness of these instructions by commenting on how precise, direct, and useful they are.

2. Find at least one example of long instructions containing an introduction, list of materials and equipment, procedural steps, statements of warning, caution, or note, and a conclusion. Bring this example to class and be prepared to show how the various steps in this set of instructions follow the principles outlined in this chapter. You can find a set of full instructions in many technical manuals and in some manuals to help consumers assemble or maintain large or complex home appliances.

3. Find a set of instructions that does not contain any visuals, but which you think should have some graphic material to make it clearer. Design those visuals yourself and indicate where they should appear in the instructions.

4. From a technical manual in your field or from an owner's manual, locate a set of instructions that you think are poorly written and illustrated. In a memo to your instructor, explain why those instructions are unclear, confusing, or badly formatted. Then revise the instructions to make them easier for the reader to carry out. Submit the original instructions with your revision.

5. Write a set of instructions in numbered steps (or in paragraph format) on one of the following relatively simple activities.
   a. tying a shoe
   b. brushing your teeth
   c. unlocking a door with a key
   d. making a call from a cellular phone
   e. planting a tree or shrub
   f. sewing a button on a shirt
   g. removing a stain from clothing
   h. pumping gas into a car

i. surfing the Internet
j. checking a book out of the library
k. parallel parking a car
l. shifting gears in a car
m. photocopying a page from a book

6. Working as part of a collaborative writing team, write a set of full instructions on one of the following more complex topics. Identify your audience. Include an appropriate introduction, a list of equipment and materials, numbered steps with necessary warning, caution, and note statements, and an effective conclusion. Also include whatever visuals you think will help your audience.
   a. programming a VCR to record a film two days from now
   b. changing a flat tire
   c. designing a computer program
   d. developing black-and-white film
   e. shaving a patient for surgery
   f. changing the oil and oil filter in a car
   g. making a blueprint
   h. installing a wind turbine on a roof
   i. filling out an income tax return
   j. surveying a parcel of land
   k. pruning hedges
   l. jumping a dead car battery
   m. using the Heimlich maneuver to help a choking individual
   n. filleting a fish
   o. arranging a footlocker for inspection
   p. taking someone's blood pressure
   q. changing a cash register tape
   r. finding and plugging a leak in a tire
   s. downloading a home page from the Web
   t. painting a car
   u. cooking a roast
   v. flossing a patient's teeth after cleaning
   w. creating a computer file

7. Write an appropriate introduction and conclusion for the set of instructions you wrote for either Exercise 5 or Exercise 6.

8. The following set of instructions is confusing, vague, and out of order. Rewrite these instructions to make them clear, easy to follow, and correct. Make sure that each step follows the guidelines outlined in this chapter.

Reupholstering a Piece of Furniture

(1) Although it might be difficult to match the worn material with the new material, you might as well try.
(2) If you cannot, remove the old material.
(3) Take out the padding.
(4) Take out all of the tacks before removing the old covering. You might want to save the old covering.

(5) Measure the new material with the old, if you are able to.
(6) Check the frame, springs, webbing, and padding.
(7) Put the new material over the old.
(8) Check to see if it matches.
(9) You must have the same size as before.
(10) Look at the padding inside. If it is lumpy, smooth it out.
(11) You will need to tack all the sides down. Space your tacks a good distance apart.
(12) When you spot wrinkles, remove the tacks.
(13) Caution: in step 11 directly above, do not drive your tacks all the way through. Leave some room.
(14) Work from the center to the edge in step 11 above.
(15) Put the new material over the old furniture.

P.S. Use strong cords whenever there are tacks. Put the cords under the nails so that they hold.

CHAPTER 15

Writing Winning Proposals

A proposal is a detailed plan of action that a writer submits to a reader or group of readers for approval. These readers are usually in a position of authority—supervisors, managers, department heads, company buyers, elected officials, civic leaders—to endorse or reject the writer's plan. Proposals are among the most important types of job-related writing. Their acceptance can lead to improved working conditions, a more efficient and economical business, additional jobs and business for a company, or a safer and more attractive environment.

Writing Successful Proposals

Proposals are written for many purposes and many different audiences; for example,

- to your boss seeking authorization to purchase a new piece of equipment for the office, as Gordon Reynolds did in Figure 2.5 (pp. 46–47) requesting that his firm buy a laser printer
- to potential customers offering a product or a service, such as to a fire chief offering to supply special firefighting gear
- to a government agency (like the Department of the Interior) seeking funds to conduct research projects (a study of the mating and feeding habits of a particular species or an investigation to determine the mercury levels at a certain lake)

Depending on the job, proposals can vary greatly in size and in scope. A proposal to your employer could easily be conveyed in a page or two. A proposal to do a research project for a class assignment can also be successfully completed in a brief memo. To propose doing a small job for a prospective client—redecorating a waiting room in an accountant's office—a letter with information on costs, materials, and a timetable might suffice. But an extremely large and costly job—constructing a ten-story office building, for example—requires a detailed report hundreds of pages long with appendixes on engineering specifications, detailed budgets, and even résumés of all key personnel working on the project.

A discussion of long, elaborate proposals is beyond the scope of this chapter. But the principles and techniques of audience analysis, organization, and drafting that this chapter does cover apply equally well to any longer project you may be called on to prepare individually or as part of a team.

Proposals Are Persuasive Plans

Proposals, whether large or small, must be highly persuasive to succeed. Without your audience's approval, your plan will never go into effect, however accurate and important you think it is. Your proposal must convince readers that your plan will help them either improve their businesses or generally make their jobs easier.

The tone of your proposal should be "Here is what I can do for you." Stress the precise benefits your plan has for the reader. Show readers how approving your plan will save them time and money or will improve their employees' morale or their customer's satisfaction.

Competition is fierce in the world of work, and a persuasive proposal frequently determines which company receives a contract. Demonstrate to your reader why your plan is better—more efficient, practical, economical—than a competitor's. In a sense, a proposal combines the persuasiveness of a sales letter (see Chapter 6), the documentation of a report (see Chapters 10 and 17), and the binding power of a contract, for if the reader accepts your proposal, he or she will expect you to live up to its terms to the letter.

Proposals Frequently Are Collaborative Efforts

Like many other types of business and technical writing, proposals often are the product of teamwork and sharing. Even a short internal proposal is often researched and put together by more than one individual in a company or agency.

Often, individual employees will pull together information from their separate areas (such as finance, marketing, sales, and transportation) and put it into a proposal that each team member then reads and revises until the team agrees that the document is ready to be released.

Types of Proposals

Proposals are classified according to how they originate and where they are sent after they are written. Distinctions are made between *solicited* and *unsolicited* proposals based on how they originate and between *internal* and *external* proposals based on where they are sent. Depending on your audience and your purpose, you may write an internal solicited or unsolicited proposal; or you may write an external solicited or unsolicited proposal.

Requests for Proposals and Solicited Proposals

When a company has a particular problem to be solved or a job to be done, it will solicit, or invite, proposals. The company will notify you and other competitors by preparing a **Request for Proposals (RFP),** which is a set of instructions that specify the exact type of work to be done along with guidelines on how and when the company wants the work completed.

Some RFPs are long and full of legal requirements and conditions. Others, like the example in Figure 15.1, are more concise. RFPs are mailed to firms with track records in the area the company wants the work done. RFPs are also printed in trade publications to attract the highest number of qualified bidders for the job. The U.S. government publishes RFPs in the *Federal Register,* while private companies sometimes send their RFPs to *Business Daily.*

FIGURE 15.1 A sample RFP.

REQUEST FOR PROPOSALS

Mesa Community College is soliciting proposals to construct and to install fifty individual study carrels in its Holmes Memorial Library. These carrels must be highly serviceable and conform to all specification standards of the ALA. Proposals should include the precise measurements of the carrels to be installed, the specific acoustical and lighting benefits, and the types and amount of storage space offered. Work on constructing and installing the carrels must be completed no later than the start of the Fall Semester, August 29, 1999. Proposals should include a schedule of when different phases of work will be completed and an itemized budget for labor, materials, equipment, and necessary tests to ensure high quality acoustical performance. Contractors should state their qualifications, including a description of similar recent work and a list of references. Proposals should be submitted in triplicate no later than May 16, 1999, to :

Mrs. Barbara Feldstein-Archer
Director of the Library
Mesa Community College
Mesa, CO 80932-0617

TECH NOTE

RFPs are also posted on the Web. Take a look at a few of them to get an idea of what government agencies and private companies look for when they request proposals. The *Federal Register,* published daily by the U.S. government, issues hundreds of legal regulations as well as requests for proposals from various government agencies. You can browse the *Federal Register* on-line via GPO access—http://www.access.gpo.gov/su_docs/aces/aces/140.html. If you have any questions, you can contact the *GPO Access* User Support Team at their e-mail address: gpoaccess@gpo.gov.

Following is a summary of a request for proposals from the Community Food Projects Program of the U.S. Dept. of Agriculture:

> Request for Proposals: Community Food Projects Program; Notice [[Page 38524]]
>
> DEPARTMENT OF AGRICULTURE
>
> Cooperative State Research, Education, and Extension Service
>
> AGENCY: Cooperative State Research, Education, and Extension Service, USDA.
>
> ACTION: Announcement of availability of grant funds and request for proposals for the Community Food Projects Program.
>
> SUMMARY: The Federal Agriculture Improvement and Reform Act of 1996 established new authority for a program of Federal grants to support the development of community food projects designed to meet the food needs of low-income people; increase the self-reliance of communities in providing for their own food needs; and promote comprehensive responses to local food, farm, and nutrition issues.
>
> This notice sets out the objectives for these projects, the eligibility for criteria for projects and applicants, and the application procedures. The legislation also allows technical assistance under the program. Therefore, the applicants may request technical assistance as a part of their proposal request in order to subcontract to consultants or other groups to provide assistance for technical needs of the applying organization.
>
> This notice contains the set of instructions needed to apply for a Community Food Project grant. To obtain application forms, please contact Proposal Services, Grants Management Branch; Office of Extramural Programs; USDA/CSREES at (202) 401-5048. When calling this office please indicate that you are requesting forms for the Community Food Projects Program.

Another source of RFPs on the Internet is *Commerce Business Daily* (CBD) which "lists notices of proposed government procurement actions, contract awards, sales of government property and other procurement information. A new edition of the CBD is issued every business day. Each edition contains approximately 500–1000 notices." The *Commerce Business Daily*—http://cos.gdb.org/repos/cbd2/cbd.intro.html—lists the government's requests for equipment, supplies, and a variety of services from assembling to maintaining equipment to fee-charging dredging.

You will need to use a server—the Community of Science Web Server—to search the *Commerce Business Daily.* That means you will need a username and a password. The Community of Science is "a consortium of research institutions and one of the largest on the Internet." Find out if your library is a member. If so, the fee may be paid for you.

Here are a few of the agencies you can search in the *Commerce Business Daily* for RFPs:

- Department of Health and Human Services
- United States Department of Energy
- United States Department of Agriculture
- United States Department of Transportation
- United States Army
- United States Navy
- United States Air Force
- United States Marine Corps
- Defense Support Center and Agencies
- Department of Defense—Advanced Research Projects Agency (ARPA)
- National Aeronautics and Space Administration
- Department of Justice and other Law Enforcement Agencies

The Community of Science Web Server also lets you search the *Federal Register, Federally-Funded Research in the U.S.,* and *Funding Opportunities Database.*

An RFP helps you to know what the customer wants. It is often extremely detailed and even tells you how the company wants the proposal prepared; for example, what information is to be included (on backgrounds, personnel, equipment, budgets), where it needs to appear, and even how many copies of the proposal you have to submit.

Your own proposal will be judged according to how well you fulfill the terms of the RFP. For that reason, follow the directions in the RFP exactly. Note that the solicited proposal in Figure 15.2 directly refers to the terms of the RFP. You should even use the language (specialized terms, specifically stated needs) of the RFP in your proposal to convince readers that you understand their needs and to get them to accept your plan.

FIGURE 15.2 An external solicited proposal in response to an RFP.

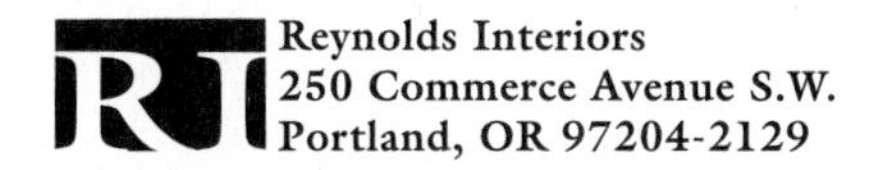

http://www.reynolds.com
503-555-8733
FAX 503-555-1629

January 21, 1998

Mr. Floyd Tompkins, Manager
General Purpose Appliances
Highway 11 South
Portland, OR 97222

Dear Mr. Tompkins:

In response to your Request 7521 for bids for an appropriate floor covering at your new showroom, Reynolds Interiors is pleased to submit the following proposal. After carefully reviewing your specifications for a floor covering and inspecting your new facility, we believe that Armstrong Classic Corlon 900 is the most suitable choice. I am enclosing a sample of the Corlon 900 so that you can see how it works.

Corlon's Advantages

Guaranteed against defects for a full three years, Corlon is one of the finest and most durable floor coverings manufactured by Armstrong. It is a heavy-duty commercial floor 0.085-inch thick for protection. Twenty-five percent of the material consists of interface backing; the other 75 percent is an inlaid wear layer that offers exceptionally high resistance to everyday traffic. Traffic tests conducted by the Independent Floor Covering Institute repeatedly proved the superiority of Corlon's construction and resistance.

Another important feature of Corlon is the size of its rolls. Unlike other leading brands of similar commercial flooring—Remington or Treadmaster—Corlon comes in 12-foot-wide rather than 6-foot-wide rolls. This extra width will significantly reduce the number of seams on your floor, thus increasing its attractiveness and reducing the dangers of the seams splitting.

Installation Procedures

The Classic Corlon requires that we use the inlaid seaming process, a technical procedure requiring the services of a trained floor mechanic. Herman Goshen, our floor mechanic, has over fifteen years of experience working with the inlaid seam process. His professional work has been consistently praised by our customers.

Continued

FIGURE 15.2 (Continued)

page 2

Installation Schedule

We can install the Classic Corlon on your showroom floor during the first week of March, which fits the timetable specified in your request. The material will take three and one-half days to install and will be ready to walk on immediately. We recommend, though, that you not move equipment onto the floor for 24 hours after installation.

The following costs include the Classic Corlon tile, labor, and tax:

750 sq. yards of Classic Corlon at $13.50	$10,125.00
Labor (28 hrs. @ $18.00/hr.)	$ 504.00
Sealing fluid (10 gals. @ $15.00/gal.)	$ 150.00
Total	$10,779.00
Tax (5 percent)	$ 538.95
GRAND TOTAL	$11,317.95

Our costs are $250.00 under those you specified in your request.

Reynolds's Qualifications

Reynolds Interiors has been in business for more than 28 years. In that time, we have installed many commercial floors in Portland and its suburbs. In the last year, we have served more than 60 customers, including the new multipurpose Tradex plant in Portland. We would be happy to furnish you with a list of satisfied customers.

Conclusion

Thank you for the opportunity to submit this proposal. We believe you will have a great deal of success with an Armstrong floor. If we can provide you with any further information, please call us or visit us at our Web site.

Sincerely yours,

Sharon Scovill

Sharon Scovill
Sales Manager

Jack Rosen

Jack Rosen
Installation Supervisor

Unsolicited Proposals

With an unsolicited proposal, you—not the reader—make the first move. You identify a problem for readers and prepare a proposal to solve it. Doing that is not as difficult as it sounds. See how the writers of the unsolicited proposal in Figure 15.3 identify a problem for their readers.

Before you write an unsolicited proposal, though, speak with an appropriate manager at the company or write an inquiry letter to determine if he or she would be interested in receiving your proposal. Many times a company is eager to learn about a problem and how the writer proposes to solve it for the company. Sometimes the company will even help by giving you information through an interview or a tour. If so, acknowledge that assistance in your proposal.

Unlike a solicited proposal in which the company knows about the problem, an unsolicited proposal has to convince readers that (1) there is a problem and (2) that you and your firm are the ones to solve it. Accordingly, your unsolicited proposal has to document the existence of the problem and demonstrate how solving it benefits readers. If they accept your identification of the problem, you have greatly increased your chances of their accepting your plan to solve it. On the other hand, if you do not convince readers that a problem exists, your solution, and hence your proposal, will be rejected or ignored.

Internal and External Proposals

An internal proposal is written to a decision maker within your own organization. As you will see on pages 556–560, this type of proposal can deal with a variety of topics, including changing a policy or procedure or requesting additional personnel or equipment.

An external proposal, on the other hand, is sent to a decision maker outside your company. It might go to a potential client you have never worked for or to a previous or current client. An external proposal can also be sent to a government funding agency such as the Department of Agriculture. External proposals tend to be more formal than internal ones.

Guidelines for Writing a Successful Proposal

Regardless of the type of proposal you are called on to write, the following guidelines will help you to persuade your audience to approve your plan. Refer to these guidelines both before and while you formulate your plan.

1. **Approach writing a proposal as a problem-solving activity.** Your goal is to solve a problem that affects the reader. Everything in your proposal should relate to the problem, and the organization of your proposal should reflect your ability as a problem-solver. Psychologically, make the reader feel confident that you can solve the problem.

FIGURE 15.3 An internal unsolicited proposal.

EQUAL HOUSING LENDER

COMMUNITY FEDERAL BANK

Powell 584-5200 *Monroe* 413-6000 *Langston* 796-3009

TO: Michael L. Sappington, Executive Vice President
Dorothy Woo, Langston Regional Manager

FROM: Tina Escobar, Oliver Jabur, ATM Services

DATE: June 12, 1997

RE: A proposal to install an ATM at the Mayfield Park branch

PURPOSE

Clearly states why proposal is being sent

We propose a cost-effective solution to what is a growing problem at the Mayfield Park branch in Langston: inefficient servicing of customer needs and rising personnel costs. We recommend that you approve the purchase and installation, within the next three to four months, of an ATM at Mayfield. Such action is consistent with Community's goals of expanding electronic banking services and promoting our image as a self-serve yet customer-oriented institution.

THE PROBLEM WITH CURRENT SERVICES AT MAYFIELD PARK

Identifies problem by giving reader necessary background information

Currently, we employ four tellers at Mayfield. However, we are spending too much on personnel/salary for routine customer transactions. In fact, as determined by teller activity reports, nearly 25 percent of the four tellers' time each week is devoted to routine activities easily accommodated by ATMs. Here is a breakdown of teller activity for the month of May:

Teller #	Total Transactions	Routine Transactions
1	6,205	1,551
2	5,989	1,383
3	6,345	1,522
4	6,072	1,518
	24,611	5,974

Continued

FIGURE 15.3 (Continued)

page 2

Divides problem into parts—volume, financial, personnel, customer service

Clearly, we are not fully using our tellers' sales abilities when they are kept busy with routine activities. To compound the problem, we expect business to increase by at least 25 percent at Mayfield in the next few months, as projected by this year's market survey. If we do not install an ATM, we will need to hire a fifth teller, at an annual cost of $20,800 ($15,500 base pay plus approximately 30 percent for fringes), for the additional 6,000 transactions we project.

Verifies that problem is widespread

Most importantly, customer needs are not being met efficiently at Mayfield. Recent surveys done for Community Federal by Watson-Perry demonstrate that customers are dissatisfied about not having the convenience of an ATM at Mayfield. Customers are unhappy about long waits in line to do simple banking business, such as deposits, withdrawals, and loan payments, and about having to drive to other branches to do their after-hours banking. Conversations we had with manager Rachael Harris-Ignara at Mayfield confirm customers' complaints.

Ultimately, the lack of an ATM at Mayfield Park hurts Community's image. With ATMs available to Mayfield residents at local stores and other banks, our institution risks having customers and potential customers disassociate Community from their banking needs. We not only miss the opportunity of selling them on our other services but also risk losing their business entirely.

A SOLUTION TO THE PROBLEM

Relates solution to individual parts of the problem

The purchase and installation of an ATM at Mayfield Park will initially result in significant savings in personnel costs and time. We will not have to hire a fifth teller and we will be able to allocate more effectively the duties of the four tellers at Mayfield. These tellers will then be able to assist customers with questions and transactions that cannot be handled through an ATM, such as purchase of savings bonds, traveler's checks, or foreign currency. Mayfield tellers, therefore, will have more opportunities for greater involvement in customer services and can spend more time cross-selling our services.

The increase in teller availability should inevitably lead to greater customer satisfaction. An ATM will allow customers the option of

Continued

FIGURE 15.3 (Continued)

page 3

meeting their banking needs electronically or through a teller. Customers can easily make a withdrawal with their ATM cards, while those who need more personal attention can use window service. As customer frustration is eased, so too will be the stress on tellers.

Shows problem can be solved and stresses how

It is feasible to install an ATM at Mayfield. This location does not pose the difficulties we face at some older branches. Mayfield offers ample room to install a drive-up ATM in the stubbed-out fourth drive-up lane. It is away from the heavily congested area in front of the bank, yet it is easily accessible from the main driveway and the side drive facing Commonwealth Avenue.

Judging from the ATM vendor's past work, the ATM could be installed and operational within two to three months. That is the amount of time it took to install ATMs at the first two locations in Powell and for Archer Avenue in Langston. Moreover, by authorizing the expenditure at Mayfield within the next month, you will ensure that ATM service is available long before the Christmas season.

COSTS

Itemizes costs

The costs of implementing our proposal are as follows:

Diebold Drive-up ATM	$28,000.00
Installation fee	2,000.00
Maintenance (1 year)	1,500.00
	$31,500.00

Interprets costs for reader

This $31,500.00, however, does not truly reflect into our annual costs. We would be able to amortize, for tax purposes, the cost of installation of the ATM over five years. Our annual expenses would, therefore, look like this:

$30,000.00 (28,000 + 2,000) divided by 5 years =
$6,000.00 + 1,500 (maintenance) or $7,500 per year.

Compared with the $20,800 a year the bank would have to expend for an additional fifth teller at Mayfield, the annual depreciated cost for the ATM ($7,500) reduces by nearly two-thirds the amount of money the bank will have to spend for much more efficient customer service.

CONCLUSION

Stresses benefits for reader and bank as a whole

Authorizing an ATM for the Mayfield Park branch is both feasible and cost-effective. Adoption of this proposal will save our bank more than $13,000 in teller services annually, reduce customer complaints, and increase customer satisfaction and approval. We will be happy to discuss this proposal with you anytime at your convenience.

2. **Regard your audience as skeptical readers.** Even though you offer a plan that you think will benefit readers, do not be overconfident that they will automatically accept it as the best and only way to proceed. To determine the feasibilty of your plan, readers will question everything you say. They will withhold their approval if your proposal contains errors, omissions, or inconsistencies. Consequently, try to examine your proposal from the readers' point of view.

3. **Research your proposal thoroughly.** A winning proposal is *not* based only on a few well-meaning, general suggestions. All your good intentions and enthusiasm will not substitute for the hard facts readers will demand. Spell out your plan or procedure. Concrete examples persuade readers; unsupported generalizations do not. To make your proposal complete and accurate, you will have to do a lot of homework; for example, reading previous correspondence or research about the problem, doing comparative shopping for the best prices, verifying schedules and timetables, interviewing customers and/or employees, making site visits.

4. **Scout out what your competitors are doing.** Closely related to number 3 above, this is an essential guideline when it comes to writing a winning proposal. Become familiar with your competitors' product lines or services, have a fair idea about their market costs, and be able to show how your company's work is better. You can accomplish all this by doing some Web browsing, looking at your competitors' home pages, and seeing what is available. You might also read print and on-line trade publications.

5. **Prove that your proposal is workable.** The bottom-line question from your reader is "Will this plan work?" Your proposal should be well thought out. It should contain no statements that say, "Let's see what happens if we do *X* or *Y*." By analyzing and, when possible, by testing each part of your proposal in advance, you can eliminate any quirks and revise the proposal appropriately before readers evaluate it. What you propose should be consistent with the organization and capabilities of the company. It would be foolish to recommend, for example, that a small company (fifty employees) triple its workforce to accomplish your plan.

6. **Be sure that your proposal is financially realistic.** This point follows from guideline 5. "Is it worth the money?" is another bottom-line question you can expect from readers. Do not submit a proposal that would require an unnecessarily large amount of money to implement. For example, it would be unrealistic to recommend that your company spend $20,000 to solve a $2,000 problem that might not ever recur. Study the economic climate, too—are you in an economic slump or in a boom time?

7. **Package your proposal attractively.** Make sure your proposal is letter-perfect, inviting, and easy to read (use plenty of headings and other visual devices discussed in Chapter 12). The appearance as well as the content of your proposal can determine whether it is accepted or rejected. Remember that readers, especially those unfamiliar with your work, will evaluate your proposal as evidence of the type of work you want to do for them. Take advantage of any software programs dealing with desktop publishing that may be available to you.

Internal Proposals

The primary purpose of an internal proposal, such as the one shown in Figure 15.3, is to offer a realistic and constructive plan to help your company run its business more efficiently and economically. On your job you may discover a better way of doing something or a more efficient way to correct a problem. You believe that your proposed change will save your employer time, money, or further trouble. (Tina Escobar and Oliver Jabur in Figure 15.3 identified and researched a more effective and less costly way for Community Federal Bank to do business and to satisfy its customers.)

Generally speaking, your proposal will be an informal, in-house message, so a brief (usually one- or two-page) memo should be appropriate. You decide to notify your department head, manager, or supervisor, or your employer may ask you for specific suggestions to solve a problem he or she has already identified. Mike Gonzalez's memo in Figure 4.3 responds to such a request from his employer.

Typical Topics for Internal Proposals

An internal proposal can be written about a variety of topics, such as

- purchasing new or more advanced equipment to replace obsolete or inefficient ones, personal computers, transducers, automobiles
- hiring new employees or retraining current ones to learn a new technique or process
- eliminating a dangerous condition or reducing an environmental risk to prevent accidents—for employees, customers, or the community at large
- improving communication within or between departments of a company or agency
- revising a policy to improve customer relations (eliminating an inconvenience, speeding up a delivery) or employee morale (offering vanpooling, adding more options for a schedule)

As this list shows, internal proposals cover almost every activity or policy that affects the day-to-day operation of a company or agency.

Your Audience and Following the Proper Chain of Command

Writing an internal proposal requires you to be aware of and sensitive to office politics. To be successful, your internal proposal should be written with the needs and likes of your audience in mind. Remember that your boss will expect you to be very convincing about both the problem you say exists and the solution you propose to correct it. You cannot assume that your reader will automatically agree with you that there is a problem or that your plan is the only way to solve it.

Anticipating and Ethically Resolving Reader Problems

Below are some difficulties you need to be aware of and some ethical ways of handling them when you prepare an internal proposal.

1. Before you write an internal proposal, consider the implications of your plan for your boss and for other offices or sections in your company. A change you propose for your department or office (transfers; new budgets or schedules) may have sweeping and potentially disruptive implications for another office or division within your company. It is wise to discuss your plan with your boss before you put it in writing. Then you might provide your boss with a draft, asking for his or her revisions or feedback.

2. Your reader in fact may feel threatened by your plan. After all, you are advocating a change. Some managers regard such changes as a challenge to their administration of an office or department. Think about the long-term effect your proposed change will have on your boss's duties, responsibilities, relationship with your co-workers or with his or her superiors. Don't step on toes or attempt to undermine existing authority.

3. Your reader may have "pet projects" or predetermined ways of doing things. Take these into account and make every attempt to acknowledge them respectfully. You may even find a way to build upon or complement such projects or procedures.

4. Keep in mind that your boss may have to take your proposal further up the organizational ladder for commentary and, eventually, approval. Again, refer to Joycelyn Woolfolk's description of the chain of command at her agency in Figure 3.4.

5. Even though you draft the proposal, it may not bear your name, or your name may be subordinate to your boss's name. It is not uncommon in the world of work to write a document so another person can sign it. Writing a proposal may mean working as a team with your supervisor, whose name goes on the document, too, for notice and credit.

6. Never submit an internal proposal that offers an idea you think will work but relies on someone else to supply the specific details on *how* it will work. For example, do not write an internal proposal that says the payroll, community relations, maintenance, or advertising department can give the reader the details and costs he or she needs about your proposal. That pushes the responsibility onto someone else, and your proposal could be rejected for lack of concrete evidence.

The Organization of an Internal Proposal

A short internal proposal follows a relatively straightforward plan of organization, from identifying the problem to solving it. Internal proposals usually contain four parts, as shown in Figure 15.3: Purpose, Problem, Solution, and Conclusion. Refer to this proposal as you read the following discussion.

Introduction or Purpose
Begin your proposal with a brief statement of why you are writing to your boss: "I propose that. . . ." State why you think a specific change is necessary now. Then succinctly define the problem and emphasize that your plan, if approved by the reader, will solve that problem. Where necessary, stress the urgency to act—within the next week? month?

Background of the Problem
In this section prove that a problem does exist by documenting its importance for your boss and your company. As a matter of fact, the more you show how the problem affects the boss's work (and area of supervision), the more likely you are to persuade him or her to act. And the more concrete evidence you cite, the easier it will be to convince the reader that the problem is significant and that action needs to be taken now.

Here are some guidelines for documenting the problem:

- Avoid vague (and unsupported) generalizations such as, "We're losing money each day with this procedure (or piece of equipment)"; "Costs continue to escalate"; "The trouble occurs frequently in a number of places"; "Numerous complaints have come in"; "If something isn't done soon, more difficulty will result."
- Provide readers with quantifiable details about the implications or consequences of the problem, such as the number of dollars a company is actually losing per day, week, or month. Emphasize the financial trouble so that you can show how your plan (described in the next section) offers an efficient and workable solution.
- Indicate how many employees (or work-hours) are involved or how many customers are inconvenienced or endangered by a procedure or condition. Notice how Escobar and Jabur include this information in their proposal in Figure 15.3.
- Verify how widespread a problem is or how frequently it occurs by citing specific occasions. Rather than just saying that a new word processor would save the company "a lot of money," document how many work-hours are lost using other equipment for the routine jobs your company now has employees perform.

The Solution or Plan
In this section describe the change you propose and want approved. Tie your solution (the change) directly to the problem you have just documented. Each part of your plan should help eliminate the problem or should help increase the productivity, efficiency, or safety you think is possible.

Your reader will again expect to find factual evidence. Do not give merely the outline of a plan or say that details can be worked out later. Supply details that answer the following questions: (1) Is the plan workable—can it be accomplished here in our office or plant? and (2) Is it cost-effective—will it really save us money in the long run or will it lead to even greater expenses?

To get the boss to say yes to both questions, supply the facts that you have gathered as a result of your research. For example, if you propose that your firm buy a new piece of equipment, do the necessary homework to locate the most efficient and cost-effective model available, as Tina Escobar and Oliver Jabur do in Figure 15.3. Supply the dealer's name, the costs, major conditions of service and training contracts, and warranties. Describe how your firm could use this equipment to obtain better results in the future. Cite specific tasks the new equipment can perform more efficiently at a lower cost than the equipment now in use.

A proposal to change a procedure must address the following questions.

1. How does the new (or revised) procedure work?
2. How many employees or customers will be affected by it?
3. When will it go into operation?
4. How much will it cost the employer to change procedures?
5. What delays or losses in business might be expected while the company switches from one procedure to another?
6. What employees, equipment, or locations are available to accomplish this change?

As these questions indicate, your boss will be concerned about schedules, working condition, employees, methods, locations, equipment, and the costs involved in your plan for change. The costs, in fact, will be of utmost importance. Make sure that you supply a careful and accurate budget so that your reader will know what the change is going to cost. Moreover, make those costs attractive by emphasizing how inexpensive they are as compared to the cost of *not* making the change, as Escobar and Jabur do in the section labeled "Costs." Double-check your math.

It is also wise to raise alternative solutions, before the reader does, and to discuss their disadvantages. Notice how Tina Escobar and Oliver Jabur do this in Figure 15.3, in their discussion of the solution to the bank's customer-service problem. They show why installing an ATM is more necessary than hiring a fifth teller. Their discussion is based on a strong persuasive strategy, demonstrating to their readers that they have examined all alternatives and chosen the best.

If you are proposing that your company hire or reassign employees, indicate where these employees will come from, when they will start, what they will be paid, what skills they must have, where they will work and for how long. If you propose to assign current employees to a job, keep in mind that their salary will still have to be paid by your company. Just because they are co-workers does not mean they will work for nothing. Note again in Figure 15.3 how Tina Escobar and Oliver Jabur raise and resolve the problem of new and profitable responsibilities for the tellers at Community Federal Bank.

The Conclusion

The conclusion to your internal proposal should be short—a paragraph or two at the most. Your intention is to remind the reader that the problem is serious, that the reason for change is justified, and that you think the reader needs to take action. Select the most important benefits and emphasize them again. In Figure 15.3, Escobar and Jabur again emphasize the savings that the bank will see by following their plan as well as the increase in customer satisfaction. Also indicate that you are willing to discuss your plan with the reader.

Sales Proposals

A sales proposal is the most common type of external proposal. Its purpose is to sell your company's products or services for a set fee. Whether short or long, a sales proposal is a marketing tool that includes a sales pitch as well as a detailed description of the work you propose to do. Figures 15.2 and 15.4 contain sample sales proposals.

The Audience and Its Needs

Your audience will usually be one or more business executives who have the power to approve or reject a proposal. Unlike readers of an internal proposal, your audience for a sales proposal may be even more skeptical since they may not know you or your work. Your proposal may also be evaluated by experts in other fields employed by your prospective customer.

Readers of a sales proposal will evaluate your work according to (1) how well it meets their needs and (2) how well it compares with the proposals submitted by your competitors. Your proposal must convince readers that you can provide the most appropriate work or service and that your company is more reliable and efficient than any other firm.

The key to success is incorporating the "you attitude" throughout your proposal. Relate your product, service, or personnel to the reader's exact needs as stated in the RFP for a solicited proposal or through your own investigations for an unsolicited proposal. You cannot submit the same proposal for every job you want to win and expect to be awarded a contract. Different firms have different needs.

Typical Questions Readers Will Ask

The most important question the reader will raise about your work is, "How does this proposal meet our company's special requirements?" Some other fairly common questions readers will have as they evaluate your sales proposal include the following.

- Does the writer's firm understand our problem?
- Can the writer's firm deliver what it promises?
- Can the job be completed on time?
- What assurances does the writer offer that the job will be done exactly as proposed?

Answer each of these questions by demonstrating how your product or service is tailored to the customer's needs.

Organizing a Sales Proposal

Most sales proposals include the following elements: introduction, description of the proposed product or service, timetable, costs, qualifications of your company, and conclusion.

Introduction

The introduction to your sales proposal can be a single paragraph in a short sales proposal or several pages in a more complex one. Basically, your introduction should prepare readers for everything that follows in your proposal. The introduction itself may contain the following sections, which sometimes may be combined.

1. **Statement of purpose and subject of proposal.** Tell readers why you are writing and identify the specific subject of your work. If you are responding to an RFP, use specific code numbers or cite application dates, as the proposal in Figure 15.2 does. If your proposal is unsolicited, indicate how you learned of the problem, as Figure 15.4 does. Briefly define the solution you propose.

2. **Background of the problem you propose to solve.** Show readers that you are familiar with their problem and why it is important. In a solicited proposal like Figure 15.2, a section outlining the problem is usually unnecessary, because the potential client has already identified the problem and wants to know how you would handle it. In that case, just point out how your company would solve the problem, mentioning your superiority over your competitors (see the third paragraph of Figure 15.2). In an unsolicited proposal, you need to describe the problem in convincing detail, identifying the specific trouble areas. However, if it is an external proposal to a current customer, such as the one in Figure 15.4, it would be unwise to point out past problems your client may have had with your company's service or products. In Figure 15.4 note how the problems of the staff in the field are described mainly as a way to sell the advantages of the Lightstar 486 notebook computer.

Description of the Proposed Product or Service

This section is the heart of your proposal. Before spending their money, customers will demand hard, factual evidence of what you claim can and should be done. Here are some points that your should cover.

1. **Carefully show potential customers that your product or service is right for them.** Stress specific benefits of your product or service most relevant to your reader. Blend sales talk with descriptions of hardware.

2. **Describe your work in suitable detail.** Specify what the product looks like, what it does, and how consistently and well it will perform in the readers' office, plant, hospital, or agency. You might include a brochure, picture, or, as the writers

FIGURE 15.4 An unsolicited sales proposal.

COMPUTER TECHNOLOGIES, INC.
VOICE 203-555-2680 FAX 203-555-6714 PHONE 203-555-1732
techsupport@interserv.com

Feb. 19, 1998

Ms. Alexandra Tyrone-Saunders
Vice President, Operations
Gemini, Inc.
Hartford, CT 06631-7106

Dear Ms. Tyrone-Saunders:

While we were servicing your Webster PC's last week, I saw several ways in which Computer Technologies might improve your communication system. Based on our assessment of Gemini's requirements for the most up-to-date automation available, we recommend that you purchase ten Lightstar 486 notebooks, one for each of your sales staff. A new generation of notebook computers can provide your staff with a portable, powerful, and affordable computer system that will meet Gemini's automation needs well into the 21st century.

PROBLEM AREAS

Each of your 10 sales staff spends up to 50 to 60 minutes at the end of each business day entering orders they have accumulated in the field into the PC's at your office. This amounts to 50 over-time hours per week. This procedure is costly and inefficient, taking your staff away from their primary responsibility of serving your clients.

ADVANTAGES OF THE LIGHTSTAR 486 NOTEBOOK

Laptop (or notebook) computers are essential complements to other computer systems. According to Felicia Gomez, writing in COMPUTER SYSTEMS WORLDWIDE, the Lightstar 486 is "one of the best computer buys of the decade, a powerhouse that will amaze the most skeptical PC user." The following description of Lightstar's features will show you how its many advantages can help Gemini.

ADVANCED TECHNOLOGY

The Lightstar is equipped with the following standard features:

* an Intel Pentium processor
* 800 x 600 SVGA active matrix color display
* 16 MB EDO RAM
* internal 6x CD-ROM (with an 8x optional)
* 1.2 GB hard drive
* built-in touchpad pointing device

Continued

FIGURE 15.4 (Continued)

2.

Your Lightstar will allow you to fax, E-mail, and connect to the Web. Please visit our Website (http://www.tech.com) for more detail about Lightstar's technical specifications.

Cost Effective
Your Lightstar 486 will cost you far less than your PC's and offers equal or greater capability. Here are some of the bottom line advantages you will be getting:

- costs $500.00 less than one of your PC's 2 years ago
- provides software upgrades at no charge
- includes everything integrated into the system —monitor, CPU, software

Assured Security
Lightstar offers a unique protection system to ensure the security of your unit:

- an encryption system requiring a special password to log onto and activate your notebook
- a removable hard drive (weighing only 1-1/2 pounds)
- a special feature known as "call home" will alert your other notebooks of unauthorized use
- a log-on procedure to protect against employee misuse

Efficient Accessibility and Customer Service
Because the Lightstar is compatible with your office PC's, your representatives can access all pertinent records at your main office while making sales calls. By having direct access to vital documentation, your staff can more efficiently verify and store information about customer orders. The Lightstar makes the virtual office a reality, since your sales force carries your office with them.

Power Supply Capabilities
Equipped with a lithium ion battery, the Lightstar 486 operates up to six hours without recharging. With a car adapter/recharger, the Lightstar can operate without standard electricity almost indefinitely, helping your staff make calls at virtually any location for extended periods of time.

Portability

- weighs only 4.4 pounds (2.2 kg)
- fits in standard briefcase
- comes with protective case

TRAINING AND SERVICE
Because all software can be preloaded, setup and installation can be brief —10 minutes. Since Lightstar is compatible with your PC's, only minimal training in how to use the Lightstar to upload sales orders or download product availability

Continued

FIGURE 15.4 (Continued)

3.

information will be necessary. Our local sales representative, Darlene Simpson, is available to instruct your staff about the operation and routine maintenance of the Lightstar.

If a problem occurs, we offer customers the latest in remote diagnostics. A modem installed in the Lightstar allows us to diagnose specific problems (and needed replacement parts).

COSTS

Below is an estimated price list for one Lightstar 486 notebook with upgraded features:

Lightstar 486 Notebook
Pentium - 166
2.3 GB HDD
32 MB EDO RAM
PCMCIA 28.8 FAX Modem
Second Lithium Ion Battery
Total Purchase Price $2795.00 total

You might also purchase one or all of these options:

8 x CD ROM Disk	$395.00
Sound Card	$215.00
External Speakers	$245.00
	$855.00 total

COMPUTER TECHNOLOGIES' REPUTATION

For the past 13 years Computer Technologies has provided quality products and fast, efficient service after the sale.

We appreciate your past confidence in Computer Technologies and look forward to expanding our service to Gemini. Please call me if you have any questions about the Lightstar or Computer Technologies. I would be happy to bring a Lightstar 486 to Gemini and let you or your sales staff use it for a few days.

If this proposal is acceptable, please sign and return a copy with this letter.

Sincerely yours,

Marion Copely

Marion Copely

I accept the proposal made by Computer Technologies.

____

for Gemini, Inc.
Encl.

of the proposals in Figures 15.2 and 15.4 do, a sample of your product for customers to study. Convince readers that your product is the most up-to-date and efficient one they could select.

3. Stress any special features, maintenance advantages, warranties, or service benefits. Highlight features that show the quality, consistency, or security of your work. For a service, emphasize the procedures you use, the terms of that service, and even the kinds of tools you use, especially any state-of-the-art equipment. Be sure to provide a step-by-step outline of what will happen and why each step is beneficial for readers.

Timetable

A carefully planned timetable shows readers that you know your job and that you can accomplish it in the right amount of time. Your dates should match any listed in an RFP. Provide specific dates to indicate

- when the work will begin
- how will the work be divided into phases or stages
- when you will be finished
- whether any follow-up visits or services are involved

For proposals offering a service, specify how many times—by the hour, week, month—customers can expect to receive your help; for example, spraying three times a month for an exterminating service or making deliveries by 10:00 A.M. for a trucking company.

Costs

Make your budget accurate, complete, and convincing. Accepted by both parties, a proposal is a binding legal agreement. Don't underestimate costs in the hope that a low bid will win you the job. You may get the job but lose money doing it, for the customer will rightfully hold you to your unrealistic figures. Neither should you inflate prices; competitors will beat you in the bidding.

Give customers more than merely the bottom-line cost. Show exactly what readers are getting for their money so they can determine if everything they need is included. Itemize costs for

- specific services
- equipment/materials
- labor (by the hour or by the job)
- transportation
- travel
- training

If something is not included or is considered optional, say so—additional hours of training, replacement of parts, and the like.

If you anticipate a price increase, let the customer know how long current prices will stay in effect. That information may spur them to act favorably now.

Qualifications of Your Company

Emphasize your company's accomplishments and expertise in using relevant services and equipment. List previous similar work. You may even want to mention the names of a few local firms for whom you have worked that would be able to recommend you. Never misrepresent your qualifications or those of the individuals who work with or for you. Your prospective client may verify if you have in fact worked on similar jobs for the last five to six years.

Conclusion

This is the "call to action" section of your proposal. As with the conclusion in an internal proposal, encourage your reader to approve your plan. Stress the major benefits your plan has for the customer. Offer to answer any questions the reader may have. Some proposals end by asking readers to sign and return a copy of the proposal thus indicating their acceptance of it, as the proposal in Figure 15.4 does.

Proposals for Research Papers and Reports

You may have to write a proposal like the one in Figure 15.5 when your instructor asks you to submit a report or research paper, a topic for an independent study, or some other major term project.

Writing for Your Teacher

The principles guiding internal and sales proposals also apply to research proposals. As with internal and sales proposals, you will be writing to convince your reader—the teacher—to approve a major piece of work. But otherwise the goals of your teacher/reader will be considerably different from those of other proposal readers. A teacher will read your proposal to help you write the best possible paper or report. In examining your proposal, the teacher will want to make sure of four things:

1. that you have chosen a significant topic
2. that you have a sufficiently restricted topic
3. that you will investigate important sources of information about that topic
4. that you can accomplish your work in the specified time

Your proposal gives your teacher an opportunity to spot omissions or inconsistencies and to provide helpful suggestions.

To prepare an effective proposal for a research project, you must do some preliminary research. You cannot pick any topic that comes to mind, or guess about procedures, sources, or conclusions. As other proposal readers do, your teacher will want convincing and specific evidence for your choice of topic and your approach to it. Be prepared to cite key facts to show that you are familiar with the topic and that you can handle it successfully.

Organization of a Proposal for a Research Paper

Your proposal for a school research project can be a memo divided into five sections, as illustrated in Figure 15.5: *introduction* (or purpose), *scope of the problem* or topic to be investigated, *methods or procedures, timetable,* and *request for approval.* However, be ready to reverse or expand these sections if your teacher wants you to follow a different organizational plan.

The Introduction

Keep your introduction short—a paragraph, maybe two, pinpointing the subject and purpose of your work.

```
I propose to research and write a report about the
"hot knife" laser used in treating port wine stain and
other birthmarks.

I intend to investigate the relationship that exists
between office design and the employee's need for
"psychological space."
```

Then briefly indicate why the topic or the problem you propose to study is significant. In other words, be prepared to explain why you have chosen that topic and why research on it is relevant or worthwhile for a specific audience or course objective. Note how Barbara R. Shoemake in Figure 15.5 states how and why her report will be useful to office managers.

Supply your teacher/reader with a few background details about your topic; for example, the importance of using a laser as opposed to conventional ways of treating birthmarks or why psychological space plays a crucial role in employee productivity and morale. Prove that you have thought carefully about a suitable topic.

The Scope of the Problem or Topic to Be Investigated

The second section, which might be entitled "Problems to Be Investigated" or "Areas to Be Studied," shows how you propose to break the topic into meaningful units. Tell your reader what specific issues, points, or areas you hope to investigate. Doing this, you show how you will limit your topic to establish the appropriate scope for your work.

Some instructors ask students to formulate a list of questions their research paper or report intends to answer. The topics included in these questions, or a list of areas or problems to be covered, might later become major sections of your paper. Make sure that the issues or questions do not overlap and that each relates directly to and supports your restricted topic. Note how the student in Figure 15.5 hopes to divide her study of electronic mail into four distinct yet related areas.

Methods or Procedures

In the third section of your proposal inform your teacher how you expect to find the answers to the questions you raised in the previous section, or how you intend

FIGURE 15.5 A proposal for a research paper.

To: Professor Leigh Felton-Parks
From: Barbara R. Shoemake B.S.
Date: February 5, 1997
Subject: Proposal for a report on the ethical and security issues involved in using e-mail

PURPOSE

For my term project I propose to research and write a report on the ethical and security issues involved in using e-mail in the workplace.

E-mail is the most frequently used form of business communication. It has been estimated that 20 million e-mail users receive 15 billion messages a year, most for internal business communications. E-mail has brought many companies closer to the paperless office. At West Industries, over 61,000 employees receive their company publications on-line. Jerry Brown has found that "E-mail gets a higher reading rate than printed material" (*Public Relations Journal* [Mar. 1996]: 25).

Despite its many advantages, e-mail presents major ethical and legal problems. Using e-mail technology is not always comfortable and uncomplicated for employees or their employers. E-mail privacy is at the heart of the controversy. Major legal battles have been waged over who can and should read an individual's e-mail at work. An understanding of the ethical considerations and security drawbacks of e-mail is important for any manager having to evaluate his or her office's internal communications. My paper will serve as a background report for that office manager.

PROBLEMS TO BE INVESTIGATED

At this stage of my research, I think my report needs to answer the following questions:

(1) How does e-mail differ from conventional communication methods (telephones, letters) in terms of confidentiality?

(2) Do employers have the right (or responsibility) to monitor employees' workplace e-mail or is this a violation of the employee's right to privacy?

(3) What can be done to establish a more secure e-mail system to protect both confidentiality and ensure company security?

(4) What types of special training programs—on business communication, Netiquette, and legal rights—are most necessary and effective for e-mail use?

Continued

FIGURE 15.5 (Continued)

I propose, therefore, to divide the body of my paper according to the four key issues of *confidentiality, monitoring, security,* and *training.*

METHODS OF RESEARCH

I will rely heavily on literature dealing with e-mail. Judging from the number of entries found on this general topic in *Business Periodical Index* (which I accessed through **FirstSearch**) and through search engines Info-Seek and Yahoo!, the subject of e-mail is both popular and significant. I found over 750 entries. Restricting my search to just security and ethical issues, I located 54 items. From a preliminary check of what was available to me at McGovern Library, I think the following may be most useful.

Aronson, Michael. "Training May Solve E-mail Worries." *Journal of Corporate Security* 5 (Sept. 1996): 70-74.

Brown, Jerry. "E-mail and Office Communications." *Public Relations Journal,* March 1996: 16-19.

Cappel, J. "A Study of Individuals' Ethical Beliefs and Perceptions of Electronic Mail Privacy." *Journal of Business Ethics* 14 (Fall 1995): 819-27.

Doss, E. and M.C. Loui. "Ethics and the Privacy of Electronic Mail." *The Information Society* 11 (July-Sept. 1995): 223-25.

Gonzalez-Zuereno, Rose. "The Problems with E-mail." *Working Woman* 19 (Oct. 1994): 61-62.

Hammonds, Keith. "E-mail: Beware of Big Brother." *Business Week* 34 Mar. 1996: 4.

"Here's Help for Your E-mail System!" Apr. 1996. <http://www.c2c.com/help> (Jan. 1997).

"Is E-mail Safe for Your Company's Health?" *Supervisory Management* 42 (Jan. 1997) : 17-18.

Maynard, Roberta. "Employee E-mail: Is It Really Private?" *Nation's Business* 84 (Feb. 1996): 10.

McGough, Nancy. "Infinite Ink's Electronic Conversations with Mail & News." Jan. 1996. <http://www.jazzie.com/ii/internet/mailnews.html> (Jan. 1997).

Ouellette, Tim. "Return to Sender?" *Computerworld* 30 Mar. 1996: 45.

2.

Continued

FIGURE 15.5 (Continued)

Pasher, Victoria S. "IMMS Members Say E-mail Is Increasingly Popular Tool." *National Underwriter* 100 (Mar. 11, 1996): 7. ABI/Inform. On-line. ProQuest. 1 Feb. 1997.

Rosner, Hillary. "Will E-mail Become J-mail?" *Brandweek* 37 (Mar. 1996) : 30.

Schneier, Bruce. *E-mail Security: How to Keep Your Electronic Message Private.* New York: Wilex, 1995.

Schwartz, Jeffery. "E-mail and the Internet: Public or Private" *Communications Week* 26 Feb. 1996: 1-9.

Sherwood, Kaitlin. "A Beginner's Guide to Effective E-mail." Dec. 1994. <http://www.webfoot.com.advice.email.top.html> (Jan. 1997).

Thompson, J., et al. "Privacy, E-mail, and Information Policy: Where Ethics Meet Reality." *IEEE Transactions on Professional Communication* 38 (Sept. 1995).

I also intend to interview two office managers in Springfield whose companies have offered employee seminars on electronic mail in the last year. My choices right now are Alice Phillips at Dodge & Spenser Hydraulic Systems and Keith Wellbridge at General Dynamics. Because of their possible schedule conflicts, I may have to interview two other individuals.

SCHEDULE

I hope to complete my library research by April 2 and my interviews by April 8. Then I will spend the following two weeks working on a draft, which I will submit by April 18, the date you specified. After receiving your comments on my draft, I will work on revisions and the final copy of my report and turn it in by May 12, the last day of class. I will submit two progress reports -- one when I finish my research and another when I decide on the final organization of my paper.

REQUEST FOR APPROVAL

I ask that you approve my topic and my approach to it. I would appreciate any suggestions on how you think I might best proceed. My e-mail address is bshoemake@whale.usm.edu, if you prefer to send them to me this way.

Thank you.

3.

to locate information about your list of subtopics. It's not enough to write, "I will gather appropriate information and analyze it." Specify what data you hope to include, where they are located, and how you intend to retrieve them.

Most students gather data from literature published about their topics. (In fact, many research papers are based exclusively on literature searches.) This literature can include books, encyclopedias or other reference materials, articles in professional publications, newspapers, bulletins, manuals, or audiovisual materials. Inform your teacher what indexes, abstracts, or Internet searches you intend to use (review pp. 329–341) as part of your search. To document your preliminary work, provide your instructor with a list of a few appropriate titles on your topic following the style of documentation used for a Works Cited page (discussed on pp. 378–383).

In addition to library materials, you might collect information from experiments you will perform in a laboratory or from a field test, interviews with experts, e-mail queries and replies, questionnaires, or a combination of any of these sources.

Timetable/Schedule

Your proposal should indicate when and in what order you expect to complete the different phases of your project. Your teacher needs this information to keep track of your progress and to make sure that you will turn in an assignment on time. Specify tentative dates for completing your research, draft(s), revisions, and final copy.

Some instructors also ask students to turn in progress reports (see pp. 589–594) at regular intervals. If you are asked to do this, indicate when those progress reports will be submitted as Barbara Shoemake does in Figure 15.5.

Request for Approval

End your proposal with a request for approval of your topic and plan of action. You might also invite suggestions from your teacher on how to restrict, research, organize, or write about your topic.

Preparing Proposals: A Final Reminder

This chapter has given you some basic information and specific strategies for writing winning proposals. Keep in mind that a proposal presents a plan to a decision maker for his or her approval. To win that approval, your proposal must be *realistic, carefully researched,* and *highly persuasive.* These essential characteristics apply to internal proposals in memo format written to your employer, more formal sales proposals sent to a potential customer, and research proposals submitted to your instructor.

✓ Revision Checklist

- ❑ Established and distinguished roles of collaborative team members involved in the preparation of the proposal.
- ❑ Appointed one individual to serve as editor or coordinator.
- ❑ Identified a realistic problem, one that is restricted and relevant to my topic and my audience's needs.
- ❑ Tried effectively to convince audience that the problem exists and needs to be solved.
- ❑ Incorporated quantifiable details demonstrating the scope and importance of problem.
- ❑ Persuasively emphasized benefits of solving the problem according to the proposal; incorporated the "you attitude" throughout.
- ❑ Investigated and overcame competing alternatives/firms.
- ❑ Offered a solution that can be realistically implemented—that is, it is both appropriate and feasible for audience.
- ❑ Wrote clearly so audience can understand how and why my proposal would work.
- ❑ Researched background of problem.
- ❑ Used specific figures and concrete details to show how proposal saves time, money.
- ❑ Double-checked proposal to catch errors, omissions, and inconsistencies.
- ❑ Avoided exaggerations.
- ❑ Organized proposal with appropriate headings for clarity and ease of reading.
- ❑ Demonstrated how proposal benefits my company and my supervisor; took into account office politics in describing problem and solution; discussed proposal with co-workers or supervisors who may be affected. (For internal proposal.)
- ❑ Related my product or service to prospective customer's needs; showed a clear understanding of those needs. (For sales proposal.)
- ❑ Prepared a comprehensive and realistic budget; accounted for all expenses; itemized costs of products and services in sales proposal.
- ❑ Provided a timetable with exact dates for implementing proposal.
- ❑ Cited other successful jobs and satisfied clients to show my company's track record.
- ❑ Concluded proposal with a summary of main benefits to readers and a call to action.
- ❑ Proved to my instructor that I researched the problem by supplying a list of possible references and sources for a proposal for a research report/paper.

Exercises

1. In two or three paragraphs identify and document a problem (in services, safety, communication, traffic, scheduling) you see in your office or plant or at your school. Make sure that you give readers—a school official (chairperson; dean) or employer (section or department head; manager)—specific evidence that a problem does exist and that it needs to be corrected.

2. Write a short internal proposal, modeled after Figure 15.3, based on the problem you identified in Exercise 1.

3. Write a short internal proposal, similar to Tina Escobar and Oliver Jabur's in Figure 15.3, recommending to a company or a college a specific change in procedure, equipment, training, safety, personnel, or policy. Make sure that your team provides an appropriate audience (a college administrator or department manager or section chief) with specific evidence about the existence of the problem and your solution of it. Possible topics include

   - providing more and safer parking
   - extending the bookstore or company credit union hours
   - purchasing new office or laboratory equipment or software
   - hiring more faculty, student workers, or office help
   - reorganizing or redesigning the school yearbook or company annual report or sales catalog
   - changing the decor/furniture in a student or company lounge
   - expanding the number of weekend or night classes in your major
   - adding more health-conscious offerings to a school or company cafeteria menu
   - altering the programming on a campus radio station
   - decreasing waiting time at school registration or in a computer lab

4. Write an unsolicited sales proposal, similar to the one in Figure 15.4, on one of the following services or products you intend to sell, or on a topic your instructor approves.

   - providing exterminating service to a store or restaurant
   - supplying a hospital with rental television sets for patients' rooms
   - keyboarding or word processing for students
   - offering temporary office help or nursing care
   - providing landscaping and lawn care work
   - testing for noise, air, or water pollution in your community or neighborhood
   - furnishing transportation for students, employees, or members of a community group
   - providing consulting service to save a company money
   - designing business forms for a local bank or hospital

- digging a septic well for a small apartment complex
- supplying insurance coverage to a small (five to ten employees) firm
- cleaning the parking lot and outside walkways at a shopping center
- selling a piece of equipment to a business
- making a work area safer
- offering a training program for employees
- increasing donations to a community or charitable fund

5. Write a solicited proposal for one of the topics listed in Exercise 4 or for a topic that your instructor approves. You might want to review Figure 15.2. Do this exercise as a collaborative effort.

6. Write an appropriate proposal—internal, solicited sales, or unsolicited sales—based on the information contained in one of the following three articles. Assume that you or your prospective customer's company or community faces a problem similar to one discussed in one of these articles. Use as much of the information in these articles as you need and add any details of your own that you think are necessary. This can be done as an individual or collaborative assignment.

Turning Schoolgrounds Green

"If our conservation district doesn't take the initiative to show our children and school leaders how to stop erosion on their playgrounds, then who will?" said Bobby Joe Ganey, Chair of the Lasalle Soil and Water Conservation District in Lasalle Parish, Louisiana.

Ganey, along with other conservation district board members, was tired of seeing bare, eroded soil outside classroom doors, so the district board initiated a project to put a cover on the schoolgrounds.

The district board members talked with school principals in the parish about erosion problems on their school campuses. The district board determined that six school campuses were suffering from a lack of vegetative cover, and erosion was keeping their playgrounds bare. Board members asked the Soil Conservation Service to prepare a plan for the schools.

"School grounds get a lot of foot traffic from the students, so it was necessary for us to establish a species of grass that could withstand this traffic," Ganey said.

The Lasalle Soil and Water Conservation District supplied the funds to buy bermuda grass seed and fertilizer.

"It was our intention from the beginning to have the students take an active part in establishing vegetative cover on the playgrounds," said Ganey. "In this way not only could they see the value of erosion control at their school but they also could learn how erosion is bad for the community and for their futures."

The district board and Soil Conservation Service introduced the students to erosion problems through a slide show. More than 650 students from the six elementary schools participated in the erosion control work on their campuses. They helped till, seed, and fertilize the eroding areas.

To be sure that the newly established grass would be properly maintained, the plan included information and training on cutting height and fertilizer requirements.

Soil and Water Conservation News (Oct. 1984): 9.

Self-illuminating Exit Signs

The Marine Corps Development and Education Center (MCDEC), Quantico, Virginia, submitted a project recently, to replace incandescent illumination exit signs with self-illuminating exit signs for a cost of $97,238. The first-year savings were anticipated to be about $37,171 with an anticipated payback time of 2.6 years—an excellent prospect. The contractor bid much lower, however, and the actual payback will be about 1.5 years.

What are the benefits of these self-illuminating exit signs? The primary benefit is that virtually all operation and maintenance expense is eliminated for the life of the device, normally from 10 to 12 years. Power failures or other disturbances will not cause them to go out.

In new construction, expensive electrical circuits can be totally eliminated. In retrofits, the release of a dedicated circuit for other use may be of considerable benefit. Initial total cost of installing circuits and conventional devices approximately equals the cost of the self-illuminating signs. Installation labor and expense for the self-illuminating signs is about that of hanging a picture.

The amount of electricity saved varies and depends on whether your existing fixtures are fluorescent (13 to 26 watts) or incandescent (50 to 100 watts). Multiply the number of fixtures × wattage/fixture × hours operated/day × days/year = KWH/year savings. For example, assume:

400 incandescent fixtures
$0.08/KWH 0.05 KW/fixture
24 hours/day 365 day/year operation
400 × 0.1 × 365 = 350400 KWH/year
350400 × 0.08 = $28,032/year for electricity

Now add in savings achieved from:

- reducing labor to change bulbs
- avoiding bulb material, stocking, and storage costs
- avoiding transportation costs involved in bulb changes
- reusing existing bulbs

The above savings can be significant. For the MCDEC Quantico project, estimates of bulb change interval and savings were 700 hours (29 days) and $13,512/year when all factors were considered.

The cost of a self-illuminating sign depends on whether one or two faces are illuminated primarily and varies between different suppliers. Single-face prices will likely be $100 to $150 while double-face prices may be $250 to $330. The contractor at Quantico found better prices than these ranges indicate. The labor cost should be about $10 per sign.

If you can use an exit-sign system with high dependability, no maintenance, and zero operations cost in your retrofit on new construction projects, try a self-illuminating exit-sign system in your economic analysis today. "Isolite" signs, by Safety Light Corp., are listed as FSC (Fire Safety Code) Group 99, Part IV, Section A, Class 9905 signs and are available through GSA contract. Contact Gerald Harnett, Safety Light Corp., P.O. Box 266, Greenbelt, MD 20070 for more information.

Lt. James F. McCollum, CEG, USN. "Self-illuminating Exit Signs Equal High Payback." *Navy Civil Engineer* (Summer 1983): 30–31.

Wheelchair-Lift Switch Covers

In order to ensure year-round access to the Springfield Armory National Historic Site (Massachusetts) museum, Michael C. Trebbe designed the cover for switches on wheelchair lifts. During the extreme New England winters, the switch buttons would freeze, thus making the lift inoperable, which in turn required several hours to thaw. The installation of these covers prevented the freezing of the switch buttons and, therefore, allowed maintenance personnel to attend to matters such as snow removal.

The covers were made of materials found on site, which resulted in the covers being almost cost free. The covers can be quickly built, and they are mounted with the same mounting screws as the switch boxes so as to not destroy any original fabric (in the case of Springfield Armory NHS, brownstone). The materials used included:

- 1/8-in. by 4 1/2-in. by 12-in. piece of rubber mat
- 3 1/2-in. by 6 1/2-in. piece of sheet metal
- three aluminum pop rivets
- primer for the sheet metal
- wheelchair symbol
- white paint for the symbol

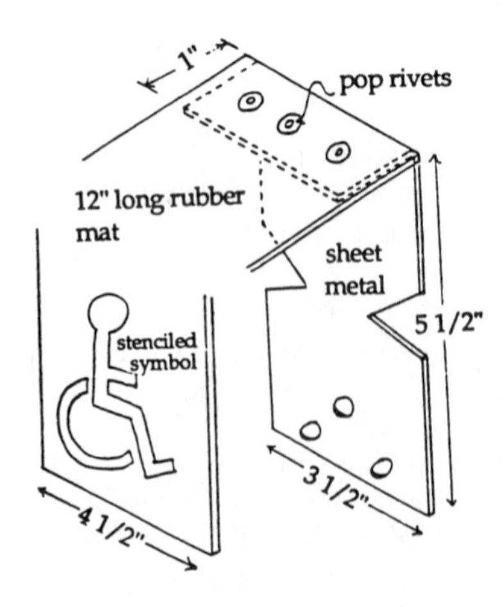

The sheet metal is bent to a 90-degree angle at the 5 1/2-inch point. The lowest two holes (see diagram on p. 576) are drilled to mount the screws of the switch box, which also secure the cover. Triangular cutouts and other holes are drilled for the clearance of the housing screws on the rear of the switch box (see diagram).

The handicap symbol is stenciled on the front of the piece of rubber mat using white paint.

Michael C. Trebbe, *Grist,* Vol. 36, No. 1 (Winter 1992). U.S. Department of the Interior, National Park Service.

---

7. Write a suitable research proposal on which the research paper on "The Advantages of Telecommuting in the Information Age" (pp. 398–413) could have been based.

8. Write a research proposal to your instructor seeking approval for a major term project. Do the necessary preliminary research to show that you have selected a suitable topic, that you have narrowed it, and that you have identified the sources of information you have to consult. Identify at least six relevant articles and two books on your topic.

CHAPTER 16

Writing Effective Short Reports

This chapter will show you how to write short reports, also called informal or semiformal reports. A short report may be defined as an organized presentation of relevant data on any topic—money, travel, time, personnel, equipment, management—that a company or agency tracks in its day-to-day operations. You will be asked to write short reports frequently on the job—on your own or in collaboration with co-workers. Short reports provide co-workers, employers, vendors, and clients with information they need to get their jobs done, too.

Why Short Reports Are Important

Business and industry cannot function without short written reports. Reports tell whether

- work is being completed
- schedules are being met
- costs are contained
- sales projections are met
- clients and patients are being effectively served
- unexpected problems have been solved

You may write an occasional report in response to a specific question, or you may be required to write a daily or weekly report on routine activities about which your readers expect detailed information. Many organizations—businesses, clinics, mass transportation systems, schools—must submit regularly scheduled reports in order to maintain their accreditation or funding by state, municipal, or federal agencies. Since a large part of your annual evaluation for raises and promotions will depend on the caliber of your short reports, it is important to know how to write them effectively.

Types of Short Reports

To give you a sense of some of the topics that you may be required to write about, here is a list of various types of short reports common in the business world.

appraisal report
audit report
construction report
design report
evaluation report
experiment report
feasibility report
incident report
inventory report
investigative report
justification report
laboratory report
manager's report
medicine/treatment error report
operations report
periodic report
production report
progress/activity report
recommendation report
research report
sales report
status report
test report
trip/travel report

This chapter will concentrate on six of the most common types of reports you are likely to encounter in your professional work:

1. periodic reports
2. sales reports
3. progress reports
4. trip/travel reports
5. test reports
6. incident reports

The first five reports can be called **routine reports** because they give information about planned, ongoing, or recurring events. The sixth category, **incident reports,** are reports that describe events the writers did not anticipate—accidents, breakdowns, environmental mishaps, delivery delays, or work stoppages. All six, however, may be termed *short* reports because they deal with current happenings rather than with long-range forecasts. Short reports focus on the "trees," not the "forest."

Guidelines for Writing Short Reports

Though there are many different kinds of short reports, they all have certain features in common. The most important point to keep in mind is that reports are written for readers who need information so that they can get a job accomplished. Never think of the reports you write as a series of short notes jotted down for *your own* convenience.

Below are six guidelines to follow to write any short report successfully.

1. Do Necessary Research

An effective short report needs the same careful planning that goes into other types of on-the-job writing. Your research may be as simple as telephoning, e-mailing, or leaving a voice mail for a colleague, or inspecting a piece of equipment. Some frequent types of research you can expect to do on the job include

- checking data in reference manuals or code books
- exploring the Web, using key search engines
- searching databases for recent discussions of a problem or procedure
- reading background information in professional journals
- reviewing a client's file
- testing equipment
- performing an experiment or procedure
- conferring with colleagues, managers, vendors, or clients
- describing a site
- attending a conference

2. Anticipate How an Audience Will Use Your Report

You will have to consider how much your audience knows about your project and in what types of information they are most interested. Employers, who will constitute the largest audience for your reports, may not always know (or be interested in) the technical details of your work. Instead, they may want you to focus on the bottom line—costs, personnel, organizational structure, problems, or delays. While co-workers may be familiar with your project, colleagues in other departments, consumers, or individuals outside your company (such as site inspectors) may not. Accordingly, these readers may require more background information, definitions, and examples. To meet your audience's needs, answer the following questions appropriately for your readers:

- *Why are you writing?* This may be your reader's most significant, and urgent, question. Always explain your reasons for writing. Is your report routine or has it been requested for a special purpose—such as justifying a new position in your department? Tell readers, too, why you think something turned out the way it did.
- *What happened?* Explain what steps you followed in a lab report, what specific events or circumstances occurred, what conclusions can be made, what prospects are likely for future business.
- *When did something happen?* Always give dates and specify the exact period the report covers. Just listing "Thursday" is not enough. Give the date, and indicate A.M. or P.M. To record time in compact and specific terms, some employers may ask you to use a twenty-four-hour clock: 1:00 A.M. is 0100 hours; 1:00 P.M. is 1300 hours. An event occurring on February 19, 1998 at 2:30 P.M. is written 98/2/19/1430—year, month, day, time (hours/minutes).
- *Where did something happen?* Give precise locations. "Highway 30" is not as helpful as "Highway 30, three miles southeast of the Morton exit."
- *Who did something or who was involved?* Give readers the names of clients, contact people, technical staff you consulted, members of your collaborative team, or individuals involved in a test or accident.

- *How did something happen?* Tell readers how a test was conducted. Describe the procedures you used. Inform them about the results you observed or how a delay, problem, or shortage affected your progress. Explain whether a presentation was relevant and effective and how it might affect your business.

3. Be Objective and Ethical

Your readers will expect you to report the facts objectively and impartially—costs, sales, weather conditions, eyewitness accounts, observations, statistics, test measurements, and descriptions. Your reports should be truthful, accurate, and complete. Your boss's decision will be based upon this information. Always avoid the following:

- *guesswork*—if you don't know or have not yet found out, say so and indicate how you'll try to find out
- *impressions*—not a substitute for careful research
- *unsupported personal opinions*—avoid using these instead of careful research
- *biased, skewed, or incomplete data*—provide a straightforward and honest account; don't exaggerate or minimize

4. Choose a Reader-Centered Format and Design

For the most part, reports for readers in your company will be written or e-mailed as memos, while those submitted to clients will be letters. But regardless of the format, help your readers easily find information by including the following:

- *a clear, precise subject line* that announces your subject and purpose
- *headings* that preview and highlight information—your report needs subdivisions; don't bombard readers with a series of uninterrupted paragraphs
- *bullets or numbers* to list and group main points
- *underscoring* to emphasize and to make it easier for readers to skim (if necessary) or review
- *visuals* to clarify and expedite. Never load readers down with numerical information in the text of your report. Note how much harder it is for readers to wade through Figure 16.1, which does not include a table, than to navigate Figure 16.2, which does. Many of the short reports in this chapter include a clarifying visual—a map (16.8), tables (16.2; 16.11).

5. Write Concisely and Clearly

Say what you need to say without wasting readers' time. Even though these reports are brief, allow time for careful revising and editing. Writing concisely—to the point and clearly—requires effort. (You may want to review pp. 54–61). Especially time-consuming for your readers are the wordy expressions below that need to be replaced with more serviceable substitutes.

Wordy	Concise
at this point in time	now
make a concerted attempt to	try
take place in such a manner that	occur

FIGURE 16.1 An example of a poorly written, organized, and formatted short report.

TO: Capt. Alice Martin
FROM: Sergeants Daniel Huxley, Jennifer Chavez, and Peter Kellog
Vague SUBJECT: Crime rate
DATE: July 11, 1998

Introduction doesn't tell reader anything about overall picture

This report will let you know what happened this quarter as opposed to what happened last quarter as far as crimes are concerned in Greenfield. This report is based on statistics the department has given us over the quarter.

Throws facts at reader without any sense of reader's needs

Irrelevant data

Here we'll let the facts speak for themselves. From Jan.-Mar. we saw 126 robberies while from Apr.-June we had 106. Home burglaries for this period: 43; last period: 36. 33 cars were stolen in the period before this one; now we have 40. Interestingly enough, last year at this time we had only 27 thefts.

No analysis or guided commentary -- just undigested numbers

Homicides were 9 this time versus 8 last quarter; assault and battery charges were 92 this time, 77 last time. Carrying a concealed weapon 11 (10 last quarter). We had 47 arrests (55 last quarter) for charges of possession of a controlled substance. Rape charges were 8, 1 less than last quarter. 319 citations this time for moving violations: speeding 158/98, and failing to observe the signals 165/102 last quarter. DUIs were good this quarter -- only 45, or 23 fewer than last quarter.

Hard to follow comparisons and contrasts

Misdemeanors this time: disturbing peace 53; vagrancy/public drunkenness 8; violating leash laws 32; violating city codes 39, including dumping trash. Last quarter the figures were 48, 59, 21, 43.

Conclusion provides no summary or recommendation

We believe this report is complete and up to date. We further hope that this report has given you all the facts you will need.

Call machines and other equipment by their precise names. Never use "thing," "gizmo," or "contraption" to refer to parts or tools.

Also, avoid vague and unsupported statements such as the following:

Vague: Sales were brisk this week with many favorable possibilities on the horizon.

Precise: During the first week of June, our sales increased by 12.25 percent over those of the last week in May. Much of the increase can be attributed to our new home page, which makes ordering easier for customers. Many of our new customers specifically mentioned our home page when placing their orders.

6. Organize Carefully

Organizing a short report effectively means that you include the right amount of information in the most appropriate places for your audience. Many times a simple chronological or sequential organization will be acceptable for your readers. Your employer may have very precise instructions on how to organize routine reports, but here is a fairly standard organizational plan to follow:

Purpose

Always begin by telling readers why you are writing and by alerting them to what you will discuss. When you establish the scope (or limits) of your report, you help readers zero in on specific times, places, procedures, or problems. Depending on your purpose and audience, you may have to start with a clear explanation or description of the problem to be studied or solved. You may also need to provide necessary background information (say, a summary of an earlier report or occurrence) to assist readers.

Findings

This is the longest part of your report. The data that you collected go under this section—facts about prices, personnel, equipment, events, locations, incidents, or experiments. Gather these data from your research, personal observations, interviews, or conversations with co-workers, employers, or clients.

Conclusion

Generally, your conclusion tells readers what your data mean. A conclusion can summarize what has happened, review what actions were taken, or explain the outcome or results of a test, a visit, or a program.

Recommendations

A recommendation informs readers what specific actions you think your company or client should take. Recommendations must be based on the data you have collected and the conclusions you reached.

The placement of recommendations in a short report can vary. Some employers prefer to see recommendations at the beginning of a report; others want them listed last. Some short reports (including those in Figures 16.2, 16.4, and 16.7) do not require a recommendation. Be sure to find out if your readers will expect you to make one.

FIGURE 16.2 A well-prepared quarterly report, revised from Figure 16.1.

GREENFIELD POLICE DEPARTMENT

"To Serve and To Protect"

Administration 555-1000 Traffic 555-1001 Drug Enforcement 555-1002

TO: Captain Alice Martin
FROM: Sergeants Daniel Huxley, Jennifer Chavez, and Peter Kellog
SUBJECT: Crime rate for the second quarter of 1998
DATE: July 11, 1998

From April 1 to June 30, 852 crimes were committed in Greenfield, representing a 5 percent increase over the 815 crimes recorded during the previous quarter.

The following report, based on the table below, discusses the specific types of crimes, organized into four categories: **robberies**, **felonies, traffic,** and **misdemeanors.**

Table 1. A comparison of the 1st and 2nd Quarter 1998 Crime Rates in Greenfield.

CATEGORY	1st Quarter	2nd Quarter
ROBBERIES		
Commercial	63	75
Domestic	36	43
Auto	33	40
FELONIES		
Homicides	16	13
Assault and Battery	77	92
Carrying a concealed weapon	10	11
Poss. of a controlled substance	55	47
Rape	9	8
TRAFFIC		
Speeding	165	158
Failure to observe signals	102	98
DUI	78	45
MISDEMEANORS		
Disturbing the peace	48	53
Vagrancy	40	48
Public drunkenness	19	40
Leash law violations	21	32
Dumping	35	37
Other	8	12

Continued

FIGURE 16.2 (Continued)

Page 2

ROBBERIES

The greatest increase in crime was in robberies, 20 percent more than last quarter. Downtown merchants reported 75 burglaries totaling more than $950,000. The biggest theft occurred on May 21 at Weisenfarth's Jewelers when three armed robbers stole more than $97,000 in merchandise. (These suspects were apprehended two days later.) Home burglaries accounted for 43 crimes, though the thefts were not confined to any one residential area. We also had 40 car thefts reported and investigated.

FELONIES

Homicides decreased slightly from last quarter—from 16 to 13. Charges for battery, however, increased—15 more than we had last quarter. Arrests for carrying a concealed weapon were nearly identical this quarter to last quarter's total. But the 47 arrests for possession of a controlled substance were appreciably down from the first quarter. The number of arrests for rape for this quarter were also less than last quarter's. Three of those rapes happened within one week (May 6-12) and have been attributed to the same suspect, now in custody.

TRAFFIC

Traffic violations for this period were lower than last quarter's figures. This quarter's citations for moving violations (335) represent a 5 percent increase over last quarter's (322). Most of these citations were issued for speeding (158) or for failing to observe signals (98). Officers issued 45 citations to motorists for DUI, an impressive decrease over the 78 DUIs issued last quarter. The new state penalty of withholding a driver's license for six months of anyone convicted of driving while under the influence appears to be an effective deterrent.

MISDEMEANORS

The largest number of arrests in this category were for disturbing the peace—53. Compared to last quarter, this is an increase of 10 percent. There were 88 charges for vagrancy and public drunkenness, an increase from the 59 charges from last quarter. We issued 32 citations for violations of leash laws, which represents a sizable increase over last quarter's 21 citations. Thirty-seven citations were issued for dumping trash at the Mason Reservoir.

CONCLUSION

Overall, while the crime rate has decreased in traffic (especially DUIs) and possession of controlled substances this quarter, we have seen a marked increase in the number of arrests for robberies and battery.

FIGURE 16.2 (Continued)

Page 3

RECOMMENDATIONS

To help deter future robberies in the downtown area, we recommend increasing our surveillance units in this area and again offering merchants our workshop on safety and security precautions, as we did during the first quarter of last year. Historically, battery arrests have risen during the second quarter, a fact that our Neighborhood Watch Group has repeatedly emphasized. We believe that this unit needs to increase foot and bicycle patrols in the neighborhoods with the highest incidence of battery reports.

The report in Figure 16.1 fails to follow these six guidelines. The revised version of the report, Figure 16.2, shows how the guidelines work to improve report writing.

Periodic Reports

Periodic reports, as their name signifies, provide readers with information at regularly scheduled intervals—daily, weekly, bimonthly (twice a month), monthly, quarterly. They help a company or agency keep track of the quantity and quality of the services it provides and the amount and types of work done by employees. Information in periodic reports helps managers make schedules, order materials, assign personnel, budget funds, and, generally speaking, determine corporate needs.

You may already be familiar with some kinds of periodic reports. For example, if you have ever punched a timecard and turned it in at the end of the week, or if you have ever taken inventory in a stockroom, you have prepared a periodic report. Delivery services require drivers to keep daily records documenting the number of packages delivered, the time, and the location; these daily records are a form of periodic report.

You may be responsible for compiling a report based on individual logs or daily or weekly employee activity reports. Figure 16.2, a report submitted to a police captain, summarizes, organizes, and interprets the data collected over a three-month period from individual activity logs. Such a report answers the reader's questions about the frequency and types of crimes committed and the work of the police force in the community. Because of this report, Captain Alice Martin will be better able to plan future protection for the community and to recommend changes in police services.

Sales Reports

Sales reports provide businesses with a necessary and ongoing record of accounts, purchases, losses, and profits over a specified period of time. Sales reports might be considered a special type of periodic report, but because of their importance in the world of business they deserve a separate category here.

Why They Are Important

Sales reports are important at various levels of business. Retail stores require a daily sales report in which purchases, classified by bar code scanners, are arranged into major categories. Salespeople often submit weekly reports on the types and costs of products sold in a given district. Branch managers write monthly reports based on the figures given them by their sales force. Higher up the business ladder, the president of a company assesses the financial health of the business for stockholders in an annual report, in which sales reports are a key feature as they relate to profits and dividends.

TECH NOTE

A report sent to someone at the same level of management as the writer (branch manager to branch manager) is known as a **lateral report.** A report sent to a higher executive level than the level of the writer (branch manager to vice president of marketing) is known as a **vertical report.**

Functions of Sales Reports

Sales reports help businesses assess past performance and plan for the future. In doing this they fulfill two functions: **financial** and **managerial.** As a financial record, sales reports list costs per unit, discounts or special reductions, and subtotals and totals. Like an accountant's spreadsheet, sales reports show gains and losses. They may also provide statistics for comparing two quarters' sales. The method or origin of a sale, if significant, can also be recorded. In selling books, for example, a publisher keeps a careful record of where sales originate—direct orders for single copies from readers, school district adoptions for classroom use, purchases at bookstores, or orders from wholesale distributors handling the book.

Sales reports are also a managerial tool because they help businesses make both short- and long-range plans. By indicating the number of sales, the report alerts store buyers and managers about which items or services to increase, modify, or discontinue. The restaurant manager's sales report illustrated in Figure 16.3

FIGURE 16.3 A sales report to a manager.

The Palace
Dayton, OH 43210
(813) 555-4000

TO: Gina Smeltzer
Frank Drew, Owners

FROM: Sam Jelinek S.J.
Manager

DATE: June 27, 1997

SUBJECT: Analysis of entrée sales, June 12-25

As we agreed at our monthly meeting on June 3, here is my analysis of entrée sales for two weeks to assist us in our menu planning. Below is a record of entree sales for the weeks of June 12-18 and June 19-25 that I have compiled into a table for easier comparisons.

	Portion	June 12-18		June 19-25		2 weeks combined	
	Size	Amount	Ratio	Amount	Ratio	Amount	Ratio
Cornish Hen	6 oz.	238	17%	307	17%	545	17%
Stuffed Young Turkey	8 oz.	112	8	182	10	294	12
Broiled Salmon Steak	8 oz.	154	11	217	12	371	13
Brook Trout	12 oz.	182	13	252	14	434	9
Prime Rib	10 oz.	168	12	198	11	366	11
Lobster Tails	2-4 oz.	147	10	161	9	308	10
Delmonico Steak	10 oz.	56	4	70	4	126	4
Moroccan Chicken	6 oz.	343	25	413	23	756	24
		1,400	100	1,800	100	3200	100

Recommendations

Based on the figures in the table above, I recommend that we do the following:

1. order at least one hundred more pounds of prime rib each two-week period to be eligible for further quantity discounts from the Northern Meat Company
2. delete the Delmonico steak entrée because of low acceptance
3. introduce a new chicken or fish entrée to take the place of the Delmonico steak; I would suggest grilled lemon chicken. This addition would help us to further honor the wishes of those of our patrons interested in tasty, low-fat, lower-cholesterol entries.

Please give me your reactions within the next week. It shouldn't take more than two weeks to implement these changes.

guides the owners in menu planning. Knowing which popular entrees to highlight and which unpopular ones to delete, the owners can increase their profits. Note how the recommendations follow logically from the figures Sam Jelinek gives to Gina Smeltzer and Frank Drew, the owners of the Palace Restaurant.

Writing a Sales Report

To write a sales report, keep a careful record of order forms, invoices, and production figures. Sales information might be arranged in list form, as in Figure 16.3. If you use a narrative format, make sure you do not overload your readers with numbers, as the poorly written short report in Figure 16.1 does. Underlining key numbers or putting them in boldface will emphasize them for readers.

Progress Reports

A progress report informs readers about the status of an ongoing project. It lets them know how much and what type of work has been completed by a particular date, by whom, how well, and how close the entire job is to being completed. A progress report emphasizes whether you are

- maintaining your schedule
- staying within your budget
- using the proper equipment
- making the right assignments
- completing the job efficiently and correctly

Almost any kind of ongoing work can be described in a progress report—research for a paper, construction of an apartment complex, preparation of a fall catalog, documentation of a patient's rehabilitation.

Audience for a Progress Report

A progress report is intended for people who generally are not working alongside you, but who need a record of your activities to coordinate them with other individuals' efforts and to learn about problems or changes in plans. For example, since supervisors (or ESL managers at an overseas office) may not be in the field or branch office or at a construction site, they will rely on your progress report for much of their information. Customers, such as a contractor's clients, often expect reports on how carefully their money is being spent. That way they can adjust schedules or alter specifications if there's a risk of going over budget.

Length of a Progress Report

The length of the progress report will depend on the complexity of the project. A short memo about organizing a time management workshop, such as that in Figure 16.4, might be all that is necessary. A report to a teacher about the progress a student is making on a research paper could be easily handled in a one-page memo, such as Barbara Shoemake's progress report in Figure 16.5 or for her research

FIGURE 16.4 A one-time progress report.

TO: Kathy Sands, Trenton
FROM: Philip Javon
DATE: September 14, 1997
SUBJECT: Preparations for Time Management Workshop

As you requested last week, I e-mailed the managers of all departments in both our Trenton and Frankfurt, Germany, offices on Thursday, September 10, to remind them of the time management workshop we will be offering on October 9 by teleconference.

I have confirmed the date and the operation of the technical links and relays with Carmen Suarez in Technical Services and have also e-mailed Jürgen Weiss in Frankfurt to make sure things are in place there.

I have reserved the corporate conference center for October 9 and have ordered DTP copies of all the packets we will need. The packets going to Germany will be Jet-Expressed, overnight delivery, on October 5 so they will be in Frankfurt two days before the teleconference.

By tomorrow, I will complete a list of all those employees scheduled to participate in the workshop and send it to you.

Plans are going according to schedule.

paper described in the proposal in Chapter 15 (pp. 568–570). Similarly, Dale Brandt's assessment of the progress his construction company is making in renovating Dr. Burke's office is given in a two-page letter in Figure 16.6.

Progress reports should contain information on (1) the work you have done, (2) the work you currently are doing, and (3) the work you will do.

Frequency of Progress Reports

Progress reports can be written daily, weekly, monthly, quarterly, or annually. Your specific job and your employer's needs will dictate how often you have to keep others informed of your progress. A single progress report is sufficient for Philip Javon's purpose in Figure 16.4. Barbara Shoemake was asked to submit two progress reports, the first of which is found in Figure 16.5. Contractor Brandt in Figure 16.6 determined that three reports, spaced four to six weeks apart, would be necessary to keep Dr. Burke posted.

FIGURE 16.5 A progress report from a student to a teacher.

TO: Professor Leigh Felton-Parks
FROM: Barbara Rene Shoemake BRS
DATE: April 8, 1997
SUBJECT: First Progress Report on Research Paper

This is the first of two progress reports that you asked me to submit about my research paper on the ethical and security issues of using electronic mail.

From March 8 until April 6, I gathered information from print and electronic resources, including library holdings, the Internet, and an interview. Of the eighteen references listed on my proposal, I found only twelve. Articles by Aronson ("Training May Solve E-mail Worries"), Doss and Loui ("Ethics and the Privacy of Electronic Mail") and "Is E-mail Safe?" (Supervisory Management) are not available in our library or on the Internet. Two of my Internet sites—"Here's Help" and Sherwood's A Beginner's Guide—are under construction. But I e-mailed Sherwood and found that her site will be open in the next few days. I'll try to replace "Here's Help."

On February 23, I had an extended interview (1½ hours) with Keith Wellbridge of General Dynamics, who gave me some seminar brochures as well as a copy of a report on electronic mail protocols that he wrote for the Society of Midwest Business Communicators—materials I hope to incorporate in my paper.

Because of a long business trip to Denver, Alice Phillips of Dodge & Spenser could not meet with me. At her suggestion, I am trying to schedule an interview with Gloria Sirkin-Dews, the Office Manager at Mid-Atlantic Power Company. Ms. Sirkin-Dews has given several seminars on e-mail security. Even if she cannot meet with me, Mr. Wellbridge gave me enough information about a business manager's view of electronic mail systems. However, not currently having the articles and Web sites listed above may slow, but not stop, my work a little.

Starting tomorrow, I will begin my paper and can submit a draft by April 27. You will receive my second progress report by April 20.

FIGURE 16.6 The second of three progress reports from a contractor to a customer.

Brandt Construction Company
Halsted at Roosevelt
Chicago, Illinois 60608-0999
http://www.brandt.com
312-555-3700 FAX 312-555-1731

April 28, 1998

Dr. Pamela Burke
1439 Grand Avenue
Mount Prospect, IL 60045-1003

Dear Dr. Burke:

Here is my second progress report about the renovation work at your new clinic at Hacienda and Donohue. Work proceeded satisfactorily in April according to the plans you had approved in March.

REVIEW OF WORK COMPLETED IN MARCH

As I informed you in my first progress report on March 31, we tore down the walls, pulled the old wiring, and removed existing plumbing lines. All the gutting work was finished in March.

WORK COMPLETED DURING APRIL

By April 8, we had laid the new pipes and connected them to the main sewer line. We also installed the two commodes, the four standard sinks, and the utility basin. The heating and air-conditioning ducts were installed by April 13. From April 18–23, we erected soundproof walls in the four examination rooms, the reception area, your office, and the laboratory. We had no problems reducing the size of the reception area by five feet to make the first examination room larger, as you had requested.

PROBLEMS WITH THE ELECTRICAL SYSTEM

We had difficulty with the electrical work, however. The number of outlets and the generator for the laboratory equipment required extra-duty power lines that had to be approved by both Con Edison and county inspectors. The approval slowed us down by three days. Also, the wholesaler, Midtown Electric, failed to deliver the recessed lighting fixtures by April 25 as promised. These fixtures and the generator are now being installed. Moreover, the cost of these fixtures will increase the material budget by $1,288.00. The cost for labor is as we had projected—$89,450.

Continued

FIGURE 16.6 (Continued)

page 2

WORK REMAINING

The finishing work is scheduled for May. By May 9, the floors in the examination rooms, laboratory, washrooms, and hallways should be tiled and the reception area and your office carpeted. By May 13, the reception area and your office should be paneled and the rest of the walls painted. If everything stays on schedule, touch-up work is scheduled for May 18–22. You should be able to move into your new clinic by May 23.

You will receive a third and final progress report by May 12. Thank you again for your business and the confidence you have placed in our company.

Sincerely yours,

Dale Brandt

Dale Brandt

How to Begin a Progress Report

In a brief introduction

- indicate why you are writing the report
- provide any necessary project titles or codes and specify dates
- help readers recall the job you are doing for them

If you are writing an initial progress report, supply background information in the opening. Philip Javon's first sentence in Figure 16.4, for example, quickly establishes his purpose by reminding Kathy Sands of their discussion last week. Similarly, Barbara Shoemake in Figure 16.5 reminds her teacher of the purpose and scope of her work in the first paragraph.

If you are submitting a subsequent progress report, your introduction should remind your reader about where your previous report left off and where the current one begins. Make sure that the period covered by each report is clearly specified. Note how Dale Brandt's first paragraph in Figure 16.6 calls attention to the continuity of his work.

How to Continue a Progress Report

The body of the report should provide significant details about costs, materials, personnel, and times for the major stages of the project.

- Emphasize completed tasks, not false starts. If you report that the carpentry work or painting is finished, readers do not need an explanation of paint viscosity or geometrical patterns.

- Omit routine or well-known details ("I had to use the library when I wanted to read the back issues of *Safety News*").
- Describe in the body of your report any snags you encountered that may affect the work in progress. See Dale Brandt's section on electrical problems in Figure 16.6. It is better for the reader to know about trouble early in the project, so that appropriate changes or corrections can be made.

How to End a Progress Report

The conclusion should give a timetable for the completion of duties or for submission of the next progress report. Give the date by which you expect work to be completed; be realistic. Do not promise to have a job done in less time than you know it will take. Readers will not expect miracles, only informed estimates. Even so, any conclusion must be tentative. Note that the good news Dale Brandt gives Dr. Burke about moving into her new clinic is qualified by the words "if everything stays on schedule." He is also well aware of the "you attitude" by thanking Dr. Burke again for her business.

How to Include a Recommendation

A recommendation may also find a place in your conclusion. Such a recommendation might advise readers, for example, of a less costly, equally durable siding than the one originally planned, suggest that a new software program would considerably improve the design of your document or your company's schedules, or show that hiring an additional part-time keyboarder would help ease the workload over a particularly busy sales period.

Trip/Travel Reports

Reporting on the trips you take is an important professional responsibility. Trips can range from a brief afternoon car ride across town to a month-long globe-hopping journey. These reports inform readers about your activities outside the office, plant, clinic, or agency. In documenting what you did and saw, these trip reports keep readers informed about your efforts and how they affect ongoing or future business. Trip reports are also written after you attend a convention or sales meeting, or call on customers.

Questions Trip Reports Answer

Specifically, a trip report should answer the following questions for your readers.

1. Where did you go?
2. When did you go?
3. Why did you go?
4. Whom did you see?
5. What did they tell you?
6. What did you do about it?

For a business trip you are also likely to have to inform readers how much the trip cost and to supply them with receipts for all your expenses.

Common Types of Trip/Travel Reports

Trip reports can cover a wide range of activities and are called by different names to characterize those activities. Most likely, you will encounter the following three types of trip/travel reports.

1. Field trip reports. These reports, often assigned in an academic course, are written after a visit to a local plant, military installation, office complex, hospital, forest, detention center, or other facility. Their purpose is to show what you have learned about the operation of these places. You will be expected to describe how an institution is organized, the technical procedures and/or equipment it uses, pertinent ecological conditions, or the ratio of one group to another. The emphasis in these reports is on the educational value of the trip, as Mark Tourneur's report in Figure 16.7 demonstrates.

2. Site inspection reports. These trip reports are written to inform managers about conditions at a branch office or plant, a customer's business, or an area directly under an employer's jurisdiction. After visiting the site, you will determine whether it meets your employer's (or customer's) needs.

Site inspection reports tell how machinery or production procedures are working or provide information about the physical plant, environment (air, soil, water, vegetation), and computer or financial operations.

A site inspection report is written for an employer or a customer who wants to relocate or build new facilities (such as a record shop, a halfway house, or a branch office) in order to assess the suitability of a particular location. Figure 16.8, which shows a report written to a district manager interested in acquiring a new site for a fast-food restaurant, begins with a recommendation (see pp. 598–599).

3. Home health or social work visits. Nurses, social workers, and probation officers report daily on their visits to patients and clients. Their reports describe clients' lifestyles, assess needs, and make recommendations. A report from a social worker to a county family services agency can be seen in Figure 16.9 on page 600. The report begins with the information the writer acquired from a family and concludes, not with a recommendation, but rather with a list of the actions this social worker has taken.

How to Gather Information for a Trip/Travel Report

Regardless of the kind of trip report you have to write, your assignment will be easier and your report better organized if you follow these suggestions.

Before you leave on the field trip, site inspection, or visit, be sure you are prepared as follows.

1. Obtain all necessary names, street and e-mail addresses, and telephone and fax numbers.

FIGURE 16.7 A student's field trip report.

TO: Katherine Holmes, RN, MSN
Director, RN Program
FROM: Mark Tourneur *M.T.*
RN Student

DATE: November 13, 1998
SUBJECT: Field Trip to Water Valley Extended Care Center

On Tuesday, November 10, I visited the Water Valley Convalescent Center, 1400 Medford Boulevard, in preparation for my internship in an extended care facility next semester.

Philosophy and Organization

Before my tour started, the director, Sue LaFrance, explained the holistic philosophy of health care at Water Valley and emphasized the diverse kinds of nursing practiced there. She stressed that the agency is not restricted to geriatric clients but admits anyone requiring extended care. She pointed out that Water Valley is a medium-sized facility (150 beds) and contains three wings: (1) the Infirmary, (2) the General Nursing Unit, and (3) the Ambulatory Unit.

Primary Client Services

My tour began with the Infirmary, staffed by one RN and two LPNs , where I observed a number of life-support systems in operation:

- IVs
- oxygen setups
- feeding tubes
- cardiac monitors

Then I was shown the General Nursing Unit, a forty-bed unit that is staffed by three LPNs and four aides. Clients can have private or semiprivate rooms; bathrooms have wide doors and lowered sinks for patients using wheelchairs or walkers. The Ambulatory unit cares for 90 clients who can provide their own daily care.

Continued

FIGURE 16.7 (Continued)

Additional Client Services

Before lunch in the main dining room, I was introduced to Doris Betz, the dietitian, who explained the different menus she coordinates. The most common are low-sodium and ADA (American Diabetic Association) restricted-calorie. Staff members eat with the clients, reinforcing the holistic focus of the agency.

After lunch, Jack Tishner, the pharmacist, discussed the agency's procedures for ordering and delivering medications. He also explained the kinds of client teaching he does and the in-service workshops he conducts.

I then observed clients in both recreational and physical therapy. Water Valley's full-time physical therapist, Tracy Cook, works with stroke and arthritic clients and helps those with broken bones regain the use of their limbs. In addition to a weight room, Water Valley has a small sauna that most of the clients use at least twice a week.

The clients' spiritual needs are not neglected, either. A small chapel is located just off the Ambulatory Unit.

Benefits for My Internship

From my visit to Water Valley, I learned a great deal about the health care delivery system at an extended care facility. I was especially pleased to have been given so much information on emergency procedures, medication orders, and physical therapy programs. My forthcoming internship will be much more useful, since I now have first-hand knowledge about these various services.

2. Check the files for previous correspondence, case studies, or terms of contracts or agreements.
3. Locate a map of the area (use the Internet research tool described on p. 498) or a blueprint of the building.
4. Gather brochures, work orders, instructions, or other documents pertinent to your visit.
5. Bring a laptop computer with you.
6. Depending on your job, you may also need to bring a camcorder, tape recorder, camera, or calculator to record important data.

FIGURE 16.8 A site inspection report using a map.

TO: Dale Gandy
FROM: Beth Armando BA
Development Department

DATE: July 1, 1999
SUBJECT: New Site for Vail's #8

RECOMMENDATION

The best location for the new Vail's Chicken House is the vacant Dairy World shop at the northeast corner of Smith and Fairfax Avenues -- 1701 Fairfax. I inspected this property on June 21 and 22 and also talked to Marge Bloom, the broker at Crescent Realty representing the Dairy World Company.

THE LOCATION

Please refer to the map below. Located at the intersection of the two busiest streets on the southeast side, the property allows us to take advantage of the traffic flow to attract customers. Being only one block west of the Cloverleaf Mall should also help business. Customers will have easy access to our location.

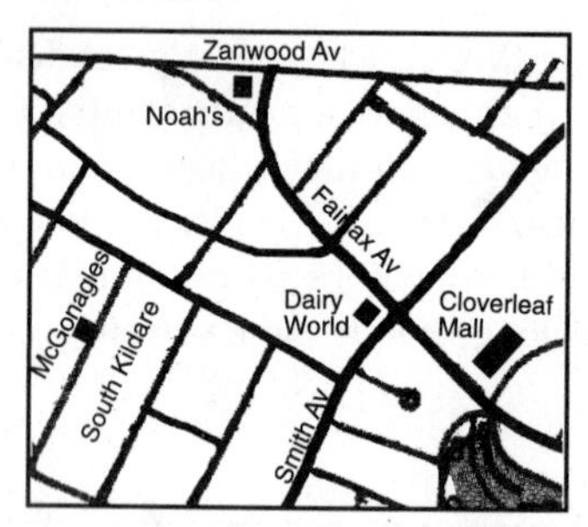

They can enter or exit the Dairy World from either Smith or Fairfax but left turns on Smith are prohibited from 7 A.M. to 9 A.M. However, since most of our business is done after 11 A.M., the restriction poses few problems.

Continued

FIGURE 16.8 (Continued)

Dale Gandy
July 1, 1999
Page 2

AREA COMPETITION

Only two other fast-food establishments are in a one-mile vicinity. McGonagles, 1534 South Kildare, specializes in hamburgers; Noah's, 703 Zanwood, serves primarily seafood entrées. Their offerings will not directly compete with ours. The closest fast-food restaurant serving chicken is Johnson's, 1.8 miles away.

PARKING FACILITIES

The parking lot has space for 45 cars. The area at the south end of the property (38 feet x 37 feet) can accommodate another 14-15 cars. The driveways and parking lot were paved with asphalt last March and appear to be in excellent condition. We will be able to make use of the drive-up window on the north side of the building.

THE BUILDING

The building has 3,993 square feet of heated and cooled space. The air-conditioning and heating units were installed within the last fifteen months and seem to be in good working order; nine more months of transferable warranty remain on these units.

The only major changes we would have to make are in the kitchen. To prepare items on the Vail's menu, we would have to add three more exhaust fans (there is only one now) and expand the grill and cooking areas. The kitchen also has three relatively new sinks and ample storage space in the sixteen cabinets.

The restaurant has a seating capacity of up to 54 persons; ten booths are covered with red vinyl and are comfortably padded. A color-coordinated serving counter could seat 8 to 10 patrons. The floor does not need to be retiled, but the walls still have to be painted to match Vail's color decor.

FIGURE 16.9 A social worker's visit report.

Green County Family Services

Randall, VA 21032
703-555-4000 FAX 703-555-4004

TO: Margaret S. Walker, Director
Green County Family Services

FROM: Jeff Bowman, Social Worker JB

SUBJECT: Visit to Mr. Lee Scanlon

DATE: October 14, 1997

PURPOSE OF VISIT

At the request of the Green County Home Health Office, I visited Mr. Lee Scanlon at his home at 113 West Diversy Drive on Tuesday, October 7. Mr. Scanlon and his three children (ages six, eight, and eleven) live in a two-bedroom apartment above a garage. Last week, Mr. Scanlon was discharged from Lutheran General Hospital after leg surgery and has asked for financial assistance.

DESCRIPTION OF VISIT

Mr. Lee Scanlon is widower with no means of support except unemployment compensation of $984 a month. He lost his job at Beaumont Manufacturing when the company went out of business four weeks ago and wants to go back to work, but Dr. Marilyn Canning-Smith advised against it for six to seven weeks. His oldest child is diabetic and the six-year-old daughter must have a tonsillectomy. Mr. Scanlon also told me the problems he was having with his refrigerator; it "was off more than it was on," he said.

Mrs. Alice Gordon, the owner of the garage, informed me that Mr. Scanlon had paid last month's rent but not this month's. She also stressed how much the Scanlons need a new refrigerator and that she had often let them use part of hers to store their food.

Continued

FIGURE 16.9 (Continued)

Margaret S. Walker
October 14, 1997
Page 2

Here is a breakdown of Mr. Scanlon's monthly bills:

Expenses	Income
$ 550 rent	$ 984 unemployment compensation
150 utilities	
350 food	
120 drugs	
80 transportation	
$ 1,250	

ACTION TAKEN

To assist Mr. Scanlon, I have done the following:

1. Set up an appointment (10/20/97) for him to apply for food stamps.
2. Talked with Blanche Derringo regarding Medicaid assistance.
3. Asked the State Employment Commission to aid him in finding a job as soon as he is well enough to work. My contact person is Wesley Sahara; his e-mail address is wsahara@empcom.gov.
4. Visited Robert Adkins at the office of the Council of Churches to obtain food and money for utilities until federal aid is available; he also will try to find the Scanlons another refrigerator.
5. Telephoned Sharon Muñez at the Green County Health Department (555-1400) to have Mr. Scanlon's diabetic daughter receive insulin and syringes gratis.

When you return from your trip, keep the following hints in mind as you compile your report.

1. Write your report promptly. If you put it off, you may forget important items.
2. When a trip takes you to two or more widely separated places, note in your report when you arrived at each place and how long you stayed.
3. Do not include everything you saw or did on the trip. Exclude irrelevant details, such as whether the trip was enjoyable, what you ate, or how delighted you were to meet people.

4. As you edit the final copy of your report, check to make sure that you have listed names and calculated figures correctly.

Test Reports

Much physical research (the discovery and documentation of facts) is communicated through reports. These reports have various names. Depending on your profession's terminology, they may be called **experiment, investigation, laboratory, operations,** or **research reports.** They all record the results of tests, whether the tests were conducted in a forest, computer center, laboratory, shopping center, or soybean field. No doubt you have already written a **test report** (or lab report) after performing an experiment in a science class.

Writing an effective test report, of course, involves specific training in a scientific or technical field. But remember that the ability to write clearly and concisely about a procedure (and the results) is as significant as the technical skills required to perform the test itself.

Style

Objectivity and accuracy are essential ingredients in a test report. Readers want to know about your empirical research (the facts), not about your feelings (the "I"). Record your observations without bias or guesswork in a laboratory journal or log book and always include precise measurements. Follow the accepted practices of your profession in documenting your findings. Use standard symbols and abbreviations.

Questions Your Report Needs to Answer

Readers will expect your test report to answer the following questions:

- why you performed the test—explain the reasons, your goals, and who may have authorized you to perform the test
- how you performed the test—under what circumstances or controls you conducted the test; what procedures and equipment you used
- what the outcomes were—your conclusions
- what implications or recommendations follow from your test—what you learned, discovered, confirmed, or even disproved or rejected

When you sign the final copy of your report, you are certifying to your readers that things happened exactly when, how, and why you say they did.

Two Sample Test Reports

Figure 16.10 contains a relatively simple and short test report in memo format regarding sanitary conditions at a hospital psychiatric unit. This report follows a direct and useful pattern of organization:

- statement of purpose—*why?*
- findings—*what happened?*
- recommendations—*what next?*

FIGURE 16.10 A test report with recommendations.

CHARLESTON CENTRAL HOSPITAL
CHARLESTON, WV 25324-0114

TO: James Dill, Supervisor
Housekeeping

FROM: Janeen Cufaude JC
Infection Control Officer

DATE: December 7, 1998

SUBJECT: Routine sanitation inspection

As part of the monthly check of the psychiatric unit (11A), the following areas were swabbed and tested for bacterial growth. The results of the lab tests of these samples are as follows:

AREA	FINDINGS
1. cabinet in patients' kitchen	1. positive for 2 colonies of strep germs
2. rug in eating area	2. positive for food particles and yeasts and molds
3. baseboard in dayroom	3. positive for particles of dust
4. medicine counter in nurses' station	4. negative for bacteria — no growth after 48 hrs.
5. corridor by south elevator	5. positive for 4 colonies of staph germs isolated

ACTIONS TO BE TAKEN AT ONCE

1. Clean the kitchen cabinet with K-504 liquid daily, 3:1 dilution.
2. Shampoo rug areas bimonthly with heavy-duty shampoo and clean visibly soiled areas with Guard-Pruf as often as needed.
3. Wipe all baseboards weekly with K-12 spray cleanser.
4. Mop heavily traveled corridors and access areas with K-504 cleanser daily, 1:1 dilution.

Submitted by an infection control officer, the report does not provide elaborate details about the particular laboratory procedures used to determine whether bacteria were present; nor does it describe the pathogenic (disease-causing) properties of the bacteria. Such descriptions are unnecessary for the audience (the housekeeping department) to do its job.

A more complex example of a short test report is found in Figure 16.11; it studies the effects of four light periods on the growth of paulownia seedlings (a flowering tree cultivated in China). This report, published in a scientific journal, is addressed to specialists in forestry.

To meet the needs of this expert audience, the writers had to include much more detailed information than Janeen Cufaude, the infection control officer who wrote Figure 16.10, did about the way the test was conducted and the types of scientific data the audience needed. The researchers did not have to define technical terms for their audience, and they can confidently use scientific symbols and formulas as well.

Such a test report follows a different, more detailed pattern of organization than the report in Figure 16.10 and includes the following parts:

- **informative abstract** to summarize the scope of the test
- **introduction** to provide background information about the importance of the test and to review previous research on the topic
- **materials and methods section** to describe the exact scientific procedures and conditions under which the test was conducted
- **results and discussion section** (sometimes separate sections) to record the data obtained from the test and to interpret the significance of the data. Many times a test report will conclude by confirming previous studies, emphasizing the need for further tests, or offering researchers new interpretations of the evidence.
- **list of references cited** in the study (follow the method of documentation recommended by the journal or society)

Notice how the needs of different audiences and the purposes of different authors control the way each report is organized and how much and what type of information is included. Janeen Cufaude's reader is a nonspecialist, less interested in scientific theory than in the practical application of that theory; Immel, Tackett, and Carpenter's readers are experts demanding scientific documentation and commentary.

As these two reports show, you should always determine how much technical knowledge an audience has about your field and how they will use your report to accomplish *their* specific job.

Incident Reports

The reports discussed thus far in this chapter have dealt with routine work. They have described events that were anticipated or supervised. But every business or agency runs into unexpected trouble that delays routine work. Employers, and on

FIGURE 16.11 A short test report published in a scientific journal.

Paulownia Seedlings Respond to Increased Daylength

M. J. Immel, E. M. Tackett, and S. B. Carpenter

Abstract

Paulownia seedlings grown under four photoperiods were evaluated after a growing period of 97 days. Height growth and total dry weight production were both significantly increased in the 16- and 24-hour photoperiods.

Introduction

Paulownia (*Paulownia tomentosa* [Thunb.] Steud.), a native of China, is a little known species in the United States. Recently, however, there has been increased interest in this species for surface mine reclamation (*1*).* Paulownia seems to be especially well adapted to harsh micro-climates of surface mines; it grows very rapidly and appears to be drought-resistant. In Kentucky and surrounding states, paulownia wood is actively sought by Japanese buyers and has brought prices comparable to black walnut (*2*).

This increased interest in paulownia has resulted in several attempts to direct seed it on surface mines, but little success has been achieved. The high light requirements and the extremely small size of paulownia seed (approximately 6,000 per gram) may be the limiting factors. Planting paulownia seedlings is preferred; but, because of their succulent nature, seedlings are usually produced and outplanted as container stock rather than bareroot seedlings. Daylength is an important factor in the production of vigorous container plants (*5*).

Our study compares the effects that four photoperiods—8, 12, 16, and 24 hours—had on the early growth of container-grown paulownia seedlings over a period of 97 days.

Materials and Methods

Seeds used in this study were stratified in a 1:1 mixture of peat moss and sand at 4° C for 2 years. Following cold storage, seeds were placed on a 1:1 potting soil-sand mix and mulched with cheesecloth. They were then placed under continuous light until germination occurred. Germination percentages were high, indicating paulownia seeds can survive long periods of storage with little loss of viability (*3*).

Thirty days after germination, 3- to 4-centimeter seedlings were transplanted into 8-quart plastic pots filled with an equal mixture of potting soil, sand, and peat

*All actual references in the Works Cited section have been omitted to save space.

Continued

FIGURE 16.11 (Continued)

moss. Seventy-five seedlings were randomly assigned to each of the four treatments. Treatments were for 4 photoperiods—8, 12, 16, and 24 hours—and were replicated three times in 12 light chambers. Each chamber was 1.2- by 1.2-meters with an artificial light source 71 centimeters above the chamber floor.

The light source consisted of eight fluorescent lights: four 40-watt plant growth lamps alternated with four 40-watt cool white lamps. Light intensity averaged 550 foot-candles (1340μ einsteins/m^2/s) at the top of each pot and the temperature averaged 23° C(± 2°C).

Seedlings were watered and fertilized after transplanting with a 6-gram 14-4-6 agriform container tablet. Beginning 1 month after transplanting, two seedlings were randomly selected and harvested from each chamber for a total of 24 trees. Height, root collar diameter, length of longest root, and oven-dry weight (at 65° C) were determined for each seedling. Harvests continued every week for 5 additional weeks.

Results and Discussion

Results indicate that early growth of paulownia is influenced by photoperiod, as shown in Table 1.

TABLE 1. Height, Diameter, Root Length, Total Dry Weight, and R/S Ratio for Paulownia Seedlings Grown Under Four Photoperiods after 97 Days

Photo period (hrs.)	Height (cm)	Diameter (cm)	Root length (cm)	Total dry weight (gm)	R/S ratio
8	13.1b	0.48	16.0	1.65c	0.18
12	17.8b	0.67	34.7	7.27b	0.32
16	27.3a	0.93	31.1	15.92a	0.39
24	29.2a	0.90	43.9	18.56a	0.33

Expanding the photoperiod from 8 to either 16 or 24 hours increased height growth by 100 percent. Height growth in the 12-hour treatment also increased, but did not differ significantly from the 8-hour treatment. Heights under photoperiods of 8, 12, 16, and 24 hours were 13.1, 17.8, 27.3, and 29.2 centimeters, respectively.

Previous studies have also shown that photoperiod affects the growth of paulownia seedlings (*4, 6*). Sanderson (*6*), for example, found that paulownia seedlings grown under continuous light averaged 27.2 centimeters in height after 101 days compared with 29.2 centimeters for our 24-hour seedlings. Other corresponding photoperiods were equally comparable. Downs and Borthwick

Continued

FIGURE 16.11 (Continued)

(*4*) also concluded that height growth of paulownia was affected by extending the photoperiod.

The greatest treatment differences were shown in total dry weight production. Refer again to Table 1. The mean weight of 1.65 grams for seedlings in the 8-hour treatment was significantly less than that of any of the other photoperiods. The 16- and 24-hour treatments did not differ significantly. In fact, they more than doubled the average weight for seedlings in the 12-hour treatment.

Root-to-shoot ratio (R/S) indicates the relative proportion of growth allocated to roots versus shoots for the seedlings in each photoperiod. In this study, shoots were developing at nearly three times the rate of the roots for seedlings in the 12-, 16-, and 24-hour photoperiods.

The 0.18 R/S ratio for seedlings in the 8-hour treatment was much lower, indicating that relative growth of the shoot is approximately five times that of the root. The shorter photoperiod therefore decreased root development relative to shoot development as well as significantly reduced total dry weight production.

Although root collar diameter and root length did not significantly differ under the different photoperiods after 97 days, there was a trend for greater diameter and root growth with longer photoperiods.

Conclusions

Results indicate that the growth of paulownia seedlings is affected by changes in the photoperiod. Increasing the photoperiod significantly increased height growth and total dry matter production. The distribution of dry matter (R/S ratio) was altered by increasing the photoperiod; the ratio was larger in the longer photoperiods. In contrast to earlier studies (*4*), we found paulownia seedlings subjected to extended photoperiods were still growing after 97 days.

Tree Planters' Notes 31 (Winter 1980): 3–5.

some occasions government inspectors, insurance agents, and attorneys, must be informed about those events that interfere with or threaten normal, safe operations.

When to Submit an Incident Report

A special type of report known as an incident report is submitted when there is, for example,

- an accident
- a law enforcement offense
- an environmental danger
- a machine breakdown
- a delivery delay

- a cost overrun
- a production slowdown

Figure 16.12 contains an incident report about a train accident submitted by the engineer on duty.

Protecting Yourself Legally

An incident report can be used as legal evidence. It frequently concerns the two topics over which powerful legal battles are waged—health and property. The report can sway the outcome of insurance claims, civil suits, or criminal cases and therefore requires careful planning and revision.

Common Errors in Writing Incident Reports

The two most common errors—for legal purposes—that writers make in submitting incident reports are that they do not give (1) enough information or (2) the right kinds of information. Here are some suggestions for avoiding these twin dangers.

- What you (or eyewitnesses) saw, heard, felt, or smelled—and when—are the essentials of your report. Make sure you report these observations completely and without bias. If you are depending on eyewitness account(s), use quotation marks to set off statements by the witness.
- So you don't omit relevant, important information, recount what happened in the order it took place. Describe what happened before or after the incident only if it is relevant—for example, environmental or weather conditions for storm damage or an automobile accident, or warning signals for a malfunctioning piece of equipment.
- Filter out irrelevant details. Delete any emotional reactions of eyewitnesses—how much an object meant to them or how surprised they were to see something happen. If you are reporting a work stoppage, it is necessary to indicate what employees did while waiting for a machine to be repaired or a facility cleaned or inspected.
- Do not duplicate information found elsewhere in a report—names, license numbers, times.

Guidelines for Writing a Legally Proper Incident Report

To ensure that what you write is legally proper, follow these guidelines.

1. **Be accurate, objective, and complete.** Never omit or distort facts; the information may surface later, and you could be accused of a cover-up. Do not just write "I do not know" for an answer. If you are not sure, state why. Also be careful that there are no discrepancies in your report.

2. **Give facts, not opinions.** Provide a factual account of what actually happened, not a biased interpretation of events. Indirect words such as "I guess," "I wonder," "apparently," "perhaps," or "possibly" weaken your objectivity. Stick to details you witnessed or that were seen by eyewitnesses. Identify witnesses or

FIGURE 16.12 An incident report in memo format.

THE GREAT HARVESTER RAILROAD
Des Moines, IA 50306-4005
http://www.ghrr.com

TO: Angela O'Brien
District Manager
James Day
Safety Inspector

DATE: March 3, 1997

FROM: Ned Roane *NR*
Engineer

SUBJECT: Derailment of Train 28 on March 3, 1997

DESCRIPTION OF INCIDENT

At 7:20 A.M. on March 3, 1997, I was driving Engine 457 traveling north at a speed of fifty-two miles an hour on the single main-line track four miles east of Ridgeville, Illinois. Weather conditions and visibility were excellent. Suddenly the last two grain cars, 3022 and 3053, jumped the track. The train automatically went into emergency braking and stopped immediately. There were no injuries to the crew. But the train did not stop before both grain cars turned at a 45° angle. After checking the cars, I found that half the contents of their load had spilled. The train was not carrying any hazardous chemical shipments.

I notified Supervisor Bill Purvis at 7:40 A.M., and within forty-five minutes he and a section crew arrived at the scene with rerailing equipment. The section crew removed the two grain cars from the track, put in new ties, and made the main-line track passable by 9:25 A.M. At 9:45 A.M. a vacuum car arrived with Engine 372 from Hazlehurst, Illinois, and its crew proceeded with the cleanup operation. By 10:25 A.M. all the spilled grain was loaded onto the cars brought by the Hazlehurst train. Bill Purvis called Barnwell Granery and notified them that their shipment would be at least three hours late.

CAUSES OF INCIDENT

Supervisor Purvis and I checked the stretch of train track where the cars derailed and found it to be heavily worn. We believe that a fisher joint slipped when the grain cars hit it, and the track broke. You can see a cracked fisher joint in the accompanying illustration.

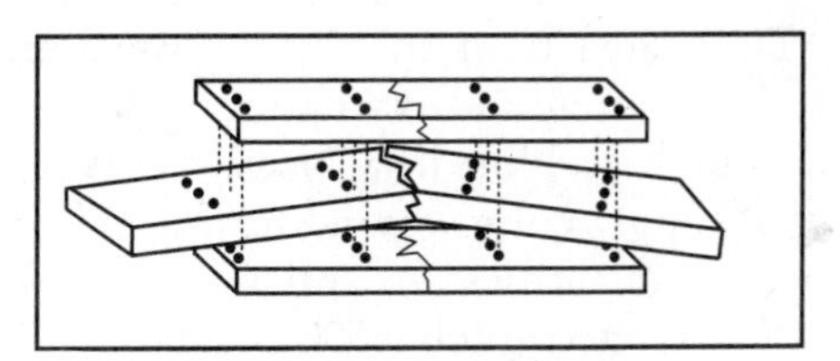

RECOMMENDATIONS

We made the following recommendations to the switch yard in Hazlehurst to be carried out immediately:

1. Check the section of track for ten miles on either side of Ridgeville for any signs of defective fisher joints.
2. Repair any defective joints at once.
3. Instruct all engineers to slow down to five to ten mph over this section of the road until the rail check is completed.

victims by giving names, addresses, places of employment, and so on. Indicate who saw what. Keep in mind that stating what someone else saw will be regarded as hearsay, and therefore not be admissible in a court of law. State only what *you* saw or heard. When you describe what happened, avoid drawing uncalled-for conclusions. Consider the following statements of opinion and fact:

Opinion: The patient seemed confused and caught himself in his IV tubing.
Fact: The patient caught himself in his IV tubing.
Opinion: The equipment was defective.
Fact: The bolt was loose.

Be careful, too, about blaming someone. Statements such as "Baxter was incompetent" or "The company knew of the problem, but did nothing about it" are libelous remarks.

In law enforcement work, identify a suspect by his or her alias and by any distinctive characteristics—for example, "jagged 4-inch scar on left forearm."

3. Do not exceed your professional responsibilities. Answer only those questions you are qualified to answer. Do not presume to speak as a detective, inspector, physician, or supervisor. Do not represent yourself as an attorney or claims adjuster in writing the report.

Parts of an Incident Report

You will use either a memo or a specially prepared form with spaces for detailed comments. See how Figure 16.12 includes the following details.

1. Personal details. Record titles, department, and employment identification numbers. Indicate if you or your fellow employees were working alone. For customers or victims, record home addresses, phone numbers, and places of employment. Insurance companies will also require policy numbers.

2. Type of incident. Briefly identify the incident—personal injury, fire, burglary, delivery delay, equipment failure. In the case of injury, identify the part(s) of the body precisely. "Eye injury" is not enough; "injury to the right eye, causing bleeding" is better. "Dislocated right shoulder," "punctured left forearm," and "twisted left ankle" are descriptive and exact phrases. A report on damaged equipment should list model numbers. Note how Ned Roane's report specifies the grain car numbers. For thefts, supply colors, brand names and/or manufacturer, model names, serial numbers, and quantities. "A stolen laptop" will not help detectives locate the right object; "a Newtech Model 5000" will.

3. Time and location of the incident. Follow the advice given at the beginning of this chapter, page 580.

4. Description of what happened. This section is the longest part of the report. Some forms ask you to write on the back, to attach another sheet, or to add a photograph or diagram. Put yourself in the reader's position. If you were not present or did not speak directly to witnesses, would you know by reading the report what happened exactly and why, how it occurred, who and what were involved, and what led up to the incident?

5. What was done after the incident. After describing the incident, describe the action you took to correct conditions, to get things back to normal. Readers will want to know what was done to treat the injured, to make the environment safer, to speed delivery of goods, to repair damaged equipment, or to satisfy a customer's demands.

6. What caused the incident. Make sure that your explanation is consistent with your description in number 4. Pinpoint the trouble. In Figure 16.12, for example, the defective fisher joint is discussed under the heading "Causes of Incident." In the following example, the two causes cited in a report of an accident involving a pipe falling from a crane are exact and helpful.

1. The crane's safety latch had been broken off and was never replaced.
2. A tag line was not used to guide the pipe onto the truckbed.

7. Recommendations. Readers will be looking for specific suggestions for solving the problem and for preventing the incident from happening again. Recommendations may involve discussing the problem at a safety meeting, asking for further training from a manufacturer, adapting existing equipment to meet customers' needs, doing some emergency planning for the next storm damage, modifying schedules, or cutting back on expenses. In the example of the falling pipe in number 6 above, the writer of the incident report listed the following two appropriate recommendations.

1. Order safety latches to replace the broken latch and have additional spare latches on hand.
2. Have a pipeshop supervisor conduct a safety meeting for employees and use a representative drawing of the incident as an aid.

Main Points to Remember About Short Reports

Before considering long reports, the subject of the next chapter, let's review the main points about short reports.

1. They are a few pages at most and get to the point quickly and concisely.
2. They can be prepared in a variety of formats—printed memos, e-mails, letters, or special forms.
3. Their content will vary with the type of report—periodic, sales, progress, trip, test, or incident.
4. In all types of short reports, the emphasis is on facts and objectivity. You need to anticipate the types of information your readers will need and how they will use it.
5. Short reports are generally organized to include information on purpose, data, conclusions, and recommendations.

✓ Revision Checklist

- ❑ Had a clear sense of how my readers will use my report.
- ❑ Consulted appropriate sources to give my audience enough information to help them make informed decisions.
- ❑ Provided significant information about costs, materials, personnel, and times so readers will know that my work consists of facts, not impressions.
- ❑ Double-checked all data—costs, figures, dates, places, and equipment numbers.
- ❑ Verified necessary sales figures, dates, costs, and references so that my report was thorough and accurate.
- ❑ Made sure all my comments and recommendations were ethical.
- ❑ Followed all agency or organization guidelines.
- ❑ Adhered to all legal requirements.
- ❑ Eliminated unnecessary details or those too technical for my audience.
- ❑ Made report concise and to the point.
- ❑ Used headings whenever feasible to organize and categorize information.
- ❑ Supplied relevant visuals to help readers understand my message and crunch any numbers.
- ❑ Employed underlining, boldface, or italics to set headings apart or to emphasize key ideas.
- ❑ Began report with statement of purpose that clearly described the scope and significance of my work.
- ❑ Incorporated tables and other pertinent visuals to display data whenever appropriate.
- ❑ Explained clearly what the data mean.
- ❑ Determined that recommendations logically follow from the data and that recommendations are realistic.

Exercises

1. Bring to class an example of a periodic report from your present or previous job or from any community, religious, or social organization to which you belong. In an accompanying memo to your instructor, indicate who the audience is and why such a report is necessary, stressing how it is organized, what kinds of factual data it contains, what visuals were used, and how it might be improved in content, organization, style, and design.

2. Assume that you are a manager of a large apartment complex (200 units). Write a periodic report based on the following information—there are 26 units vacant, 38 ready to become vacant, and 27 soon to be leased (by June 1). Also add a section of recommendations to your boss (the head of the real estate management company for which you work) on how vacant apartments might be leased more quickly and perhaps at an increased rent. Consider such important information as decorating, advertising, and installing a new security system.

3. Assume you work for a household appliance store. Prepare a sales report based on the information contained in the following table. Include a recommendation section for your manager.

	Numbers Sold	
Product	*October*	*November*
Kitchen Appliances		
Refrigerators	72	103
Dishwashers	27	14
Freezers	10	36
Electric Ranges	26	26
Gas Ranges	10	3
Microwave Ovens	31	46
Laundry Appliances		
Washers	50	75
Dryers	24	36
Air Treatment		
Room Air-conditioners	41	69
Dehumidifiers	7	2

4. Write a progress report on the wins, losses, and ties of your favorite sports team for last season. Address the report to the director of publicity for the team and stress how the director might use these facts for future publicity. As part of your report, indicate what might be an effective lead for a press release about the team's efforts.

5. Submit a progress report to your writing teacher on what you have learned in his or her course so far this term, which writing skills you want to develop in greater detail, and how you propose doing so. Mention specific memos, letters, instructions, reports, or proposals you have written or will soon write.

6. Compose a site inspection report on any part of the college campus or plant, office, or store in which you work that might need remodeling, expansion, rewiring for computer use, or new or additional air-conditioning or heating work.

7. You and your collaborative team have been asked to write a short preliminary inspection report on the condition of a historic building for your state historical

society. Inspecting the building, the home of a famous late nineteenth-century governor, you discover the following problems. Include all these details in your report. Also supply recommendations for your readers—a director of the state historical society, a state architect, and four representatives of the subcommittee on finance from your state legislature. Design two appropriate visuals to include in your report.

- Eight front columns are all in need of repair; two of them in fact may have to be replaced.
- The area below the bottom window casements needs to be excavated for waterproofing.
- The slate tile on the roof has deteriorated and needs immediate replacement.
- The front stairs show signs of mortar leaching requiring attention at once.
- Sections of gutter on the northwest and northeast sides of the house must be changed; other gutters are in fair shape.
- Wood louvers need to be repainted; four of the twelve may even need to be replaced.
- All trees around the house need pruning; an old elm in the backyard shows signs of decay.
- The siding is in desperate need of preparation and painting.
- The brick near the front entrance is dirty and moss-covered.

8. Write a report to an instructor in your major about a field trip you have taken recently—to a museum, laboratory, health care agency, correctional facility, radio or television station, agricultural station, or office. Indicate why you took the trip, name the individuals you met on the trip, and stress what you learned and how that information will help you in course work or on your job.

9. Submit a test report on the purpose, procedures, results, conclusions, and recommendations of an experiment you conducted on one of the following subjects:

a. soil
b. machinery
c. water
d. automobiles
e. textiles/clothing
f. animals
g. recreational facilities
h. computer hardware or software
i. forests
j. food
k. air quality
l. transportation
m. blood
n. noise levels

10. Write an incident report about a problem you encountered in your work or at home in the last year. Document the problem and provide a solution. Use the memo format in Figure 16.12.

11. Write an incident report about one of the following problems. Assume that it has happened to you. Supply relevant details and visuals in your report. Identify the audience for whom you are writing and the agency you are representing or trying to reach.

a. After hydroplaning, your company car hits a tree and has a damaged front fender.

b. You have been the victim of an electrical shock because an electrical tool was not grounded.

c. You twist your back lifting a bulky package in the office or plant.

d. Your boat capsizes while you are patrolling the lake.

e. The crane you are operating breaks down and you lose a half-day's work.

f. The vendor shipped the wrong replacement part for your computer, and you cannot complete a job without buying a more expensive software package.

g. An electrical storm knocked out your computer; you lost 1,000 mailing labels and will have to hire additional help to complete a mandatory mailing by the end of the week.

12. Choose one of the following descriptions of an incident and write a report based on it. The descriptions contain unnecessary details, vague words, insufficient information, unclear cause-and-effect relationships, or all of these errors. In writing your report, correct the errors by adding or deleting whatever information you believe is necessary. You may also want to rearrange the order in which information is listed. Use a memo format as in Figure 16.12 to write the report.

a. After sliding across the slippery road late at night, my car ran into another vehicle, one of those fancy imported cars. The driver of that car must have been asleep at the wheel. The paint and glass chips were all over. I was driving back from our regional meeting and wanted to report to the home office the next day. The accident will slow me down.

b. Whoever packed the glass mugs did not know what he or she was doing. The string was not the right type, nor was it tied correctly. The carton was too flimsy as well. It could have been better packed to hold all those mugs. Moreover, since the bus had to travel across some pretty rough country, the package would have broken anyhow. The best way to ship these kinds of goods is in specially marked and packed boxes. The value of the box contents was listed at $575.

CHAPTER 17

Writing Careful Long Reports

This chapter will introduce you to long reports—how they differ from short reports, how they are written, and how they are organized. It is appropriate to discuss long reports in one of the last chapters of *Successful Writing at Work.* The long report is often assigned last in class because writing one gives you an opportunity to use and combine many of the writing skills and strategies you have already learned. In business, a long report is the culmination of many weeks or months of hard work on an important company project.

The following skills and strategies will be most helpful to you as you prepare to study long reports; appropriate page numbers appear below where these topics have already been discussed.

- assessing and meeting your audience's multiple needs (pp. 4–11)
- gathering and summarizing information, especially from print and on-line reference works (pp. 325–362; 417–438)
- generating, drafting, revising, and editing your ideas (pp. 36–65)
- reporting the results of your research accurately and concisely (pp. 54–61; 581–583)
- creating and introducing visuals (pp. 469–503)
- using an appropriate method of documentation (pp. 374–397)
- preparing an informative abstract (pp. 438–440)

Having improved these skills, you should be ready to write a successful long report.

How a Long Report Differs from a Short Report

Both long and short reports are invaluable tools in the world of work. Basic differences exist, though, between these two types of reports. A short report is not a watered-down version of a long report; nor is a long report simply an expanded version of a short one. These reports differ in purpose, scope, format, and, many times, audience.

The following section explains some of the key differences between these two reports. By understanding these differences, you will be better able to follow the rest of Chapter 17 as it covers the process of writing a long report and the organization and parts of such a report. A model long report is included in Figure 17.3.

1. Scope

A long report is a major study that provides an intensive and in-depth view of the problem or idea. For example, a long report written for a course assignment may be eight to twenty pages long; a report for a business or industry may be that long or, more likely, much longer, depending on the scope of the subject. The implications of a long report are wide-ranging for a business or industry—relocating a plant, adding a new line of equipment, changing a computer programming operation.

While the long report examines a problem or idea in detail, the short report covers just one part of the problem. Unlike a short report, a long report may discuss not just one or two current events, but rather a continuing history of a problem or idea (and the background information necessary to understand it in perspective). For example, the short test report on paulownia in Figure 16.11 would be used with many other test reports for a long report for a group of industrialists or a government agency on the value of planting these trees to prevent soil erosion at mine sites.

The following titles of some typical long reports further suggest their extensive (and in some cases exhaustive) coverage:

- A Master Plan for the Recreation Needs of Dover Plains, New York
- The Transportation Problems in Kingford, Oregon, and the Use of Monorails
- Building and Technology Requirements for Rivera Plastics Over the Next Five Years
- The Use of Virtual Reality Attractions in Theme Parks
- Public Policy Implications of Expanding Health Care Delivery Systems in Tate County
- The Contributions of the Internet in Providing Health Care in Rural Areas

2. Research

A long, comprehensive report requires much more extensive research than a short report does. Such research can be gathered over time from the Internet and other research sources, questionnaires, laboratory experiments, on-site visits and tests, interviews, and the writer's own observations. For a course report, you will have to do a great deal of research and possibly interviewing to identify a major problem or topic, to track down the relevant background information, and to discover what experts have said about the subject and what they propose should be done.

TECH NOTE

The "virtual library" of the Internet offers unparalleled access both to primary research and to long reports published by scientists and mathematicians (.sci), university researchers (.edu), government agencies (.gov), the military (.mil), and every kind of organization (.org) and business (.com). Below is a "short list"—very short—of government information sources that have Web sites.

Specific URLs (Internet "addresses") change constantly and lead to ever-expanding links on the Web, making it impractical to list exact forms. But today most agencies (and even many restaurants!) regularly include URLs in their contact information, along with telephone and fax numbers. To get you started, here are some examples that are likely to stand:

http://www.epa.gov will lead you to press releases, test guidelines, and information about grants, contracts, and job opportunities at the Environmental Protection Agency.

http://www.osha.gov and **http://www.osha-slc.gov/osha.html** are Web sites of the Occupational Safety and Health Agency; you'll find information about OSHA standards, news releases and fact sheets, publications, technical information, and safety links.

http://www.astd.org is the home page of the American Society for Training and Development, which oversees research and maintains databases of information about employee training in business, industry, education, and government.

http://www.aaas.org is the site of the largest general scientific organization, the American Association for the Advancement of Science, which publishes *Science* magazine and many research reports.

http://english.ttu.edu/acw/ leads you to the Alliance for Computers and Writing; its members include educators and researchers who collaborate on the methods and technology of electronic communication.

A Short List of U.S. Government Web Sites

Advanced Research Projects Agency, Bureau of Labor Statistics, Central Intelligence Agency, Department of Agriculture, Department of Education, Department of Energy, Department of Interior, Federal Bureau of Investigation, Federal Communications Commission, General Services Administration, Library of Congress, National Performance Review, Small Business Administration, Social Security Administration, U.S. Business Advisor, U.S. Census Bureau, U.S. House of Representatives, U.S. Patent and Trademark Office, U.S. Senate Gopher Site, White House

Information gathered over many short reports can also be used to help prepare a long report. In fact, as the example of paulownia shows, a long report can use the experimental data included in a short report to arrive at a conclusion. Also for a long report writers often supply one or more progress reports (one type of short report).

Finally, preparing a proposal can lead to writing a long report. You might suggest a change to an employer, who then would request you to write a long report containing the research necessary to implement that change. Your instructor may ask you to write a proposal on doing a long report for a class project, as Barbara Shoemake did in Figure 15.5.

3. Format

A long report is too detailed and complex to be adequately organized in a memo or letter format. The product of thorough research and analysis, the long report gives readers detailed discussions and interpretations of large quantities of data. To present this information in a logical and orderly fashion, the long report contains more parts, sections, headings, subheadings, documentation, and supplements (appendixes) than would ever be included in a short report. The long report often includes many graphs, spreadsheets, charts, and tables to provide readers with extensive background information.

4. Timetable

The two types of reports also differ in the time it takes to prepare them and in the way in which they are written. Writers of these two types of reports are working under different expectations from their readers and under different kinds of deadlines. A long report is generally commissioned by a company or agency to explore in detail some subject involving personnel, locations, costs, safety, or equipment. Many times a long report is required by law—for example, investigating the feasibility of a project that will affect the ecosystem. A short report is often written as a matter of routine duty, with the writer given little or no advance notice. The long report may take weeks or even months to write. Over this period the researchers have to gather the extensive documentation readers will expect. When you prepare a long report for a class project, make sure that you select a topic that really interests you, for you will spend a good portion of the term working on it.

5. Audience

The audience for a long report is generally broader—and goes higher in an organization's hierarchy—than that for a short report. Your short report may be read by co-workers, a first-level supervisor, and possibly that person's immediate boss, but a long report is always intended for people in the top levels of management—presidents, vice presidents, superintendents, directors—who make executive, financial, and organizational decisions. These individuals are responsible for long-range planning and for always seeing the big picture, so to speak. In addition, copies of long reports may be sent to appropriate department heads for their information and commentary. A long report written about a campus issue or problem may at first be read by your teacher and then sent to an appropriate decision maker, such as a dean of students, a business manager, a director of athletics, or the head of campus security.

6. Collaborative Effort

Unlike many short reports, the long report in the business world and industry quite often is not the work of one employee. Rather, it may be a collaborative effort, the product of a committee or group whose work is reviewed by one main editor to make sure that the final copy is consistently and accurately written. Individuals in many departments within a company—art, computer programming, engineering, public relations, industrial safety—may cooperate in planning, researching, drafting, revising, and editing a long report. Your instructor may ask you to work in a group (or alone) in preparing your long report (review Chapter 3 on collaborative writing).

The Process of Writing a Long Report

As we just saw, writing a long report requires a lot of time and effort. Your work will be spread over many weeks, and you need to see your report not as a series of static or isolated tasks but as an evolving and accomplishable project. Before you embark on that project, review the information on the writing process found in Chapter 2. You may also want to study the flow chart found in Chapter 3, which illustrates the different stages in writing a research paper. Follow these guidelines:

1. **Identify a broad yet significant topic.** You'll have to do some preliminary research—general reading, on-line searching, conferring with and interviewing experts—to get an overview of main problems, key ideas and individuals involved, and implications for your company and/or community. Many times in the business world you are likely to be assigned a given subject area and will be expected to arrive at a focused topic only after doing some preliminary research. Note the kinds of research Amy Dolejs did for her long report in Figure 17.3.

2. **Expect to confer regularly with your supervisor.** In these meetings, plan to ask a lot of questions in order to pin down exactly what your boss wants. Your questions will focus on the company's use of your report, how the company wants you to express certain ideas, and the amount of information it wants you to present. Your supervisor may want you to submit an outline before you draft the report, and may expect you to submit a certain number of drafts for his or her approval before you proceed.

3. **Revise your work often.** In fact, your revisions may sometimes be extensive, depending on what your boss, teacher, or collaborative team recommends. You may have to consult new sources and arrive at a *new interpretation* of those sources. Be sure to share major changes in your thinking with coauthors or the supervisor who assigned the project. Also, as you narrow your purpose and scope, you may find

yourself deleting information or modifying its place in your work. At the earlier stages of outlining and preparing drafts, you may move material around a great deal to avoid unnecessary duplications and to ensure adequate coverage. At the later stages, you will be revising and editing your words, sentences, and paragraphs.

4. **Keep the order flexible at first.** Even as you work on your drafts and revisions, keep in mind that a long report is not written in the order in which the parts will finally be assembled. You cannot write in "final" order—abstract to appendix. Instead, expect to write in "loose" order to reflect the process in which you gathered information and assembled it for the final copy of the report. Accordingly, the body section is usually written first, for authors must obtain material included here in order to construct the rest of the report. Introductions are usually written later because authors then can make sure they have not left anything out. The abstract, which appears very early in the report, is always written last, after all the facts have been recorded and the recommendations made or the conclusions drawn. The title page and table of contents, too, are always prepared after you write the body of the report.

5. **Prepare both a work calendar and a checklist.** Keep both posted where you do your work—above your desk or computer—so you can track your progress. Make sure that every member of your collaborative team is following the same calendar and using the same checklist. The calendar should mark the dates by which each stage of your work must be completed. Match the dates on your calendar with the dates your instructor or employer may have given you to submit an outline, for progress report(s), and for the final copy. Your checklist should present the major parts of your report. As you complete each section, check it off. Before assembling the final copy of your report for readers, use the checklist to make sure that you do not accidentally omit something.

Parts of a Long Report

A long report may include some or all of the following twelve parts (pp. 621–629), which form three categories: *front matter, report text,* and *back matter.* The entire report may be placed in a clear plastic folder or other suitable cover.

Front Matter

As the name implies, the front matter of a long report consists of everything that precedes the actual text of the report. Such elements introduce, explain, and summarize to help the reader locate various parts of the report. Use lowercase roman numerals for numbering front matter elements.

1. **Letter of transmittal.** This three- or four-paragraph letter states the purpose, scope, and major recommendation of the report. If written to a teacher, the letter should additionally note that the report was done as a course assignment. Sometimes a letter of transmittal is bound with the report as part of it; most often it comes before the report, serving as a cover letter. Figure 17.1 is a sample letter of

FIGURE 17.1 A letter of transmittal for a long report.

ALPHA CONSULTANTS
1400 Ridge
Evanston, California 97213-1005
805-555-9200 FAX 805-555-9221
www.alpha.com

August 3, 1998

Dr. K. G. Lowry, President
Coastal College
San Diego, CA 93219-2619

Dear Dr. Lowry:

We are happy to offer you the enclosed report, "A Study to Determine New Directions in Women's Athletics at Coastal College," that you commissioned us to prepare. The report contains our recommendations about strengthening existing sports programs and creating new ones at Coastal College.

Our recommendation is that Coastal should engage in more active recruitment to establish a competitive women's baseball team, start to offer athletic activities in women's track and field by August 2000, and create a new interdisciplinary program between the Athletic Department and the Women's Studies Program.

We hope that you find our report useful in meeting students' needs at Coastal College. If you have any questions or if you would like to discuss any of our recommendations, please call us.

Sincerely yours,

Barbara Gilchrist

Lee T. Sidell

Encl. Report

transmittal for a business report; Figure 17.3 contains a letter of transmittal for a student's long report.

2. Title page. The title, which should be printed in all capital letters, indicates a specific subject and how or why you studied it. Select your title carefully so that it clearly tells the reader what your topic is and how you have restricted it in time, space, or method of research. Your title says that the topic of your report is focused and significant. The title page also gives the name of the company or agency preparing the report, the name(s) of the report writer(s), the date, any agency or order numbers, and the name of the firm for which the report was prepared. For a report for a class, you will indicate your instructor's name and the specific course for which you prepared the report.

It is important that your title page look professional. Center your title and logically and graphically subordinate any subtitles. Experiment with different type sizes and fonts. Figure 17.3 shows a typical title page for a long report.

3. Table of contents. The contents page tells readers on which pages they can find different parts of your report and shows how you organized your report. By looking at the table of contents in Figure 17.2, for example, the reader sees at a glance how the report, "A Study to Determine New Directions in Women's Athletics at Coastal College," is divided into four chief parts (Introduction, Discussion, Conclusion, and Recommendations). A table of contents emerges from the many outlines and drafts based on the outlines that you prepared for your report. The items on these outlines frequently expand, shrink, and move around until you decide on the formal divisions and subdivisions of your report. Include front matter components in your table of contents, but never list the contents page itself, the letter of transmittal, or the title page in your table of contents, and never have just one subheading under a heading. You cannot divide a single topic by one.

Incorrect: EXPANDING THE SPORTS PROGRAM
Basketball
BUILDING A NEW ARENA
The West Side Location

Correct: EXPANDING THE SPORTS PROGRAM
Basketball
Track and Field
BUILDING A NEW ARENA
The West Side Location
Costs

4. List of illustrations. This list of all the visuals indicates where they can be found in your report. Figure 17.3 shows a list of illustrations.

5. Abstract. As discussed in Chapter 11, an abstract presents a brief overview of the problem and conclusions; it summarizes the report. An informative abstract is far more helpful to readers of a report than is a descriptive one, which gives no conclusions or results.

FIGURE 17.2 A table of contents for a long report.

CONTENTS

Not every member of your audience will read your entire report. But almost everyone will read the abstract of your long report. For example, the president of the corporation or the director of an agency may use the abstract as the basis for approving the report and passing it on for distribution. Thus the abstract may be the most important part of your report.

Abstracts may be placed at various points in long reports. They may be placed on the title page, on a separate page, or on the first page of the report text.

Text of the Report

6. Introduction. The introduction may constitute as much as 10 or 15 percent of your report but should not be any longer. If it were, the introduction would be disproportionate to the rest of your work, especially the body section. The introduction is essential because it tells readers why your report was written and thus helps them to understand and interpret everything that follows. Do not regard the introduction as one undivided block of information. It includes the following related parts, which should be labeled with subheadings. Keep in mind, though, that your instructor or employer may ask you to list these parts in a different order.

a. Background. To understand why your topic is significant and hence worthy of study, readers need to know about its history. This history may include information on such topics as who was originally involved, when, and where; how someone was affected by the issue; what opinions have been expressed on the issue; what the implications of your study are. See how the long report on e-cash (Figure 17.3) provides useful background information on when, where, and how e-cash was used, why, and by whom.

b. Problem. Identify the problem or issue that led you to write the report. Since the problem or topic you investigated will determine everything you write about in the report, your statement of it must be clear and precise. That statement may be restricted to a few sentences. Here is a problem statement from a report on how construction designs have not taken into account the requirements of a growing number of older and disabled Americans.

> The construction industry has not satisfactorily met the needs for accessible workplaces and homes for all age and physical ability groups. The industry has relied on expensive and specialized plans to modify existing structures rather than creating universally designed spaces that are accessible to everyone.

c. Purpose statement. The purpose statement, crucial to the success of the report, tells readers why you wrote the report and what you hope to accomplish or prove. In explaining why you gathered information about a particular problem or topic, indicate how such information might be useful to a specific audience, company, or group. Like the problem statement, the purpose statement does not have to be long or complex. A sentence or two will suffice. You might begin simply by saying, "The purpose of this report is. . . ."

d. Scope. This section informs readers about the specific limits—number and type of issues, times, money, locations, personnel, and so forth—you have placed on your investigation. The long report on e-cash in Figure 17.3 concentrates on the economic and security benefits of e-cash in an Internet society, not on e-cash technology—two completely different topics. To cite another example, a report on waste disposal might include a scope statement, such as

> This report examines the recent techniques involved in the disposal of liquid and solid wastes; gaseous wastes are not discussed in this report.

In your report, you may not have studied individuals in an adjacent town or county because of time or may not have reviewed certain types of electronic equipment because of their costs or availability. If so, indicate this in your scope section. You also might limit your report by directing it to a particular audience or by writing at a particular technical level for that audience.

> This report is intended for nonengineering managers to acquaint them with recent investigations in unit mechanization design.

A careful statement of the scope should tie in with the purpose of your report.

7. The body. Also called the *discussion,* this section is the longest part, possibly making up as much as 70 percent of your report. Everything in this and all the other sections of your report grows out of your purpose and how you have limited your scope. The body of your report should supply readers with statistical information, details about the environment, and physical descriptions, as well as the various interpretations and comments of the authorities whose work you consulted as part of your library research. (Follow one of the methods of documentation discussed in Chapter 10.)

The body of your report should be carefully organized to reveal a coherent and well-defined plan. Separate the material in the body of your report into meaningful parts to make sure that you identify the major issues as well as subissues in your report. These must be clearly related to each other. Headings help your reader identify major sections more quickly.

Your organization should be carefully reflected in the different headings (and even subheadings) included in your report. Use them throughout your report to make it easy to follow. These organizational headings will also enable someone skimming the report to find specific information quickly. These headings, of course, will be included in the table of contents page. (Note how Figure 17.3 is carefully organized into sections.)

In addition to headings, use transitions to reveal the organization of the body of your report. At the beginning of each major section of the body, tell readers what they will find in that section and why. Summary sentences at the end of a section will tell readers where they have been and prepare them for any subsequent discussions. The report in Figure 17.3 does an effective job of providing these internal summaries.

8. Conclusion. The conclusion should tie everything together for readers by presenting the findings of your report. Findings, of course, will vary depending on the type of research you engage in. For a research report based on a study of sources located through various reference searches, the conclusion should summarize the main viewpoints of the authorities whose works you have cited. Perhaps your instructor will ask you to assess in your conclusion which resource materials were most thorough and helpful, and why. For a marketing report done for a business or industry, you must spell out the implications for your readers in terms of costs, personnel, products, location, and so forth.

Regardless of the type of research you do, your conclusions should be based on the information and documentation in the body of the report. Be careful that your conclusions do not contradict the evidence/information you gave in the body of your report.

Also, your conclusions should not stray into areas that your report did not cover. In essence, to write an effective conclusion, you will have to summarize carefully a great deal of information accurately and concisely. Notice how the following conclusion of a long report on well casing materials concisely summarizes the findings of the report by reviewing the literature at this crucial stage of the report.

Conclusions

Presently, many materials are available for coating monitoring wells. These casing materials, when they interact with ground water, are affected by pH levels, composition of the ground water, and the casing–ground water contact time. The complex and varied nature of ground water makes it difficult to predict the sorption and leaching potential of these various casing materials. Consequently, selecting the proper casing material for a particular monitoring application is difficult. Researchers disagree about which is the best material.

The two main classes of casing materials are metals and synthetics. SS304 and SS316 are the preferred metals, whereas PTFE and PVC are the two preferred synthetic polymer casing materials.

Research offers no clear choice as to which material is best for sampling organic or inorganic material. However, we can draw the following conclusions from our review of the literature:

1. If metals are to be tested, metallic casing of any type should **not** be used.
2. If organic materials in high concentrations are to be tested, SS metals are preferred, since PVC and PTFE are questionable.
3. If metals and low levels of organic materials or compounds are to be tested, PVC and PTFE are the acceptable choices.[1]

---

[1]Adapted from *Ground-Water Issue: The Effects of Well Casing Material on Ground-Water Quality*, Oct. 1991, p. 12, U.S. Environmental Protection Agency, Office of Solid Waste and Emergency Response.

Let's look at another effective example of a conclusion. The following conclusion to a long report on the Japanese tuna market clearly summarizes the market opportunities explained in the report.

Conclusion

The U.S. tuna industry has great potential to expand its role in the Japanese market. This market, presently 400,000 tons a year and growing rapidly, is already being supplied by imports that account for 35 percent of all sales. Our report indicates that not only will this market expand but its share of imports will continue to grow. The trend is alarming to Japanese tuna industry leaders, because this important market, close to a billion dollars a year, is increasingly subject to the influence of foreign imports. Decreasing catches by Japan's own tuna fleet as well as an increased preference for tuna by affluent Japanese consumers have contributed significantly to this trend.[2]

[2]Adapted from Sunee C. Sonu, *Japan's Tuna Market,* Sept. 1991, U.S. Department of Commerce, NOAA Technical Memorandum NMFS.

9. Recommendations. A research report for a course may not require a recommendation section. But for a business or scientific report, the most important part of the report, after the abstract, is the recommendation(s) section, which tells readers what should be done about the findings recorded in the conclusion. Your recommendation section shows readers how you want them to solve the problem your report has focused on.

The report on the Japanese tuna market mentioned above, uses a numbered list to make its recommendations.

Recommendations

Based on our analysis of the Japanese tuna market, we recommend five marketing strategies for the U.S. tuna industry:

1. Farm greater supplies of bluefin tuna to export.
2. Market our own value-added products.
3. Sell fresh tuna directly to the Tokyo Central Wholesale Market.
4. Sell wholesale to other Japanese markets.
5. Advertise and supply to Japanese supermarket chains.[3]

[3]Adapted from Sonu, *Japan's Tuna Market.*

Back Matter

Included in this section of the report are all the supporting data that, if included in the text of the report, would bog the reader down in details and cloud the main points the report makes.

10. Glossary. An alphabetical list of the specialized vocabulary with its definitions appears in the glossary. A glossary might be unnecessary if your report does not use highly technical vocabulary or if all members of your audience are familiar with the specialized terms you do use.

11. References cited. Any sources cited in your report—Web sites, books, articles, television programs, interviews, reviews, and audiovisuals—are usually listed in this section (see Chapter 10 on preparing a Works Cited or References page). Also, ask your instructor or employer about how he or she wants information to be documented. Sometimes, in the world of work, employers prefer all information to be documented in footnotes or cited parenthetically in the text (without a formal Works Cited page). Note that the long report in Figure 17.3 uses the American Psychological Association (APA) system of documentation.

12. Appendix (plural Appendixes). The appendix contains all the supporting materials for the report—tables and charts too long to include in the discussion, sample questionnaires, budgets and cost estimates, correspondence about the preparation of the report, case histories, transcripts of telephone conversations. Group like items together in the appendix, as the examples under "Appendixes" in Figure 17.2 show.

A Model Long Report

The long report in Figure 17.3 is a research report written by a student for a business communication class. It deals with e-cash as future currency, a topic about which the student gathered relevant data primarily through reading print and electronic sources and interviewing an expert. It contains all the parts of a long report discussed earlier in this chapter except a recommendation, glossary, and appendix. Intended for a general audience unfamiliar with e-cash technology, the report consequently avoids the technical terms and descriptions for which a glossary, might be necessary. Because the student is surveying authorities' opinions, she does not supply a recommendation section of her own.

The student's main task is to investigate what has been said about the use of electronic cash in American society and to report the results of her research in a logically organized discussion. Notice how the body section of her report is divided into three closely related areas—"What Is E-cash," "How Long and Difficult Will the Transition to E-cash Be," and "Pros and Cons of Using E-cash." Each of these areas is further subdivided, as you will see from the table of contents and the subheadings in the report.

Study the report to learn how one writer researched, organized, and discussed a major problem (or topic) suitable for a long report.

FIGURE 17.3 A long report.

345 Spruce Lane
Apt 34-C
Gunderson, NE 68345

December 5, 1996

Professor Janice Ranzel
Business Communications Department
Fairmont Central College
Fairmont, NE 68339-6717

Dear Professor Ranzel:

With this letter I am enclosing my research report on the topic of electronic cash, which you approved six weeks ago. My report discusses various aspects of electronic cash, including Internet usage and smart card development.

E-cash offers significant and widespread benefits. Though it will not replace owner responsibility, e-cash will allow greater freedom and safety than credit cards, checks, or paper money. E-cash will further reduce bank storage and security costs while providing consumers and merchants with the security of encryption. Daily purchases, both in retail outlets and over the Internet, will be faster and less expensive for the consumer and the merchant. Both Internet e-cash and smart cards have the potential to transform the way the world buys and sells.

I hope that you will find this report carefully organized and researched. If you would like to discuss it with me, I can be reached at 555-3400 or you might write me at my e-mail address acdolejs@fcc.edu.

Sincerely yours,

Amy C Dolejs

Amy C. Dolejs

Enclosure

THE E-CASH REVOLUTION:
New Currency for an Internet Society

Amy C. Dolejs

December 5, 1996

Business Communication 204

Professor Janice Ranzel

Table of Contents

LIST OF ILLUSTRATIONS

ABSTRACT

After decades of research, a global electronic cash economy is now becoming possible. Some consumers fear that e-cash means the end of anonymity, a limit to personal transactions, or random technical errors resulting in a loss of money. E-cash is safer to carry than paper money because only the owner can use it. It is more economical for merchants and consumers to use, and it offers the same anonymity associated with cash purchases. E-cash has additional benefits impossible with physical cash, including the ability to be earmarked and secured for specific purchases. However, as with any new technology, individuals need to be responsible about their financial obligations.

INTRODUCTION

Background

For decades computer programmers have been trying to develop faster, safer ways to transmit important information electronically (Kirkman, 1987; Schneier, 1996). Such a goal in fact goes back centuries, since consumers and merchants have repeatedly looked for more efficient, safer ways to conduct business transactions. Thanks to developing Internet technology two fields of research--banking and the electronic transferral of information--have been combined to create a convenient, secure type of money called electronic cash, or e-cash. On-line, e-cash works very much like e-mail, which allows people to send information from one computer to other computers around the world almost instantaneously and inexpensively by phone lines. E-cash, encoded on a plastic card, performs the same jobs as regular cash.

E-cash has seriously held the attention of the general public only since early 1994, when Mondex International, a subsidiary of the British clearing bank National Westminster (NatWest), began to develop e-cash technology by introducing a potentially viable alternative to cash--the Mondex smart card ("Smart Card Cashes In," 1995). This "smart card" was not meant to be a credit card or even a check card. Its designers intended it to replace cash in many, but not all, commercial and personal transactions. The current smart card is only a stepping stone in the potentially revolutionary e-cash technology.

Another company, Digicash, worked toward the same objective but from a different starting point. Cooperating with Mark Twain Bancshares in Ladue, Missouri, Digicash, in Amsterdam, developed software allowing users with a World Currency Access account to pay bills over the Internet (Lunt, 1996a). Bankers, economists, credit card companies, consumers, and even the IRS are all following the development of e-cash--technology that might very easily define the way the entire world buys and sells in the next century.

So far over 2,000 accounts have been opened with Digicash and are used daily for relatively small payments over the Internet--usually amounts under $500. Mondex, geared more toward getting whole communities to adopt their smart card, has opened trials with Wells Fargo in San Francisco (where 550 employees have been issued smart cards that can be used at 23 area merchants) and in Swindon, England, where over 10,000 people, 700 merchants, and 250 pay phones use and accept smart cards for everyday purchases (Lunt, 1996b). The possibility of widespread use of some form of e-cash grows stronger every day. Economists like Patiwat Panurach (1996) maintain that while electronic methods of payment may not be immediately accepted by all consumers, "the trends of modern commerce, driven by the weaknesses of traditional payment systems, point to the eventual rise of electronic payments" (p. 50).

Problem

Many consumers are wary of e-cash's implications. Some frequently voiced worries are that money may be lost by some technical error, that e-cash will not be accepted everywhere, that personal transactions will be limited, or that e-cash results in the loss of the anonymity associated with paper money. Moreover, with such a convenient spending option available, consumers have to be careful to avoid bankruptcy.

Yet, if used responsibly, e-cash will allow for the same anonymity as cash, while guaranteeing the merchant against counterfeited cash, stolen credit cards, or bounced checks, and providing the consumer with security against theft. E-cash may also help protect the thousands of smaller merchants who conduct business over the Internet by introducing a freer, less expensive way of collecting payments.

Purpose of Report

The purpose of this report is to discuss the benefits e-cash will offer consumers all over the world, especially the rapidly growing number of consumers who regularly do business over the Internet.

Scope

This report includes information on the relationship of e-cash to paper money, including how e-cash works and what similar technology we already use. The report also surveys the causes of several consumer problems and demonstrates that e-cash, if not abused, offers many advantages over our current paper-bound monetary system.

DISCUSSION

What Is E-cash?

An Alternative to Credit Cards

Programmers and entrepreneurs became interested in developing something resembling e-cash when Internet consumers and merchants wanted a safer, faster, cheaper method of payment than credit cards afforded. When a consumer used a credit card to buy from merchants on the Internet, he or she exposed his or her card number to possible theft. The system for transmitting the card number from consumer to merchant was not entirely protected, and sophisticated and dishonest computer users often found ways to steal credit card numbers and run up huge balances on other people's accounts (Freeman, 1996).

Since a large percentage of Internet sales deal with information, a merchant who unknowingly accepts a stolen credit card number and then allows the user access to the information may never be paid. The user can quickly take the information and escape without leaving a traceable path to his or her identity. Also, accepting credit cards costs the merchant money because of the vertification process necessary with every transaction. As Levy (1995) points out, ". . . credit cards cost merchants a fee averaging two percent of the charge" (p.63).

Even more problems can arise for credit card users who surf the Net. Credit cards reveal the identity of the buyer, and credit card companies must know the identity of the merchant. Kleiner (1995) emphasizes that some consumers are unsettled by the idea that whatever they buy can be traced. Even law-abiding consumers report that they are uncomfortable with companies knowing too much about them, if only because these companies trace buying habits and subsequently annoy purchasers with telephone, mail, or e-mail offers (Sanders, 1996).

Credit cards also come close to undermining what some consider to be the real value of the Internet. A "merchant" on the Internet can be anyone with a minimum amount of computer knowledge--a fitness instructor, a chef, a florist. Any individual with something even marginally marketable--a good joke or a beer can collection--can sell his or her information or items on-line. However, most unincorporated individuals are not authorized to accept credit cards, which causes problems when selling an item to an anonymous buyer. Credit cards, according to Holland and Cortese (1995), are simply "not suited for the grassroots economy the Internet makes possible" (p. 69). They require too much expense for small merchants, and they are not safe enough for the merchant or the consumer to warrant such expense.

However, smart cards or e-cash in other forms would eliminate the problems associated with credit card use.

David Chaum, CEO of Digicash, has developed a way for a bank to certify electronic money without knowing to whom it was issued. Chaum's system keeps track of which customers withdrew which money, but does not trace where the money goes after it leaves the bank. In this way, merchants are assured of getting paid, and consumers can buy anonymously, as if they used paper money. Under Chaum's system, a bank customer simply notifies the bank of how much money to credit to his or her electronic account; after the money is credited to the account, the consumer is free to spend it in cyberspace on anything he or she wants. The consumer reaps an added benefit: the account number is encrypted, which makes it very difficult for anyone to steal and use.

The Mondex smart card also offers the security of encryption, though further refinements in the system will be necessary. Mondex designer Tim Jones admits that the system is not 100 percent secure, but assures consumers that the encrypted code is so tamper-reistant, it would probably cost thieves more money and time to break it than they would make from it (Palmer, 1994). As one computer expert cautioned, "The arguments for encryption do have drawbacks sometimes. Only recently two industrious individuals cracked an encryption algorithm that Worldscape has been using to secure financial transactions" (Freeman, 1996).

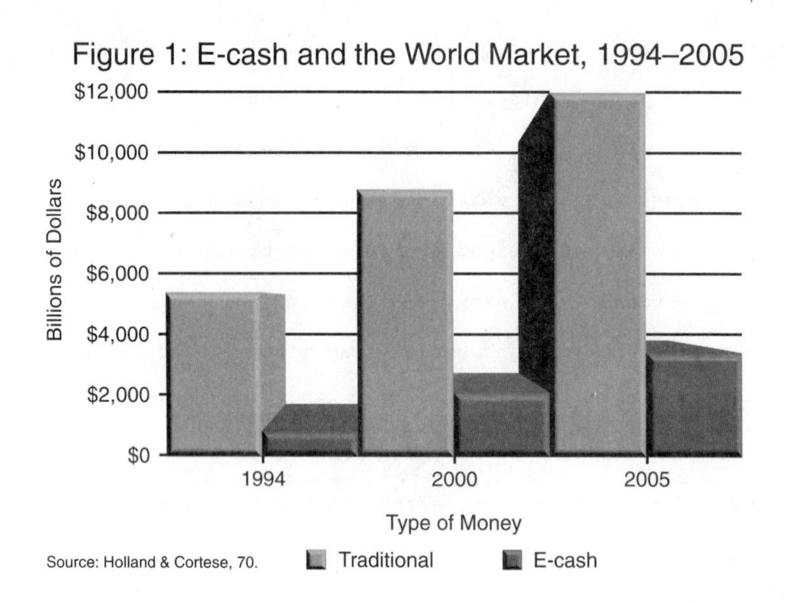

Simple Supply and Demand

Some consumers might argue that implementing such a large-scale change in our monetary system simply to cater to the desires of a few Internet consumers would be foolhardy. Yet the group of people who understand and operate computers daily is no longer marginal. As Figure 1, above, shows, in 1994, Internet purchases accounted for 4.5 percent of all commercial transactions, which means that $245 billion worth of merchandise was bought and sold from computer terminals around the world (Holland & Cortese, 1995). Experts predict that by the year 2002, that number will increase by nearly five times, making Internet transactions 18.6 percent of the market. By 2005, 20 percent of all purchases will be made over the Net, with a monetary value of nearly $3 trillion (Holland &

Cortese, 1995). Other economists predict a much faster adoption of e-cash, suggesting that it will claim 25 percent of the market by 2000 (Quittner, 1995). Such a large percentage of consumers looking for an easier way to spend money is bound to advance many possible solutions. Even so, e-cash stands out as the most feasible answer to solve currency transaction problems both on- and off-line (Maidment, 1995).

An Electronic Alternative to Cash and Checks

Many consumers have heard about and welcome the concept of e-cash, but widespread misunderstanding of what it actually is and what it can do impedes greater acceptance of e-cash. E-cash is not a credit system. In other words, a consumer using a smart card does not accrue debts to be paid off in a month or a year as with a credit card; rather, the consumer depletes his or her store of ready cash, which happens to be coded on a card instead of rattling and rustling around in a pocket.

Nor is e-cash the equivalent of a debit card. The cash on the card is actually on the card, not resting in a bank account. When the consumer makes a purchase, it is as if he or she had paid with paper money or coins. The merchant does not check with a bank or credit card company for verification, because all the verification the merchant needs is right there, encoded on the smart card. Mondex founder Tim Jones explains, "Imagine it's the same as physical money, and you won't be far off" (Palmer, 1994, p. 69). The only difference between e-cash and paper money is that e-cash is much more efficient.

If e-cash is the same as physical money, why do we need it? For one thing, physical cash is expensive for banks to move around and to keep secure. In Britain alone, the annual cost for banks of moving cash safely around is over three billion dollars ("Smart Card Cashes In," 1995). The three billion is spent on armored trucks, vaults to store the cash, people to count and recount the cash, guards, and security cameras and alarms. That cost eventually reaches bank customers through larger service charges, lower interest rates on savings accounts, and higher interest rates on loans. E-cash has built-in security for banks and consumers alike. It weighs nothing and takes up no space, so the need for large vaults is eliminated. The savings on security and storage alone would be enough to make bankers want to develop and promote e-cash systems.

E-cash would also replace checks, which cost both banks and consumers billions of dollars a year: "Once you factor in processing, the 30 billion checks written each year cost banks (and ultimately, us) up to a buck apiece" (Levy, 1995, p. 63). Checks, like cash, present a storage problem, which could be eliminated if there were a safer way to carry money around. Checks present the merchant with another problem as well: insufficient funds. E-cash would eliminate the possibility of bounced checks, saving merchants, consumers, and banks millions of dollars a year.

How Long and Difficult Will the Transition to E-cash Be?

Cards in Motion

Many countries already have an established infrastructure which, with modifications, could support a global e-cash economy. Credit cards are accepted nearly everywhere, and merchants worldwide have machines that read and verify credit card purchases. Smart card manufacturers are working closely with credit card companies to develop a standard that would allow credit cards, e-cash, insurance information, driver's license data, and almost any other information to be encoded on the same card. Switching over to a partial or even a full e-cash economy would require almost no effort on the part of merchants or consumers except minor timesaving adjustments. But even e-cash pioneers had not predicted an immediate worldwide change from physical cash to electronic cash. Tim Jones believes that within a few decades, his smart card will be used in 40 percent of cash transactions. He is not so brave as to think it will replace cash altogether. On the Internet, however, e-cash will probably soon be used in a majority of transactions (Kelly, 1995; Taylor, 1995).

Many consumers around the world already use cards much like the smart card. Many toll booths in Europe and

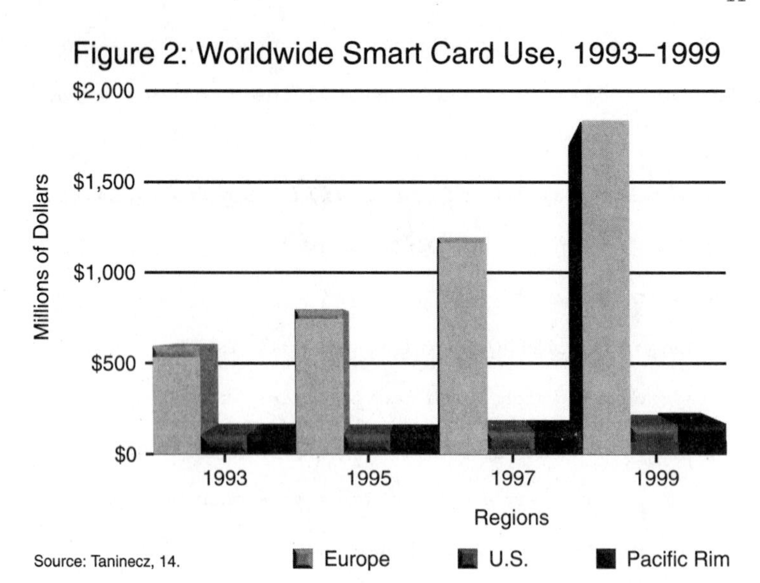

pay phones around the world accept prepaid cards in place of cash. These cards can be used as many times as it takes to spend all the money on them. Americans have already seen some fairly common uses of the prepaid phone card. However, as we can see from the chart above in Figure 2, Europeans are far ahead of us in using phone and toll cards and currently spend over a billion dollars a year with them as compared to our $38 million (Taninecz, 1994). And, though e-cash is not the same as a credit card, global consumers already are comfortable with using plastic to pay for purchases.

American Use of Electronic Cash

Some areas in the United States are working toward setting up electronic money systems. In 1993, Maryland began issuing plastic cards credited with food stamps and other benefits to federal aid recipients. Within a year, the state had reduced its paperwork costs by 10 percent. The cards also helped reduce welfare fraud and made some recipients feel more confident when paying for groceries with plastic instead of paper food stamps. According to Stix (1994), Maryland's system has been duplicated around the country, with millions of U.S. consumers becoming familiar and comfortable with the idea of cash on a card.

In July 1996, Visa International provided prepaid debit cards for visitors to the Olympic games in Atlanta. The cards reduced the possibility of theft or loss of money and were eagerly accepted by most area merchants and all Olympic affiliations (Woods, 1995). Visitors to the Olympics reported that the prepaid cards were easy to use and provided a sense of security not available to those carrying paper money (Curtis, 1996).

College students around the country may be the most loyal consumers with the prepaid phone card and the prepaid copy card. Ten-, twenty-, or thirty-minute phone cards can be bought at vending machines on most college campuses. Copy cards purchased in campus libraries can be used in all the copy machines in that library, giving users savings of up to 50 percent (Sinclair, 1996). The difference between these cards and a smart card is that when the minutes or copies are exhausted on a prepaid card, the card is worthless; yet a smart card can be recharged with cash over and over again. Moreover, a phone card is restricted to phone charges and a copy card to copy machine fees. But a smart card can be used anywhere for anything, proving that these prepaid cards point the way toward widespread e-cash use. When consumers see the many ways electronic cash is already used, the concept of e-cash does not seem as frightening or foreign.

Pros and Cons of E-cash

Privacy Concerns

Potential e-cash users may fear that governments or companies would be able to find out every purchase they make. They reason that if we do away with anonymous paper money, every transaction can be electronically cataloged. Understandably, that idea is frightening. But e-cash proponents are working to ensure consumer privacy. David Chaum of Digicash considers privacy a major priority, and his company is a pioneer in devising protections for consumers: "We call it one-way privacy. When you buy something you aren't forced to reveal who you are, but you know who they are and have proof they took your money" (Levy, 1995, p. 65). Encrypted information on the cards can be used only one way and cannot be reversed to find out who or where the buyer is.

Theft or Loss

Because a smart card is just like cash, an owner who loses e-cash risks the same fate as if he or she had lost paper money. However, the danger of losing a card loaded with cash is no more likely than dropping a wallet or a wad of bills in the street. The smart card delivers one thing cash doesn't, however: identification. While someone could pick up a lost twenty-dollar bill and spend it easily with no one being the wiser, no one else can use a lost smart card because it requires the owner's personal code to be activated. A smart card, like a credit card, would have the name of the issuing bank on it, so anyone finding a lost card and being unable to use it can return it to the bank, which can then contact the person who lost it (Mondex, 1995). Moreover, if a card is damaged but still holds money, the bank can reissue a new card to the customer, who loses nothing in the long run.

Financial institutions also have to take precautions. Until a few years ago, banks almost always lost money when they were held up by robbers. Yet today most of the money banks lose is from an inside computer theft. Any computer system needs to have strong safeguards against insider abuse. As one expert notes,

> The move to e-cash has implicit in it a move to an increase in the level of integration between banks and other financial institutions, i.e., a move toward putting all of our eggs in one basket. We will see a more monolithic banking structure. If one bank dies, all banks will die. (Freeman, 1996)

Personal Transactions

Again just like cash, e-cash can be used to make personal transactions. On-line, one person can simply send bits of e-cash to a friend's e-mail account. Off-line, transactions will require that one person have a device called a wallet, which works much like a miniature ATM. The person giving money puts his or her card into the wallet, indicates the amount, and then takes the card out. The person receiving money then puts his or her card into the wallet, and the money is automatically credited to that card. A balance reader, about the size of a keychain, can show the smart card user his/her card's balance as well as the last ten transactions.

Earmarked Money

The possibilities of e-cash are almost limitless. Parents of college students may like the option of designating what items their kids can buy, since smart cards can be programmed to pay for certain items and reject others. For example, a parent could provide a college student with a loaded card that could be used only for rent, food, and books--no beer (Levy, 1995). Impulse buyers may even be able to program their cards not to let them spend money on certain items or at certain types of stores--electronic willpower.

CONCLUSION

The steady increase in prepaid card use and Internet purchases is leading us toward a cashless society. Though paper money is not likely to disappear as soon as some economists predict, a definite rise in the use of e-cash is imminent. E-cash will free merchants, consumers, and bankers from the shackles of paper money security and costly storage and allow open trade on the Internet to flourish. Though legitimate concerns may at first keep some consumers from adopting electronic spending habits, the slow proliferation of smart cards and Internet bank accounts, along with even better encrypted secured privacy, may convince many consumers about the flexibility and freedom of electronic cash. Ultimately, though, consumers must be careful not to overspend and thus lose their financial independence and stability in the cashless society developing around the world.

17

REFERENCES

Curtis, J. Y. (1996, September 2). Electronic cash at the Olympics. *Cash Advances Monthly, 5,* 52-53.

Freeman, C. (1996, November 12). Professor of Computer Science. Midwest Technical Institute. Personal Interview.

Holland, K., & Cortese, A. (1995, June 12). The future of money. *Business Week,* 66-78.

Kelly, K. (1995). Program capitalism. *New Perspectives Quarterly, 12,* 32.

Kirkman, P. (1987). *Electronic funds transfer systems.* Oxford: Basil Blackwell.

Kleiner, K. (1995, April 8). Banking on electronic money. *New Scientist,* 26-30.

Levy, S. (1995, October 30). The end of money? *Newsweek,* 62-65.

Lunt, P. (1996a, January). E-cash becomes reality, via Mark Twain and Digicash. *ABA Banking Journal, 88,* 62.

Lunt, P. (1996b, February). Mondex spreads its wings. *ABA Banking Journal, 88,* 50-54.

Maidment, P. (1995, December 26). The age of cybercash. *Newsweek,* 128.

18

Miller, J. (1996). E-money mini-FAQ (release 2.0). [On-line], Internet. http://www.ex.ac.uk/~RDavies/arian/emoneyfaq.html

Mondex at a glance. (1995). *Mondex Home Page.* [On-line], Internet. http://www.mondex.com/mondex/home.htm

Palmer, G. (1994, August). Tim Jones: the man behind Mondex. *Banking World, 12,* 6-10.

Panurach, P. (1996, June). Money in electronic commerce: Digital cash, electronic fund transfer, and e-cash. *Communications of the ACM, 39,* 45-50.

Quittner, J. (1995, June 12). How to send real money over the Internet. *Time,* 64.

Sanders, F. (1996). E-money: Paradise or prison? *Swiss America Trading Corporation.* [On-line], Internet. <http://com.primenet.com/callme/coins/emoney.html>

Schneier, B. (1996). *Applied cryptography: Protocols, algorithms, and source code in C.* (2nd ed.). New York: John Wiley & Sons.

Security truce? (1995, July). *Byte, 20,* 80.

Sinclair, M. (1996, October 31). Reference Librarian. Hardy Community College. Personal Interview.

The smart card cashes in. (1995, January). *The Economist, 29,* 73-74.

Smart cash cards proliferate. (1995, September). *Electronics Now,* 4-10.

Stix, G. (1994, August). Welfare plastic. Electronic benefits may become a big business. *Scientific American,* 84-86.

Taninecz, G. (1994, April 11). Visa to unleash power of the purse. *Electronics,* 14.

Taylor, W. (1995, April). They're going for your wallet! *PC Computing, 8,* 150.

Woods, W. (1995, June 12). Cashless society becomes a reality. *Fortune,* 24.

Final Words of Advice About Long Reports

Perhaps no piece of writing you do on the job—or as a course assignment for your instructor—carries more weight than your work on a long report. Your readers will inevitably place a great deal of emphasis on your work. As we have seen, these reports deal with major issues affecting long-range planning and decision making. The long report requires you to use all the researching, organizing, drafting, revising, and editing skills you have learned.

The preparation of a long report may appear at first to be a formidable task. But you can simplify your job and increase your chances for success by following the guidelines offered in this chapter. Among those guidelines, always keep these four points in mind:

1. Plan and work early—do not postpone work on identifying, researching, and drafting until a deadline draws near.
2. Divide your workload into meaningful units—reassure yourself that you do not have to write the report or even an entire section of a report in a day or two.
3. Set up mini-deadlines for each phase of your work and then meet them.
4. If you are preparing the report as part of a team, confer often and carefully with others on your team.

✓ Revision Checklist

- ❑ Concentrated on a major problem—one with significant implications for my major, neighborhood, city, or employer.
- ❑ Identified, justified, and described the significance of the main problem as opposed to focusing on a minor side issue.
- ❑ Did sufficient research—in the library, on the Internet, through interviewing, from personal observation and/or testing—to convince my readers that I am knowledgeable about this problem, its scope and effects, and likely solution.
- ❑ Became familiar with key terms, major researchers in the field, major changes, trends, and accomplishments.
- ❑ Anticipated how various readers will use and profit from my report for their long-range planning.
- ❑ Made sure I understand what employer/teacher/reader is looking for.
- ❑ Followed company's/instructor's guidelines for the format and documentation of work.

- ☐ Followed company's/instructor's schedule for completing various stages of long report.
- ☐ Divided and labeled the parts of long report to make it easy for readers to follow and to show careful plan of organization.
- ☐ Supplied an abstract that leaves no doubt in readers' minds about what report deals with and why.
- ☐ Designed attractive title page that contains all the basic information—title, date, for whom the report is written, my name—readers require.
- ☐ Gave readers all the necessary introductory information about background, problem, purpose of report, and scope. Made sure that introduction is neither too long nor too short.
- ☐ Included in body of report the weight of all my research—the facts, statistics, and descriptions—that my readers need in order to know that I have done my homework on the topic well.
- ☐ Included subheadings to reflect the major divisions into which I have organized the research that forms the nucleus of the text.
- ☐ Wrapped up report in succinct conclusion. Told readers what the findings of my research are and accurately interpreted all data.
- ☐ Supplied a recommendations section (if required) that tells readers concretely how they can respond to the problem using the data. Recommendations make sense—are realistic and practical and related directly to the research and topic. Ensured that recommendations are persuasive.
- ☐ Included in the final copy of report all the parts listed in table of contents.
- ☐ Supplied a one-page letter of transmittal or cover letter informing readers why the report was written and describing its scope and findings.

Exercises

1. Send an e-mail to your instructor on how one of the short reports in Chapter 16 could be useful to someone who has to write a long report.
2. Using the information contained in Figures 17.1 and 17.2, draft an introduction for the report "A Study to Determine New Directions in Women's Athletics at Coastal College." Add any details you think will be relevant.
3. What kinds of research did the student do to write the long report in Figure 17.3? As part of your answer, include the titles of any specific reference works you think the writer may have consulted. (You may want to review Chapter 9.)
4. Study Figure 17.3 and answer the following questions based on it.
   - **a.** Why can the abstract be termed informative rather than descriptive?
   - **b.** How has the writer successfully limited the scope of the report?

c. Where does the writer use internal summaries especially well?
d. Where and how has the writer adapted her technical information for her audience of general readers?
e. What visual devices does the writer employ to separate parts of the report and divisions within each part?
f. How does the writer introduce, summarize, and draw conclusions from the expert opinions she cites in order to substantiate the main points?
g. What are the ways in which the writer documents information she has gathered?
h. What functions does the conclusion serve for readers? Cite specific examples from the report.

5. Come to class prepared to discuss at least two major problems that would be suitable topics for a long report. Consider an important community problem—traffic, crime, air and water pollution—or a problem at your college. Then write a letter to a consulting firm or other appropriate agency or business, requesting a study of the problem and a report.

6. Write a report outline for one of the problems you decided on in Exercise 5. Use major headings and include the kinds of information discussed in the Front Matter section of this chapter (pp. 621–625).

7. Have your instructor look at and approve the outline you prepared for Exercise 6. Then write a long report based on the outline, either individually or as part of a collaborative writing team.

CHAPTER

18

Making Successful Presentations at Work

Almost every job requires employees to have and to use carefully developed speaking skills. Your oral presentations can be just as important to your career as your written ones are. In fact, to get hired, you have to be a persuasive speaker at your job interview. And to advance up the corporate ladder you will have to continue to be a confident, well-prepared, and persuasive speaker.

Types of Oral Presentations at Work

On the job you will have numerous presentation responsibilities that will vary in the amount of preparation they require, the time they last, and the audience and occasion for which they are intended. Here are some frequent types of presentations you may have to make as part of your job:

- sales pitches to prospective clients
- evaluations of products or policies
- progress reports to members of your collaborative team and your boss
- reports to superiors about your job accomplishments
- justifications of your own position or perhaps even that of an entire section or group
- appeals and/or explanations before elected officials about some aspect of your company or work

Whatever your oral communication responsibilities, this chapter gives you practical advice on how to become a better, more assured communicator in both informal briefings and formal speeches.

TECH NOTE

To get helpful experience delivering both impromptu briefings and formal speeches, consider joining Toastmasters International, an educational organization that has transformed many frightened speakers into accomplished speechmakers. Members of Toastmasters meet at least once a month (and sometimes as often as every week) to listen to and evaluate one another's speeches. From feedback and strategic training, you will grow to be a more confident public speaker thanks to Toastmasters. Their Web site is http://www.toastmasters.org.

Informal Briefings

If you have ever given a book report or explained laboratory results in front of a class, you have given an informal briefing. Such semiformal reports are a routine part of many jobs. Here is a list of some of the typical informal briefings you may need to deliver at work:

- a status report on your current project
- an update or end-of-shift report, as nurses and police officers give
- an explanation of a policy to co-workers
- a report on a conference you attended, as Judith Kim and Lee Schoppe do in Figure 11.8.
- a demonstration of a new piece of equipment or software
- an introduction of a guest speaker, co-worker, supervisor, or inspector
- a summary of a meeting you attended

These informal reports are usually short (one to seven minutes, perhaps), and you won't always be given advance notice. When the boss tells you to "say a few words about the new Web site (or the new accounting procedure)," you will not be expected to give a lengthy formal speech. For example, the personnel officer informing employees about recently extended insurance coverage does not read them the fine print in the policy, but rather covers its key points, saving detailed questions for private conferences.

Guidelines for Preparing Informal Briefings

Follow these guidelines when you have to make an informal briefing:

- Make your comments brief and to the point.
- Jot down a few rough notes on a 3" × 5" note card that you can easily hold in the palm of your hand.

FIGURE 18.1 A note card for an informal briefing to introduce an engineer to a group of safety directors.

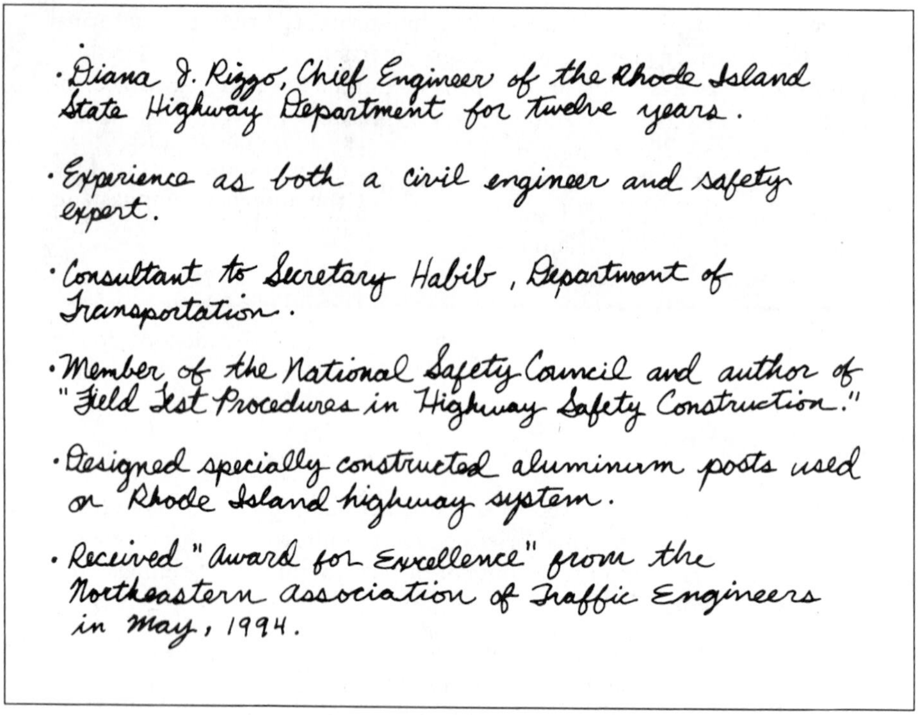

- Diana J. Rizzo, Chief Engineer of the Rhode Island State Highway Department for twelve years.
- Experience as both a civil engineer and safety expert.
- Consultant to Secretary Habib, Department of Transportation.
- Member of the National Safety Council and author of "Field Test Procedures in Highway Safety Construction."
- Designed specially constructed aluminum posts used on Rhode Island highway system.
- Received "Award for Excellence" from the Northeastern Association of Traffic Engineers in May, 1994.

- Highlight key phrases and terms that you need to stress.
- Include in your notes only the major points you want to mention.
- Arrange your points in chronological order or from cause to effect.

A note card with key facts used by an employee who is introducing Diana Rizzo, a visiting speaker, to a monthly meeting of safety directors might look like Figure 18.1.

Formal Speeches

Whereas an informal briefing is likely to be short, generally conversational, and intended for a limited number of people, a formal speech is much longer, far less conversational, and intended for a wider audience. Therefore, it involves more preparation and more sustained interaction between speaker and audience; it is, in other words, more "formal."

Many of us are uncomfortable in front of an audience because we feel frightened or embarrassed. Much of this anxiety can be eased, though, if you know what to expect. The two areas you should investigate thoroughly before you begin to prepare your speech are (1) who will be in your audience and (2) why they are there.

Analyzing Your Audience

The more you learn about your audience, the better prepared you will be to give them what they need. Just as you do for your written work, for your oral presentation you will have to do some research about the audience, emphasizing the "you attitude," and establishing your own credibility.

Consider Your Audience as a Group of Listeners, Not Readers

While many elements of audience analysis pertain both to readers of your work and to listeners of your speech, keep in mind that there are several fundamental differences between these two groups. Unlike a reader of your report, the audience for your speech

- is a captive audience
- has only one chance to get your message
- has less time to digest what you say
- has a shorter attention span
- can't go back to review what you said or jump ahead to get a preview
- is more easily distracted by noise in the room—interruptions, chairs being moved, people coughing, and so on
- cannot absorb as many of the technical details as you would include in a written report

Take all of these differences into account as you plan your speech and assess your audience.

Remember that everything in your speech is being delivered for the benefit of that audience. Relate everything in your talk to them. Do this by selecting only details that are relevant to your audience and your purpose. Choose concrete (not abstract or general) examples; look for memorable stories, analogies, or events that an audience can easily recall after your speech is completed.

Analyzing Your Listening Audience

Here are four key rules you need to follow when analyzing the audience for your talk.

1. **Find out what unites them as a group.**

   - members of the same profession
   - customers using the same products
   - employees of the company you work for
   - supporters of the same club or organization
   - members of the same ethnic, religious, or political group

- united in their emotional response to a topic—such as a campaign to curb property taxes in their area

2. **Determine how much they know about your topic.**
   - all consumers who have little or no technical knowledge of your subject
   - all technical individuals who understand the terms, jargon, and background of your subject
   - a mixed audience

3. **Establish how interested they are in your topic.**
   - highly motivated, willing audience eager to hear what you have to say and happy to endorse your conclusions
   - mildly interested but not totally convinced of your point of view
   - neutral—waiting to be informed, entertained, or persuaded
   - uninterested in your topic—their attendance is mandatory and they are listening indifferently or reluctantly
   - hostile—opposed to your opinion

4. **Anticipate their most likely response to you.**
   - positive and with great interest
   - open-minded and uncommitted
   - mildly skeptical
   - uncooperative and antagonistic

Special Considerations for an ESL Audience

Given the international make-up of audiences at many business presentations today, it is not unlikely that you will have to address a group of ESL listeners or even make a presentation before individuals in a country other than your own. While it is difficult to generalize about specific communication guidelines, consider your audience's particular cultural taboos and protocols. Do they accept your looking at them directly, or do they frown on eye contact? Will they expect you to stand in one place, or will they be comfortable if you move about the room while you speak?

As you prepare a talk before an ESL audience, keep the following points in mind.

1. Brush up on your audience's culture, especially accepted ways they communicate with each other (see pages 163–170).
2. Find out what constitutes an appropriate length for a talk before your audience. (German listeners might be accustomed to hearing someone read a thirty- to forty-page paper while members from another culture would regard this practice as improper.)
3. Be especially careful about introducing humor—avoid anything that is based on nationality, dialect, religion, or race.
4. Avoid U.S. clichés and jargon.

5. Think twice about injecting anything autobiographical into your speech. Some cultures regard such intimacy as an inappropriate invasion of privacy.
6. Steer clear of politics; you risk losing your audience's confidence.
7. Choose visuals with universally understood icons and logos.

Speaking for the Occasion

Understanding why your audience has gathered will help you deliver a successful speech. An audience may be present for a variety of reasons—for a social gathering, a business meeting, or an educational forum. Shape your remarks to fit the occasion.

Your Allotted Time

Consider also how much time is allotted for your speech and never exceed it. In fact, your audience may be even more likely to respond enthusiastically if you finish a little early.

Also factor in whether you are the only speaker scheduled. It makes a big difference in your preparation if you are the first and only speaker at a breakfast meeting or the last of four speakers at an evening meeting. Are you speaking on early Monday morning or late Friday afternoon? Take into account your audience's interest level and attention span at various times of the day or week. Also find out whether someone will introduce you or whether you will begin on your own. If someone introduces you, it would be embarrassing to repeat (or contradict!) the information from that introduction in your speech.

Number of People in Your Audience

The number of people in your audience is also significant. A formal presentation to a small group—five or six supervisors or buyers—seated around a conference table can nonetheless be made more personal; you can walk around the table or interrupt your talk for a few times to answer questions. You will obviously have much less flexibility when addressing a large group—seventy or eighty people—in an auditorium.

Ways to Present a Speech

Your effectiveness depends directly on the extent of your preparation. Of the four approaches below, the extemporaneous is best suited to most individuals and occasions. But first we will examine three other possibilities and their advantages and disadvantages.

1. Speaking "off the cuff." The professional speechmaker may be comfortable with an off-the-cuff approach, but for most of us, the worst way to deliver a speech is to speak without any preparation whatsoever. You may know a subject very well and think that your experience qualifies you for an on-the-spot performance. But you only fool yourself if you think you have all the necessary details and explanations in the back of your head. It is equally dangerous to believe that once you start talking, everything will fall into place smoothly. The "everything works out for the best" philosophy, unaided by a lot of hard work, does not operate in public speaking. Without preparation, you are likely to confuse important points or forget them entirely. Mark Twain's advice is apt here: "It takes three weeks to prepare a good impromptu speech."

2. Memorizing a speech. The exact opposite of the off-the-cuff approach, a memorized speech does have some advantages for certain individuals—tour bus drivers, guides, or salespeople—who must deliver the same speech verbatim many times over. But for the individual who has to deliver an original speech just once, a memorized speech contains pitfalls.

- The hours spent memorizing exact words and sentences would be better devoted to organizing your speech or gathering information for it.
- If you forget a single word or sentence, you may lose the rest of the speech.
- A memorized delivery can make you appear stiff and mechanical; rather than adjusting to an audience's reactions, you will be obligated to speak the exact words you wrote before you ever saw your audience.

3. Reading a speech. Reading a speech may be appropriate if you are presenting information on company policy or legal issues on which there can be no deviation from the printed word. Most speeches, however, will not require this rigid adherence to a text. Your speeches will be more acceptable, socially and professionally, when you interact with the audience. Reading a speech is also likely to bore your audience and suggest to them that you don't know they are there. In reading, you set up a barrier between yourself and the audience by not establishing eye contact for fear of losing your place.

4. Delivering a speech extemporaneously. An extemporaneous delivery is the most widely used method for a variety of occasions. Unlike a speaker using a memorized or written speech, you do not come before your audience with the entire speech in hand. *By no means, though, is an extemporaneous delivery an off-the-cuff performance.* It requires a great deal of preparation. But what you prepare is an outline of the major points of your speech, as discussed later in this chapter. You will rehearse using that outline, but the actual words you use in your speech will not necessarily be those you have rehearsed. Stand confidently in front of the audience with your outline reminding you of major points but not the precise language for expressing them. In this way you are free to establish contact with your audience.

The rest of this chapter discusses various effective ways of preparing and delivering an extemporaneous speech.

The Parts of a Speech

As you read this section, refer to Marilyn Ford's speech outline in Figure 18.2.

The Introduction

The most important part of a speech, your introduction should capture the audience's attention by answering these questions: (1) Who are you? (2) What are your qualifications? (3) What specific topic are you speaking about? and (4) How is the topic relevant to us?

Don't lose sight of the fact that, as you accomplish all four points above, your first and most immediate goal in your introduction is to establish rapport with your audience. You have to win their confidence and elicit their cooperation. Your audience is probably at their most attentive during the first few minutes of your talk. Accordingly, they will pay very close attention to everything about you and what you say. Seize the moment and build momentum.

An effective introduction is proportional to the length of your speech. A ten-minute speech requires no more than a sixty-second introduction; a twenty-minute speech needs no more than a two- or three-minute introduction.

Ways to Begin a Speech

You can begin by introducing yourself, emphasizing your professional qualifications and interests. (This self-introduction is unnecessary if someone else has introduced you or if you know everyone in the room.)

Never begin by

- apologizing for taking up the audience's time
- pointing out your limitations as a professional speaker
- criticizing the room, the lighting, the furniture, or the time and date of your speech
- finding fault with the audience for any past decisions or actions

Instead, be positive and thank the audience for the opportunity of addressing them. Below are some more specific guidelines to follow to deliver a successful speech.

Give Listeners a Road Map

Give listeners a "road map" at the beginning of your talk so that they will know where you are, when you are there, and what they have to look forward to or to recall. Indicate what your topic is and how you have organized what you have to say about it.

> Tonight I will discuss the three problems a food store manager faces when deciding to stock inventory: (1) the difficulty in predicting needs, (2) the inability of the production department to record current levels of stock accurately, and (3) the uncertainty of marketplace conditions.

The most informative speeches are the easiest to follow. Restrict your topic to ensure that you will be able to organize it carefully and sensibly—a tasty diet under 1,000 calories a day, for example, or a course in canoeing.

Capture Audience Attention

Moving from your announcement of the topic to your actual presentation requires skill at inducing an audience to listen. Use any of the following strategies to get your audience to "bite the hook":

Question:	Do you know how much actual meat there is in a hot dog?
Quotation:	Winston Churchill said, "We get things to make a living but we give things to have a life."
Interesting Statistic:	In 1998, 2 million heart attack victims will live to tell about it.
Anecdote:	Be sure it is relevant and in good taste; make your audience feel at ease and friendly toward you by establishing a bond with them.
Background Information:	Provide information about some local history, event, or tradition to show your knowledge of your audience's past.
Compliment:	Find some way to praise or recognize your audience (or someone prominent in the audience) for an accomplishment related to your talk.

The Body of the Speech

The body is the longest part of your speech, just as it is in a long report. It supplies the substance of your speech by (1) explaining a process, (2) describing a condition, (3) telling a story, (4) arguing a case, or (5) doing all of the above. See how the body of Marilyn Ford's speech outlined in Figure 18.2 is organized around the customer benefits of desktop videoconferencing.

To get the right perspective, recall your own experiences as a member of an audience. How often did you feel bored or angry because a speaker tried to overload you with details or could not stick to the point? The cartoon in Figure 18.3 points out the consequences of boring an audience.

Ways to Organize the Body of Your Speech

Here are a few helpful ways you can present and organize information in the body of your speech. In writing a report, you assist readers by designing your document to help them visually, supplying headings, underscorings, bullets, necessary white space, and headers and footers. In a speech, switch from these purely visual devices to aural ones, such as the following.

1. Give signals (directions) to show where you are going or where you have been. These signals will convince an audience that your speech does not ramble. Enumerate your points: *first, second, third.* Emphasize cause and effect relationships with *subsequently, therefore, furthermore.* When you tell a story, follow a chronological sequence and fill your speech with signposts: *before, following, next, then.* In the body of her speech, Marilyn Ford tells readers her company can offer GTP Industries "Three Communication Benefits" (IV).

2. Comment on your own material. Tell the audience if some point is especially significant, memorable, or relevant. "This next fact is the most important

FIGURE 18.2 A speech outline.

An Outline for a Speech Promoting Desktop Videoconferencing to a Company

Audience: Executives of GTP Industries
Purpose: To convince management to invest in videoconferencing technology
Speaker: Marilyn Ford, Southwest Telecommunications

Introduction

I. Southwest's Desktop Videoconferencing can increase the efficiency of GTP's communications by 50 to 100 percent — and dramatically cut costs. **[VISUAL AID: Show map of GTP's sites with airplanes connecting them.]**

- A. GTP spent over $350,000 for business travel last year (much more than necessary).
- B. Conducting staff meetings among branch offices poses several problems: extensive preparations, scheduling conflicts, travel delays.
- C. Desktop Videoconferencing solves these problems; employees at different sites interact as though all in the same room (a competitive advantage).
- D. Desktop Videoconferencing transforms existing computers into interactive, multimedia conference rooms.

Body

II. Southwest's technology for Desktop Videoconferencing is easy to use and cost-effective. **[VISUAL AID: Show employees connecting with distant sites by dialing the Microlink.]**

- A. Desktop Videoconferencing is as simple as a telephone call.
  1. Arrange a meeting time with colleagues at other sites.
  2. Use your computer to access Southwest's Microlink.
  3. Dial in to the conference (see and hear all participants access the Microlink).
  4. Talk directly with all participants by telephone.
- B. Reap great benefits with an inexpensive or existing computer system.
  1. Technology is based on a 486-class computer with 8 MB of memory; if new, cost ranges from $900 to $1,500.
  2. Existing computer system can be upgraded quickly for less than the cost of a new computer.
  3. Cost of additional equipment (software, camera, modem, video and sound card) is under $1,400.
- C. Integrated service data network (ISDN)—a dedicated phone line that transfers digital data—is widely available and costs only $900/year to rent.
- D. Desktop Videoconferencing can cut data-processing costs by 60 percent or more.

Continued

FIGURE 18.2 (Continued)

page 2

III. Desktop Videoconferencing is a strategic weapon in today's global competitive environment.

- A. Business today is information-driven, not product-driven.
  1. GTP can access the globe via the Internet.
  2. Desktop Videoconferencing provides flexible, low-cost networking.
- B. Telecommunications is transforming the traditional office.
  1. Desktop Videoconferencing consolidates your communications network into a single point of contact for all types of information—data, image, voice, and video.
  2. For less than the cost of a personal computer, GTP can create an interactive, multimedia conference room.

IV. Southwest Telecommunications offers three communication benefits.

- A. Desktop Videoconferencing enhances employees' collaboration and interaction.
  1. Communication improves when employees can see each other's facial expressions and body language.
  2. Southwest's "whiteboard" allows employees to work simultaneously on a document (they see each other's revisions while sharing ideas on the telephone). **[VISUAL AID: Show employees using the whiteboard at different sites to edit a document simultaneously.]**
- B. Desktop Videoconferencing will increase GTP's productivity.
  1. Employees transmit both audio and video information, sharing more data with more people—from 50 to 100 percent more effectively.
  2. Employees interact directly, streamlining collaborative projects by 35 to 45 percent.
  3. Employees work more effectively in their own offices.
- C. Desktop Videoconferencing makes better use of time and saves money.
  1. Schedule meetings quickly and avoid conflicts.
  2. Desktop Videoconferencing brings the right people together, no matter where they are.
     - a. Important discussions will no longer depend on one person's travel schedule.
     - b. Schedule emergency meetings quickly when problems arise.

FIGURE 18.2 (Continued)

page 3

Conclusion

V. Southwest Telecommunications can bring the communication benefits of Desktop Videoconferencing to GTP Industries. **[VISUAL AID: Show map of GTP's sites with Microlink web connecting them.]**
  A. Desktop Videoconferencing will save GTP both time and money.
  B. Desktop Videoconferencing will enhance the professional development of GTP's employees.
  C. Southwest Telecommunications will tailor Desktop Videoconferencing to GTP's needs, ensuring a competitive advantage in today's complex business world.

thing I'll say today." "The best determiner of pressure is the volume of liquid present in the chamber."

3. Repeat key ideas. Concisely restate your main idea or topic. You can repeat a sentence or a word to emphasize its importance and to help the audience remember it. But do this sparingly; repeating the same point over and over again bores an audience.

FIGURE 18.3 How to find out if you are a boring person.

4. **Provide internal summaries.** Spending a few seconds to recap what you have just covered will reassure your audience and you as well.

> We have already discussed the difficulties in establishing a menu repertory, or the list of items that the food service manager wants to appear on the menu. Now we will turn to ways of determining which items should appear on a menu and why.

The Conclusion

Conclusions, even for long speeches (twenty minutes), should never run more than sixty to ninety seconds. Plan your conclusion as carefully as your introduction. Stopping with a screeching halt is as bad as trailing off in a fading monotone. An effective conclusion leaves the audience feeling that you have come full circle and accomplished what you promised. See how Marilyn Ford ends her speech emphasizing how GTP Industries can ensure having a competitive edge, a point with which she began.

What to Put in Your Conclusion

A conclusion should contain something lively and memorable. Never introduce a new subject in your conclusion or simply repeat your introduction. A conclusion can contain the following:

- a fresh restatement of your three or four main points
- a call to action, just as in a sales letter—to buy, to note, to agree, to volunteer
- a final emphasis on a key statistic—for example, "The installation of the stainless steel heating tanks has, as we have seen, saved our firm 32 percent in utility costs, since we no longer have to run the heating system all day."

Note how Marilyn Ford ends her speech (Figure 18.2) with a concise summary of her main points and urges her listeners to invest company money in her product, Southwest Telecommunications Desktop Videoconferencing system.

Mean It When You Say "Finally"

When you tell your audience that you are concluding your talk, make sure that you mean it. An audience will resent a speaker who gives them a false ending. Saying, "In conclusion," and then talking for another ten minutes only frustrates your listeners and makes them less receptive to your message.

The Speech Outline

Using information from the previous discussion of the parts of a speech, you will need to construct a speech outline to represent the introduction, the body, and the conclusion. In an extemporaneous speech, remember, you do not write out your entire speech in manuscript form. Instead you will put together a speech outline similar to Marilyn Ford's in Figure 18.2.

Your speech outline has to be far more substantial than the rough notes found on the cards for informal briefings shown in Figure 18.1 . To construct an effective speech outline, you will have to do research, analyze your audience's needs, and collect suitable data (and visuals) for them. Your outline has to be detailed enough to give your readers the three Ds:

- direction you will follow
- development of your topic
- documentation of your topic

Your speech outline will not only assist your readers but it also has great psychological value for you. The outline gives you enough facts to handle your speaking engagement confidently, yet it is not so detailed that it places you in a straitjacket.

The speech outline illustrated in Figure 18.2 contains an appropriate amount of detail to represent the introduction, body, and conclusion. A roman numeral designates each major point; capital letters indicate appropriate supporting facts. Be careful about crowding too much into a speech outline. You do not need an outline as highly developed as the one below; an audience would get lost trying to follow five levels of subordination.

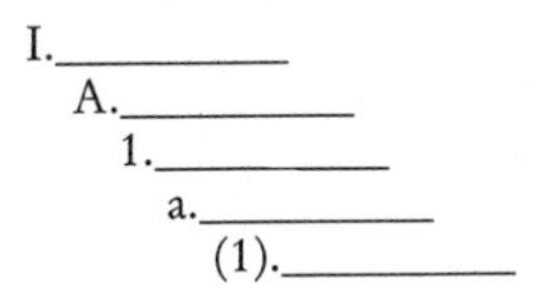

Nor will every main point will require four or five capital letters. But each point, whether indicated by a roman numeral or a capital letter, should be written as a complete sentence. Since your outline must be easy to read and follow, leave wide margins and double-space the entire outline. Mark, perhaps with color or in capital or boldfaced letters, where visuals appear in your speech. That way, you won't overlook or omit them.

Using Visuals

The more successful a speech, the greater the chances that the speaker has effectively incorporated visuals into it. Many listeners even judge a speech by the quality of the visuals the speaker presents. Visuals have numerous benefits for speakers and their audiences. Visuals can

- arouse an audience's interest
- add variety
- explain information quickly
- summarize a great deal of information
- reinforce and enhance the main points of your talk

Keep in mind that visuals in any presentation must be constructed even more carefully than for a written report. Unlike a reading audience, a listening audience cannot refer to a visual again and may not have time to study the visual in detail. Keep your visuals clear and simple.

Types of Presentation Visuals

You may use a variety of visuals during your talk: photographs, maps, chalkboards, models, pasteboards (large, 2' × 3' pieces of stiff white cardboard), diagrams, transparencies, slides, and computer presentations taking advantage of text, sound, animation, and even virtual reality.

TECH NOTE

You can create and display electronic visuals for your audience directly from your desktop software program. On your computer screen you can design an appropriate visual and then transfer it from your computer file to an especially equipped overhead projector. This way you will be spared the difficulty of having to make transparencies and, furthermore, you will not have to worry about adapting or enlarging existing graphics and photocopying them for your audience.

Make Sure Visuals Are Readable

The size and shape of your visuals are important. Make sure that they are large and easy to read, even from the back row or from the far corner of a long conference table. If they cannot be seen clearly from a distance and understood at once, they are not very useful. If possible, before using any visual, place it in the front of the room and sit in the last row of chairs to see whether your audience will be able to decipher it. You might decide to enlarge your visual or to show it to different sections of your audience by walking from one side of the room to another. All photocopy machines have enlargement features, and computer software programs also allow you to blow up a figure, table, or other visual.

Make Your Visuals Easy to Understand and Relevant

Each visual should be clear and simple, easy to understand the very first time an audience sees it. If you suspect that your audience may have trouble, even momentarily, knowing what the visual is or why and how it works in your presentation, redesign it or delete it from your talk. For example, using technical (or excessively complex) computer flow charts may detract from rather than add to your talk if an audience does not recognize what the flow charts show. And don't make the mistake of taking five minutes to explain the content of one visual; the visual itself should explain the content of your talk. Use that time more profitably to deliver your speech.

Make Sure Your Visuals Are Relevant

Also, be careful that each visual relates directly to your topic and that it does not interfere with your message. Will it lead an audience astray on a tangent? Will it contradict the message you are attempting to deliver? If so, delete or revise it.

Determining How Many Visuals to Use

Use visuals only when your audience needs them. Even a very long speech may need only two or three visuals. Their purpose is to clarify (or supplement), not to compete with, what you say. Nor are they a substitute for your speech. (Note Marilyn Ford's effective use of visuals in the speech outlined in Figure 18.2.) As a general rule, too many visuals will distract your audience.

Getting the Most from Your Visuals

The following practical suggestions will help you get the most from your visuals.

1. **Do not set up your visuals before you begin speaking.** The audience will be wondering how you are going to use them and so will not give you their full attention. When you are finished with a visual, put it away. Your audience will not be distracted by it or tempted to study it instead of listening to you.

2. **Firmly anchor any maps or illustrations.** Having a map roll up or a picture fall off an easel during a presentation is embarrassing.

3. **Never obstruct the audience's view by standing in front of your visuals.** Use a pointer or a laser pointer (a pen-sized tool that projects a bright red spot up to 150 feet) to direct audience attention to your visual.

4. **Avoid crowding too many images onto one computer screen that you may transfer to a projector or onto one pasteboard.** Use different screens or pasteboards instead.

5. **Do not put a lot of writing on a visual.** Elaborate labels or wordy descriptions defeat your reason for using the visual. Your audience will spend more time trying to decipher the writing than attempting to understand the visual itself. If any writing must appear on one of your visuals, enlarge it so your audience can read it quickly and easily.

6. **Be especially cautious with a slide projector.** Check beforehand to make sure that all your slides are in the order in which you are going to discuss them and that they are right-side-up. Most important, make sure the projector is in good working order. And test your tape recorder if you are using one as part of your talk.

Using Videos in Your Talk

Many business presentations rely on video players and large-screen televisions to sell a product, service, or idea. Animation, sound, color—all can engage an audience's interest. There is no doubt that a film (or video) can show things that could never be duplicated in a conference room or that could never be described in such vivid detail. If you intend to show a film or video to your audience, follow these guidelines.

1. **Prepare your audience for viewing the video.** Don't just walk into the room and turn on the VCR. Also, don't stop your talk and say, "Now let's view this." Help your audience by telling them why you are going to show them the video, what they should look for, and how long the video will last.
2. **Never substitute a video for a talk.** Don't show a twenty- or thirty-minute film or video and think it alone will sell your product or make all your points. Preview the video or film and show only those sections that are essential to your talk. The film or video should become a selected part of your presentation, not the presentation itself. For a twelve- to fifteen-minute talk, plan on devoting no more than two or three minutes of your time to showing a video.
3. **Always identify the source of the video** (where it comes from, even if from your own company), and indicate whether your audience can obtain a copy of the edited or full video, if they wish.

Rehearsing Your Speech

An efficient writer never submits a rough draft of a paper as final copy to an instructor or employer. Rather, the draft is revised, edited, and carefully checked before a final copy is printed and sent. Similarly, a careful speaker does not write a speech and march off to deliver it. Between the time you write a speech and deliver it, rehearse it several times. Going over your ideas aloud may help you to spot poor organization and insufficient or inaccurate content.

Effective Rehearsal Strategies

Rehearsing will help you become more familiar with your topic and overall message, building your confidence. It will also help you to acquire more natural speech rhythms—pitch, pauses, and pacing. Here are some strategies to use as you rehearse your speech.

- Speak in front of a full-length mirror for at least one rehearsal to see how an audience might view you.
- Talk into a tape recorder to determine whether you sound friendly or frantic, poised or pressured. You can also catch and correct yourself if you are speaking too quickly or too slowly.
- Time yourself so you will not exceed your allotted time or fall far short of your audience's expectations.
- Practice with the visuals or equipment you intend to use in your speech for valuable hands-on experience.

- Videotape (on a camcorder) your final rehearsal and show it to a colleague or instructor for feedback.

Delivering a Speech

A poor delivery can ruin a good talk. You will be evaluated by your style of presentation just as you are in your written work. When you speak before an audience, you will be evaluated on the image you project—

- how you look
- how you talk
- how you move (your body language)

Do you mumble into your notes, never looking at the audience? Do you clutch the lectern as if to keep it in place? Do you shift nervously from one foot to another? All these actions betray your nervousness and detract from your presentation.

The following suggestions on how to present a speech will help you to be a well-prepared speaker.

Before You Speak

You name is called, and within a minute or two you will have to begin addressing the audience. You will be nervous. Accept the fact and even allow a few seconds for a "panic time." But then put your nervous energy to work for you. Chances are, your audience will have no idea how anxious you are; they cannot see the butterflies in your stomach. See your audience as friends, not enemies.

If you have to go to a lectern, walk slowly so you do not trip. Once you are before the lectern, or conference table, remove any pitcher or glasses of water if you are worried about spilling them. Always have a wristwatch with you. Before speaking, lay it on the table or lectern so that you can occasionally glance down to see how much time you have left. This unobtrusive act is far preferable to reminding the audience that you are running long by noticeably raising your arm to look at your watch.

During the Speech

Begin your speech slowly. Give listeners a chance to sit back in their chairs and establish a mental connection with your topic. Rushing into your speech may be startling, causing you to lose the audience from the start. To speak effectively, pay attention to the following points.

1. Establish eye contact with your listeners. Look at your audience to establish a relationship with them. Never bury your head in notes; you will signal your lack of interest in the audience, or your fear of public speaking. Some timid speakers think that if they look only at some fixed place or object in the back of the room, the audience will still regard this as eye contact. But this kind of cover-up does not work.

Another tactic poor (or frightened) speakers use is to look at only one member of the audience or to focus, with frequent sidewise glances, on the individual next to them on the stage, perhaps the person who has introduced the speakers.

Establish a pattern of glancing at your notes and then looking up at various individuals in the audience. If the group you are addressing is small (five to ten people), look at each person in the course of your talk. When you speak to a large audience (fifty or more individuals), visually divide this group into four or five sections, and look at each section a number of times as you speak. Or try to focus on the heads of your listeners in the last row.

2. Use a friendly, confident tone. Speak in a natural, conversational voice, but avoid such "verbal tics" as "you know" or "I mean" repeated several times each minute. Such nervous habits will make your audience nervous too and make your speech less effective.

3. Vary the rate of your delivery. Talking in a monotone, never raising or lowering your voice, will lull your audience to sleep, or at least inattention! Use your rate to help you emphasize key points and make transitions in your speech. Talk slowly enough for your audience to understand you, yet quickly enough so that you don't sound as if you are belaboring or emphasizing each word.

4. Adjust your volume appropriately. Talk loudly enough for everyone to hear, but be careful if you are using a microphone. Your voice will be amplified, so if you speak loudly, you will boom rather than project. Watch out for the other extreme—speaking so softly that only the first two rows can hear you.

TECH NOTE

As you rehearse your speech, look up in a dictionary the pronunciation of any words you are unsure about. Pay special attention to the pronunciation of individuals' names and company and city names. Consult a dictionary (*Webster's Collegiate Dictionary*, 10th Edition, or *The American Heritage Dictionary of the English Language*, 3rd Edition). Whenever you are in doubt, ask ahead of time. It's far less embarrassing than having someone in the audience stop you in the middle of your speech to correct you or tell you *after* your speech that you pronounced the CEO's name wrong ten times!

5. Watch your posture. If you stand motionless, looking as if rigor mortis has set in, your speech will be judged cold and lifeless, no matter how lively your words are. Don't stand in one spot as if your legs were set in concrete. Be natural; move, and let your body react to what you are saying. Smile, nod your head, move your arms, point at an object, stand back a little from the lectern. This is not to say you should be a moving target. Never sit on a desk or lean on a lectern in front of your audience. Listeners will be waiting to see if you fall off your newly discovered perch.

6. Use appropriate body language. Be natural and consistent. Do not startle an audience by suddenly pounding on the lectern for emphasis. Any quick, unexpected movement detracts from what you are saying. Let your material suggest appropriate movements. If you are itemizing two or three points, hold up the appropriate number of fingers to indicate which point you are discussing. Use your hands and arms to indicate direction, size, or relationships.

Avoid any gesture that will distract your audience. For example, don't fold your arms as you talk, a gesture that signals you are unreceptive (closed) to your audience's reactions. Also, avoid the nervous habits that can divert the audience's attention: scratching your head, rubbing your nose, twirling your hair, pushing up your glasses, fumbling with your notes, or tapping your foot.

7. Dress professionally. Do not wear clothes or jewelry that call attention to themselves. Be conservative and dress formally. Women should wear a business-like dress or suit. Men should wear a dark business suit, white shirt, and a tasteful tie.

When You Have Finished Your Speech

Don't just sit down, walk back to your place on the platform or in the audience, or, worse yet, march out of the room. Thank your listeners for their attention, and stay at the lectern for audience applause or questions. If appropriate, give the person who introduced you a chance to thank you while you are still in front of the group.

If a question-and-answer session is to follow your speech, give your audience a time limit for questions. For example, you might say, "I'll be happy to answer your questions now before we break for lunch in ten minutes," or "I have set aside the next ten minutes for questions." By setting limits, you reduce the chances of engaging in a lengthy debate with members of the audience, and you also can politely leave after your time elapses.

Evaluating Your Speech

A large portion of Chapter 18 has given you information on how to construct and deliver a formal speech. As a way of reviewing that advice, study Figure 18.4, an evaluation form similar to those used by instructors in speech classes. Note that the form gives equal emphasis to the speaker's performance or delivery and to the organization and content of the speech.

FIGURE 18.4 A speech evaluation form.

Name of speaker ____________ Date of speech ____________

Title of speech ____________ Length of speech ____________

PART I: THE SPEAKER (circle the appropriate number)

1. Appearance:	1 sloppy	2	3	4	5 well groomed
2. Eye contact:	1 poor	2	3	4	5 effective
3. Voice:	1 monotonous	2	3	4	5 varied
4. Posture:	1 poor	2	3	4	5 natural
5. Gestures:	1 disturbing	2	3	4	5 appropriate
6. Self-confidence:	1 nervous	2	3	4	5 poised

PART II: THE SPEECH (circle the appropriate number: 1 = poor; 5 = superior):

A. Overall performance

1. Speaker's knowledge of the subject—carefully researched; factual errors; missing details:

   1 2 3 4 5

2. Relevance of the topic for audience—suitable for this group:

   1 2 3 4 5

3. The speaker's language—too technical; filled with cliches or slang expressions; or crisp and descriptive:

   1 2 3 4 5

✓ Revision Checklist

- ❑ Anticipated audience's background, interest, or even potential resistance to message.
- ❑ Prepared introduction to provide "road map" of speech and to arouse audience interest.
- ❑ Started with interesting and relevant statistics, a question, an anecdote, or similar "hook" to capture audience attention.
- ❑ Limited body of speech to main points.
- ❑ Arranged main points logically, and made connections among them.
- ❑ Used supporting examples and illustrations appropriate to audience.
- ❑ Made sure conclusion contains summary of main points of my speech and/or specific call for action.
- ❑ Prepared speech outline and identified and corrected any weak or redundant areas.
- ❑ Designed visuals that are clear and easy to read.
- ❑ Double-spaced speech outline with wide margins to make it easy to read.
- ❑ Rehearsed speech thoroughly so I am familiar with its organization and am comfortable using visuals.
- ❑ Monitored volume, tone, and rate to vary delivery and emphasize major points.
- ❑ Double-checked gestures to make them relevant and nonintrusive.
- ❑ Timed speech, complete with visuals, so as to run close to allotted time.

Exercises

1. Prepare a three- to five-minute talk explaining how a piece of equipment that you use on your job works. If the equipment is small enough, bring it with you to class. If it is too large, prepare an appropriate visual or two for use with your talk.

2. Find an article on a technical subject from a professional journal in your field. Prepare a five- to seven-minute briefing on the topic of this article but adapt your remarks for a general audience of consumers.

3. You have just been asked to talk about the students at your school. Narrow this topic and submit a speech outline to your instructor, showing how you have limited the topic and gathered and organized evidence. Use two or three appropriate visuals (tables, photographs, charts). Follow the format of the outline in Figure 18.2.

4. Prepare a ten-minute talk on a controversial topic that you would present before a civic group—the local PTA, a local chapter of an organization, a post of the Veterans of Foreign Wars, a synagogue or church club. Submit a speech outline similar to that in Figure 18.2 to your instructor, together with a one-page statement of your specific call to action and its relevance for your audience.
5. Using the information contained in the research paper on telecommuting in Chapter 10 (pp. 398–413) or the long report on e-cash in Chapter 17 (pp. 631–652), prepare a short speech (five to seven minutes) for your class.
6. Using the evaluation form in Figure 18.4, evaluate a speaker—a speech class student, a local politician, or a co-worker delivering a report at work. Specify the time, place, and occasion of the speech.
7. Deliver a formal speech (fifteen to eighteen minutes) on the various uses and advantages of the Internet for individuals in your chosen career field. Use at least three visuals with your talk. Submit a speech outline to your instructor.

A Writer's Brief Guide to Paragraphs, Sentences, and Words

To write successfully, you must know how to create effective paragraphs, write clear sentences, and use words correctly. This guide succinctly explains some of the basic elements of clear and accurate writing.

Paragraphs

Writing a Well-Developed Paragraph

A paragraph is the basic building block for any piece of writing. It is (1) a group of related sentences (2) arranged in a logical order (3) supplying readers with detailed, appropriate information (4) on a single important topic. Each of these points is discussed below.

A paragraph expresses one central idea, with each sentence contributing to the overall meaning of that idea. The paragraph does this by means of a *topic sentence,* which states the central idea, and *supporting information,* which explains the topic sentence.

Supply a Topic Sentence

The topic sentence is the most important sentence in your paragraph. Carefully worded and restricted, it helps you to generate and control your information. An effective topic sentence also helps readers grasp your main idea quickly. As you draft your paragraphs, pay close attention to the following three guidelines:

1. Make sure you provide a topic sentence. In their rush to supply readers with facts, some writers forget or neglect to include a topic sentence. The following paragraph, with no topic sentences, shows how fragmented such writing can be.

No topic sentence

Sensors found on each machine detect wind speed and direction and other important details such as ice loading and potential metal fatigue. The information is fed into a small computer (microprocessor) in the nacelle (or engine housing). The microprocessor automatically keeps the blades turned into the wind, starts and stops the machine, and changes the pitch of the tips of the blades to increase power under varying wind conditions. Should any part of the wind turbine suffer damage or malfunction, the microprocessor will immediately shut the machine down.

Only when a suitable topic sentence is added—"The MOD-2 wind turbine is designed to be operated completely by computer"—can readers understand what the technical details have in common.

2. Put your topic sentence first. Place your topic sentence at the beginning—not the middle or end—of your paragraph because the first sentence occupies an emphatic position. Burying the key idea in the middle or near the end of the paragraph makes it harder for readers to comprehend your purpose or act on your information.

3. Be sure your topic sentence is focused. If restricted, a topic sentence discusses only one central idea. A broad or unrestricted topic sentence leads to a shaky, incomplete paragraph for two reasons.

- The paragraph will not contain enough information to support the topic sentence.
- A broad topic sentence will not summarize or forecast specific information in that paragraph.

The following example of a carefully constructed paragraph, contains a clear topic sentence in an appropriate position (highlighted in color) and adequate supporting details.

> Fat is an important part of everyone's diet. It is nutritionally present in the basic food groups we eat—meat and poultry, dairy products, and oils—to aid growth or development. The fats and fatty acids present in these foods ensure proper metabolism, thus helping to turn what we eat into the energy we need. These same fats and fatty acids also act as carriers for important vitamins like A, D, E, and K. Another important role of fat is that it keeps us from feeling hungry by delaying digestion. Fat also enhances the flavor of the food we eat, making it more enjoyable.

Three Characteristics of an Effective Paragraph

Effective paragraphs have **unity, coherence,** and **completeness.**

Unity

A unified paragraph sticks to one topic without wandering. Every sentence, every detail, **supports, explains,** or **proves** the central idea. A unified paragraph includes only relevant information and excludes unnecessary or irrelevant comments.

Coherence

In a coherent paragraph all sentences flow smoothly and logically to and from each other like the links of a chain. Use the following four techniques to achieve coherence:

1. Use transitional words and phrases. Some useful connective words, along with the relationships they express, are listed in Table A.1.

Paragraph with connective words

Advertising a product on the radio has many advantages over using television. *For one thing,* radio rates are much cheaper. *For example,* a one-time 60-second spot on television can cost $750. *For that money,* advertisers can purchase nine 30-second spots on the radio. *Equally attractive* are the low production costs for radio advertising. *In contrast,* television

TABLE A.1 Transitional, or Connective, Words and Phrases

Addition	again	furthermore	next
	also	first, second, third	too
	and	in addition	what's more
	as well as	many	
	besides	moreover	
Cause/Effect	and so	consequently	on account of
	accordingly	due to	since
	as a result	hence	therefore
	because of	if	thus
Comparison/ Contrast	but	in contrast	on the other hand
	conversely	in the same way	similarly
	equally	likewise	still
	however	on the contrary	yet
Conclusion	all in all	in conclusion	on the whole
	at last	to conclude	to put into perspective
	finally	in short	to summarize
	in brief	in summary	
Condition	although	granted that	to be sure
	even though	if	unless
		of course	
		provided that	
Emphasis	above all	as I said	of course
	after all	for emphasis	obviously
	again	indeed	surely
	as a matter of fact	in fact	to repeat
		in other words	unquestionably
Illustration	for example	in other words	that is
	for instance	in particular	to illustrate
	in effect	specifically	
Place	across from	below	over
	adjacent to	beyond	there
	alongside of	here	under
	at this point	in front of	where
	behind	next to	wherever
Time	afterwards	earlier	presently
	at length	later	soon
	at the same time	meanwhile	then
	at times	next	until
	beforehand	now	when
	currently	once	while
	during		

advertising often includes extra costs for models and voice-overs. *Another* advantage radio offers advertisers is immediate scheduling. *Often* the ad appears during the same week a contract is signed. *On the other hand,* television stations are *frequently* booked up months in advance, so it may be a long time *before* an ad appears. *Furthermore,* radio gives advertisers a greater opportunity to reach potential buyers. *After all,* radio follows listeners everywhere—in their homes, at work, and in their cars. *Although* television is very popular, its viewers cannot do that.

2. Use pronouns and demonstrative adjectives. Such words as *he, she, him, her, they,* and so on contribute to paragraph coherence and increase the flow of sentences.

Paragraph with pronouns

Traffic studies are an important tool for store owners looking for a new location. These studies are relatively inexpensive and highly accurate. They can tell owners how much traffic passes by a particular location at a particular time and why. Moreover, they can help owners to determine what particular characteristics these individuals have in common. Because of their helpfulness, these studies can save owners time and money and possibly prevent financial ruin.

3. Use parallel (coordinated) grammatical structures. Parallelism means using the same *kind of* word, phrase, clause, or sentence to express related concepts.

Orientation sessions accomplish four useful goals for trainees. First, they introduce trainees to key personnel in accounting, data processing, maintenance, and security. Second, they give trainees experience logging into the data base system, selecting and appropriate menu, editing core documents, and getting off the system. Third, they explain to trainees the company policies affecting the way supplies are ordered, used, and stored. Fourth, they help trainees understand their responsibilities in such sensitive areas as computer security and use.

Parallelism is at work on a number of levels in this paragraph, among them,

- the four sentences about the four goals start in the same way grammatically ("... they introduce/give/explain/help trainees ...") to help readers categorize the information.
- Within individual sentences, the repetition of *present participles* (logg*ing*, select*ing*, edit*ing*, gett*ing*) and of *past participles* (order*ed*, us*ed*, stor*ed*) helps the writer to coordinate information.
- Transitional words—*first, second, third, fourth*—provide a clear-cut sequence.

Completeness

A complete paragraph provides readers with sufficient information to **clarify, analyze, support, defend,** or **prove** the central idea expressed in the topic sentence. The reader feels satisfied that the writer has given sufficient details.

Skimpy paragraph

Farmers can turn their crops and farm wastes into useful, cost-effective fuels. Much grown on the farm can be converted to energy. This energy can have many uses and save farmers a lot of money in operating expenses.

Fully developed paragraph

Farm crops and wastes can be turned into fuels to save farmers on their operating costs. Alcohol can be distilled from grain, sugar beets, potatoes—even from blighted crops. Converted to gasohol (90 percent gasoline, 10 percent alcohol), this fuel can run such farm equipment as irrigation pumps, feed grinders, and tractors. Similarly, through a biomass digestion system, farmers can produce methane from animal or crop wastes as a natural gas for heating and cooking. Finally, cellulose pellets, derived from plant materials, become solid fuel that can save farmers money in heating barns.

Sentences

Constructing and Punctuating Sentences

The way you construct and punctuate your sentences can determine whether you succeed or fail in the world of work. You sentences reveal a lot about you. They tell readers how clearly or how poorly you can convey a message. And any message is only as solid and thoughtful as the sentences that make it up.

What Makes a Sentence

A **sentence** is a complete thought, expressed by a subject and a verb that can make sense standing alone.

Subject Verb

Web sites sell products.

The first step toward success in writing sentences is learning to recognize the difference between phrases and clauses.

The Difference Between Phrases and Clauses

A **phrase** is a group of words that does not contain a subject and a verb; phrases cannot make sense standing alone. Phrases cannot be sentences.

in the park	**No subject:** Who is in the park? **No verb:** What was done in the park?
for every patient in intensive care	**No subject:** Who did something for every patient? **No verb:** What was done for the patients?

A **clause** does contain a subject and a verb, but *not every clause is a sentence.* Only **independent** (or **main**) **clauses** can stand alone as sentences. Here is an example of an independent clause that is a complete sentence.

subject verb object

The president closed the college.

A **dependent** (or **subordinate**) **clause** also contains a subject and a verb, but does not make complete sense and cannot stand alone. Why? A dependent clause contains a subordinating conjunction—*after, although, as, because, before, even though, if, since, unless, when, where, whereas, while*—at the beginning of the clause. Such conjunctions subordinate the clause in which they appear and make the clause dependent for meaning and completion on an independent clause.

After
Before
Because
Even though
Unless } the president closed the college

"After the president closed the college" is not a complete thought. This dependent clause leaves us in suspense. It needs to be completed with an independent clause telling us what happened "after."

dependent clause	independent clause		
	subject	verb	phrase
After the president closed the college,	we	played	in the snow.

Avoiding Sentence Fragments

An incomplete sentence is called a **fragment.** Fragments can be phrases or dependent clauses. They either lack a verb or a subject or have broken away from an independent clause. A fragment is isolated: it needs an overhaul to supply missing parts to turn it into an independent clause or to glue it back to an independent clause to have it make sense.

To avoid writing fragments, follow these rules. Note that incorrect examples are preceded by a minus sign, corrected revisions by a plus sign.

1. Do not use a subordinate clause as a sentence. Even though it contains a subject and a verb, a subordinate clause standing alone is still a fragment. To avoid this kind of sentence fragment, simply join the two clauses (the independent clause and the dependent clause containing a subordinating conjunction) with a comma—*not* a period or semicolon.

– Unless we agreed to the plan. (What would happen?)
– Unless we agreed to the plan; the project manager would discontinue the operation. (A semicolon cannot set off the subordinate clause.)
+ Unless we agreed to the plan, the project manager would discontinue the operation.

– Because safety precautions were taken. (What happened?)
+ Because safety precautions were taken, ten construction workers escaped injury.

Sometimes subordinate clauses appear at the end of a sentence. They may be introduced by a subordinate conjunction, an adverb, or a relative pronoun (*that, which, who*). Do not separate these clauses from the preceding independent clause with a period, thus turning them into fragments.

– An all-volunteer fire department posed some problems. Especially for residents in the western part of town.
+ An all-volunteer fire department posed some problems, especially for residents in the western part of town. (The word *especially* qualifies posed, referred to in the independent clause.)

2. Every sentence must have a subject telling the reader who does the action.

– Being extra careful not to spill the water. (Who?)
+ The aide was being extra careful not to spill the water.

3. Every sentence must have a complete verb. Watch especially for verbs ending in *-ing*. They need another verb (some form of *to be*) to make them complete.

– The machine running in the computer department. (Did what?)

You can change this fragment into a sentence by supplying the correct form of the verb.

+ The machine *is running* in the computer department.
+ The machine *runs* in the computer department.

Or you can revise the entire sentence, adding a new thought.

+ The machine running in the computer department processes all new accounts.

4. Do not detach prepositional phrases (beginning with *at, by, for, from, in, to, with,* and so forth) **from independent clauses.** Such phrases are not complete thoughts and cannot stand alone. Correct the error by leaving the phrases attached to the sentence to which they belong.

– By three o'clock the next day. (What was to happen?)
+ The supervisor wanted our reports by three o'clock the next day.

Avoiding Comma Splices

Fragments occur when you use only bits and pieces of complete sentences. Another common error that some writers commit involves just the reverse kind of action. They weakly and wrongly join two complete sentences (independent clauses) with a comma as if those two sentences were really only one sentence. This error is called a **comma splice.** Here is an example.

– Gasoline prices have risen by 10 percent in the last month, we will drive the car less often.

Two independent clauses (complete sentences) exist:

+ Gasoline prices have risen by 10 percent in the last month.
+ We will drive the car less often.

A comma alone lacks the power to separate independent clauses. As the preceding example shows, many pronouns—*I, he, she, it, we, they*—are used as the subjects of independent clauses. A comma splice will result if you place a comma instead of a semicolon between two independent clauses where the second clause opens with a pronoun.

– Rosa approved the plan, she liked its cost-effective approach.
+ Rosa approved the plan; she liked its cost-effective approach.

However, relative pronouns (*who, whom, which, that*) are preceded by a comma, not a period or a semicolon, when they introduce subordinate clauses.

– She approved the plan. Which had the cost-effective approach.
+ She approved the plan, which had the cost-effective approach.

Four Ways to Correct Comma Splices

1. **Remove the comma separating two independent clauses and replace it with a period.** Then capitalize the first letter of the first word of the newly reinstated sentence.

+ Gasoline prices have risen by 10 percent in the last month. We will drive the car less often.

2. **Insert a coordinating conjunction (*and, but, or, nor, for, yet*) after the comma.** Together, the conjunction and the comma properly separate the two independent clauses.

+ Gasoline prices have risen by 10 percent in the last month, and we will drive the car less often.

3. **Rewrite the sentence** (if it makes sense to do so). Turn the first independent clause into a dependent clause by adding a subordinate conjunction; then insert a comma and add the second independent clause.

+ Because gasoline prices have risen by 10 percent in the last month, we will drive the car less often.

4. **Delete the comma and insert a semicolon.**

+ Gasoline prices have risen by 10 percent in the last month; we will drive the car less often.

(Of the four ways to correct the comma splice, sentences 3 and 4 are equally suitable, but sentence 3 reads more smoothly and so is the better choice.)

The semicolon is an effective and forceful punctuation mark when the two independent clause are closely related, that is, when they announce contrasting or parallel views, as the two following examples reveal.

+ The union favored the new legislation; the company opposed it. (contrasting views)
+ Night classes help the college and the community; more students can take more credit hours. (parallel views)

How Not *to Correct Comma Splices*

Some writers mistakenly try to correct comma splices by inserting a conjunctive adverb (*also, consequently, furthermore, however, moreover, nevertheless, then, therefore*) after the comma.

– Gasoline prices have risen by 10 percent in the last month, consequently we will drive the car less often.

Because the conjunctive adverb (*consequently*) is not as powerful as the coordinating conjunction (*and, but, for*), the error is not eliminated. If you use a conjunctive adverb—*consequently, however, nevertheless*—you still must insert a semicolon or a period before it, as the following examples show.

+ Gasoline prices have risen by 10 percent in the last month; consequently, we will drive the car less often.

+ Gasoline prices have risen by 10 percent in the last month. Consequently, we will drive the car less often.

Avoiding Run-on Sentences

A **run-on sentence** is the opposite of a sentence fragment. The fragment gives the reader too little information, the run-on too much. A run-on sentence forces readers to digest two or more grammatically complete sentences without the proper punctuation to separate them.

Run-on The Internet is unquestionably a major source of information and students and other researchers are right to call it a virtual library this library is not like the collections of books and magazines that are carefully shelved always waiting for students to check and recheck them too often an electronic document on the Internet disappears or changes considerably and without a back-up file or a hard copy the researcher has no document to quote from and no exact citation to prove that he or she consulted an authentic source.

Revised The Internet is unquestionably a major source of information. Students and other researchers are right to call it a virtual library, although this library is not like the collections of books and magazines that are carefully shelved, always waiting for students to check and recheck them out. But too often an electronic document on the Internet disappears or changes considerably. Without a back-up file or a hard copy the researcher has no document to quote from and no exact citation to prove that he or she consulted an authentic source.

As the revision above shows, you can repair a run-on by (1) dividing it into separate, correctly punctuated sentences, and (2) by adding coordinating conjunctions (*and, but, yet, so, or, nor*) between clauses.

Making Subjects and Verbs Agree in Your Sentences

A subject and verb must agree in number. A singular subject takes a singular verb whereas a plural subject requires a plural verb.

Singular Subject	Plural Subjects
the engineer calculates	engineers calculate
a memo requests	memos request
a policy changes	policies change

You can avoid subject-verb agreement errors by following several simple rules.

1. **Disregard any words that come between the subject and its verb.**

   Faulty: The customer who ordered three parts want them shipped this afternoon.
   Correct: The customer who ordered three parts wants them shipped this afternoon.

2. **A compound subject** (two parts connected by *and*) **takes a plural verb.**

   Faulty: The engineering department and the safety committee prefers to develop new guidelines.
   Correct: The engineering department and the safety committee prefer to develop new guidelines.

3. When your compound subject contains *neither . . . nor, either . . . or,* the verb agrees with the subject closest to it.

Faulty: Either the residents or the manager are going to file the complaint.
Correct: Either the residents or the manager is going to file the complaint.
Correct: Either the manager or the residents are going to file the complaint.

4. Use a singular verb after collective nouns (like *committee, crew, department, group, organization, staff, team*) **when the group functions as a single unit.**

Correct: The crew was available to repair the machine.
Correct: The committee asks that all recommendations be submitted by Friday.

BUT NOTE

Correct: The staff were unable to agree on the best model.
(The staff acted as individuals, not a unit, so a plural verb is required.)

5. Use a singular verb with indefinite pronouns (such as *anyone, anybody, each, everyone, everything, no one, somebody, something*).

Each of the programmers has completed the seminar.
Somebody usually volunteers for that duty.

Similarly, when *all, most, more,* or *part* is the subject, it requires a singular verb.

Most of the money is allocated.
Part of the equipment was salvageable.

6. Words like *scissors* and *pants* are plural when they are the true subject.

Correct: The trousers were on sale.
Faulty: A pair of trousers was available in his size. (*Pair* is the subject.)

7. Some foreign plurals (*curricula, data, media, phenomena, strata, syllabi*) **always take a plural verb.**

The data conclusively prove my point.
The media are usually the first to point out a politician's weak points.

8. Use a singular verb with fractions.

Three-fourths of her research proposal was finished.

Writing Sentences That Say What You Mean

Your sentences should say exactly what you mean, without doubletalk, misplaced humor, or nonsense. Sentences are composed of words and word groups that influence each other.

Writing Logical Sentences

Sentences should not contradict themselves or make outlandish claims. The following examples contain errors in logic; note how easily the suggested revision handles the problem.

Illogical: Steel roll-away shutters make it possible for the sun to be shaded in the summer and to have it shine in the winter. (The sun is far too large to shade; the writer meant that a room or a house, much smaller than the sun, could be shaded with the shutters.)

Revision: Steel roll-away shutters make it possible for owners to shade their living rooms in the summer and to admit sunshine during the winter.

Sentences Using Contextually Appropriate Words

Sentences should use the combination of words most appropriate for the subject.

Inappropriate: The members of the Nuclear Regulatory Commission saw fear radiated on the faces of the residents. (The word *radiated* is obviously ill advised; use a neutral term.)

Revision: The members of the Nuclear Regulatory Commission saw fear reflected on the faces of the residents.

Writing Sentences with Well-Placed Modifiers

A **modifier** is a word, phrase, or clause that describes, limits, or qualifies the meaning of another word or word group. A modifier can consist of one word (a *green* car), a prepositional phrase (the man *in the telephone booth*), a relative clause (the woman *who won the marathon*), or an *-ing* or *-ed* phrase (*walking three miles a day,* the student was in good shape; *seated in the first row,* we saw everything on stage).

A **dangling modifier** is one that cannot logically modify any word in the sentence.

– When answering the question, his notebook fell off the table.

One way to correct the error is to insert the right subject after the *-ing* phrase.

+ When answering the question, he knocked his notebook off the table.

You can also turn the phrase into a subordinate clause.

+ When he answered the question, his notebook fell off the table.
+ His notebook fell off the table as he answered the question.

A **misplaced modifier** illogically modifies the wrong word or words in the sentence. The result is often comical.

– Hiding in the corner, growling and snarling, our guide spotted the frightened cub. (Is our guide growling and snarling in the corner?)
– All travel requests must be submitted by employees in green ink. (Are the employees covered in green ink?)

The problem with both of the examples above is word order. The modifiers are misplaced because they are attached to the wrong words in the sentence. Correct the error by moving the modifier where it belongs.

+ Hiding in the corner, growling and snarling, the frightened cub was spotted by our guide.
+ All travel requests by employees must be submitted in green ink.

Misplacing a relative clause (introduced by such relative pronouns as *who, whom, that, which*) can also lead to problems with modification.

- – The salesperson recorded the merchandise for the customer that the store had discounted. (The merchandise was discounted, not the customer.)
- – The salesperson recorded the merchandise that the store had discounted for the customer. (The salesperson recorded for the customer; the store did not discount for the customer.)
- + The salesperson recorded for the customer the merchandise that the store had discounted.

Always place the relative clause immediately after the word it modifies.

Correct Use of Pronoun References in Sentences

Sentences will be vague if they contain a faulty use of pronouns. When you use a pronoun whose **antecedent** (the person, place, or object the pronoun refers to) is unclear, you risk confusing your reader.

Unclear: After the plants are clean, we separate the stems from the roots and place them in the sun to dry. (Is it the stems or the roots that lie in the sun?)

Revision: After the plants are clean, we separate the stems from the roots and place the stems in the sun to dry.

Unclear: The park ranger was pleased to see the workers planting new trees and installing new benches. This will attract more tourists. (The trees or the benches or both?)

Revision: The park ranger was pleased to see the workers planting new trees and installing new benches, for the new trees will attract more birds.

Words

Spelling Words Correctly

Your written work will be judged in part on how well you spell. A misspelled word may seem like a small matter, but on an employment application, memo, incident report, or letter, it stands out to your discredit. You will look careless, or even worse, uneducated to a client or supervisor. Readers will inevitably question your other skills if your spelling is incorrect.

The Benefits and Pitfalls of Spell-checkers

Spell-checkers can be very handy for flagging potential problem words. But beware! Spell-checkers only recognize words that have been listed in them. A proper name or infrequently used word may be flagged as an error even though the word is spelled correctly. Moreover, it will not differentiate between such homonyms as *too* and *two* or *there* and *their*. A spell-checker identifies only misspelled words, not misused words. In short, do not rely exclusively on spell-checkers to solve all of your spelling and word-choice problems.

Using Apostrophes Correctly

Apostrophes cause some writers special problems. Basically, apostrophes are used for four reasons: (1) contractions, (2) possessives, (3) plurals, and (4) abbreviations. The guidelines below will help you to sort out these uses:

1. To form a **contraction,** the apostrophe takes the place of the missing letter or letters: *I've* = *I have; doesn't* = *does not; he's* = *he is; it's* = *it is.* (*Its* is a possessive pronoun (the dog and its bone), not a contraction. There is no such form as *its'*.)

2. To form a **possessive,** follow these rules.

a. If a singular or plural noun does not end in an *-s,* add *'s* to show possession.

Mary's locker
children's books
the staff's dedication
the woman's jacket
the women's jackets
the company's policy

b. If a singular noun ends in *-s,* add *'s* to show possession.

The class's project
the boss's schedule

c. If a plural noun ends in *-s,* add just the ' to indicate possession.

employee's benefits
lawyers' fees
computers' speed

d. If a proper name ends in *-s,* use just the ' to form the possessive.

Jones' account
the Williams' house
Keats' poetry
James' contract

e. If it is a compound noun, add an ' or an *'s* to the end of the word.

brother-in-law's business
Ms. Allison Jones-Wyatt's order

f. If you wish to indicate shared possession, add just *'s* to the last name.

Warner and Kline's Computer Shop
Sue and Anne's major

If you want to indicate separate possession, add an *'s* to each name.

John's and Mary's transcripts
Shakespeare's and Byron's poetry

3. To form the plural of numbers and capital letters used as nouns, including abbreviations without periods, just add *s.* To avoid misreading some capital letters, however, you may need to add the apostrophe.

during the 1980s
his SATs
the 3 Rs
all perfect 10s
several local YMCAs
straight A's

4. For abbreviations with periods, however, and for lowercase letters used as nouns, form the plural by adding *'s.*

his *p*'s and *q*'s
Ph.D.'s

Matching the Right Word with the Right Meaning

The words in the following list frequently are mistaken for one another. Some are true homonyms; others are just similar in spelling, pronunciation, or usage.

accept (v) to receive, to acknowledge: *We accept your proposal.*
except (prep) excluding, but: *Everyone attended the meeting except Neelou.*

advice (n) a recommendation: *I should have taken Leroy's advice.*
advise (v) to counsel: *Our lawyers advised us not to sign the contract.*

affect (v) to change, to influence: *Does the detour on Route 22 affect your travel plans?*
effect (n) a result: *What was the effect of the new procedure?*
effect (v) to bring about: *We will try to effect a change in company policy.*

all ready (adj) two-word phrase *all + ready;* to be finished; to be prepared: *We are all ready for the inspector's visit.*
already (adv) previously, before a given time: *Our supply department had already ordered the replacement valve.*

attain (v) to achieve, to reach: *We attained our sales goal this month.*
obtain (v) to get, to receive: *You can obtain a job application via their Web site.*

cite (v) to document: *Please cite several examples to support your claim.*
site (n) place, location: *They want to build a parking lot on the site of the old theater.*
sight (n) vision: *His sight improved with bifocals.*

complement (v) to add to, enhance: *His graphs and charts complemented my proposal.*
compliment (v) to praise: *The customer complimented us on our courteous staff.*

continually (adv) frequently and regularly: *This answering machine continually disconnects the caller in the middle of the message.*
continuously (adv) constantly: *The air conditioning is on continuously during the summer.*

council (n) government body: *The council voted to increase salaries for all city employees.*
counsel (n) advice: *She gave the trainee pertinent counsel.*

discreet (adj) showing respect, being tactful: *The manager was discreet in answering the complaint letter.*
discrete (adj) separate, distinct: *Put those figures into discrete categories for processing.*

dual (adj) double: *A clock-radio serves a dual purpose.*
duel (n) a fight, a battle: *The argument almost turned into a real duel.*

eminent (adj) prominent, highly esteemed: *Dr. Rollins is the most eminent pediatrician in our community.*
imminent (adj) about to happen: *A hostile takeover of that company is imminent.*

foreword (n) preface, introduction to a book: *The foreword outlined the author's goals and objectives in her research study.*
forward (adv) toward a time or place; in advance: *We moved the time of the visit forward on the calendar so we could meet the overseas manager.*
forward (v) to send ahead: *We forwarded his mail to his new address.*

imply (v) to suggest: *Mr. Chin implied that the mechanics had taken too long for their lunch break.*
infer (v) to draw a conclusion: *We can infer from these sales figures that the new advertising campaign is working.*

it's (noun + verb) contraction of *it* and *is*: *Do you think it's too early to tell?*
its (adj) possessive form of *it*: *That old typewriter is on its last legs.*

lay/laid/laid (v) to put down: *Lay aside that project for now. He laid aside the project. He had already laid aside the project twice before.*
lie/lay/lain (v) to recline: *I think I'll lie down for a while. He lay there for only two minutes before the firefighter rescued him. She has lain out in the sun too often.*

lose (v) to misplace, to fail to win: *Be careful not to lose my calculator. I hope I don't lose my seat on the planning board.*
loose (adj) not tight: *The printer ribbon was too loose.*

passed (v) went by (past tense of *pass*): *He passed me in the hall without recognizing me.*
past (n) time gone by: *We've never used their services in the past.*

personal (adj) private: *The manager closes the door when she discusses personal matters with one of her staff.*
personnel (n) staff of employees: *All personnel must participate in the 401(k) retirement program.*

perspective (n) view: *From the customer's perspective, we are a fair and courteous company.*
prospective (adj) expected, likely to happen or become: *E-mail the new prospective budget to district managers.*

precede (v) to go before: *A slide show will precede the open discussion.*
proceed (v) to carry on, to go ahead: *Proceed as if we had never received that letter.*

principal (adj) main, chief: *Sales of new software constitute their principal source of revenue.*
principal (n) the head of a school: *She was a high school principal before she entered the business world.*
principal (n) money owed: *The principal on that loan totaled $21,900.*
principle (n) a policy, a belief: *Sales reps should operate on the principle that the customer is always right.*

quiet (adj) silent, not loud: *He liked to spend a quiet afternoon surfing the Net.*
quite (adv) to a degree: *The officer was quite encouraged by the recruit's performance.*

stationary (adj) not moving: *Sam rides a stationary bicycle for an hour every morning.*
stationery (n) writing supplies, such as paper and envelopes: *Please stop off at the stationery store and buy some more address labels.*

than (conj) as opposed to (used in comparisons): *He is a faster keyboarder than his predecessor.*
then (adv) at that time: *First she called the vendor; then she summarized their conversation in an e-mail to her boss.*

their (adj) possessive form of *they: All of the lab technicians took their vacations during June and July.*
there (adv) in that place: *Please put the printer in there.*
they're (noun + verb) contraction of *they* and *are: They're our two best customer service representatives.*

who's (noun + verb) contraction of *who* and *is: Who's up next for a promotion?*
whose (adj) possessive form of *who: Whose idea was that in the first place?*

you're (noun + verb) contraction of *you* and *are: You're going to like their decision.*
your (adj) possessive form of *you: They agree with your ideas.*

Index